# PHYSICAL CHEMISTRY

*By the same Author*

———————

NEW ORGANIC CHEMISTRY
*Second Edition*
384 pages, illustrated.

CHEMISTRY EXPERIMENTS AT
HOME FOR BOYS AND GIRLS
237 pages, illustrated.

AN INTRODUCTION TO ELECTRONIC
THEORY OF ORGANIC COMPOUNDS
236 pages, illustrated.

Cover photograph by Paul Brierley
Reverse osmosis, courtesy of Patterson Candy Ltd

# PHYSICAL CHEMISTRY

H. L. HEYS M.A. (CAMBRIDGE)

*Lately Senior Science Master*
*Liverpool Collegiate School*

### FIFTH EDITION

Fully revised in SI units

and with modern nomenclature

HARRAP LONDON

*To*
# E. H.

First published in Great Britain 1935
as "Physical Chemistry for Schools and Colleges"
by GEORGE G. HARRAP & CO. LTD
182–184 High Holborn, London WC1V 7AX

Reprinted: 1939; 1943; 1945; 1946; 1947; 1948

Second Edition, revised 1952

Reprinted: 1953; 1954; 1955; 1957; 1959;
1960; 1961; 1962; 1963; 1964

Third Edition, completely revised and reset 1966

Reprinted: 1968; 1969

Fourth Edition, fully revised in SI units 1970

Reprinted: 1971; 1972; 1973; 1974

Fifth Edition, revised, with SI units and modern
nomenclature 1975

Reprinted 1976; 1977 (twice)

ISBN 0 245 52675 7

Printed in Great Britain by offset lithography by
Billing & Sons Ltd, Guildford, London and Worcester

# Preface to the Third Edition

This is the third edition of *Physical Chemistry for Schools and Colleges* and the opportunity has been taken to make the form of title less unwieldy. More important, drastic changes have been made in the subject matter and its arrangement. The modern approach to the teaching of physical chemistry is based fairly and squarely on the particle theory of matter and on energy concepts, and these have provided the guiding principles for the new edition. The latter includes nuclear reactions, ionization energies. bond strengths, intermediate character of bonds, hydrogen bonding, shapes of simple molecules and ions, the redox series, activation energies, and a number of related topics. A more detailed treatment of the metallic state has been given, and a complete chapter devoted to the Periodic Table. As in previous editions, the author has endeavoured to grade the standard of the work by leading up carefully to more difficult portions of the subject matter. In a few cases more advanced material has been printed in smaller type to indicate that it may be passed over by students reading the book for the first time. A certain amount of 'conventional' physical chemistry (*e.g.*, limiting densities and the phase rule) has now been omitted. The book meets all the requirements of the new syllabuses in physical chemistry at 'A' level and 'S' level of the various examining boards.

A new feature of this edition is the inclusion of a number of fundamental experiments which can be carried out under school laboratory conditions. The sign conventions used in thermochemistry have been altered so that heat evolved in a reaction is expressed as a negative quantity, and symbols generally have been brought into line with those recommended by the British Standards Institution.

The author's thanks are due to Professor E. G. Cox, F.R.S., and the Royal Society, John Murray Ltd., George Newnes Ltd., and the Syndics of the Cambridge University Press for permission to reproduce illustrations or details of experiments from their publications. Fuller acknowledgments of these are made in the text.

The author is indebted to the following bodies for permission to reproduce questions from their G.C.E. examination papers at 'A' level and 'S' level:

Joint Matriculation Board of the Universities of Manchester, Liverpool,
    Leeds, Sheffield, and Birmingham
Welsh Joint Education Committee
University of Cambridge Local Examinations Syndicate
Oxford Delegacy of Local Examinations
Oxford and Cambridge Schools' Examination Board
Senate of the University of London
Southern Universities' Joint Board for School Examinations

The author also wishes to express his gratitude to Dr. W. J. Hughes of
Liverpool University Education Department for reading the book in
proof and making many valuable suggestions for its improvement.

H. L. H.

# Preface to the Fourth Edition

The only major change in this edition is the substitution of SI units,
symbols, and abbreviations in place of traditional ones. The latter have
been kept only in one or two cases where practical considerations are
likely to lead to the older units being retained (at least for the present).
    The author wishes to thank the examining bodies listed in the Preface
to the Third Edition for permission to alter their examination questions
to conform with the new system. Any errors arising from these changes
are the sole responsibility of the author.

H. L. H.

# Preface to the Fifth Edition

Another edition of this book has been rendered necessary by the
changes in chemical nomenclature which have recently come into general
use. Advantage has been taken of the opportunity to incorporate certain
revisions in SI units and to make a few corrections and improvements in
the text.

The author wishes to thank the G.C.E. examining boards listed in the
Preface to the Third Edition for permission to substitute the new nomenclature in questions reproduced from their 'A' Level examinations in
Chemistry. Any errors resulting from these changes are entirely the responsibility of the author.

H. L. H.

# CONTENTS

# CHAPTER 1

# Chemical evidence for the particle theory of matter

## What is Physical Chemistry?

"We come here to be philosophers, and I hope you will always remember that whenever a result happens, especially if it be new, you should say, 'What is the cause? Why does it occur?' and you will in the course of time find out the reason."

Michael Faraday, *The Chemical History of a Candle*

There are three chief objects in studying chemistry. These are:

- To discover as much as we can about the behaviour of different kinds of matter.
- To find out the reasons for this behaviour and so obtain a deeper understanding of nature.
- To put the knowledge gained to practical use.

For convenience the subject is usually divided into physical, inorganic, and organic chemistry. Physical chemistry, however, is not so much a separate branch of chemistry as a method of studying the behaviour of inorganic and organic substances. It has been called the intellectual side of chemistry. It attempts to discover general patterns in the behaviour of matter, summarizes these patterns in the form of *laws*, and then tries to explain them by means of *theories*. We may say that inorganic and organic chemists supply the facts, which physical chemists try to understand.

**Laws.** A *law* is a statement which summarizes some general feature of substances or their behaviour. Thus, from experiments with a large number of gases, it is found that at constant temperature, changes of pressure affect the volumes of all gases in the same way: the volume is halved when the pressure is doubled, reduced to one third when the pressure is trebled, and so on. Boyle's law (see later) expresses this general feature of gases. Other physical laws concerning gases are Charles's law and Graham's law of diffusion (see later).

Chemical laws describe the ways in which chemical changes take place. For example, when elements combine together to form compounds they do so in fixed proportions by mass. This is stated in the law of definite proportions.

Many laws in physics and chemistry are not strictly true. Thus, as explained later, neither the law of definite proportions nor Boyle's law holds in all cases. However, a law must be approximately true if it is to be of any use. In everyday life people are constantly proclaiming unjustifiable 'laws' such as "It always rains on Sundays" or "Boys with red hair are hot tempered". In science laws are accepted only when experiments show that they are 100 per cent accurate, or nearly so. The failure of a law to hold in certain circumstances often has valuable results because it stimulates research into reasons for the failure. Thus, the fact that gases do not obey Boyle's law at high pressures led to discoveries about the size of gaseous molecules and the forces acting between them.

**Theories.** Why are the volumes of different gases affected in the same way by changes of pressure? The explanations which we advance for scientific laws are called *theories*. Now theories can be invented by almost anybody—for example, by the father of the boy who is bottom of the form ("My theory is that John doesn't do enough homework"). A scientist *tests* his theories by finding whether predictions based on them work out in practice. (Occasionally a possible explanation has to be provisionally accepted without experimental testing because the latter is impossible. In this case the explanation is called a *hypothesis*.) Thus, if John is bottom of the form only because he does insufficient homework, we can predict that his position will improve if he spends more time on the task. If his father ensures that John does this and John's place in the form improves, the theory is confirmed—as far as the evidence goes. A theory can never be *proved*, because some unknown factor may affect the situation. Thus John's apparent improvement might be caused by some of the very bright boys in the form being promoted and replaced by very dull boys (a fact which John, perhaps, carefully conceals from his father).

The method used by scientists to test theories is called the *scientific method*. Take, for example, the way in which a doctor deals with a patient. He first makes a careful note of all the facts (the symptoms) of the case. Some of these facts may be of a qualitative nature (*e.g.*, a headache), while others may be quantitative (*e.g.*, a high temperature or a low pulse rate). The doctor next makes a diagnosis or theory, to explain the illness, and from his theory he *predicts* that a certain treatment will help the patient. Finally he prescribes a treatment—that is, the test by which his diagnosis stands or falls.

Experiments used to test scientific theories often involve constructing a model in accordance with the theory. Thus an aircraft designer tests his theories of flight by making models and placing them in a wind tunnel. In chemistry we often make models of crystals to check our theories about the size and arrangement of the particles which they contain. Sometimes we cannot construct actual models, and we have to be satisfied with imaginary ones. For example, we cannot make a satisfactory model of a gas or an electric current. Nevertheless, our mental pictures of these enable us to carry out experiments to test the soundness of our theories about them.

Just as a doctor may have to change his diagnosis if a patient develops new symptoms, so in science we sometimes have to change our theories when new facts are brought to light. Theories are seldom completely wrong, but from time to time they have to be modified or expanded. This has happened with the particle theory of matter. The theory that matter consists of extremely small particles is of fundamental importance not only in chemistry, but in almost every other branch of science. (For this reason we shall be much concerned with the theory in this book.) During the nineteenth century nearly all the facts known about substances and their reactions were explained by Dalton's theory (see later), according to which the smallest particles of matter are the individual atoms of the elements. Early this century Dalton's atomic theory had to be modified to explain discoveries connected with radioactivity and the discharge of electricity through gases at low pressure. The new facts could be accounted for only by supposing that atoms are composed of still smaller particles —protons, neutrons, and electrons.

**Practical Uses of Physical Chemistry.** About 1954 a series of disasters occurred involving the *Comet* aircraft. Each time the same component had failed. The cause was traced to changes in structure of the metal brought about by continued stress, and steps were taken to guard against any more disasters. The investigation of the structure of materials is one of the most important tasks of physical chemistry because the properties and uses of matter in bulk depend on the arrangement of the tiny invisible particles which make it up.

The manufacture of chemicals is one of Britain's most important industries. British products have to be sold abroad in the face of foreign competition, which means that they have to be produced as cheaply and efficiently as possible. Efficiency in the manufacture of chemicals requires a thorough understanding of the processes used. Chemical engineers and works chemists use their knowledge of physical chemistry to arrange the best working conditions. Many modern processes—for example, the production of plastics like poly(ethene)—require careful adjustment of physical conditions.

Physical chemistry has many other applications. For instance, we need it to understand the fertility of soil, cracking of paint films, the setting of concrete, the corrosion of metals, and the removal of dirt by detergents. It thus has great practical importance, and this is one of the chief reasons for studying it.

## Classical Atomic Theory

The idea that matter is not continuous, but is composed of very small individual particles, originated over 2 000 years ago and persisted down the centuries. The 'granular theory' of matter formed part of the philosophy of such early Greek thinkers as Democritus. In the seventeenth century we find Sir Isaac Newton writing: "It seems probable to me that God in the beginning formed matter in solid, massy, hard, impenetrable, movable particles . . . so very hard as never to wear or break in pieces."

At the beginning of the nineteenth century mere speculation was replaced by a definite theory based on experimental evidence. Dalton's atomic theory was inspired by the results of experiments in which the masses of substances taking part in chemical reactions were measured. By the beginning of the last century two important gravimetric ('measurement by mass') laws had become established. These laws still form the basis of chemistry.

**Law of Conservation of Mass (or Indestructibility of Matter).** This law states:

*Matter can be neither created nor destroyed by chemical change.*

A better form for expressing its application to chemical reactions is as follows:

*In any chemical reaction the total mass of the primary substances is equal to the total mass of the products of the reaction.*

This law was put forward by Lavoisier, who showed in 1774 that when tin was calcined in a sealed vessel no change in mass took place.

Since Lavoisier's time the law has been tested on many occasions. Between 1893 and 1908 Landolt carried out a series of experiments designed to test the law to an accuracy of 1 part in 10 million. He sealed solutions of silver(I) nitrate(V) and potassium(I) chromate(VI) in the two limbs of a glass $\lambda$-tube (Fig. 1–1). He carefully weighed the tube, counterpoising a similar tube from the other side of the balance to eliminate error due to the buoyancy effect of the air. The solutions were made to react by tilting the tube.

$$2AgNO_3 + K_2CrO_4 = Ag_2CrO_4 \downarrow + 2KNO_3$$

The heat of the reaction caused a slight expansion of the tube, which led to an increase in the volume of air displaced and therefore a slight decrease in mass. To avoid the error thus caused, Landolt waited for several days before reweighing the tube, but on reweighing he found no difference in mass. Altogether he investigated fifteen different chemical reactions, all of which testified to the validity of the law.

### Law of Definite Proportions (or Constant Composition):

*The same chemical compound, however prepared, always contains the same elements in the same proportions by mass.*

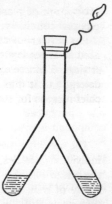

Fig. 1–1. Landolt tube

This law was deduced by Proust in 1799. If pure calcium(II) carbonate, whether in the form of chalk, marble, or Iceland spar, is analysed it is found that 100 parts by mass always contain 40 parts by mass of calcium, 12 parts by mass of carbon, and 48 parts by mass of oxygen. At first the truth of the law was disputed by Berthollet on the ground that when lead was heated in air the mass of lead oxide obtained was not constant. Proust, however, showed that the inconstancy was due to the formation of a mixture of lead oxides whose proportions varied with the temperature.

**Dalton's Explanation of the Two Basic Gravimetric Laws.** In 1808 John Dalton, a Manchester schoolmaster and keen amateur scientist, published a book called *A New System of Chemical Philosophy*. In this he explained the laws of conservation of mass and definite proportions by his 'atomic' theory. The chief points of the theory were the following:

- Matter cannot be subdivided indefinitely, because each element consists of ultimate, chemically indivisible particles called *atoms* (Greek, *atomos*, indivisible).
- The atoms of an element are indestructible.
- Atoms of the same element are alike in every respect, having the same mass, volume, chemical properties, etc.
- Atoms of different elements have different mass, volume, chemical properties, etc.
- Chemical combination occurs between different elements by atoms joining together in simple definite numbers (one to one, one to two, etc.) to give 'compound atoms' (*e.g.*, HCl, $CO_2$, $NH_3$). The latter were later called *molecules*.

By using this theory we can explain the two basic gravimetric laws. When a reaction occurs in a closed vessel all the atoms present before the reaction are still present after the reaction. They have merely become arranged differently. Since the masses of the atoms are constant, the total masses of the substances before and after the reaction should be the same.

Similarly the law of definite proportions is easily understood. Suppose an atom A of one element has a mass $x$ gram and an atom B of another element a mass $y$ gram, and that one atom of the first element combines with one atom of the second element to give the compound AB. If only one atom of each combines, the proportion by mass of the elements in the compound is $x:y$. If $n$ atoms of each combine the proportion by mass is $nx:ny$, which is the same as $x:y$. Thus any pure sample of the compound must contain the two elements in the fixed ratio by mass of $x:y$. If different numbers of atoms of the two elements combine together there will still be a constant ratio for the masses of the two elements in the compound. Thus when the compound formed is $AB_2$ the ratio by mass of the elements will always be $x:2y$.

**Testing Dalton's Atomic Theory.** We have already said that before a theory can be accepted it has to be checked by further experiment. Practical confirmation for Dalton's theory was provided by cases in which two elements combine to form *more* than one compound. An example of this is the two oxides of carbon.

Suppose that in one oxide of carbon one atom of carbon is combined with one atom of oxygen, while in the other oxide one atom of carbon is combined with two atoms of oxygen. Also, let the mass of one carbon atom be $x$ gram and the mass of one oxygen atom be $y$ gram. Then in the first oxide the ratio of the mass of carbon to that of oxygen will be $x:y$, while in the second oxide the ratio will be $x:2y$. Thus the two masses of oxygen which combine with a fixed mass of carbon should be in the simple proportion of $y:2y$, or $1:2$. Analysis of the two oxides of carbon shows this to be the case. Thus

In carbon monoxide 1 g of carbon is combined with 1·33 g of oxygen.
In carbon dioxide 1 g of carbon is combined with 2·66 g of oxygen.

These results are in agreement with one atom of carbon combining with one atom of oxygen in one oxide (CO) and with two atoms of oxygen in the other ($CO_2$). They do not *prove* that the elements combine in this way. The same ratio for the two masses of oxygen would also be obtained if one atom of carbon combined with two and four atoms of oxygen respectively ($CO_2$ and $CO_4$). The practical results do not tell us the number of atoms which combine together. They do show, however, that twice as many atoms of oxygen are united with one atom of carbon in one oxide as in the other, and, therefore, confirm Dalton's theory that atoms of different elements join together in simple fixed numbers.

Many more examples of combination between two elements support

Dalton's theory. The general results of these experiments are summarized in the third gravimetric law, the **Law of Multiple Proportions**:

*If two elements combine to form more than one compound the different masses of the one which combine with a fixed mass of the other are in a simple whole-number ratio.*

**Law of Reciprocal Proportions.** If a football team, A, beats a second team, B, by 3 goals to 2, and team B beats a third team, C, by 2 goals to 1, we might expect that when A and C meet A will win by 3 goals to 1 in view of their respective performances against B. We know how fallacious such reasoning is in football. In chemistry, however, the relation (or a very similar one) holds if we replace football teams by elements and goals by masses. Thus

1 g of hydrogen (A) combines with 23 g of sodium (B)
23 g of sodium (B) combine with 35·5 g of chlorine (C)
1 g of hydrogen (A) combines with 35·5 g of chlorine (C)

Sometimes, instead of the masses corresponding, we find that a simple multiple of one of the masses has to be taken. Thus

16 g of sulphur combine with 20 g of calcium
20 g of calcium combine with 8 g of oxygen

But 16 g of sulphur combine with 2 × 8 g of oxygen to form sulphur dioxide and with 3 × 8 g of oxygen to form sulphur trioxide.

These facts are generalized in the fourth gravimetric law, called the *Law of Reciprocal Proportions* (Berzelius, 1812):

*The masses of two elements A and B which combine with a fixed mass of a third element C are the same as the masses in which A and B themselves combine, or simple multiples thereof.*

The experimental results embodied in the law of reciprocal proportions provide additional evidence to support Dalton's theory of combination by atoms of constant mass. Suppose that three elements A, B, and C combine in pairs. If one atom of A combines with one atom of B, one of B with one of C, and one of A with one of C, the masses of A and C which combine with a fixed mass of B will be the same as the masses in which A and C themselves combine. This occurs with sodium, hydrogen, and chlorine, which form compounds NaH, HCl, and NaCl.

In another instance the three compounds formed may be of the types AB, BC, and $AC_2$. Again a certain mass of A will combine with a certain mass of B. The mass of C, however, combining with this mass of A will be double that combining with the corresponding mass of B. This is found for sulphur, calcium, and oxygen (in compounds SCa, CaO, and $SO_2$).

The law of reciprocal proportions states the manner of combination by mass only for three elements which combine in pairs. However, we need not limit the law to three elements. We can assign to each element a characteristic mass, which gives the ratio by mass in which it will combine with other elements. Dalton chose hydrogen as the standard (H = 1) for expressing the combining proportions of different elements because hydrogen has the lowest combining proportion. Thus 1 g of hydrogen unites with approximately 23 g of sodium, 20 g of calcium, 35·5 g of chlorine, or 8 g of oxygen, and these masses of the different elements also combine with each other. As these quantities were equal as regards combining power, they were called 'chemical equivalent weights' of the elements. A more accurate term is *chemical equivalent mass*. Some elements can combine in more than one proportion, and can therefore have more than one equivalent mass.

**Inexactness of the Gravimetric Laws.** Later research has shown that the four gravimetric laws are only approximately correct.

*Law of Conservation of Mass.* We assume the truth of the law of conservation of mass when we use equations to calculate the masses of substances taking part in chemical reactions, but, in spite of Landolt's experiments, we now know that the law is not rigidly true. This is due to energy changes which accompany chemical reactions. Matter and energy are interconvertible, the relation between them being given by Einstein's equation

$$E = mc^2$$

where $E$ represents energy (in joule), $m$ the mass (in kilogram), and $c$ is the velocity of light (in metre per second). Since, however, $c$ has the value $3 \times 10^8$ m s$^{-1}$, a huge amount of energy corresponds to the 'destruction' of even a tiny amount of matter.

Conversely, in all chemical reactions where heat is given out, the amount of mass which disappears is far too small to be detected by a balance.[1] Strictly speaking, it is the sum of the mass and energy which remains the same after a chemical reaction. Alternatively, we can regard the mass of a body as a form of energy, in which case the law of conservation of mass becomes merged in the *law of conservation of energy*, which states that *in any isolated system the total amount of energy remains constant*.

*Law of Definite Proportions.* Exceptions to this law are the oxides and sulphides of a number of metals. Iron(II) sulphide is given the formula

---

[1] It is simple to calculate the loss of mass in a chemical reaction due to the heat evolved. When 12 g of carbon are burned in air or oxygen $406 \times 10^3$ joule are given out.

$$\text{Loss of mass} = \frac{E}{c^2} = \frac{406 \times 10^3}{9 \times 10^{16}} \text{ kg}$$
$$= 4·5 \times 10^{-12} \text{ kg} = 0·000\ 000\ 004\ 5 \text{ g}$$

FeS, but its composition seldom agrees with this. There is usually a deficiency of iron, the composition of the compound varying between $Fe_6S_7$ and $Fe_{11}S_{12}$ (or between $Fe_{0.86}S$ and $Fe_{0.92}S$). A similar deficiency of iron occurs in iron(II) oxide, which is usually represented as FeO. Compounds like iron(II) sulphide and iron(II) oxide, which have a variable composition, are called *berthollide compounds* after Claude Louis Berthollet, who disputed the truth of the law of definite proportions (see above). Compounds of this type can be indicated by writing the mathematical sign for 'approximately' before the formula (*e.g.*, $\approx$ FeS).

The great majority of compounds, however, have a fixed composition and obey the law of definite proportions. These are called *daltonide compounds* after Dalton. (Compounds of fixed composition are also described as *stoichiometric* (pronounced 'sto-ick-iometric'), those of variable composition as *non-stoichiometric*.) Metals which can exist in two valency states often form berthollide oxides and sulphides—for example, other compounds of this type are copper(I) oxide, $\approx Cu_2O$, and the lower oxides of manganese, nickel and lead.

Berthollide oxides are crystalline and consist of metal ions and oxide ions. For example, an ideal FeO crystal would contain equal numbers of $Fe^{2+}$ and $O^{2-}$ ions. (An iron(II) ion ($Fe^{2+}$) is an iron atom which has lost two electrons. An oxide ($O^{2-}$) ion is an oxygen atom which has gained two electrons.) Berthollide character arises because metal ions of higher valency may replace some of lower valency, although if this is to happen the sizes of the ions must not differ greatly. In iron(II) oxide some of the iron(II) ions are replaced by iron(III) ions ($Fe^{3+}$). For the crystal to be electrically neutral the number of positive charges must equal the number of negative charges of the $O^{2-}$ ions. The introduction of, for example, two $Fe^{3+}$ ions into the crystal in place of two $Fe^{2+}$ ions produces an excess of two positive charges, but electrical neutrality can still be maintained if another $Fe^{2+}$ ion is omitted completely. In this way a 'defect' is caused in the crystal (Fig. 1–2).

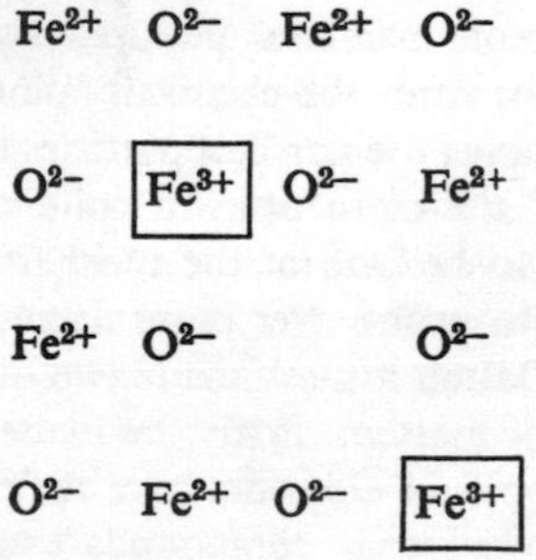

$$Fe^{2+} \quad O^{2-} \quad Fe^{2+} \quad O^{2-}$$

$$O^{2-} \quad \boxed{Fe^{3+}} \quad O^{2-} \quad Fe^{2+}$$

$$Fe^{2+} \quad O^{2-} \qquad\qquad O^{2-}$$

$$O^{2-} \quad Fe^{2+} \quad O^{2-} \quad \boxed{Fe^{3+}}$$

*Fig.* 1–2. Berthollide character in iron(II) oxide. If the arrangement of ions shown were repeated throughout the crystal the oxide would have the formula

$$Fe_7O_8, \text{ or } Fe_5^{2+}Fe_2^{3+}O_8^{2-}$$

Ionic crystals which are daltonide in composition are non-conductors of electricity. Iron(II) oxide, copper(I) oxide, and similar berthollide compounds are semiconductors. This is because electrons can pass through the crystal via the metal ions of different valency in adjacent layers. Thus

$$Fe^{3+} + e \rightarrow Fe^{2+} \text{ and } Fe^{2+} - e \rightarrow Fe^{3+}$$
$$Fe^{3+}Fe^{2+}Fe^{2+} \rightarrow Fe^{2+}Fe^{3+}Fe^{3+} \rightarrow Fe^{2+}Fe^{2+}Fe^{3+}$$

Berthollide character can also arise in metal oxides in another way. If zinc(II) oxide, $Zn^{2+}O^{2-}$, is heated it gives off a little oxygen, as neutral atoms, and excess electrons are left in the crystal; these convert some of the zinc ions into atoms ($Zn^{2+} + 2e \rightarrow Zn$). The crystal now contains excess of zinc, some of the oxide ion sites being vacant. Owing to the presence of the metal atoms the crystal again becomes a semiconductor. As the oxide changes from a daltonide composition, ZnO, to a berthollide one, $\approx$ ZnO, its colour changes from white to yellow. The change, however, is reversed on cooling, when it absorbs oxygen once more.

*Law of Multiple Proportions.* When two elements unite to form more than one compound, the different masses of one element which combine with a fixed mass of the other are not always in a simple whole-number ratio. The hydrides of carbon are a well-known exception to the law of multiple proportions. Thus in the alkane hydrocarbons heptane ($C_7H_{16}$) and octane ($C_8H_{18}$) the hydrogen ratio for a fixed mass of carbon is 64:63, which is a whole-number ratio, but certainly not a simple one.

Further exceptions to the laws of definite, multiple, and reciprocal proportions arise from *isotopes*. These are forms of the same element in which the atoms have different masses because they contain different numbers of neutrons in the nucleus. The above gravimetric laws hold only when compounds formed from the same isotopes are involved, or when the compounds are made from a mixture of isotopes in constant proportion. The latter condition usually exists in nature and the laws are obeyed.

**Usefulness of Dalton's Atomic Theory.** Although we now know that atoms are not the simple particles pictured by Dalton, this has not affected their importance from the chemical point of view; it is still true that when elements combine the smallest particles involved in the combination are the atoms of the elements. In spite of its imperfections the atomic theory proved to be one of the most fruitful scientific theories. It prompted scientists to probe ever more deeply into the structure and behaviour of matter. Dalton himself made two outstanding contributions to the particle theory of matter: firstly, he made it clear that the fundamental particles, or atoms, of elements have their own individual properties; secondly, he showed that compounds usually form by atoms of different elements combining together in fixed numbers.

One important result of Dalton's theory was that for the first time it became possible to represent a chemical compound by a formula and a chemical reaction by an equation. In Dalton's time, however, no accurate method was known for finding the number of atoms in a molecule. In deriving his chemical formulæ Dalton made certain unjustified assumptions about the ways in which atoms combine, and, as a result, many of his formulæ were wrong. Thus, he gave water the formula $\infty$, or HO. In other cases he was more fortunate—for example, he deduced correctly the formulæ of carbon monoxide and carbon dioxide.

A few years after the publication of the atomic theory, the Swedish chemist Berzelius (1814) invented the modern system of chemical symbols and the method of using numbers to show the occurrence of more than one atom of an element in a molecule. With the introduction of Cannizzaro's method (see later) for finding relative atomic masses and the right number of atoms in a molecule, it became possible to show the composition of quite large molecules—*e.g.*, $C_4H_{10}O$, ethoxyethane ('ether')—by a chemical formula. This led to an understanding of the structure of organic compounds and helped a great deal in the enormous expansion of organic chemistry which took place during the nineteenth century.

**Problem of Relative Atomic Masses.** Dalton was unable to find the actual masses of atoms, but he saw that it might be possible to *compare* the masses of atoms of different elements. As hydrogen had the lowest combining proportion by mass of all the elements, he assumed, correctly, that hydrogen atoms had the smallest mass. The masses of atoms of other elements compared with that of a hydrogen atom were called their 'atomic weights'. An 'atomic weight' was not actually a weight, but a ratio, and the term has now been replaced by *relative atomic mass*.

In trying to assess the relative atomic masses of elements Dalton encountered an immediate difficulty. The only information available was the proportion by mass in which the elements combined together, and this was insufficient to give the relative masses of the combined atoms, as the *number* of atoms which united was unknown. To get over this difficulty Dalton assumed that if two elements combined to form one compound only, their combination was the simplest possible—namely, one atom of the first element with one atom of the second. If two compounds were formed, one atom of one element would be joined to one atom and two atoms respectively of the other element in the different compounds. Now, Dalton knew that in water one gram of hydrogen is combined with (approximately) eight gram of oxygen. Let us assume with Dalton that one atom of hydrogen combines with one atom of oxygen to form water (HO). Then any sample of pure water will contain $N$ atoms of hydrogen to $N$ atoms of oxygen. We have

1 g of H : 8 g of O

= mass of $N$ atoms of H : mass of $N$ atoms of O
= mass of 1 atom of H : mass of 1 atom of O

That is, an oxygen atom has eight times the mass of a hydrogen atom.

If, however, two atoms of hydrogen combine with one atom of oxygen to form water ($H_2O$) we have

1 g of H : 8 g of O

= mass of $2N$ atoms of H : mass of $N$ atoms of O

= mass of 2 atoms of H : mass of 1 atom of O

It follows that

Mass of one H atom : mass of one O atom = $\frac{1}{2}$ : 8

= 1 : 16

Clearly, to use this method for comparing the masses of atoms we need to know (i) the combining proportions by mass of the elements, and (ii) the relative numbers of atoms which join together. Dalton, Berzelius, and others spent much time measuring the first, but they were unable to measure the second. Hence many of their relative atomic masses (such as O = 8 and C = 6) were wrong. Actually, the key to solving the problem lay at hand, but went unrecognized by Dalton and his fellow workers. This key was contained in the molecular theory of gases, which was invented to explain the manner in which gases combine.

## Molecular Theory of Gases

**Gay-Lussac's Law of Combining Volumes.** In the same year (1808) that Dalton published his atomic theory the French chemist Gay-Lussac announced the results of a series of experiments on the volumes of combining gases. The results were summarized in a law:

*When gases combine together at constant temperature and pressure they do so in volumes which bear a simple ratio to each other, and to the volume of the product if gaseous.*

Examples of this are the combination of hydrogen with chlorine, hydrogen with oxygen, and nitrogen with hydrogen.

1 volume of hydrogen + 1 volume of chlorine → 2 volumes of
hydrogen chloride

2 volumes of hydrogen + 1 volume of oxygen → liquid water

2 volumes of hydrogen + 1 volume of oxygen → 2 volumes of steam

1 volume of nitrogen + 3 volumes of hydrogen → 2 volumes of
ammonia

Gay-Lussac's law can be verified practically by combining hydrogen and chlorine as described below.

**Experiment.** The apparatus (Fig. 1–3) consists of two glass bulbs of equal capacity, approximately 50 cm³, fitted with two-way taps A and B. The bulbs are joined by a short tube fitted in the middle with a three-way tap, C, which can also be connected to a short exit tube D, at right angles to the first tube. Dry the apparatus carefully before use.

Fill one of the bulbs with hydrogen from a Kipp's apparatus. Dry the gas by bubbling it through two Dreschel bottles containing concentrated sulphuric(VI) acid. Pass the hydrogen slowly through the bulb for about 20 minutes, allowing it to escape through the three-way tap and exit tube D. Close this bulb and fill the second one with dry chlorine in a similar manner. Prepare the chlorine by dropping concentrated hydrochloric acid on to potassium(I) manganate(VII), and dry the gas with concentrated sulphuric(VI) acid. Again pass the gas in slowly for about 20 minutes. Close the second bulb, and put the two bulbs into communication by means

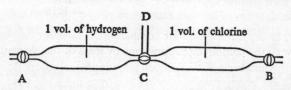

*Fig.* 1–3. Apparatus used to find the composition of hydrogen chloride by volume

of the three-way tap. Note the laboratory temperature and pressure, and leave the apparatus for two days in diffused daylight to allow the gases to mix and combine. Then proceed as follows:

1. Holding the apparatus vertically, open the bottom tap under mercury. Providing the temperature and pressure have not altered appreciably, no mercury enters the apparatus and no gas is expelled. This shows that the total volume after the reaction is the same as before the reaction.

2. Open one tap under water. The latter rises quickly up the apparatus and fills it completely. As hydrogen is insoluble in water and chlorine has only a small solubility the residual gas must be hydrogen chloride which is very soluble.

From the results of this experiment we can conclude that *one volume of hydrogen combines with one volume of chlorine to give two volumes of hydrogen chloride.*

**Explanation of Gay-Lussac's Law.** Gay-Lussac's law put the early nineteenth century chemists into a quandary because it appeared to be at variance with the atomic theory. To understand their problem we must remember that gaseous elements like hydrogen and chlorine were thought to consist of single atoms. There were two things to be reconciled: (i) according to Dalton's theory atoms of one gaseous element combined with atoms of another gaseous element in simple whole numbers, and (ii) the volumes of the gases which combined were also in simple whole numbers. To Dalton and Berzelius the only explanation seemed to be that equal volumes of gases contain the same number of atoms. Then

1 volume of hydrogen combines with 1 volume of chlorine
∴ $N$ atoms of hydrogen combine with $N$ atoms of chlorine
∴ 1 atom of hydrogen combines with 1 atom of chlorine

This seemed reasonable enough until the volume of the hydrogen chloride formed was taken into account.  Thus

$\quad$ 1 volume of hydrogen $+$ 1 volume $\quad\rightarrow$ 2 volumes
$\qquad\qquad\qquad\qquad\quad$ of chlorine $\qquad$ of hydrogen chloride
$\therefore\quad N$ atoms of hydrogen $+\ N$ atoms $\quad\rightarrow 2N$ particles
$\qquad\qquad\qquad\qquad\quad$ of chlorine $\qquad$ of hydrogen chloride
$\therefore\quad$ 1 atom of hydrogen $\quad+$ 1 atom $\quad\rightarrow$ 2 particles
$\qquad\qquad\qquad\qquad\quad$ of chlorine $\qquad$ of hydrogen chloride

Now, each particle of hydrogen chloride must contain some hydrogen and some chlorine.  Therefore each atom of hydrogen and chlorine must split into at least two parts, which is impossible since by definition an atom is indivisible chemically.

The same difficulty was encountered with oxygen atoms in the formation of steam from hydrogen and oxygen, and with nitrogen atoms in the formation of ammonia.  Apparently the atomic theory and Gay-Lussac's law could not be reconciled with each other.  Dalton's reaction was to deny the truth of the law and assert that Gay-Lussac's experiments were inaccurate.  In 1810 he wrote, "The truth is, I believe, that gases do not unite in equal or exact measures in any one instance."

The difficulty was resolved by the Italian scientist Avogadro in 1811. Avogadro suggested that in gaseous elements like hydrogen and chlorine the atoms were joined together into larger particles containing two or more atoms.  For the larger particles he used the name *molecule*.  (The molecule is now defined as *any group of atoms, like or unlike, chemically combined together*.)  Avogadro pointed out that, if 'molecule' were substituted for 'atom' in the explanation of Dalton and Berzelius, Gay-Lussac's law and the atomic theory were no longer inconsistent.  Avogadro thus arrived at the following statement, known as **Avogadro's hypothesis**:

*Equal volumes of all gases at the same temperature and pressure contain the same number of molecules.*

The explanation of the combination of hydrogen and chlorine then became as follows:

$\quad$ 1 volume of hydrogen $\quad+$ 1 volume $\quad\rightarrow$ 2 volumes
$\qquad\qquad\qquad\qquad\qquad$ of chlorine $\qquad$ of hydrogen chloride
$\therefore\quad$ 1 molecule of hydrogen $+$ 1 molecule $\quad\rightarrow$ 2 molecules
$\qquad\qquad\qquad\qquad\qquad$ of chlorine $\qquad$ of hydrogen chloride

It follows that each molecule of hydrogen and chlorine must split into at least two parts and must therefore contain at least two atoms.  In the same way it could be shown that the molecules of oxygen and nitrogen were at least diatomic.  Evidence that the molecules of these gases consisted of *only* two atoms was not obtained until later.

With the application of Avogadro's hypothesis it was obvious that Gay-Lussac's law, far from disagreeing with the atomic theory, actually supplied experimental evidence to confirm it. In fact, the law and the hypothesis together carried the atomic theory a stage further: they showed that many gaseous elements consisted, not of single atoms, but of composite particles made up of two or more atoms. In view of this support for his atomic theory, it was ironic that Dalton refused to believe in Gay-Lussac's law.

One difficulty was that at the time Avogadro's hypothesis could not be verified by experiment; in 1811 it was not possible to find the number of molecules in a given volume of a gas. The hypothesis therefore remained a reasonable, but unconfirmed, assumption. Many years later, when practical evidence of its truth was obtained and its status was raised to that of a fully fledged scientific theory, it became known as Avogadro's theory. Nowadays it is regarded as so well established that it is often called Avogadro's law, although it is still commonly referred to by its original title.

**Cannizzaro solves the Problem of Relative Atomic Masses.** Avogadro himself first pointed out that, if his hypothesis were true, it led to a simple method of comparing the masses of different gaseous molecules. If equal volumes of two gases, A and B, contain the same number of molecules the relative densities of the two gases must equal the relative masses of their molecules. Thus

$$\frac{\text{Mass of 1 m}^3 \text{ of A}}{\text{Mass of 1 m}^3 \text{ of B}} = \frac{\text{mass of } N \text{ molecules of A}}{\text{mass of } N \text{ molecules of B}}$$

$$= \frac{\text{mass of 1 molecule of A}}{\text{mass of 1 molecule of B}}$$

In 1858, the Italian Cannizzaro showed how the above relationship could be used to compare atomic masses of elements. As hydrogen was the lightest gas he took hydrogen as the standard (H = 1) for comparing the densities of different gases. Then the relative density, formerly called 'vapour density', of any gas was *the ratio of the mass of any volume of the gas to the mass of an equal volume of hydrogen at the same temperature and pressure.*

Cannizzaro first applied his method to hydrogen itself. He assumed that amongst the many gaseous hydrogen compounds there must be at least one which contained only one hydrogen atom in its molecule. For such a compound, the mass of hydrogen present in one 'molecular weight' of the compound would be a minimum. For compounds containing two or three hydrogen atoms in their molecules, the mass of hydrogen in one 'molecular weight' would be double or treble the minimum value; this can be seen from the formulæ $HCl$, $H_2S$, and $NH_3$.

For reasons given shortly, Cannizzaro also assumed the *relative molecular mass* ('molecular weight') of hydrogen to be 2, that is, twice its relative density. Then, to obtain the relative molecular masses of the compounds he had to multiply their densities relative to hydrogen also by 2; this follows from the relationship given at p. 23, the substance B being hydrogen. The relative molecular masses thus obtained are shown in the third column of Table 1–1. Finally, he analysed the various hydrogen

**Table 1–1.** APPLICATION OF CANNIZZARO'S METHOD TO HYDROGEN

| Substance | Relative density of gas | Relative molecular mass ($M_r$) | Mass/g hydrogen in $M_r$ g of substance |
|---|---|---|---|
| Hydrogen | 1 | 2 (assumed) | 2 |
| Hydrogen chloride | 18·25 | 36·5 | 1 |
| Hydrogen bromide | 40·5 | 81 | 1 |
| Steam | 9 | 18 | 2 |
| Hydrogen sulphide | 17 | 34 | 2 |
| Ammonia | 8·5 | 17 | 3 |
| Phosphine | 17 | 34 | 3 |
| Ethene | 14 | 28 | 4 |
| Methane | 8 | 16 | 4 |
| Ethanol vapour | 23 | 46 | 6 |

compounds and determined the mass in gram of hydrogen in 1 *mole* (relative molecular mass in gram) of the compounds. These masses are shown in column 4.

Several conclusions can be drawn from the results given in the table which could be extended to include some thousands of hydrogen compounds. Thus we find:

- The minimum number of gram of hydrogen contained in one mole of any of the hydrogen compounds is *one half the relative molecular mass of hydrogen*. This is true whatever value we assume for the latter. Thus, if we take the relative molecular mass of hydrogen to be 4, instead of 2, the numbers in columns 3 and 4 are merely doubled.
- This minimum mass of hydrogen, or whole multiples of it, are contained in the molar quantities of all the hydrogen compounds—for example, half a molecule of hydrogen is present once in the hydrogen chloride molecule, twice in the steam molecule, and three times in the ammonia molecule. It is reasonable to conclude that this common unit must be an atom of hydrogen. Since the hydrogen molecule has two units it must contain two atoms—that is, the molecule is $H_2$.
- *If* we take the relative molecular mass of hydrogen to be 2, the relative atomic mass of hydrogen must be 1. Conversely, *if* we take the relative

atomic mass of hydrogen to be 1 (Dalton) the relative molecular mass of hydrogen must be 2.

- The relative molecular mass of any gas or vapour—that is, the mass of one molecule relative to that of a hydrogen atom—is obtained by multiplying its density relative to that of hydrogen by two.

Knowing that the molecule of hydrogen contains two atoms, we can deduce the relation between relative molecular mass and relative gaseous density more formally as follows:

Relative density of gas A

$$= \frac{\text{mass of } x \text{ cm}^3 \text{ of A}}{\text{mass of } x \text{ cm}^3 \text{ of hydrogen}}$$

$$= \frac{\text{mass of } N \text{ molecules of A}}{\text{mass of } N \text{ molecules of hydrogen}}$$

$$= \frac{\text{mass of 1 molecule of A}}{\text{mass of 1 molecule of hydrogen}}$$

Taking the mass of one atom of hydrogen as the unit for relative molecular masses and knowing that one molecule of hydrogen contains two atoms, we have:

$$\text{Relative density of gas A} = \frac{\text{relative molecular mass of A}}{2}$$

Hence,

$$\text{Relative molecular mass of gas A} = \text{relative density} \times 2$$

*Note.* For more accurate working the standard of reference for gaseous density was later changed from H = 1 to O = 16. On the oxygen standard the relative density of hydrogen is 1·008. The change did not affect the relation, relative molecular mass of a gas = relative density × 2, but on the oxygen standard

$$\text{Relative density} = 16 \times \frac{\text{mass of the gas}}{\text{mass of equal vol. of oxygen}}$$

Unless great accuracy is required, however, relative density is usually determined with respect to hydrogen.

Cannizzaro proceeded to apply his method of finding relative atomic masses to other elements such as chlorine, and carbon. In each case he determined the least mass of the element which occurred in one mole of any of its volatile compounds, finding the relative molecular mass by doubling the relative density of the gas or vapour. Table 1–2 shows how the method applies to carbon.

**Table 1–2.** APPLICATION OF CANNIZZARO'S METHOD TO CARBON

| Compound | Relative gaseous density | Relative molecular mass | Mass/g C in 1 mole of compound |
|---|---|---|---|
| Carbon dioxide | 22 | 44 | 12 |
| Ethoxyethane | 37 | 74 | 48 |
| Ethyne | 13 | 26 | 24 |
| Methane | 8 | 16 | 12 |
| Propane | 22 | 44 | 36 |
| Pentane | 36 | 72 | 60 |

Only 12 or multiples of 12 occur in the last column. Less than 12 g of carbon are never found in one mole of any carbon compound. Hence the relative atomic mass of carbon is 12. In the same way the relative atomic mass of chlorine was shown to be 35·5 and that of oxygen 16.

Note that relative atomic masses obtained by Cannizzaro's method were only approximate. Because hydrogen is a very light gas and is difficult to weigh accurately, the values of the relative densities and relative molecular masses of the volatile compounds were only rough values. Nevertheless, as we shall see shortly, Cannizzaro's approximate results could form a basis for accurate determination of relative atomic masses.

Cannizzaro's method is not suitable for finding the relative atomic masses of metals, because the latter form few volatile compounds. However, we can deduce the *probable* relative atomic mass of a metal from its equivalent mass and the relative gaseous density of a volatile compound.

**Example.** *The equivalent mass of a metal M is* 25·1. *The relative density of the vapour of the anhydrous chloride is* 90·7. *Find the probable relative atomic mass of the metal.*

Let the valency of M be $n$. Then the relative atomic mass $= 25·1n$. The formula of the chloride is $MCl_n$, and the relative molecular mass, obtained by adding together the relative atomic masses, is

$$25·1n + 35·5n = 60·6n$$

But the relative molecular mass of the chloride equals the relative gaseous density multiplied by two.

$$\therefore \quad 60·6n = 90·7 \times 2 = 181·4$$

and
$$n = 3$$

$$\therefore \quad \text{Relative atomic mass of M} = 75·3$$

If the volatile compound consists of associated molecules (such as $Fe_2Cl_6$ and $Al_2Cl_6$) in the vapour state, the relative atomic mass obtained by this method will be a simple multiple of the correct value.

## Relative Atomic Masses after Cannizzaro

Cannizzaro's work started a period of intense activity in the determination of relative atomic masses. Before we describe these developments, however, we shall look at some of the ways in which the newly discovered relative atomic masses were useful.

**Significance of 'Mole' as Applied to Elements.** If we express the relative atomic mass of an element in gram, we get the quantity known as a 'mole' of the element. Thus a mole of carbon consists of 12 g of carbon, and a mole of magnesium of 24 g of magnesium. Now we see from the relative atomic masses of the elements that a magnesium atom has double the mass of a carbon atom. Hence it follows that 24 g of magnesium contain the same number of atoms as 12 g of carbon. This can be extended to elements in general, in which case we find that:

- One mole of all elements contains the same number of atoms (this is true whether the element is a solid, a liquid, or a gas).[1]
- If the masses of two elements are divided by their relative atomic masses we obtain the relative number of atoms of each element.

**Finding Empirical Formulæ.** The relative masses of the elements in a compound depend partly on the masses of the atoms and partly on the relative numbers of atoms of each element which have joined together. If the masses of the elements in a known mass of compound have been found by analysis, we can deduce the relative numbers of atoms by dividing the masses of the elements by their respective relative atomic masses, which need only be approximate.

**Example.** *1·32 g of magnesium were dissolved in dilute hydrochloric acid and the solution was then heated in a stream of hydrogen chloride. 5·26 g of the anhydrous metal chloride remained. Find the simplest formula for the metal chloride.* ($Mg = 24$, $Cl = 35·5$.)

1·32 g of magnesium combined with (5·26 − 1·32) g of chlorine

= 3·94 g of chlorine.

---

[1] 2·016 g of hydrogen ($H_2$) contain $6·023 \times 10^{23}$ molecules. Hence 1·008 g of hydrogen (or one mole of any other element) contain $6·023 \times 10^{23}$ atoms.

Number of Mg atoms : number of Cl atoms

$$= \frac{1 \cdot 32}{24} : \frac{3 \cdot 94}{35 \cdot 5}$$

$$= 0 \cdot 055 : 0 \cdot 111$$

$$= 1 : 2 \text{ (to the nearest whole number)}$$

The simplest formula which shows the above ratio is $MgCl_2$, although the same ratio would be expressed by $Mg_2Cl_4$, $Mg_3Cl_6$, etc. The first of these formulæ is called the 'empirical formula' of the metal chloride.

*The* **empirical formula** *of a compound is the simplest formula which shows the ratio of the atoms of the different elements in the compound.*

The formula which shows the number of atoms of each element present in one *molecule* of a compound is called its *molecular formula*. Many compounds do not exist as molecules, but as crystalline solids composed of oppositely charged ions. Such compounds have to be represented by their empirical formulæ. Thus the formulæ used for anhydrous magnesium(II) chloride ($MgCl_2$), calcium(II) carbonate ($CaCO_3$), and anhydrous copper(II) sulphate(VI) ($CuSO_4$) are empirical formulæ.

**Valency** (or **Valence**). When we examine the formulæ, empirical or molecular, of a series of binary compounds of the same element it is clear that different combining capacities are associated with atoms of different elements. Thus in the chlorides $NaCl$, $MgCl_2$, $AlCl_3$, and $CCl_4$ we have one atom of the first element united with one, two, three, and four atoms of chlorine. The term *valency* (Latin *valere* = to be worth) was introduced by Frankland in 1852 to denote the relative combining powers of different atoms. It is customary to measure the combining powers in terms of hydrogen atoms.

*The* **valency** *of an element is the number of atoms of hydrogen which combine with, or are displaced by, one atom of the given element.*

We can, however, deduce the valencies of most atoms from the compounds which they form with atoms or groups equivalent to hydrogen. Thus $Cl$, $Br$, $CH_3$, and $C_2H_5$ have the same combining power as one atom of hydrogen. Again, most elements form oxides, and, assuming, from the formula $H_2O$ for water, that oxygen has a valency of two, we can often infer the valency of an element from the formula of its oxide. Thus the monovalency of chlorine can be deduced from any of the formulæ $HCl$, $C_2H_5Cl$, and $Cl_2O$.

Some elements have two, or even more, valencies. The different valencies are indicated by roman numerals, as in iron(II) chloride ($FeCl_2$) and iron(III) chloride ($FeCl_3$). Strictly speaking, the roman numerals stand for 'oxidation numbers' of the elements (see p. 437).

Another definition of valency is given by the relation

$$\text{Valency} = \frac{\text{relative atomic mass (in g)}}{\text{equivalent mass (in g)}}$$

This interpretation is fundamentally the same as the previous one. Thus suppose that an element M has a relative atomic mass $A_r$ and an equivalent mass $E$. Then by the second definition valency of M $= A_r\,$g$/E\,$g. But, by Dalton's definition of equivalent mass, $E\,$g of M combine with 1 g of hydrogen. Division by relative atomic masses gives the atom ratio. Thus

$E/A_r$ atoms of M combine   with $1/1$ atoms of hydrogen

    $E$ atoms of M combine   with $A_r$ atoms of hydrogen

$\therefore$     1 atom   of M combines with $A_r/E$ atoms of hydrogen

Consequently the first definition agrees with the second. It is obvious that, if an element has more than one equivalent mass, it will have more than one valency.

**General Procedure in Chemical Determination of Relative Atomic Masses.** In the post-Cannizzaro period of chemistry, and, indeed, right up to modern times, the second definition of valency given above was frequently used for finding relative atomic masses accurately; that is, most of the methods were based on the relationship

relative atomic mass (in g) $=$ equivalent mass (in g) $\times n$,

where $n$, the valency, is a simple whole number.

In finding the accurate relative atomic mass of an element by chemical methods, the first step is to determine the equivalent, or combining, mass *as accurately as possible*. The second step is to ascertain the valency; the accurate equivalent mass is then multiplied by the valency. Methods of finding the equivalent masses and valencies of elements are described below.

**Combining Ratio of Hydrogen and Oxygen by Mass.** Several accurate determinations have been made of the combining proportions by mass of these elements. The first was by Dumas in 1842, who passed carefully purified hydrogen over hot copper(II) oxide, and measured the loss in mass of the metal oxide and the mass of water formed. Dumas found that 8 g of oxygen combined with 1·007 g of hydrogen. Other determinations were carried out by Morley in 1895 and by Burt and Edgar in 1915.

In Morley's method, hydrogen was prepared by electrolysis of dilute sulphuric(VI) acid and purified from possible traces of sulphur dioxide by potassium(I) hydroxide, from oxygen by hot copper, and from moisture by phosphorus(V) oxide. It was then stored in metallic palladium, which absorbs about 930 times its own volume of hydrogen at room temperature, and gives it off again when heated. Oxygen was made from potassium(I) chlorate(V) without manganese(IV) oxide, to avoid traces of chlorine. After being dried with phosphorus(V) oxide, the gas was stored in a large glass globe. The hydrogen and oxygen were then admitted to the apparatus (Fig. 1–4), which had previously been exhausted of air and weighed. The hydrogen was ignited by an electric spark and burned at a platinum jet. The apparatus was in a cooling bath so that the water vapour formed condensed to liquid.

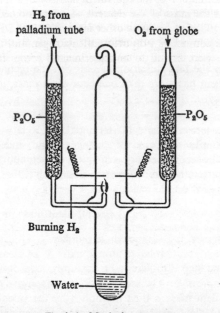

*Fig.* 1–4. Morley's apparatus

At the end of the experiment, when some 40 dm³ of hydrogen had burned, the apparatus was placed in a freezing bath to freeze the water to ice. The gases remaining in the apparatus were pumped off (the phosphorus(V) oxide retained any water vapour), and were weighed and analysed. From these masses and the loss in mass of the palladium tube and the oxygen globe, the combining proportion by mass of hydrogen and oxygen was calculated. As a check the mass of water produced was also measured. The average result obtained by Morley from twelve experiments was: 8 g of oxygen to 1·0076 g of hydrogen.

Burt and Edgar exploded pure hydrogen with pure oxygen at s.t.p. and measured very accurately the volumes of gases which combined. From fifty-nine different experiments they found an average combining proportion of 2·002 9 volumes of hydrogen to 1 volume of oxygen. They were then able to calculate from the densities of the gases that 8 g of oxygen combined with 1·007 7 g of hydrogen.

**Oxygen (O = 16) Standard of Relative Atomic Masses.** Dalton used hydrogen (H = 1) as his standard for relative atomic masses. From about 1860 onwards, the hydrogen standard was gradually replaced by the oxygen standard (O = 16). This was far more suitable because few elements combine with, or displace, hydrogen quantitatively. Consequently, in the preliminary determination of the equivalent mass of an element it was often necessary to find the mass of the element which combined with a certain mass of oxygen, chlorine, or some other intermediate element and then to make use of the combining proportion of hydrogen and the intermediate element. There were bound to be experimental errors in both of these stages. To minimize experimental errors, it is desirable to find the equivalent mass of an element by direct reference to the standard element. Oxygen is, therefore, a more suitable standard then hydrogen, because most elements combine with oxygen and the oxides produced are easily analysed. The oxygen standard also has the advantage that it depends on measurement of mass, whereas the hydrogen standard usually involves measurement of the volume of gas produced; we can measure mass much more accurately than volume. For these reasons Dalton's definitions of equivalent mass and relative atomic mass were replaced by those given below.

*The* **equivalent,** *or* **combining, mass** *of an element is the number of parts by mass of the element which combine with eight parts by mass of oxygen.*

*The* **relative atomic mass** *of an element is the ratio of the mass of one atom of the element to 1/16 of the mass of an oxygen atom.*

By 1900 the oxygen standard for relative atomic masses had been universally accepted. It remained in force until 1962, when another change was made (see page 36). This, however, has made little practical difference to relative atomic masses. Thus both on the oxygen standard and on the more recent standard the relative atomic mass of hydrogen is 1·008—the value, to three decimal places, found for the equivalent mass of hydrogen by Morley and by Burt and Edgar.

**Determination of Equivalent Masses.** Various methods have been used to measure equivalent masses of elements. A list of the commoner methods is given below.

(Equivalent masses have lost most of their former importance because relative atomic masses are now easily and accurately measured by mass spectrometer.)

METHOD 1. *Displacement of Hydrogen.* A known mass of a metal is used to liberate hydrogen from hydrochloric acid (*e.g.*, zinc, iron, magnesium, aluminium), from sodium(I) hydroxide solution (*e.g.*, aluminium), from water (*e.g.*, calcium), or from ethanol (*e.g.*, sodium and potassium). The volume of hydrogen is measured, reduced to standard temperature and pressure (see later) and its mass is calculated. The mass of metal needed to displace 1 g (more accurately, 1·008 g) of hydrogen is found by proportion.

METHOD 2. *Oxidation of the Element.* A known mass of the element (Cu, Sn, Pb, Fe, Al, C, S) is converted to the oxide by suitable means. The mass of combined oxygen is found, and the mass of element which unites with 8 g of oxygen is calculated.

METHOD 3. *Reduction of the Oxide.* A known mass of the oxide (*e.g.*, $CuO$, $Fe_2O_3$) is heated and reduced in a current of hydrogen until no further loss in mass occurs. The equivalent mass is calculated as in Method 2.

METHOD 4. *Combination with Chlorine.* A known mass of a metal (*e.g.*, silver) is converted to the chloride by passing pure dry chlorine over it while it is heated. The mass of combined chlorine is found, and the mass of metal which combines with 35·5 g (the equivalent mass) of chlorine is calculated.

METHOD 5. *Electrolytic Method based on Faraday's Laws.* This depends on the fact that many elements are liberated from combination by passing an electric current through an aqueous solution—*e.g.*, metals from their salts. The laws of electrolysis were established by Faraday, and are dealt with later. If we pass the same current through acidified water, aqueous copper(II) sulphate(VI), and aqueous silver(I) nitrate(V), the masses of hydrogen, oxygen, copper, and silver set free are proportional to their equivalent masses. Hence we can determine the equivalent mass of copper or silver by finding the masses of these elements which are liberated in the same time as 1 g of hydrogen or 8 g of oxygen.

METHOD 6. *'Conversion Ratio' Method.* In hydrochloric acid, nitric(V) acid, etc., the masses of the acid radicals which are combined with 1 g of hydrogen can be regarded as 'equivalent masses' of these radicals. Thus:

| HCl | HNO$_3$ | H$_2$SO$_4$ | H$_2$CO$_3$ |
|---|---|---|---|
| (1 + 35·5) | (1 + 62) | (2 + 96) | (2 + 60) |
| Cl = 35·5 | NO$_3$ = 62 | $\frac{1}{2}$SO$_4$ = 48 | $\frac{1}{2}$CO$_3$ = 30 |

We can then determine the equivalent mass of a metal by finding how much combines with the equivalent mass of an acid radical. It is usually more convenient, however, to convert a known mass of one compound into a second compound which can be weighed. Both compounds must be pure. The conversion of a soluble salt to an insoluble salt is frequently used. The equivalent mass of the metal can then be found by simple proportion.

**Example 1.** *3·31 g of lead(II) nitrate(V) gave 3·03 g of the sulphate(VI) with sulphuric(VI) acid. Calculate the equivalent mass of lead.*

Let $m$ = the equivalent mass of lead. Then we have the following ratios for lead(II) nitrate(V) and lead(II) sulphate(VI):

$$\frac{m + 62 \text{ g}}{m + 48 \text{ g}} = \frac{3 \cdot 31 \text{ g}}{3 \cdot 03 \text{ g}}$$

This gives 103·5 g for the equivalent mass of lead.

**Example 2.** *1·50 g of anhydrous calcium(II) chloride gave 3·89 g of silver(I) chloride on precipitation with silver(I) nitrate(V) solution. Calculate the equivalent mass of calcium.*

Let $m$ = the equivalent mass of calcium. Then, assuming the equivalent mass of silver to be 108, we have

$$\frac{m + 35 \cdot 5 \text{ g}}{108 + 35 \cdot 5 \text{ g}} = \frac{1 \cdot 50 \text{ g}}{3 \cdot 89 \text{ g}}$$

Solving this equation, we get

$$m = 19 \cdot 8 \text{ g}$$

**Accurate Determination of Relative Atomic Mass.** When the equivalent mass of an element has been found accurately its relative atomic mass can be obtained numerically by multiplying the equivalent mass by the valency. The valency of an element can be determined by the methods given below.

METHOD 1. *From Cannizzaro's Method.* As explained previously, Cannizzaro's method yielded only rough values for the relative atomic masses of elements. However, if the numerical value of the accurate equivalent mass is divided into the rough value, the quotient *taken to the nearest whole* number represents the valency. Thus Cannizzaro found that the relative atomic mass of carbon was approximately 12. By oxidation of carbon to carbon dioxide the equivalent mass of carbon can be shown to be 3·003 g. Hence the valency is 4, and the accurate relative atomic mass is 12·012.

METHOD 2. *From Dulong and Petit's Law.* Put in modern form, this law states that *the molar heat capacities of solid elements are approximately the same, being equal to about 26 joule per mole kelvin.* This means that about 26 joule of heat energy raise the temperature of one mole of different solid elements by one kelvin ($=1°C$). The molar heat capacity is obtained by multiplying the heat capacity per gram by the relative atomic mass. Table 1–3 shows the molar heat capacities of some typical solid elements.

We see from Table 1–3 that an approximate value can be deduced for the relative atomic mass of an element by measuring its heat capacity per gram and dividing this into 26 J K$^{-1}$. Then, if the accurate equivalent mass is known, we can find the valency and the accurate value of the relative atomic mass as in Method 1.

Table 1–3. MOLAR HEAT CAPACITIES

| Element | Relative atomic mass | Heat capacity/ J g$^{-1}$ K$^{-1}$ | Molar heat capacity/ J mol$^{-1}$ K$^{-1}$ |
|---------|----------------------|-------------------------------------|---------------------------------------------|
| Arsenic | 75 | 0·347 | 26·0 |
| Copper | 63·6 | 0·353 | 22·5 |
| Lead | 207 | 0·130 | 26·9 |
| Nickel | 58·7 | 0·451 | 26·5 |
| Tin | 118·7 | 0·230 | 27·3 |

**Example.** *The equivalent mass of magnesium is* 12·16 *g, and its heat capacity per gram is* 1·04 J K$^{-1}$. *What is its accurate relative atomic mass?*

Approximate relative atomic mass = 26 J K$^{-1}$ ÷ 1·04 J K$^{-1}$ = 25.

Valency = relative atomic mass (in g) ÷ equivalent mass (in g)

$$= 25 \text{ g} \div 12\cdot16 \text{ g}$$

= 2, *to the nearest whole number*

∴    Accurate relative atomic mass = 12·16 × 2 = 24·32.

The heat capacities of elements vary with temperature. At ordinary temperatures they are approximately constant, but at lower temperatures they decrease and tend towards zero as the absolute zero of temperature is approached. Carbon, silicon, boron, and beryllium are exceptions to Dulong and Petit's law. At ordinary temperatures the molar heat capacities of these elements are considerably lower than 26 J mol$^{-1}$ K$^{-1}$. If the elements are heated to higher temperatures, however, their molar heat capacities gradually increase and finally become constant, although the value of the constant is still less than 26 J mol$^{-1}$ K$^{-1}$, being about 23 J mol$^{-1}$ K$^{-1}$ for carbon, silicon, and beryllium.

Dulong and Petit put forward their law in 1819, but it made little impact at the time because of the chaotic state existing in regard to relative atomic masses. The validity of the law was first demonstrated by Cannizzaro, using the relative atomic masses which he himself had found. He then used the law to deduce the relative atomic masses of a number of metals such as sodium, copper, silver, and gold.

METHOD 3. *From Isomorphism of Salts.* The term 'isomorphous' means literally 'of the same shape'. When applied to crystals, however, it means more than just similarity of crystalline form. When compounds are isomorphous they usually show the following characteristics:

- They crystallize in the same form.
- They form 'mixed crystals'—that is, if solutions of the two compounds are mixed, homogeneous crystals, containing both compounds, are produced on crystallizing out. If the two compounds are not isomorphous—*e.g.* copper(II) sulphate(VI)-5-water and sodium(I) chloride—separate kinds of crystals are deposited.
- They form 'overgrowths'—that is, if a crystal of one compound is suspended in a saturated solution of the other, the crystal continues to grow and becomes covered with a layer of the second compound, the same crystalline form being maintained.
- They have a similar kind of chemical formula.

There are many cases of salts showing the above characteristics, some well known examples being those given below.

*The Alums*

Aluminium(III) potassium(I) sulphate(VI)-12-water ('alum')

$$AlK(SO_4)_2 . 12 H_2O$$

Iron(III) potassium(I) sulphate(VI)-12-water ('iron alum')

$$FeK(SO_4)_2 . 12 H_2O$$

General formula: $RX(SO_4)_2 . 12 H_2O$, where R represents a trivalent metal (Al, Fe, Cr, Ga, etc.) and X a monovalent metal or radical (Li, Na, K, $NH_4$, etc.). The cations are named in alphabetical order. Alums form characteristic octahedral crystals (Fig. 9–8).

*Hydrated Sulphates(VI) of Mg, Mn, Fe, Ni, Co, Zn*

Magnesium(II) sulphate(VI)-7-water      $MgSO_4 . 7H_2O$
Manganese(II) sulphate(VI)-7-water      $MnSO_4 . 7H_2O$
etc.

General formula:    $MSO_4 . 7H_2O$

where M stands for the divalent metal ion.

*Potassium(I) Sulphate(VI), Chromate(VI) and Selenate(VI)*

$$K_2SO_4 \qquad K_2CrO_4 \qquad K_2SeO_4$$

Isomorphism was discovered by Mitscherlich, in the early part of the last century, and was used in several instances for fixing relative atomic masses. It was assumed that if two compounds had similar crystals and formed mixed crystals and overgrowths they were chemically similar and had the same type of chemical formula. Thus the valency of gallium was taken to be three because gallium forms alums, *e.g.*, $Ga(NH_4)(SO_4)_2 . 12H_2O$.

The relative atomic mass of gallium was therefore equal to the numerical value of its equivalent mass multiplied by three.

The term isomorphism has a somewhat deeper meaning in modern chemistry. It is applied to compounds which are built up from a similar kind of crystal unit, or 'unit cell'. Such compounds do not always have the features of isomorphism given above. Calcium(II) carbonate, $CaCO_3$, and sodium(I) nitrate(V), $NaNO_3$, have similar types of unit cell. They form similar crystals because the relative sizes of the $Ca^{2+}$ ion and the $CO_3^{2-}$ ion are about the same as the relative sizes of the $Na^+$ ion and $NO_3^-$ ion. The two compounds, however, are chemically dissimilar and do not form an overgrowth; they cannot form mixed crystals. They are isomorphous in the modern sense, but not in the old one.

### Physical Determination of Relative Atomic Masses.

*Relative Molecular Mass and Atomicity.* The *atomicity* of an element is the number of atoms in one molecule. If the relative molecular mass of nitrogen is 28, and there are two atoms in the molecule, the relative atomic mass of nitrogen must be 14.

$$\text{Relative atomic mass} = \frac{\text{relative molecular mass}}{\text{atomicity}}$$

As we have seen, the relative molecular mass of a gas can be found from its relative density. The atomicity of a gas can be deduced from the ratio of the two values of the molar heat capacity. As explained later (page 57), the latter has two values according to whether the gas is heated at constant pressure or constant volume. The first value is always the larger. The ratio $\gamma$ (gamma) of the first value to the second depends on the number of atoms in the molecule.

$\gamma = 1·66$ approximately, for monatomic gases (helium, argon, and other rare gases).

$\gamma = 1·40$ approximately, for diatomic gases ($O_2$, $N_2$, $Cl_2$, etc.).

$\gamma = 1·30$ approximately, for triatomic gases ($O_3$, $CO_2$, $N_2O$, etc.).

Until the mass spectrometer was invented early in the present century, this was the only method for determining the relative atomic masses of the rare gases. This was because no compounds of the rare gases were known, and, therefore, the equivalent masses could not be obtained.

*Mass Spectrometer.* The use of this instrument has now superseded all other methods for measuring relative atomic masses because it allows such a high degree of accuracy. The mass spectrometer and its use for this purpose are dealt with later.

A further change has also been made in the standard used for relative atomic masses. For reasons given at p. 129 the oxygen standard ($O = 16$) has been replaced by one based on a particular isotope of carbon. This isotope is the one containing six protons and six neutrons in the nucleus and is represented by the symbol $^{12}C$. One atom of this isotope is given a mass of 12 units, so that the new standard is expressed by $^{12}C = 12$. The modern definition of relative atomic mass is therefore as follows:

*The* **relative atomic mass** *of an element is the average mass of its atoms on a scale on which one atom of the $^{12}C$ isotope of carbon has a mass of 12 units.*

The latest change has made only a very slight difference to the values cf relative atomic masses previously expressed on the O = 16 standard.

**Summary of Chemical Evidence for the Particle Theory of Matter.** The law of conservation of mass and the law of definite proportions are most reasonably explained by Dalton's theory that elements are composed of different kinds of very small particles called atoms. For any one element these appear to have a constant mass. When atoms of different elements combine to form a compound they usually do so in fixed proportions by numbers. Dalton's theory is supported by the laws of multiple and reciprocal proportions.

In gaseous elements atoms are usually joined together into larger particles called molecules. Gay-Lussac's law of combining volumes can only be explained by Avogadro's hypothesis, which requires the smallest particles of gases like hydrogen and chlorine to be made up of two, or more, atoms. The ratio of molar heat capacities indicates, however, that the rare-gas elements do consist of single atoms.

Further discoveries have shown that, very often, atoms of an element are not exactly alike. They can have different masses, and it is the average mass of the atoms which is generally constant. The relative, average, masses of atoms of different elements are called their relative atomic masses. These can be measured by both chemical and physical methods. In the course of time the scale of relative atomic masses has been based on three different standards—firstly H = 1, secondly O = 16, and now the carbon standard $^{12}C = 12$.

### EXERCISE 1[1]
(*Relative atomic masses are given at the end of the book*)

**The Gravimetric Laws**

1. Describe fully, using diagrams and equations, an experiment which you could carry out in the laboratory to verify the Law of Constant Proportions.

---

[1] The letters in brackets after some of the questions indicate the examining bodies from whose G.C.E. papers questions have been taken. The abbreviations used are as follows: *J.M.B.*—Joint Matriculation Board of the Universities of Manchester, Liverpool, Leeds, Sheffield, and Birmingham; *W.J.E.C.*—Welsh Joint Education Committee; *O.L.*—Oxford Delegacy of Local Examinations; *C.L.*—University of Cambridge Local Examinations Syndicate; *O. and C.*—Oxford and Cambridge Schools' Examination Board; *Lond.*—Senate of the University of London; *S.U.*—Southern Universities' Joint Board for School Examinations.

Explain what is meant by isotopes. To what extent does the existence of isotopes affect the exactness of this law as originally stated?    (S.U.)

**2.** When hydrogen was passed over hot copper (II) oxide 0·63 g of water was formed, while the mass of the copper oxide decreased by 0.56 g. In an electrolysis of water experiment 20 $cm^3$ of hydrogen were evolved at the same time as 10 $cm^3$ of oxygen. Oxygen is 16 times as heavy as hydrogen. Do these results agree with any of the gravimetric laws?

**3.** 0·862 g of copper when converted into oxide through the nitrate(V) left 1·079 g of copper(II) oxide. In another experiment 1·716 g of copper(I) oxide when reduced in hydrogen left 1·524 g of copper. Show that these results agree with the law of multiple proportions.

**4.** The two chlorides of gold contain 15·2 per cent and 35·1 per cent respectively of chlorine. Show that these figures agree with the law of multiple proportions.

**5.** (Part question.) A metal forms three oxides containing 23·53, 31·58, and 48 per cent of oxygen. Deduce a possible value for the relative atomic mass of the metal.    (J.M.B.)

### Equivalent Masses and Relative Atomic Masses

**6.** 1·49 g of a metal chloride gave 1·74 g of the sulphate(VI) on treatment with sulphuric(VI) acid. Find the equivalent mass of the metal.

**7.** 1·875 g of a metal nitrate(V) left on ignition 0·795 g of the metal oxide. What is the equivalent mass of the metal?

**8.** 1·72 g of the carbonate of a metal were dissolved in hydrochloric acid and hydrogen sulphide passed through the solution. The precipitated sulphide of the metal was filtered off, dried, and weighed. Its mass was found to be 1·44 g. Calculate the equivalent mass of the metal.

**9.** (Part question.) When 1·000 g of a metallic oxide $M_xO_y$ was completely converted into the corresponding sulphate(VI) it gave 2·579 g of the latter. Given that the relative atomic mass of M = 52, what was the formula of the oxide?    (J.M.B.)

**10.** Give two methods by which the accurate relative atomic mass of an element may be determined.

5·21 g of the chloride of a divalent metal were dissolved in distilled water and to the solution an excess of a solution of silver(I) nitrate(V) was added. The mass of silver(I) chloride obtained was 7·17 g. Calculate the relative atomic mass of the metal. Ag = 107·9; Cl = 35·5.    (O. and C.)

**11.** (Part question.) A solution of 0·312 4 g of the anhydrous chloride of a metal M yielded a quantitative precipitate (0·350 1 g) of anhydrous sulphate(VI) of M on treatment with dilute sulphuric(VI) acid. The heat capacity of the metal per gram is 0·188 $JK^{-1}$. Calculate its accurate relative atomic mass. O = 16; S = 32·06; Cl = 35·46.    (W.J.E.C.)

**12.** The heat capacity per gram of a metal X is 0·226 $JK^{-1}$ and it forms two oxides containing 11·88 per cent and 21·23 per cent of oxygen. It forms a volatile chloride the relative gaseous density of which is approximately 130. Calculate the relative atomic mass of X; assign formulæ to the two oxides and deduce the molecular formula of the chloride, stating the principles on which your calculations are based. O = 16·00; Cl = 35·46.    (W.J.E.C.)

**13.** (*a*) When are two substances said to be isomorphous? In this connection explain the terms *mixed crystals* and *overgrowths*. State the law of isomorphism.

The chloride of a metal R is isomorphous with potassium(I) chloride. The percentage of R in its chloride is 70·66; what is its relative atomic mass?

(*b*) The heat capacity of a metal M is 0·46 J $g^{-1}$ $K^{-1}$. On suitable treatment, 2 g of M gave 2·857 g of oxide. What is the relative atomic mass of M?    (O.L.)

**14.** Describe briefly the methods for the determination of relative atomic masses associated with Cannizzaro, Mitscherlich, and Dulong and Petit.

0·345 g of the oxide of a metal was exactly neutralized by 45·0 cm³ of M/10 HCl solution. The carbonate of the metal was isomorphous with aragonite (CaCO₃). Calculate the relative atomic mass of the metal. O = 16. (Lond.)

### Gay-Lussac's Law and Avogadro's Hypothesis

15. State Gay-Lussac's law of combining volumes.

Illustrate this law by reference to the combining volumes of carbon monoxide and oxygen, and indicate how it can be shown practically that carbon dioxide contains its own volume of oxygen.

15 cm³ of a mixture of methane and ethane were completely oxidized by 45 cm³ of oxygen. The volumes were measured at the same temperature and pressure. What was the composition of the mixture? (O.L.)

16. (*a*) Describe, with the aid of a diagram, an experiment to find the volume composition of hydrogen chloride. Explain how the result of this experiment is used to deduce (i) the atomicity of hydrogen, (ii) the formula of hydrogen chloride. What further evidence is necessary?

(*b*) Taking coal gas to consist by volume of 50 per cent hydrogen, 30 per cent methane, 10 per cent carbon monoxide, 6 per cent carbon dioxide, and 4 per cent nitrogen, calculate the percentage composition by volume of the product formed at 227°C and atmospheric pressure when coal gas is burnt with an equal volume of oxygen. (S.U.)

## MORE DIFFICULT QUESTIONS

17. If you were given the oxide of an unknown metal outline the experiments you would do in order to find the valency of the metal in the oxide.

18. "Although in the development of chemical theory Dalton's work stands pre-eminent, almost equal prominence must be given to the combined contribution of Avogadro and Cannizzaro since this provided the first correct methods for determining

(i) the relative molecular masses of gaseous substances,

(ii) the relative atomic masses of gaseous elements,

(iii) the relative atomic masses of certain non-gaseous elements,

(iv) the empirical formulae of many compounds and the molecular formulae of many of these."

Elaborate and justify this statement. (J.M.B.)

19. (Part question.) A metal M has a heat capacity per gram of 0·31 J K⁻¹. It forms a volatile methyl derivative of formula $M(CH_3)_x$, the relative molecular mass of which is 132·8. Calculate (i) the value of $x$, (ii) the accurate relative atomic mass of M. (O.L.)

20. An element forms two bromides containing 88·56 per cent and 92·81 per cent of bromine respectively. What information about the element can be deduced from these figures? Br = 80.

21. Explain briefly two methods for deciding the relative atomic mass of an element the equivalent mass of which is known. 3·264 g of the chloride of an element were converted by hydrolysis into 1·909 g of the oxide. The oxide is acidic and yields salts isomorphous with chromates(VI). What conclusions do you draw from these data regarding the equivalent mass and relative atomic mass of the element and the formula of the oxide? (O. and C.)

22. What is meant by the Principle of Isomorphism? Give two examples of groups of isomorphous compounds and show how the principle may be used to find the relative atomic mass of an element. Assuming that the relative atomic mass of calcium is 40·1, find the relative atomic mass of magnesium, using the following data of percentage composition:

Calcium(II) carbonate: Ca = 40·06; C = 11·99; O = 47·95.

Magnesium(II) carbonate: Mg = 28·91; C = 14·22; O = 56·87. (J.M.B.)

# CHAPTER 2

# Physical evidence for the particle theory of matter

## Physical Gas Laws

The chemical behaviour of gases is described by Gay-Lussac's law of combining volumes. Four laws describe the physical behaviour of gases: Boyle's law, Charles's law, Dalton's law of partial pressures, and Graham's law of diffusion. In the first part of this chapter we shall see how these four laws are rationally explained by the particle theory of matter, thus providing experimental support for the latter.

**Boyle's Law.** Boyle's law states that *at a given temperature the volume of a given mass of gas is inversely proportional to the pressure.*

That is,

$$v \propto \frac{1}{p}$$

Or, $\qquad\qquad pv = \text{a constant}$

Thus, if a gas has a volume $v_1$ at a pressure $p_1$ and the pressure is changed to $p_2$, the new volume, $v_2$, at constant temperature is given by the equation

$$p_1 v_1 = p_2 v_2$$

Boyle's law can equally well be stated in the form:

*At a given temperature the pressure of a given mass of gas is inversely proportional to the volume.*

That is, the pressure is doubled when the volume is halved, etc. As we shall see shortly, this form of the law is more useful for discussing the nature of gaseous pressure.

Graphs are convenient for representing how the magnitude of one quantity varies with that of another. With a given mass of gas we have three variables to consider, volume, pressure, and temperature; the magnitude of any one depends on the magnitude of the other two. A single two-dimensional graph cannot show the relation between three dependent variables. If one of the variables is constant, however, we can show the relation between the other two. Thus, if temperature is constant we can plot the volume of a gas against its pressure and thus obtain a graph of the form shown in Fig. 2–1 (a rectangular hyperbola).

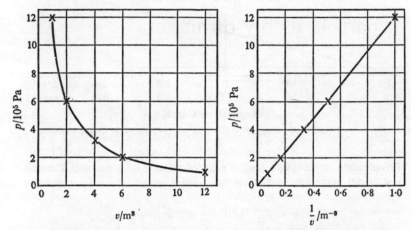

*Fig.* 2–1. Variation of volume of a gas with pressure (at constant temperature)

*Fig.* 2–2. Variation of reciprocal of volume of a gas with pressure (at constant temperature)

If we plot the pressure against the reciprocal of the volume we obtain a straight-line graph passing through the origin (Fig. 2–2). This is a more useful form of graph; it always indicates that *one plotted variable is directly proportional to the other*. Thus, from Fig. 2–2, we can immediately deduce that $p \propto 1/v$.

**Charles's Law:** *At a given pressure the volume of a given mass of gas is directly proportional to the absolute, or kelvin, temperature.*

Charles's law followed from the discovery that, at constant pressure, the volume of a given mass of gas increases (or decreases) by 1/273 of its volume at 0° Celsius for each degree rise (or fall) in temperature. Thus, if a gas has a volume of 273 m$^3$ at 0°C, the volume would be 272 m$^3$ at −1°C, 271 m$^3$ at −2°C, and zero at −273°C. The last temperature, at which all gases would theoretically have zero volume, is the *absolute zero*. As explained later, −273°C, or more accurately −273·16°C, is theoretically the lowest possible temperature. In practice we cannot reduce the temperature of a gas to absolute zero because all gases liquefy above this temperature.

Fig. 2–3 shows how Charles's law can be expressed graphically, the pressure being fixed.

The scale of temperature which has −273°C as zero is called the *thermodynamic*, or *kelvin*, scale. Degrees Celsius are converted to kelvin (symbol K with no degree sign) by adding 273. Thus 15°C is 288 K.

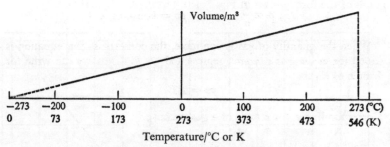

| | −273 | −200 | −100 | 0 | 100 | 200 | 273 (°C) |
| | 0 | 73 | 173 | 273 | 373 | 473 | 546 (K) |

Temperature/°C or K

*Fig.* 2–3. Variation of volume of a gas with temperature (at constant pressure)

According to Charles's law, if $v$ is the volume of a gas and $T$ is the kelvin temperature, we have at constant pressure

$$v \propto T \qquad \text{or} \qquad \frac{v}{T} = \text{a constant}$$

We see that if a gas has a volume $v_1$ at a temperature $T_1$, and the temperature is changed to $T_2$, the new volume, $v_2$, at constant pressure, is given by the equation

$$\frac{v_1}{T_1} = \frac{v_2}{T_2}$$

$$\text{or} \quad v_2 = v_1 \times \frac{T_2}{T_1}$$

It follows from the last equation that if a gas has a certain volume at 0°C the volume will be doubled at 273°C, if the pressure remains constant. To reduce the volume to its previous value the pressure would have to be doubled (Boyle's law). From this example we see that Charles's law can be put in the form:

*The pressure of a given mass of gas is directly proportional to the kelvin temperature if the volume is kept constant.*

This again is a more useful form of the law for discussing the nature of gaseous pressure.

**Ideal Gas Equation.** Boyle's law and Charles's law can be combined to give a single equation which represents the relation between the pressure, volume, and kelvin temperature of a given mass of gas under different conditions. Thus,

$$p \propto \frac{1}{v} \text{ at constant temperature (Boyle's law)}$$

$$p \propto T \text{ at constant volume (Charles's law)}$$

$$\therefore \quad p \propto \frac{T}{v} \quad \text{or} \quad pv = \text{constant} \times T$$

When the quantity of gas is one mole, the constant in this equation is called the *molar gas constant*, represented by $R$. Thus we can write for 1 mole of a gas

$$pv = RT$$

Correspondingly, for $n$ mole of a gas we have

$$pv = nRT$$

This is known as the *ideal gas equation*, because it holds only when gases are behaving as 'ideal', or 'perfect', gases. The conditions under which gases behave as ideal gases are discussed later.

Since for a given mass of gas we can write $pv/T = $ a constant, we have

$$\frac{p_1 v_1}{T_1} = \frac{p_2 v_2}{T_2}$$

where $p_1$, $v_1$, and $T_1$ refer to one set of conditions, and $p_2$, $v_2$, and $T_2$ refer to a different set of conditions. If we know any five of the six values in this equation, we can calculate the sixth. This is often very useful—for example, in comparing the densities of two gases we calculate the volumes of known masses of the gases under the same conditions of temperature and pressure. The values chosen for these are usually 0°C, or 273 K, and 101 325 N m$^{-2}$, or 101 325 Pa, where Pa stands for the pressure unit called the *pascal*. The values given above are known as *standard temperature and pressure* (s.t.p.). The standard pressure is equal to 760 mmHg pressure (see next section). If we know the volume of a gas at a certain temperature and pressure we can calculate its volume at s.t.p. as now shown.

**Example.** *A certain mass of a gas has a volume of 241 m$^3$ at 18°C and 100 400 Pa pressure. What would its volume be at s.t.p.?*

*Note.* In calculations of this type it is important to remember to change the temperature from degrees Celsius to the kelvin scale.

Let $p_1$, $v_1$, $T_1$ refer to the first set of conditions and $p_2$, $v_2$, $T_2$ to the second. Then, taking standard pressure to be 101 300 N m$^{-2}$ approximately, we have

$$\frac{p_1 v_1}{T_1} = \frac{p_2 v_2}{T_2}$$

$$\frac{100\,400 \times 241}{(18 + 273)} = \frac{101\,300 \times v_2}{(0 + 273)}$$

$$v_2 = \frac{100\,400 \times 241}{291} \times \frac{273}{101\,300} \text{ m}^3$$

$$= 224 \text{ m}^3$$

**Calculation of the Molar Gas Constant R.** The value of $R$ depends on the units adopted for the quantities in the equation $pv = nRT$. The units now used are those of the Système International d'Unités (usually abbreviated to SI). In this system the basic units of mass, length, and time are the kilogram (kg), the metre (m), and the second (s). (Note that the letter s is never added to the name of an SI unit or to its symbol to express a plural.)

For 1 mole of substance $n = 1$, and

$$R = \frac{pv}{T}$$

Since $T$ has no dimensions of mass, length, and time we have

$$R = \frac{\text{force}}{\text{area}} \times \text{volume} = \text{force} \times \text{length}$$

In SI the derived unit of force is the *newton* (N), the force which, acting on a mass of 1 kg, gives the latter an acceleration of 1 metre per sec$^2$. We thus have

Force in newton (N) $= \text{kg m s}^{-2}$

The product of force and length represents *work* or *energy*. In SI the unit for these is the *joule* (J), which is the work done when a force of 1 newton acts over a distance of 1 metre.

Work in joule (J) $= \text{force} \times \text{length}$

$$= \text{N m} = \text{kg m}^2 \text{s}^{-2}$$

$R$ must therefore be expressed in joule per mole kelvin (J mol$^{-1}$ K$^{-1}$).

The value of $R$ can be calculated from the fact that 1 mole of any gas occupies a volume of 22·4 dm$^3$, or 0·002 4 m$^3$, at s.t.p. A pressure of 760 mmHg is the pressure exerted by a column of 760 mm, or 0·760 m, of mercury of density 13 600 kg m$^{-3}$ at a place where the acceleration due to gravity is 9·81 m s$^{-2}$. Hence

Standard pressure $= 0·76 \times 13\,600 \times 9·81$ N m$^{-2}$

$$= 101\,325 \text{ N m}^{-2}$$

$$R = \frac{pv}{T} = \frac{101\,325\,\text{N m}^{-2} \times 0.022\,4\,\text{m}^3}{273}\,\text{mol}^{-1}\,\text{K}^{-1}$$

$$= 8.31\,\text{N m mol}^{-1}\,\text{K}^{-1}$$

$$= 8.31\,\text{J mol}^{-1}\,\text{K}^{-1}$$

### Dalton's Law of Partial Pressures

This law states that *for a mixture of gases the total pressure is equal to the sum of the partial pressures of the constituent gases.*

The partial pressure of each of the constituent gases is the pressure which it would exert if it alone occupied the gaseous volume. It is difficult to verify this law directly. It was deduced by Dalton from experiments on the volumes of gases dissolved from a mixture of gases by a solvent; the volumes dissolved depended on the partial pressures of the gases (see later).

Dalton's law of partial pressures is important in correcting the volume of a gas to s.t.p. when the gas has been collected in a graduated vessel over water. The pressure inside the vessel is made equal to that of the atmosphere by 'levelling' the water inside and outside the vessel. The volume is then measured and the barometer reading noted. The gas in the vessel, however, is saturated with water vapour, and this is responsible for part of the pressure. To obtain the pressure of the gas alone we must therefore subtract from the atmospheric pressure the pressure of saturated water vapour at the given temperature. Some values of the pressure of saturated water vapour at different temperatures are given later.

**Graham's Law of Diffusion.** The diffusion, or intermingling, of the particles of substances when placed in contact is not confined to gases. It occurs also with liquids and solids. We therefore postpone consideration of the fourth of our physical gas laws until we discuss diffusion in general.

**Kinetic Theory of Gases.** Boyle's law, Charles's law, and Dalton's law of partial pressures are all concerned with gaseous pressure. These laws are most reasonably explained by supposing that a gas consists of particles in a state of rapid and constant movement, and that the pressure is caused by bombardment of the walls of the containing vessel by the particles. In Chapter 1 we saw that the particles of which gases are composed are usually *molecules* consisting of two or more atoms. The extension of the particle theory which postulates that the molecules of a gas are in a state of movement is called the *kinetic theory* (Greek *kinesis*, motion).

The molecules in a gas must be more spread out in space than they are in a liquid or solid. This is shown both by the ease with which gases can be compressed as compared with liquids and solids, and by the large increase of volume which takes place when a liquid changes into vapour:

thus 1 cm$^3$ of water at 100°C and standard pressure gives 1 650 cm$^3$ of steam. According to the kinetic theory, the gaseous molecules travel in straight lines with velocities which are constantly changing as a result of the collisions of the molecules with each other. The molecules are assumed to be perfectly elastic, so that they rebound from each other and from the walls of the containing vessel without loss of momentum. This gives us a mental model of a gas which we can use to explain the laws of Boyle, Charles, and Dalton.

*Boyle's law.* Pressure is measured by force per unit area. The pressure exerted by a gas depends on the mass of the molecules, on the number striking a unit area of the walls per second, and on the average molecular velocity. Although the mass of a single molecule is very small this is compensated for by the large number of molecules and by their high velocity. If the volume of a given mass of gas is halved, at constant temperature, the number of molecules striking unit area of the walls in one second is doubled; therefore, the pressure is doubled. We see that the pressure is proportional to the number of molecules in a given volume of the gas—that is, to the concentration of the gas.

*Charles's law.* The temperature of a gas depends on the average kinetic energy of its molecules. Thus, if $m$ is the mass of a single molecule and $\bar{u}$ is the average molecular velocity, the average kinetic energy of the molecules is determined by $\frac{1}{2}m\bar{u}^2$. Hence the temperature depends on the average velocity of the molecules.[1] The speed of the molecules increases when the gas is heated, and decreases when it is cooled. Since an increase in average molecular velocity results in more molecules striking the walls of the container per second, the pressure of a gas at constant volume increases with rise of temperature, and decreases with fall of temperature.

If a gas could be cooled sufficiently and still kept as a gas, its molecules would eventually come to rest and have no kinetic energy. The gas would then be at zero temperature and would exert zero pressure. Now, by experiment it is found that, at constant volume, the pressure decreases by 1/273 of its value at 0°C for each degree fall in temperature. Hence a gas would have zero pressure at −273°C, which is therefore the absolute zero of temperature.

*Dalton's law of partial pressures.* If two or more gases are contained in a closed vessel the pressure is caused by impacts of the molecules of the different gases against the walls of the vessel. Each gas acts independently of the others, and its contribution to the total pressure is, therefore, the pressure which it would exert if it alone occupied the whole of the volume. The total pressure is the sum of the individual pressure of the gases.

---

[1] The average kinetic energy of the molecules is actually given by $\frac{1}{2}mu_{r.m.s.}^2$ where $u_{r.m.s.}$ is the *root mean square velocity* (see p. 55). However, the average molecular velocity, $\bar{u}$, is proportional to the root mean square velocity, and therefore the average kinetic energy is proportional to $\bar{u}^2$.

## Diffusion

**Diffusion of Gases.** If a light gas is brought into contact with a heavier gas the two gases mix together despite the difference in their densities. This can be shown by the following experiment.

**Experiment.** Fill a gas jar with bromine vapour by allowing the heavy vapour to roll down the side of the inclined jar and displace the air. Cover the jar with a glass plate. Fill a second jar with hydrogen and cover this also with a glass plate. Place the jars mouth to mouth, with the lighter hydrogen uppermost, and withdraw the plates (Fig. 2–4). The brown colour of the bromine vapour gradually becomes evenly spread through the two jars, showing that, despite its greater density, the bromine vapour has mixed with the hydrogen.

To test whether any hydrogen has travelled into the lower jar remove the plate from this jar and apply a lighted taper to the mouth of the jar. There will be a small explosion owing to the combination of hydrogen and bromine to form colourless hydrogen bromide ($H_2 + Br_2 \rightarrow 2HBr$).

The intermingling, or *diffusion*, of two gases when placed in contact is readily understood in the light of the kinetic theory. The molecules of both gases are moving. At the beginning of the experiment the partial pressure of the hydrogen in the lower jar and that of the bromine vapour in the upper jar are both zero. The molecules of both travel from the region of higher pressure to the one of lower pressure until their partial pressures become the same in both jars. This takes an appreciable time owing to the constant collisions between the molecules. When the gases become uniformly mixed a *dynamic equilibrium* is established. The rate at which the hydrogen molecules pass into the bottom jar equals the rate at which they return to the top jar; the same applies to the bromine molecules.

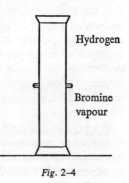

*Fig. 2–4*

Hydrogen

Bromine vapour

**Graham's Law of Diffusion.** The rates at which different gases diffuse are not equal; in general, a lighter gas diffuses more rapidly than a heavier one. This can be demonstrated by the experiment given below.

**Experiment.** Set up the apparatus shown in Fig. 2–5. P is a porous pot containing air. It allows gaseous molecules to pass through the very small pores in its walls, and normally, in air, molecules of oxygen and nitrogen are travelling in both directions through the walls. The porous pot is closed with a rubber stopper, through which a short glass tube passes. The latter is connected by rubber tube carrying a spring clip, to a pressure gauge made from two 25-cm³ pipettes. These have had part of the stems cut off, and are joined by rubber tubing. Coloured water is put into the pressure gauge. The porous pot is enclosed in a glass jar covered with a layer of cardboard, through which pass a short delivery tube for gases lighter than air and a long delivery tube for gases heavier than air.

First introduce hydrogen into the glass jar. Open the clip and note that the water in the pressure gauge moves away from the porous pot, showing that the pressure has increased inside the pot; the increase in pressure is caused by the hydrogen molecules diffusing into the pot more quickly than the oxygen and nitrogen molecules diffuse outwards.

Remove the hydrogen from the apparatus. This is most rapidly done by temporarily disconnecting the rubber tubing above the porous pot and blowing air through the longer delivery tube with a hand pump. Repeat the experiment using carbon dioxide in the glass jar. This time the coloured

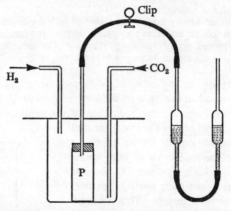

*Fig.* 2–5. Experiment to show that a lighter gas diffuses more rapidly than a heavier gas

water will move towards the porous pot, showing that there is a decrease in pressure inside the pot; this is because the molecules of oxygen and nitrogen diffuse outwards more quickly than the heavier carbon dioxide molecules diffuse inwards.

As a result of his experiments on the rates at which gases mix together Graham (1832) established the following law:

> *The relative rates at which two gases diffuse are inversely proportional to the square roots of their densities.*

That is,

$$\frac{R_1}{R_2} = \frac{\sqrt{\rho_2}}{\sqrt{\rho_1}}$$

where $R_1$ and $R_2$ are the rates of diffusion and $\rho_1$ and $\rho_2$ the densities of the gases ($\rho$ is Greek letter rho). *The density of a gas is the mass of one cubic metre of the gas at a given temperature and pressure* (usually at s.t.p., when it is called *normal density*).

According to the kinetic theory the relative rates of diffusion of two gases are determined by the average velocities of their molecules. If the gases are at the same temperature the average kinetic energies of their molecules are the same. If $m$ is the mass of one molecule and $\bar{u}$ is the average molecular velocity

$$\tfrac{1}{2}m_1\bar{u}_1^2 = \tfrac{1}{2}m_2\bar{u}_2^2$$

$$\text{or} \quad \frac{\bar{u}_1^2}{\bar{u}_2^2} = \frac{m_2}{m_1}$$

Thus the smaller their mass the faster molecules move, the average molecular velocity being inversely proportional to the square root of the relative molecular mass. We saw in Chapter 1 that the ratio of the masses of the molecules is the same as the ratio of the densities, $\rho$, of the gases at the same temperature and pressure. Hence

$$\frac{\bar{u}_1^2}{\bar{u}_2^2} = \frac{m_2}{m_1} = \frac{\rho_2}{\rho_1}$$

$$\text{or} \quad \frac{\bar{u}_1}{\bar{u}_2} = \frac{\sqrt{\rho_2}}{\sqrt{\rho_1}}$$

Thus the average velocities of the molecules, and therefore their relative rates of diffusion, are inversely proportional to the square roots of their densities. The experimental method of testing Graham's law is described in Chapter 3.

**Uses of Gaseous Diffusion.** The different rates of diffusion of gases can be used to separate them, although complete separation by this method is difficult. One process used in the U.S.A. for extracting helium from natural gas is based on the rapid diffusion of the light helium (as compared with that of the hydrocarbon gases) through a very thin sheet of Pyrex glass.

Diffusion has also been used to separate isotopes (forms of an element with different atomic mass). The isotopes of neon and chlorine have been partially separated in this way. The most important example of separating isotopes by diffusion occurs with uranium, which is used in the nuclear reactors of atomic energy plants. Uranium required for the spontaneous nuclear fission reaction has a relative atomic mass of 235. This occurs to the extent of only one part in 140 in the naturally occurring metal, most of which consists of atoms of relative atomic mass 238. The lighter atoms are partially separated from the heavier ones by converting the metal to uranium hexafluoride, $UF_6$, and allowing the vapour of this compound to diffuse repeatedly through porous barriers made of nickel. Subsequently the diffused vapour is reconverted to the metal, yielding suitably enriched uranium.

**Diffusion of Liquids and Dissolved Substances.** Diffusion is not confined to gases; it may also occur when two liquids are in contact.

> **Experiment.** Fill a dry test tube one third full of phenylamine. Slope the tube and slowly run some ethoxyethane on to the phenylamine. The lighter ethoxyethane forms a separate colourless layer on top (Fig. 2–6). Close the test tube with a tightly fitting stopper to prevent evaporation, and leave the tube standing in a place where it can be observed. After a day or two the boundary surface between the two liquids becomes indistinct as the liquids slowly diffuse into each other. Only after some weeks, however, has one homogeneous liquid formed.

The diffusion of liquids shows that their molecules, like those of gases, are moving. The slower diffusion of liquids as compared with gases is explained by the closer packing of the molecules and the greater frequency of collision. An important difference between gases and liquids is that, whereas any two gases form a homogeneous mixture as a result of diffusion, this happens only with certain pairs of liquids. Thus if we add paraffin or benzene to water the oily liquid floats on the water without mixing with it (except to a minute extent). This is explained later.

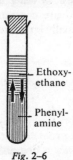

Ethoxy-
ethane

Phenyl-
amine

*Fig.* 2–6

Further simple experiments show that the particles of a dissolved substance can also undergo diffusion, and hence must be constantly moving, like the molecules of the solvent itself.

> **Experiment.** Introduce about 5 cm$^3$ of bromine water into a test tube and cover the solution with a layer of benzene. Stopper the tube and leave it on one side. Some of the reddish-brown bromine molecules will diffuse into the upper benzene layer.

**Diffusion of Solids.** In solids movement of particles is usually restricted to vibration about a mean position, but under suitable conditions the particles can also move from place to place. Particles on the outside of a solid have more freedom of movement than those inside, and if their thermal energy becomes sufficiently high they may leave the main body of the solid. Thus if a block of lead is clamped to a block of gold, and left for some months, the gold diffuses into the lead, the rate of diffusion increasing with rise of temperature. Diffusion is possible because the smaller gold atoms are able to move into vacant spaces between the larger lead atoms.

Most crystals are imperfect and contain 'holes', or vacant sites. These can be filled by migration of neighbouring particles, thus creating new vacant sites, which are filled in turn. In this way both particles and vacant sites may travel considerable distances through a solid.

## Brownian Movement

**Brownian Movement in Liquids.** In 1827 Robert Brown, a botanist, while examining a suspension of pollen grains in water under a microscope, observed that the grains were not stationary, but moved in zigzag fashion over small distances. This phenomenon, which is called *Brownian movement*, can be demonstrated as now described.

> **Experiment.** Various kinds of very small particles can be used, such as colloidal carbon (graphite) or a suspension of fine magnesium(II) oxide in water. Excellent results can be obtained with toothpaste. Place a drop of the suspension on a microscope slide, cover the liquid with a cover slip, and observe under a high-power microscope (magnification 450–500). The smaller the particles, the more vigorous is their movement.

Brownian movement occurs in many other liquids besides water and provides direct evidence that the molecules of a liquid are constantly moving. The solid particles move because they are continually bombarded

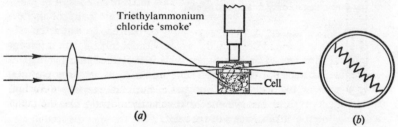

*Fig.* 2–7. (*a*) Method of showing Brownian movement of particles in air;
(*b*) Zigzag track of a particle

by molecules of the liquid. The particles would not move if at any instant they received an equal number of impacts on all sides from liquid molecules having the same velocity. This does not happen. At one moment the particle is struck by a greater number of molecules on one side than on another, and also the molecules move with different velocities. A momentum in a definite direction is therefore given to the particle, but this is altered the next moment by further impacts. Larger particles receive a greater number of impacts and the effects of the impacts on different sides tend to cancel out, so that movement is less marked.

**Brownian Movement in Gases.** Brownian movement also occurs when small particles are suspended in a gas. It can be shown to take place in air by the method described below.

> **Experiment.** In this experiment 'smoke' particles are suspended in air in a small glass cell. The particles are viewed through a *low-power* microscope while the cell is illuminated by a powerful beam of light at right angles to the direction of observation (Fig. 2–7*a*). The chief difficulty

arises from convection currents, which cause very fine particles to move too rapidly across the field of view for them to be observed satisfactorily. This difficulty can be overcome by having a sufficiently small cell and using coarser particles than in the previous experiment. Particles of suitable size can be obtained by mixing the vapour of a secondary or tertiary amine with hydrogen chloride (ammonium chloride 'smoke' is less satisfactory).

Cut off the open end of an ordinary test tube about 1 cm from the end, and cement the flanged end to a microscope slide. Dip small strips of filter paper into a 33 per cent solution of triethylamine and concentrated hydrochloric acid respectively, and place the wet ends close together inside the cell. 'Smoke' particles of triethylammonium chloride are formed. Cover the cell with a disc of cellophane and fasten this to the outside of the cell with a rubber band. Concentrate a beam of light from a 250-watt projector lamp by means of a lens so that the light is brought to a focus just below the cellophane cover. *Using a low power lens* (a magnification of 100 is suitable) focus the microscope just below the cellophane. Wait for a few minutes to allow the convection currents in the cell to die down and the coarser particles to settle. The smaller particles in suspension can then be observed moving across the field of view in zigzag fashion (Fig. 2–7b). The irregular motion is caused by uneven bombardment of the particles by molecules of oxygen and nitrogen.

## Diffraction of X-Rays by Crystals

**Internal Structure of Solids.** Solids are often divided into two classes: *crystalline* substances, like common salt; and *amorphous*, or non-crystalline, substances, like sealing wax and glass. Even before modern methods of investigation there was evidence that substances in the first class have a definite internal arrangement of particles, while those in the second class have a more haphazard distribution of particles. Thus crystals form in characteristic shapes bounded by plane surfaces that meet at definite angles. When crystals are struck they usually fracture along particular planes called *cleavage planes*. Again, physical properties, such as refractive index and thermal and electrical conductivity, often vary in different directions in a crystal. Finally, when a crystalline solid is heated, it melts at a definite temperature, suggesting a sudden change in the distribution of its particles.

Amorphous solids have no regular outline, possess no cleavage planes, and usually show no variation in properties in different directions. They do not melt sharply at a definite temperature, but, when heated, gradually soften and so reach the liquid stage. This indicates that there is no abrupt change in particle arrangement. Amorphous substances are often regarded not as true solids, but as supercooled liquids because, like liquids, they do not possess the orderly internal structure associated with crystals.

In 1912 it was suggested by von Laue that X-rays might be used to obtain direct evidence of the orderly arrangement of particles in crystals. When ordinary monochromatic light (light of a single wavelength, such as sodium light) is passed through a diffraction grating (a glass plate on which parallel lines are ruled very close together) a series of alternating light and dark lines is produced. Von Laue suggested that if a crystal consists of particles arranged close together in parallel planes it should be possible to use it as a diffraction grating. He considered that ordinary light would be useless for the purpose because its wavelength is much longer than the distance then thought to exist between the crystal planes. A necessary condition for diffraction is that the distance and the wavelength should be of the same order of magnitude. Von Laue suggested

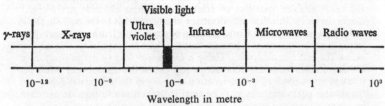

| γ-rays | X-rays | Ultra violet | Visible light | Infrared | Microwaves | Radio waves |
|---|---|---|---|---|---|---|

| $10^{-12}$ | | $10^{-9}$ | | $10^{-6}$ | | $10^{-3}$ | | 1 | | $10^{3}$ |

Wavelength in metre

*Fig.* 2–8. The electromagnetic spectrum (the different regions overlap to some extent)

that X-rays might be suitable because their wavelengths are thousands of times smaller than those of visible light. (Fig. 2–8.)

His predictions were strikingly confirmed. When a narrow pencil of X-rays was passed through various kinds of crystals and then allowed to fall on a photographic plate, a number of bright spots were formed on the plate (Fig. 2–9). The spots had a regular distribution that was

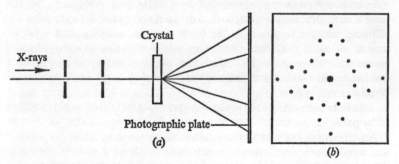

*Fig.* 2–9. (*a*) Diffraction of X-rays by a crystal; (*b*) Laue pattern from a crystal of sodium(I) chloride

characteristic for the particular substance. They could be explained only if the crystal acted as a three-dimensional diffraction grating formed by parallel layers of particles. Furthermore, the fact that diffraction took place showed that the distance between the layers (and hence between the particles) was of the same order as the wavelength (about 100 pm, or $10^{-10}$ m) of the X-rays.

X-ray analysis reveals that nearly all solids are crystalline, even when their appearance would suggest otherwise. Thus, powders like chalk and magnesium(II) oxide consist of crystals, as do all metals when solid. The kind of diffraction pattern obtained with amorphous solids is different from that given by crystalline substances; it consists of a series of diffuse light and dark circles, and resembles that obtained with liquids. Formerly it was thought that the particles in amorphous solids and in liquids have a completely disorderly arrangement, but more detailed investigation has shown that the structural difference between these and crystalline substances is really the degree of order. In a crystalline solid like sodium(I) chloride there is 'long-range' order, the regular arrangement of particles extends over a distance that is large compared with the distance between the particles. In amorphous solids there is only 'short-range' order. Thus 'amorphous' carbon (charcoal) is microcrystalline; it consists of extremely small crystals of graphite which have a random distribution. As we shall see later, there is also evidence that in some liquids at least the molecules can arrange themselves to a limited extent.

X-rays have become one of the most important tools for investigating the structures of substances. We can now measure not only the distances between particles, but the sizes of the particles themselves. A more detailed description of the use of X-rays is given in Chapter 9.

**Summary of Physical Evidence Given for the Particle Theory.** To explain the physical gas laws, diffusion, and Brownian movement we must take the particle theory of matter a stage further than the simple theory of atoms and molecules developed by Dalton and Avogadro. In the extended theory, the kinetic theory, we regard these particles as continually moving, whether they exist in the form of a gas, a liquid, or a solid, or are dissolved in a solvent. From the rates of diffusion of different gases at the same temperature, we deduce that the average velocities of gaseous molecules vary with the density of the gas, being larger for a lighter gas than a heavier one.

A further extension of the particle theory explains the physical properties of crystals—in particular, how they act as diffraction gratings for X-rays. The diffraction patterns indicate that the particles in solids are usually arranged in a *regular manner*, which varies with the nature of the solid. The distances between the particles in a crystal must correspond roughly with the wavelength of the X-rays.

In this chapter we have given only some of the physical evidence which supports the particle theory of matter. Other evidence, involving further amplification of the theory, will appear in later chapters.

**Some Molecular Statistics.** Since molecules are not usually spherical their sizes vary in different directions. By measuring the thickness of oil films on water we can obtain some indication of molecular size. These films are found experimentally to have a thickness of about $10^{-9}$m. Since any of the films must be at least one molecule thick, the size of the oil molecules in the measured direction cannot be larger than about one nanometre (1 nm = $10^{-9}$ m). The size of simple diatomic molecules such as $O_2$ is usually expressed by the distance between the nuclei of the two atoms. This distance can be measured by X-ray analysis of the element in the solid form.

According to Avogadro's hypothesis the number of molecules in equal volumes of different gases at the same temperature and pressure is the same. *The number of molecules in 22·4 $dm^3$ of any gas at s.t.p. is called the Avogadro constant* ($N_A$ or $L$). 22·4 $dm^3$ is chosen because this is the volume which is occupied by one mole of a gas at s.t.p. Several methods (described later) are available for measuring the Avogadro constant. Strong evidence in support of Avogadro's hypothesis is that, although the experimental procedures used in these methods are quite different, the results obtained agree closely. The accepted value of the Avogadro constant is $6·023 \times 10^{23} \, mol^{-1}$.

We can determine the mass of one molecule of a gas by dividing the mass of 1 $cm^3$ of the gas by the number of molecules which it contains. Other interesting deductions concerning molecules can be made with the help of the fundamental gas equation (see next section). The following data apply to oxygen molecules at s.t.p.:

| | |
|---|---|
| Number of molecules in 1 $cm^3$ | $27 \times 10^{18}$ |
| Mass of one molecule/g | $53 \times 10^{-24}$ |
| O–O distance/m | $0·132 \times 10^{-9}$ |
| Average distance between molecules/m | $3 \times 10^{-9}$ |
| Average molecular velocity/m s$^{-1}$ | $0·42 \times 10^{3}$ |
| Number of collisions per second | $4·2 \times 10^{9}$ |

We have seen earlier that the smaller the mass of the molecules the larger is their average velocity. Thus for hydrogen at s.t.p. the average molecular velocity (1 680 m s$^{-1}$) is four times that of oxygen molecules.

## Deduction of the Gas Laws from the Kinetic Theory

**Fundamental Gas Equation.** According to the kinetic theory the temperature of a gas is determined by the average kinetic energy of its molecules. The average kinetic energy ($\frac{1}{2}$ mass $\times$ velocity$^2$) depends on the average of the squares of the individual molecular velocities. If the latter at any instant are represented by $u_1, u_2, u_3, \ldots u_N$, the total number of molecules being $N$, we can write

$$u_{\text{r.m.s.}}^2 = \frac{u_1^2 + u_2^2 + u_3^2 + \ldots u_N^2}{N}$$

where $u_{\text{r.m.s.}}$ is the *root mean square velocity*. This is not the same as the average molecular velocity, $\bar{u}$, but it can be shown mathematically to be proportional, and nearly equal, to the average molecular velocity ($\bar{u} = 0.921\ u_{\text{r.m.s.}}$).

Applying to the gas molecules the laws of dynamics obeyed by ordinary colliding bodies, and assuming that the molecules are perfectly elastic, we can derive the following equation:

$$pv = \tfrac{1}{3}mNu^2_{\text{r.m.s.}},$$

where $p$ is the pressure of the gas, $v$ its volume, $m$ the mass of a single molecule, $N$ the number of molecules, and $u_{\text{r.m.s.}}$ the root mean square velocity. This is the *fundamental gas equation*. (Its deduction is usually given in text books of physics, and for this reason has been omitted here.) From it we can deduce the gas laws and the ratio of the specific heat capacities of a monatomic gas such as helium. This close agreement between theory and experiment strongly supports the kinetic explanation of the properties of gases.

### Deduction of the Gas Laws.

*Boyle's law.* According to the kinetic theory the kelvin temperature, $T$, of a gas is proportional to the average kinetic energy of its molecules, or, for a given mass of gas, to the total kinetic energy, $\tfrac{1}{2}mNu^2_{\text{r.m.s.}}$. If the temperature remains constant, $\tfrac{1}{2}mNu^2_{\text{r.m.s.}}$ is constant, and hence

$$pv = \tfrac{1}{3}mNu^2_{\text{r.m.s.}} = \text{a constant}$$

*Charles's law.* Again, the kelvin temperature, $T$, of a given mass of gas is proportional to the total kinetic energy, $\tfrac{1}{2}mNu^2_{\text{r.m.s.}}$. $T$ is thus proportional to $u^2_{\text{r.m.s.}}$. But $pv = \tfrac{1}{3}mNu^2_{\text{r.m.s.}}$. Hence for a given mass of gas at a given pressure $v$ is proportional to $u^2_{\text{r.m.s.}}$ and thus to $T$. That is, the volume of a given mass of gas is proportional to the kelvin temperature when the pressure is fixed, or, alternatively, the pressure is proportional to the kelvin temperature when the volume is fixed.

*Avogadro's hypothesis.* Consider equal volumes of two gases A and B at the same temperature and pressure. Since the pressures and volumes are equal $pv$ for A $= pv$ for B.

$$\therefore \quad \tfrac{1}{3}mNu^2_{\text{r.m.s.}} \text{ for } A = \tfrac{1}{3}mNu^2_{\text{r.m.s.}} \text{ for } B \qquad (1)$$

Since the gases are at the same temperature, the average kinetic energy of a molecule of gas A is the same as the average kinetic energy of a molecule of gas B.

That is,

$$\tfrac{1}{2}mu^2_{\text{r.m.s.}} \text{ for } A = \tfrac{1}{2}mu^2_{\text{r.m.s.}} \text{ for } B \qquad (2)$$

Combining (1) and (2), we have

$$N \text{ for } A = N \text{ for } B$$

Therefore equal volumes of two gases at the same temperature and pressure contain the same number of molecules.

*Graham's law.* Since $pv = \frac{1}{3}mNu^2_{r.m.s.}$, we have $u^2_{r.m.s.} = 3pv/mN$. The density, $\rho$, of a gas is the mass per unit volume, that is, $mN/v$.

$$\therefore \quad u^2_{r.m.s.} = \frac{3p}{\rho}$$

If the pressure is fixed

$$u_{r.m.s.} \propto \frac{1}{\sqrt{\rho}}$$

We have already stated that the root mean square velocity, $u_{r.m.s.}$, is proportional to the average molecular velocity, $\bar{u}$. Therefore

$$\bar{u} \propto \frac{1}{\sqrt{\rho}}$$

That is, the average velocity of the molecules, and hence their rate of diffusion, is inversely proportional to the square root of the density of the gas. This is Graham's law of diffusion.

**Ratio of Molar Heat Capacities of a Gas.** If a gas is heated at constant pressure energy is absorbed in increasing the kinetic energy of the molecules, and in increasing the volume against the external pressure. If the gas is heated at constant volume no energy is absorbed in expansion. Hence a gas has two specific heat capacities, one at constant pressure, and the other at constant volume, the former being the greater. In the case of liquids and solids expansion is small and its effect can be ignored.

The heat capacities of gases are usually expressed per mole, and are called *molar heat capacities*. Since a mole of different substances contains the same number of molecules, molar heat capacities represent the heat required to raise the temperature of equal numbers of molecules of the substances by one degree kelvin. The molar heat capacity at constant pressure is denoted by $C_p$, and that at constant volume by $C_V$. The ratio of $C_p$ to $C_V$ is expressed by $\gamma$.

Consider the increase in kinetic energy when 1 mole of a gas is heated through 1 kelvin.

$$pv = RT = \frac{1}{3}mNu^2_{r.m.s.}$$

But        Average kinetic energy $= \frac{1}{2}mNu^2_{r.m.s.} = \frac{3}{2}RT$

Therefore the increase in average kinetic energy for 1 kelvin rise in temperature is $\frac{3}{2}R$, which equals $12 \cdot 46$ J mol$^{-1}$ K$^{-1}$, since $R$ is approximately $8 \cdot 31$ J mol$^{-1}$ K$^{-1}$.

Now consider the energy absorbed in increasing the volume when 1 mole of gas is heated through 1 kelvin at constant pressure. Let the increase in volume be $v'$.

Before heating,

$$pv = RT \tag{1}$$

After heating,

$$p(v + v') = R(T + 1)$$
$$\therefore \quad pv + pv' = RT + R \tag{2}$$

Subtracting (1) from (2), we have

$$pv' = R$$

The product (pressure × increase in volume) represents the work done in increasing the volume and is equal to $R$ or $8.31$ J mol$^{-1}$ K$^{-1}$.

Thus, $C_p$ (the energy absorbed in raising the temperature of 1 mole of gas by 1 kelvin at constant pressure) equals $\frac{3}{2}R + R$, or $20.77$ J mol$^{-1}$ K$^{-1}$, while if the heating is done at constant volume the energy absorbed ($C_V$) is only $\frac{3}{2}R$ or $12.46$ J mol$^{-1}$ K$^{-1}$.

$$\therefore \quad \gamma = \frac{C_p}{C_V} = \frac{20.77 \text{ J mol}^{-1}\text{K}^{-1}}{12.46 \text{ J mol}^{-1}\text{K}^{-1}} = 1.66$$

This will only be true if energy is absorbed in the ways described. Gaseous molecules, however, may absorb energy in other ways. When a molecule consists of two or more atoms, energy may be absorbed *internally* by the molecule. Thus if the atoms are vibrating energy may be used to increase the speed of vibration. This internal absorption of energy will take place whether the heating occurs at constant pressure or constant volume. Representing the energy absorbed internally by $E$, we can write

$$\gamma = \frac{C_p}{C_V} = \frac{20.77 \text{ J mol}^{-1}\text{K}^{-1} + E}{12.46 \text{ J mol}^{-1}\text{K}^{-1} + E}$$

The larger the value of $E$, the smaller is the ratio. If no energy at all is absorbed internally $E = 0$ and $\gamma = 1.66$. The ratio of the molar heat capacities of the monatomic rare gases helium, neon, etc., approximates closely to $1.66$. For diatomic gases like hydrogen $\gamma$ decreases to about $1.40$, and for triatomic gases like carbon dioxide to about $1.30$. The ratio of the molar heat capacities therefore furnishes evidence as to the atomicity of the gas. The value of $\gamma$ can be determined by ascertaining the velocity of sound in the gas and substituting in the expression

$$c = \sqrt{(\gamma p/\rho)}$$

where $c$ = velocity of sound, $p$ = the pressure, and $\rho$ = the normal density.

### EXERCISE 2
*(Relative atomic masses are given at the end of the book)*

**1.** Calculate the volumes of the following gases when reduced to s.t.p.: (*i*) 1 500 m$^3$ of hydrogen at 27°C and 102 700 Pa pressure, (*ii*) 241 m$^3$ of air at 18°C and 100 400 Pa pressure. (Use 101 300 Pa as standard pressure.)

**2.** A certain gas has a volume of 75 m³ at 15°C and 104 000 Pa pressure. What would be the volume at 27°C and 98 700 Pa pressure?

**3.** A closed bulb contains a certain volume of gas at 21°C and 100 700 Pa pressure. Find (*i*) the pressure of the gas if the temperature is raised to 51°C, and (*ii*) the temperature to which the bulb must be heated to double the pressure.

**4.** A sealed flask contains oxygen at 17°C and 99 300 Pa pressure. What would be the pressure of the oxygen if the temperature were lowered to −23°C?

**5.** 0·12 g of a metal liberated from an acid 118 cm³ of hydrogen collected over water at 15°C and 101 700 Pa pressure. Find the equivalent mass of the metal. H = 1. Saturated aqueous vapour pressure at 15°C = 1 700 Pa. 1 g of hydrogen at s.t.p. occupies 11·2 dm³.

**6.** At s.t.p. 1 m³ of oxygen has a mass 1·54 × 10³ g and 1 m³ of carbon dioxide 1·98 × 10³ g. How much faster does oxygen diffuse than carbon dioxide?

**7.** A gas X diffuses four times as rapidly as sulphur dioxide under the same conditions. If the density of sulphur dioxide at the given temperature and pressure is 2·88 × 10³ g/m³ what is the density of X?

**8.** What is the kinetic theory? How does the theory account for (*i*) gaseous pressure, (*ii*) the different rates of diffusion of gases? What evidence is there that the molecules of liquid water are in a state of constant movement?

## MORE DIFFICULT QUESTIONS

**9.** (Part question.) Two gas burettes, one containing 10 cm³ of sulphur dioxide and the other 30 cm³ of hydrogen sulphide, both at atmospheric pressure, are separated by a stopcock. The connecting stockcock is opened and the gases allowed to mix. Calculate the final pressure (in atmospheres) after reaction has ended and the apparatus has regained room temperature. (W.J.E.C.)

**10.** 2·20 g of carbon dioxide occupy 1 166 cm³ at 15°C and 102 700 Pa pressure. Calculate the value of the gas constant $R$ in joule per mole kelvin.

**11.** State (*a*) Charles's law and (*b*) Dalton's law of partial pressures. Describe concisely an experiment by which Charles's law could be verified.

(*i*) A volume of 100 cm³ of dry hydrogen, initially at 0°C and 1 atm pressure, is heated at constant pressure to 60°C. What is the final volume?

(*ii*) A volume of 100 cm³ of moist hydrogen at 0°C and 1 atm total pressure, confined over water, is heated (together with the water) at a constant total pressure to 60°C. Calculate the final volume of the gas.

Comment on any difference you find between the increases in (*i*) and (*ii*), making clear, with reasons, whether Charles's Law is or is not valid in the case of (*ii*).

Vapour pressure of water at 0°C = 700 Pa.

Vapour pressure of water at 60°C = 20 000 Pa. (J.M.B.)

CHAPTER 3

# Relative molecular masses
# in the gaseous state

**Relative Molecular Mass.** Since atoms of the same element may exist as isotopes of different mass it follows that molecules of the same element or compound may have different masses. However, as mentioned previously, isotopes of an element usually occur in a constant ratio, so that a constant average value is obtained when relative atomic masses or relative molecular masses are measured by the ordinary methods of the laboratory. Using the modern standard $^{12}C = 12$ for relative atomic masses and relative molecular masses, we define the latter as follows:

*The* **relative molecular mass** *of an element or compound is the average mass of its molecules on a scale on which one atom of the $^{12}C$ isotope of carbon has a mass of* 12 *units.*

The adoption of the $^{12}C = 12$ standard in place of the $O = 16$ standard has made little difference in the values of relative molecular masses, and the difference can usually be ignored. The chief use of relative molecular masses is to help in establishing molecular formulæ, and approximate values are satisfactory for this purpose. It is therefore still the practice to refer relative molecular masses found by ordinary methods to oxygen or hydrogen. This is done in the following pages.

The relative molecular mass of an element or compound is equal to the sum of the relative atomic masses of the atoms in the molecule. Thus for oxygen and sulphur dioxide we have

|  | $O_2$ | $SO_2$ |
|---|---|---|
|  | $16 \times 2$ | $32 + 32$ |
| Relative molecular mass: | 32 | 64 |

Note, however, that in practice molecular formulæ are derived from relative molecular masses, and not vice versa.

Many compounds, *e.g.*, sodium(I) chloride, consist, not of molecules, but of ions. They therefore do not possess a relative molecular mass.

**Meaning of 'Mole' and 'Molar'.** In Chapters 1 and 2 we used the term 'mole' provisionally for the relative atomic mass of an element or the relative molecular mass of a compound when these are expressed in gram. We also deduced (p. 27) that one mole of different elements contains the same number of atoms. We can show in the same way that one mole of different substances composed of molecules contains the same number of molecules. Thus a mole of oxygen ($O_2$) is 32 g, and a mole of sulphur dioxide is 64 g. But we see from the relative molecular masses of these gases that a molecule of sulphur dioxide has double the mass of an oxygen molecule. Hence the 64 g of sulphur dioxide must contain the same number of molecules as the 32 g of oxygen.

Again, since a sulphur dioxide molecule has twice the mass of a sulphur atom, the number of molecules in 64 g (1 mole) of sulphur dioxide must be the same as the number of atoms in 32 g (1 mole) of sulphur (S). We can also extend the term 'mole' to substances in which the elementary particles, or units, are ions or groups of ions, and again we can show that the number of particles in one mole is the same.

*A* **mole** *is the amount of a substance which contains the same number of specified particles, or units, as there are carbon atoms in 12 gram of carbon*—12 ($^{12}C$).

*Examples* (approximate masses of 1 mole)

1 mole of S          has a mass of 32 g
1 mole of $SO_2$      has a mass of 64 g
1 mole of $Na^+$      has a mass of 23 g
1 mole of $Cl^-$      has a mass of 35·5 g
1 mole of $Na^+Cl^-$ has a mass of 58·5 g

The term 'mole' replaces previously used terms such as 'gram-atom', 'gram-molecule', and 'formula-weight in gram'. It embraces all these terms.

'Molar' usually means 'per mole' ($mol^{-1}$). Thus the molar heat capacity of sulphur (S) is the heat capacity of 32 g of sulphur. The use of the term in 'molar solution' (a solution of concentration 1 mole per $dm^3$) is anomalous since concentration should be in $mol\ m^{-3}$.

**Molar Volume of a Gas.** The density of oxygen at s.t.p. is $1·429\ kg\ m^{-3}$, or $1·429\ g\ dm^{-3}$. Thus a mole (32 g) of oxygen ($O_2$) occupies at s.t.p. a volume of $32 \div 1·429\ dm^3 = 22·4\ dm^3$ approximately. But by Avogadro's hypothesis this volume of oxygen, chlorine, or any other gas at s.t.p. contains the same number of molecules. Hence the volume occupied by a mole of any gas at s.t.p. is approximately $22·4\ dm^3$. This is called the *molar volume* ($V_m$) of a gas at s.t.p.

As stated earlier, the actual number of molecules in 22·4 dm³ of a gas at s.t.p. is called the Avogadro constant ($N_A$), and is equal to 6·023 × 10²³ mol⁻¹. Thus 64 g of sulphur dioxide contain 6·023 × 10²³ $SO_2$ molecules. Since the number of units in one mole is always the same, it follows that 32 g of sulphur (S), 23 g of sodium ions (Na⁺), and 35·5 g of chloride ions (Cl⁻) also contain 6·023 × 10²³ particles.

## Relative Densities and Relative Molecular Masses of Gases and Vapours

Since the relative molecular mass of a gas is twice its relative density, we can find the former by comparing the mass of a known volume of the gas with the mass of an equal volume of hydrogen at the same temperature and pressure, and doubling the result obtained. It is not, however, necessary to weigh the hydrogen; this is a difficult operation to carry out accurately owing to the extreme lightness of the gas, and we know that a mole (2 g) of hydrogen ($H_2$) occupies 22·4 dm³ at s.t.p., or 1 g of hydrogen 11·2 dm³ at s.t.p. Hence by weighing a known volume of a gas we can find its relative density or relative molecular mass from the following values at s.t.p.:

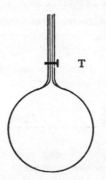

*Fig.* 3–1. A density globe

T

Relative density = mass in g of 11·2 dm³ ÷ 1 g

Relative molecular mass = mass in g of 22·4 dm³ ÷ 1 g

The method can also be applied to the vapours of substances which are liquids or solids at ordinary temperatures, providing the substances can be vaporized readily.

METHOD 1. *Direct Weighing of Gases* (*Regnault's Method*). In this method a strong glass globe (Fig. 3–1) which has a capacity of about one dm³ is used. The globe is fitted with a glass tap, T, and has a neck made of capillary tubing. The globe is evacuated by means of a suction pump, and it is then suspended from one side of a balance and weighed ($m_1$). The true mass of the globe is diminished by the mass of air displaced, which varies if the temperature or pressure alters during the experiment. Therefore a second globe of approximately the same size, and which contains air throughout the experiment, is suspended from the other side of the balance, so that the mass of air displaced on both sides of the balance is always the same.

After the first globe has been weighed completely empty it is filled with the gas, which should be pure and dry. The globe is filled and emptied (through the pump) several times to ensure that it does not contain any air from the connecting tubes. It is then weighed again. If this mass is $m_2$, the mass of gas filling the globe is $(m_1 - m_2)$.

To find the volume of the globe it is weighed when full of water $(m_3)$. Hence, if we assume the density of water to be 1 g/cm³, the volume in cm³ is numerically equal to $(m_3 - m_1)$. Temperature and pressure are noted,

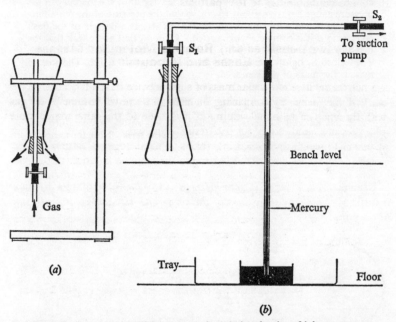

*Fig.* 3–2. Apparatus used to measure the relative density of laboratory gas

and the volume of the gas is reduced to s.t.p. (1mm Hg pressure = 133·3 Pa. The ratio of the mass of 11·2 dm³ or 22·4 dm³ of the gas at s.t.p. to 1 g then gives the relative density or relative molecular mass of the gas.

The next experiment is a simpler method of finding the relative density or relative molecular mass of a gas by direct weighing, and is more suitable for the school laboratory. It is a convenient method of measuring the relative density of laboratory gas, which is used later in testing Graham's law of diffusion for gases.

**Experiment.** Fit a dry 350-cm³ conical flask with a rubber stopper through which passes a glass tube reaching nearly to the bottom of the flask. Connect to the upper end of the tube a short length of PVC tubing (or rubber pressure tubing) carrying a screw clip. Poly(chloroethene) (PVC) tubing, which is flexible but does not cave in under reduced pressure, is more suitable than thick pressure tubing for making the short gas-tight connections. Invert the flask in a ring (Fig. 3–2a) and place the apparatus in the fume cupboard. Pass a slow stream of the gas into the flask for 2 or 3 minutes, loosening the stopper so that the gas escapes round the sides. Shut off the gas, insert the stopper firmly, and after disconnecting from the gas supply immediately screw up the clip tightly. This leaves the flask full of gas at room temperature and pressure. Weigh the flask to three places of decimals.

Attach the flask to the apparatus shown in Fig. 3–2b. With the screw clips $S_1$ and $S_2$ open, draw out gas from the flask until the mercury in the pressure gauge has risen about 50 cm, so that the pressure of the remaining gas is about one third of an atmosphere. (There is a danger of the flask collapsing if it is completely exhausted.) Close $S_2$ first and check that the mercury level remains constant, thus ensuring that the flask is air-tight. Then close $S_1$ tightly, detach the flask, and weigh it again. The loss in mass is the mass of gas removed.

Invert the flask and open the screw clip under water in a large trough. Water enters the flask. Equalize the water levels as far as possible and close the clip again. Remove the flask and measure the volume of water which has entered. This is the volume, at laboratory temperature and pressure, of the gas which was displaced from the flask. Note the temperature and pressure and correct the volume to s.t.p. Find the relative density of the gas as illustrated below.

**Example**. *The mass of 267 cm³ of a gas at* 18 °C *and* 100 400 Pa *pressure is* 0·162 g. *Calculate the values of the normal density and relative density of the gas.*

Volume of gas converted to s.t.p.

$$= 267 \times \frac{273}{291} \times \frac{100\ 400}{101\ 300}\ \text{cm}^3$$

$$= 248\ \text{cm}^3$$

Mass of 248 cm³ of gas at s.t.p. = 0·162 g.

Mass of 1 dm³ of gas at s.t.p. $= 0 \cdot 162 \times \dfrac{1\ 000}{248}\ \text{g dm}^{-3}$

$= 0 \cdot 653\ \text{g dm}^{-3} = 0 \cdot 653\ \text{kg m}^{-3}$ (normal density).

Mass of 11·2 dm³ of gas at s.t.p. $= 0 \cdot 653 \times 11 \cdot 2\ \text{g} = 7 \cdot 31\ \text{g}$

Relative density $= 7 \cdot 31\ \text{g}/1\ \text{g} = 7 \cdot 31$.

METHOD 2. *Dumas' Method.* This method is suitable for liquids like trichloromethane or solids like sulphur, which can be easily turned into vapour by heat. The mass of vapour which fills a bulb of known volume at a measured temperature and pressure is found.

A bulb with a capacity of about 200 cm³ has the neck drawn out into a fine capillary tube (Fig. 3–3). The bulb is weighed ($m_1$) when full of air. Some of the liquid is introduced, and the bulb is immersed as completely as possible in a bath containing boiling water or some liquid maintained at a temperature ($t$) at least 20 degrees higher than the boiling point of the liquid. The liquid is rapidly vaporised, and the air contained in the bulb is expelled along with excess of vapour. When the vapour ceases to be expelled, and the bulb remains filled with vapour at atmospheric pressure ($p$), the end of the capillary tube is rapidly sealed by a blow-pipe flame. The bulb is cleaned, dried, and weighed. This gives the mass ($m_2$) of bulb + vapour. The neck of the bulb is broken under water, which enters to take the place of the now condensed vapour. The combined mass ($m_3$) of the bulb full of water and the fragments of glass broken off is obtained.

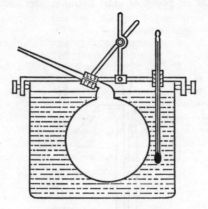

*Fig.* 3-3.  Dumas' apparatus

Assuming the density of water to be 1 g/cm³, we have

$$\text{Volume of bulb} = (m_3 - m_1) \div 1 \text{ g cm}^{-3} = v$$

The mass of air originally in the bulb is too small to make an appreciable difference in the above calculation but not in the next one.

$$
\begin{aligned}
\text{Mass of bulb} &= m_1 - \text{mass of contained air} \\
&= m_1 - \text{mass of a volume } v \text{ of air} \\
&= m_4
\end{aligned}
$$

(assuming the density of air to be known).

Mass of vapour filling bulb = $m_2 - m_4$.

Therefore a volume $v$ of vapour at temperature $t$ and pressure $p$ has a mass of $m_2 - m_4$. As before, the relative density is numerically equal to the mass of 11·2 dm³ of vapour at s.t.p.

METHOD 3.  *Victor Meyer's Method.*  In this method the vapour from a known mass of liquid displaces its own volume of air, which is collected in a graduated tube and measured at atmospheric temperature and pressure.

The apparatus consists of two tubes (Fig. 3–4). The outer one, a boiling tube, contains a liquid which boils at a temperature at least 20°C above the boiling point of the liquid to be vaporized. The inner tube is fitted with a stopper and with a side tube near the top of the narrower part of the tube. Thus any gas which is evolved passes into a graduated tube filled with water and inverted in a trough of water. The graduated tube is not placed over the delivery tube until everything is ready.

It is important that during the experiment the temperature should remain as constant as possible. The flame is therefore shielded from

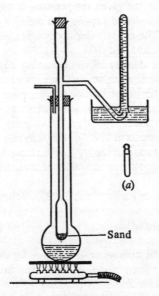

(a)

Sand

*Fig.* 3–4. Victor Meyer's apparatus

draughts, which would cause a temporary cessation of boiling and consequent sucking back. Care is also taken that the inner tube does not touch the heating liquid as the latter is liable to superheating, which would result in uneven temperature and irregular expansion and contraction of the air in the inner tube.

The liquid in the boiling tube is heated to boiling point, and air is expelled from the inner tube. When the evolution of air bubbles from the delivery tube has ceased the graduated tube is placed in position. The stopper at the top of the inner tube is removed and a small 'Hofmann' bottle (Fig. 3–4 a) containing a known mass of the liquid is dropped into the tube. The stopper is at once replaced to prevent air being driven out through the top of the inner tube. The bottle falls on to a little sand which prevents it from breaking, and immediately vaporization of the liquid occurs and the stopper is blown out of the bottle. Air is expelled into the graduated tube. Providing that the vaporization of the liquid has been rapid—and this we have ensured by the use in the boiling tube of a liquid of appreciably higher boiling point—the volume of air collected is equal to the volume of the vapour, not as it is, but as it would be if it were at the same temperature and pressure as the air in the graduated tube.

The volume of air collected in the graduated tube is measured after placing the tube in a deep vessel of water and levelling. The barometric pressure is noted, and, since the air is collected over water, the pressure due to water vapour at the given temperature is subtracted. The volume occupied by a certain mass of vapour at a definite temperature and pressure is now known. The relative density is again obtained from the mass of $11 \cdot 2$ dm$^3$ of vapour at s.t.p.

The relative vapour density of the following liquids can be found by Victor Meyer's method: trichloromethane, benzene, phosphorus trichloride, etc. (using water in the boiling tube); water, methylbenzene, etc. (using phenylamine, b.p. 184°C). For determinations at still higher temperatures the apparatus is made of silica glass. In this case a liquid like molten tin, which melts at 232°C, or boiling sulphur (b.p. 445°C) must be used in the boiling tube and the bulb of the inner tube must be immersed in the liquid. Relative vapour densities of substances vaporizing up to 2000°C have been determined by using a porcelain or iridium bulb with electrical heating. In this way the relative density of sodium(I) chloride vapour has been shown to correspond to NaCl particles and that of copper (I) chloride to $Cu_2Cl_2$ particles.

METHOD 4. *Diffusion or Effusion Method.* According to Graham's law of diffusion, the relative rates of diffusion of two gases are inversely proportional to the square roots of their densities. The ratio of the densities of two gases, A and B, is the same as the ratio of their relative densities. This is easily seen from the following:

Relative density of A : relative density of B

$$= \frac{\text{mass of 1 m}^3 \text{ of A}}{\text{mass of 1 m}^3 \text{ of hydrogen}} : \frac{\text{mass of 1 m}^3 \text{ of B}}{\text{mass of 1 m}^3 \text{ of hydrogen}}$$

$= $ mass of 1 m$^3$ of A : mass of 1 m$^3$ of B

$= $ density of A : density of B

Hence Graham's law can also be written in the form

$$\frac{R_1}{R_2} = \frac{\sqrt{d_2}}{\sqrt{d_1}}$$

where $d_1$ and $d_2$ are the relative densities of the gases.

In testing or applying Graham's law we make use, not of diffusion, but of *effusion*, which is closely related. Diffusion is the free intermingling of

two gases when placed in contact. Effusion refers to the driving of a gas
by pressure through a small opening. Graham's law applies equally well
to effusion as to diffusion, so that the relative density of a gas can be determined by comparing its rate of effusion with that of a gas of known relative
density. The relative rates of effusion are found from the times taken for
equal volumes of the gases to be driven through a fine aperture under the
same conditions. Since the rates of effusion are inversely proportional to
the times, $t$, we have

$$\frac{R_1}{R_2} = \frac{t_2}{t_1} = \frac{\sqrt{d_2}}{\sqrt{d_1}}$$

Or    $$\frac{t_1}{\sqrt{d_1}} = \frac{t_2}{\sqrt{d_2}} = \text{a constant (for different gases)}$$

The effusion times depend on the size of the aperture and on the
temperature.

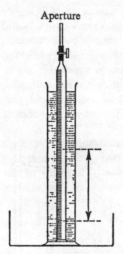

Fig. 3–5. Simple form of
effusiometer

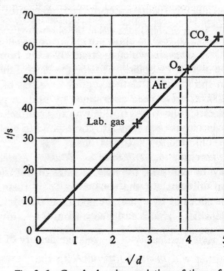

Fig. 3–6. Graph showing variation of time of
effusion with square root of relative density

**Experiment.** A simple effusiometer can be made as described below.
Cement a disc of thin aluminium foil to one end of a short length (5 cm)
of glass tubing. Pierce a minute hole in the foil with a very fine needle.
Join the tube by a rubber connection to a burette, from which the spout
has been removed. Place a spring clip on the rubber connection.

First find the effusion time for 20 cm³ of laboratory gas. Close the clip,
fill the burette with water, and clamp it upside down in a trough of water.
Fill the burette with the gas, taking care to remove air from the delivery
tube first. Stand a large measuring cylinder or a tall levelling jar full of
water in a small trough to catch the overflow (Fig. 3–5). Invert the burette
into the water. Open the clip and with a stop watch measure the time

required for the water to rise from the 5-cm³ mark to the 25-cm³ mark (it is convenient to have rubber bands placed at these levels). Repeat the measurement until three consistent times are obtained, and find the average time.

Repeat the experiment with oxygen and carbon dioxide in the burette. (If an oxygen cylinder is not available oxygen is obtained by adding a little manganese(IV) oxide to 20 cm³ of 10-volume hydrogen peroxide in a conical flask fitted with a stopper and delivery tube.) Assuming the relative densities of oxygen and carbon dioxide are 16 and 22 respectively, and using the relative density already found practically for the laboratory gas, plot a graph of times of effusion, $t$, against the square roots of the relative densities, $d$ (Fig. 3–6). The three points should lie on a straight line passing through zero $t$ and zero $\sqrt{d}$. A straight-line graph indicates that $t/\sqrt{d}$ is constant.

Use the graph obtained to find the relative density of air, measuring the time of effusion when the burette contains air.

*Notes.* (1) Graham's law does not apply strictly to a mixture of gases such as fuel gas or air. A lighter constituent effuses more rapidly than a heavier one and therefore the composition of the mixture alters. The law is approximately obeyed, however, if the effusion time is short.

(2) This method is not suitable for hydrogen, because the presence of water vapour in the gas in the burette makes a large difference to the effective relative density of hydrogen; as a result the effusion time for hydrogen is always appreciably larger than the theoretical value. Water vapour does not change the effective relative densities of fuel gas, oxygen, air, or carbon dioxide.

Fig. 3–7 shows an accurate effusiometer used in more advanced work. Mercury is used instead of water to avoid errors due to the presence of water vapour in the gas. The two cylindrical glass bulbs, P and Q, are of different sizes; the capacity of P is about 160 cm³ and that of Q about 50 cm³. Marks are etched on the glass at $M_1$ and $M_2$. By means of a three-way tap P can be connected either with a gas supply or with the tube R. Across the top of R is fixed a disc of platinum foil containing the aperture. A mercury reservoir, S, is attached to the lower end of Q by rubber tubing.

First, the air is expelled from the apparatus by raising S until the mercury reaches the three-way tap. A sample of pure dry gas is drawn

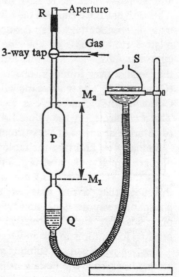

*Fig.* 3–7. An accurate form of effusiometer

into the apparatus by lowering S and is then expelled by again raising S. This is repeated several times to ensure that all the air has been removed.

Finally gas is drawn in until it fills P and most of Q. P and R are connected and S is raised to a height which is kept constant for experiments with different gases. Gas is driven through the aperture and the mercury rises in the bulbs. The purpose of the second bulb, Q, is to slow down the initial rate of rise so that the observer has time to clamp S and start a stop watch as soon as the mercury reaches the first mark $M_1$. The time required for the mercury to reach the second mark $M_2$ is noted. The experiment is repeated three or four times and the average value of the time is found.

**Relative Molecular Masses by Mass Spectrometer.** The most accurate method of measuring the relative molecular masses of gases or vapours is by mass spectrometer. This is described in Chapter 5.

## Molecular Formulae of Gases and Vapours

**Atomicity of Elements.** For elements the atomicity, or number of atoms in one molecule, is given by

$$\text{Atomicity} = \frac{\text{relative molecular mass}}{\text{relative atomic mass}}$$

We have already shown how this relationship can be used to fix relative atomic mass when the relative molecular mass and atomicity are known. Alternatively, if we can find the relative atomic mass and relative molecular mass independently, we can deduce the atomicity of the element. Thus the relative atomic mass of chlorine obtained by the mass spectrometer is approximately 35·5, and this number of gram is also the least mass of the element ever found in one mole of the volatile chlorine compounds (Cannizzaro). The relative molecular mass found by direct weighing is 71. Consequently the molecule is diatomic and the molecular formula is $Cl_2$. This is confirmed by the value, 1·4, of the ratio of the molar heat capacities of chlorine at constant pressure and constant volume.

Similarly we find that the molecular formulæ of hydrogen, oxygen, and nitrogen are $H_2$, $O_2$, and $N_2$. The relative molecular masses of the rare gases (helium, argon, etc.) are the same as the relative atomic masses; this shows that the rare gases exist as single atoms. The relative density of phosphorus vapour is 62, and hence the relative molecular mass of phosphorus is 124; since the relative atomic mass of phosphorus is 31, there must be four atoms in the molecule ($P_4$).

The relative density of sulphur vapour decreases with increase of temperature. Just above the boiling point (445°C) it is rather too low to correspond to $S_8$ molecules. At 900°C it corresponds to $S_2$ molecules, and at 2000°C to single atoms. Sulphur vapour furnishes an illustration of *thermal dissociation* (see later). It is possible that

$$S_8 \rightleftharpoons 2S_4 \rightleftharpoons 4S_2 \rightleftharpoons 8S$$

Iodine vapour similarly dissociates on heating $(I_2 \rightleftharpoons 2I)$. Mercury vapour and metals generally in the vapour state are monatomic.

**Molecular Formulæ of Compounds.** We can establish the molecular formulæ of compound gases and vapours in two ways. One method is to find first the empirical formula of the compound by dividing the relative masses of the elements present by their relative atomic masses. Then

$$\text{Molecular formula} = \text{empirical formula} \times n$$

where $n$ is (usually) a simple whole number such as 1, 2, 3, etc. The correct multiple is determined by the relative molecular mass.

Water contains 1 g of hydrogen to 8 g of oxygen. Dividing these masses by the relative atomic masses we obtain the ratio of the atoms:

$$\frac{1}{1} \text{ H atoms} : \frac{8}{16} \text{ O atoms}$$

$$= 2 \text{ H atoms} : 1 \text{ O atom}$$

The empirical formula of water (or steam) is thus $H_2O$. The relative density of steam (found by Victor Meyer's method) is 9. We now have

$$\text{Molecular formula of steam} = (H_2O)_n$$

$$\text{Relative molecular mass of steam} = (2 + 16)n = 18n$$

But from relative density the relative molecular mass of steam is 18

$$\therefore \quad 18n = 18$$
$$\text{and} \quad n = 1$$

The molecular formula of steam is thus $H_2O$.

Note that we have not proved that the molecular formula of liquid water is $H_2O$. The formula deduced by this method applies only when the substance is in the same state as when the relative molecular mass is found. There is strong evidence that in liquid water the $H_2O$ molecules are partially combined into larger aggregates (see Ch. 10).

The above method can be applied to gases like carbon dioxide, but it is usually easier to establish the molecular formulæ of gases from the volume relationships in certain chemical reactions. In some cases we must also know the relative density or relative molecular mass.

*Dinitrogen oxide* ($N_2O$). When strongly heated the gas decomposes at about 700°C, yielding only nitrogen and oxygen. The oxide, therefore, is composed of these two elements. When a known volume of the gas is heated with iron, copper, or sodium the oxygen combines with the metal and the volume of nitrogen left is equal to the original volume of dinitrogen oxide.

1 volume of dinitrogen oxide contains 1 volume of nitrogen

∴   1 molecule of the oxide contains 1 molecule of nitrogen (by Avogadro's hypothesis)

If we assume that one molecule of nitrogen contains two atoms, the formula must be $N_2O_x$.

The relative density = 22 and relative molecular mass = 44.

Since $N_2 = 28$ and the relative atomic mass of oxygen = 16.

$$x = 1$$

That is, the molecular formula = $N_2O$

By similar reasoning the molecular formulæ NO, $CO_2$, $SO_2$, $H_2S$, $PH_3$, and $AsH_3$ can be proved.

*Carbon monoxide (CO)*. The gas is known to consist of carbon and oxygen because it can be synthesized from these two elements. By exploding a known volume of carbon monoxide with a known volume (excess) of oxygen over mercury in a eudiometer and absorbing the carbon dioxide produced in aqueous potassium(I) hydroxide, it is found that:

2 volumes of carbon monoxide combine with 1 volume of oxygen to give 2 volumes of carbon dioxide

∴   2 molecules of carbon monoxide combine with 1 molecule of oxygen to give 2 molecules of carbon dioxide (by Avogadro's hypothesis)

1 molecule of carbon monoxide combines with $\frac{1}{2}$ molecule of oxygen to give 1 molecule of carbon dioxide

Assuming that oxygen is diatomic and that the formula of carbon dioxide is $CO_2$, we have

1 molecule of carbon monoxide = $CO_2 - O = CO$

*Note*. The relative density (14) of carbon monoxide is not used in the above proof, but it provides confirmatory evidence of the formula CO.

*Ammonia (NH₃)*. When ammonia contained over mercury in a eudiometer is subjected to electric sparks it is decomposed almost entirely into nitrogen and hydrogen (which are therefore its constituent elements). If a known volume is treated in this way, it is found that 2 volumes of ammonia yield approximately 4 volumes of nitrogen and hydrogen. By adding excess of oxygen and passing an electric spark, the hydrogen is removed as water. Since 2 volumes of hydrogen combine with 1 volume of oxygen, the volume of hydrogen in the gaseous mixture is equal to two thirds of the contraction produced by the explosion. It is thus shown that 2 volumes of ammonia give 1 volume of nitrogen and 3 volumes of hydrogen. By reasoning similar to that used in the case of carbon monoxide, we can then prove that the formula for ammonia is $NH_3$.

*Steam.* The volumetric composition of steam can be found by exploding together a mixture of two volumes of hydrogen and one volume of oxygen. The explosion is carried out in a U-type eudiometer containing mercury (Fig. 3–8). The explosion limb is surrounded by a jacket, through which passes the vapour of an alcohol of fairly high boiling point, *e.g.*, pentan-1-ol (b.p. 138°C). The volume is noted at this temperature and at atmospheric pressure after 'levelling' by running out mercury at the tap at the bottom of the open limb. Before the mixture is exploded a considerable part of the mercury is removed in this way to reduce the pressure of the gases and lessen the force of the explosion. After the explosion mercury is poured back into the open limb until the mercury is again at the same level in both limbs. It is then found that a contraction in volume of one third (represented by the cross-hatched portion in Fig. 3–8) has occurred.

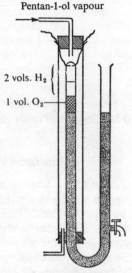

Fig. 3–8. Apparatus used to show the volumetric composition of steam

Thus

2 volumes of hydrogen combine with 1 volume of oxygen to give 2 volumes of steam

Again using Avogadro's hypothesis and the known atomicities of hydrogen and oxygen, we can show that the formula of steam is $H_2O$.

*Hydrocarbons.* A known volume of the gas is exploded with excess of oxygen over mercury in a eudiometer and the volume of carbon dioxide produced is determined by absorption in alkali. The following illustration shows how the formula of the hydrocarbon is deduced from the experimental results.

**Example.** *20 $cm^3$ of a gaseous hydrocarbon X were exploded with 120 $cm^3$ of oxygen. After the explosion the volume of gases remaining was 90 $cm^3$, and this decreased to 50 $cm^3$ on treatment with aqueous potassium(I) hydroxide. Calculate the molecular formula of X.*

Let the formula of X be $C_xH_y$. Then the equation for the combustion is:

$$C_xH_y + (x + \tfrac{1}{4}y)O_2 \rightarrow xCO_2 + \tfrac{1}{2}y\,H_2O$$

$$1\ cm^3 \quad (x + \tfrac{1}{4}y)\ cm^3 \quad x\ cm^3$$

$$20\ cm^3\ 20(x + \tfrac{1}{4}y)\ cm^3\ 20x\ cm^3$$

But

Volume of carbon dioxide produced $= 90 - 50 = 40 \text{ cm}^3$

and        Total volume of oxygen used $= 120 - 50 = 70 \text{ cm}^3$

$$\therefore \quad 20x = 40 \quad \text{from which } x = 2$$

and        $20\left(x + \dfrac{y}{4}\right) = 70 \quad \text{from which } y = 6$

$$\therefore \quad \text{Formula of } X = C_2H_6$$

Notice the prominent part played by Avogadro's hypothesis in the determination of molecular formulæ in the gaseous state.

## Abnormal Relative Densities of Vapours

**Association.** Sometimes the relative density found for a vapour is too high to agree with the simplest chemical formula of the substance. Thus, the vapour of iron(III) chloride just above its boiling point has a relative density about double that corresponding to $FeCl_3$. We must therefore conclude that the simple molecules are associated into $Fe_2Cl_6$ molecules. Other examples of association include aluminium(III) chloride ($Al_2Cl_6$), hydrogen fluoride ($H_2F_2$), arsenic (III) oxide ($As_4O_6$), and the oxides of phosphorus ($P_4O_6$ and $P_4O_{10}$). When the vapours are heated beyond the boiling points their relative densities usually decrease until they agree with the simpler formula—for example,

$$Fe_2Cl_6 \rightleftharpoons 2FeCl_3$$

**Thermal Dissociation.** Sometimes the value found for the relative density of a vapour is too low to agree with the simplest formula of the substance. Thus, the relative density of ammonium chloride vapour calculated from the formula $NH_4Cl$ is 26·75. If the relative density is determined by Dumas' method just above its sublimation temperature (350°C) the value obtained is only half the calculated value. The actual value corresponds to a molecule $N_{\frac{1}{2}}H_2Cl_{\frac{1}{2}}$, which is impossible. The low relative density is explained by dissociation of the salt into ammonia and hydrogen chloride.

$$NH_4Cl \rightleftharpoons NH_3 + HCl$$
(vapour)

Since one volume of ammonium chloride vapour would produce on dissociation two volumes of products the relative density would be halved if all the ammonium chloride were split up. The value obtained for the relative density shows that this does occur and that almost no ammonium chloride particles exist in the vapour. Ammonium chloride can be re-formed, however, by cooling the dissociated vapour, and therefore the reaction is written with the usual reversible reaction sign.

> **Experiment.** To show the thermal dissociation of ammonium chloride vapour put a little of the solid into a dry test-tube. About half way up the tube insert a porous plug of glass-wool or asbestos (Fig. 3–9). Use this to wedge a piece of damp blue litmus paper so that the latter is suspended in the lower part of the tube. Introduce a piece of damp red litmus paper into the tube above the plug. Gently heat the ammonium chloride. In a few moments the blue litmus paper will turn red, and the red litmus paper will turn blue. *Both* gases diffuse through the porous plug, but ammonia, having the smaller density, diffuses more rapidly than the hydrogen chloride (Graham's law). Thus above the plug we get an excess of ammonia, while below the plug there is an excess of hydrogen chloride.

**Further Examples of Thermal Dissociation.**

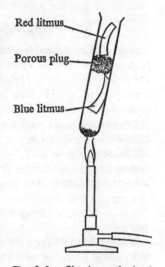

*Mercury(I) Chloride (Calomel).* Baker showed that if this substance is first subjected to prolonged drying by phosphorus(V) oxide the relative density of the vapour agrees with the formula $Hg_2Cl_2$. For the undried compound the relative density is only half of the value for the dried substance. This can be explained by dissociation in two ways, as represented by the equations

$$Hg_2Cl_2 \rightleftharpoons 2HgCl$$

$$Hg_2Cl_2 \rightleftharpoons Hg + HgCl_2$$

We can show that dissociation occurs into mercury and mercury(II) chloride, in accordance with the second equation, if we warm mercury(I) chloride in a flask containing a small piece of gold leaf; the mercury forms a grey amalgam on the surface of the gold. The manner of dissociation has been confirmed from the ultraviolet absorption spectrum of the vapour. Baker's experiment showed that the dissociation is prevented by intensive drying. He found that vigorous drying was equally effective in stopping the dissociation of ammonium chloride.

*Fig.* 3–9. Simple method of showing the dissociation of ammonium chloride vapour

*Phosphorus Pentachloride and Phosphorus Pentabromide.* The relative density of the vapour of both these substances decreases progressively as the temperature of the vapour is increased. Finally it reaches a constant value equal to half that calculated from the formulæ $PCl_5$ and $PBr_5$.

Thus, with these compounds the fraction of the vapour dissociated increases with rise of temperature until dissociation is complete.

$$PCl_5 \rightleftharpoons PCl_3 + Cl_2$$
$$PBr_5 \rightleftharpoons PBr_3 + Br_2$$

The presence of free chlorine or bromine in the vapour can be shown by warming a little of the substance in an open flask. The green colour of chlorine or the red colour of bromine appears and deepens as the temperature is increased. If a piece of starch iodide paper is enclosed in the flask the paper turns blue owing to liberated iodine.

*Dinitrogen Tetraoxide* ($N_2O_4$). At low temperatures this substance forms colourless crystals. These melt at $-11°C$ to a faintly yellow liquid. If the temperature is allowed to rise the colour of the liquid deepens to an orange colour, until at $22°C$ the liquid boils, giving off a reddish brown vapour. When the vapour is heated in a closed bulb the colour grows more intense until at $140°C$ it is brownish black. After this the colour diminishes and at $620°C$ it has completely disappeared. These colour changes are explained by dissociation, which takes place as follows:

$$N_2O_4 \rightleftharpoons 2NO_2 \rightleftharpoons 2NO + O_2$$

colourless  deep brown    colourless

At low temperatures the substance consists of colourless $N_2O_4$ molecules. As the temperature rises the coloured nitrogen dioxide ($NO_2$) molecules increase in number until at $140°C$ the gas is composed entirely of these molecules. Above $140°C$ further dissociation occurs into a colourless mixture of nitrogen oxide (NO) and oxygen, this dissociation becoming complete at $620°C$. The dissociations are accompanied by a progressive decrease in the relative density. The relative density corresponding to $N_2O_4$ is 46, while nitrogen dioxide consisting entirely of $NO_2$ molecules has a relative density 23. The value for the final mixture of nitrogen oxide and oxygen is 15·3 (one third of 46). The variation of the relative density with temperature between $25°C$ and $205°C$ is shown in Fig. 3–10.

**Calculation of Degree of Dissociation.** The degree of dissociation is the fraction of the original molecules which have undergone dissociation. It is usually expressed as a decimal fraction or as a percentage.

In all cases where dissociation takes place with increase in volume the density (or relative density) decreases and the amount of decrease depends on the degree of dissociation.

$$N_2O_4 \rightleftharpoons 2NO_2$$

1 volume   2 volumes

Let the degree of dissociation be $\alpha$. Then the fraction of $N_2O_4$ molecules remaining is $1 - \alpha$. If there were before dissociation $N$ molecules of $N_2O_4$, there will be after dissociation $N(1 - \alpha)$ molecules of $N_2O_4$.

Since each molecule of $N_2O_4$ which dissociates furnishes two molecules of $NO_2$, there will be after dissociation $2N\alpha$ molecules of $NO_2$.

Total number of molecules after dissociation
$$= N(1 - \alpha) + 2N\alpha$$
$$= N(1 - \alpha + 2\alpha)$$
$$= N(1 + \alpha)$$

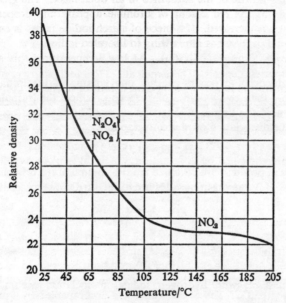

*Fig.* 3–10

Now for a given mass of gas the density is inversely proportional to the number of molecules present. Let $\rho_1$ = density if no dissociation occurs and $\rho_2$ = the observed density. Then

$$\frac{\rho_1}{\rho_2} = \frac{N(1 + \alpha)}{N} = \frac{1 + \alpha}{1}$$

Hence
$$\alpha = \frac{\rho_1}{\rho_2} - 1$$

(Knowing that $\rho_1/\rho_2$ must always be greater than 1, it is easy to remember which of the densities should be the numerator and which the denominator in this fraction.)

Using this formula we can calculate the degree of dissociation in the vapour of dinitrogen tetraoxide, ammonium chloride, phosphorus pentachloride, or any vapour where the volume is increased from one volume to two volumes on complete dissociation. Since densities and relative densities ($d$) are proportional to each other, the latter can be substituted for the former in the expression $\alpha = (\rho_1/\rho_2) - 1$.

**Example.** *The relative density of partially dissociated dinitrogen tetraoxide at 26·7°C is 38·3. The relative density of undissociated dinitrogen tetraxide* $N_2O_4$, *is 46. Find the degree of dissociation.*

$$\alpha = \frac{d_1}{d_2} - 1 = \frac{46}{38·3} - 1$$

$$= 1·20 - 1 = 0·20$$

$$= 20 \text{ per cent}$$

When dissociation occurs so that one volume of the initial vapour yields more than two volumes of gaseous products the formula for the calculation of the degree of dissociation must be modified accordingly. Thus in the dissociation of methanol vapour

$$CH_3OH \rightleftharpoons CO + 2H_2$$

3 volumes of products are obtained from 1 volume of the original vapour. In general, if $N$ gaseous molecules are formed from one molecule,

$$\frac{\rho_1}{\rho_2} = \frac{1 - \alpha + N\alpha}{1}$$

Hence

$$\alpha = \frac{\rho_1 - \rho_2}{\rho_2(N - 1)}$$

Sometimes, as with hydrogen iodide, dissociation occurs with no change of volume:

$$2HI \rightleftharpoons H_2 + I_2$$

The degree of dissociation cannot then be obtained by observation of the relative density, since the latter does not alter. Instead, the extent of dissociation is measured by a chemical method (see Ch. 15).

### EXERCISE 3
*(Relative atomic masses are given at the end of the book)*

**1.** Explain why one mole of different gases (*i*) contains the same number of molecules, (*ii*) occupies 22·4 dm$^3$ at s.t.p.

**2.** Dalton's view of the constitution of water may be represented by the formula HO, while nowadays the formula $H_2O$ is adopted. Discuss this change of view, and give an account of the principles underlying it.          (J.M.B.)

**Relative Gaseous Density and Relative Molecular Mass**

(Take the molar volume of a gas at s.t.p. as 22·4 dm$^3$.)

**3.** What is the mass of 1 dm$^3$ of carbon dioxide at s.t.p.?

**4.** State Boyle's law and Charles' law. From these laws, and given that 1 mole of an ideal gas occupies 22·4 dm$^3$ at s.t.p., deduce the ideal gas equation.

A 1 000-cm$^3$ bulb weighs 50·883 g when evacuated and 52·080 g when filled with a gas at standard pressure and 20°C. Assuming the gas behaves ideally, calculate its relative molecular mass. Suggest the identity of the gas.

(O. and C.)

5. (Part question.) In a determination of the relative vapour density of a volatile compound X by the method of Dumas, the volume of the vessel used was 152 cm$^3$, and the mass of X needed to fill it at a temperature of 100°C to the prevailing atmosphere pressure of 100 000 Pa was 0·450 g. Calculate the relative molecular mass of X. (O.L.)

6. State Avogadro's hypothesis. Explain what is meant by the relative gaseous density of a substance, and describe one method for determining it.

0·2 g of a liquid when vaporized in a Victor Meyer's apparatus displaced 31·2 cm$^3$ of air, measured over water at 13°C and 100 300 Pa. Find its relative molecular mass. (Pressure of saturated aqueous vapour at 13°C = 1 500 Pa.) (Lond.)

7. What volume of air (measured at s.t.p.) should be expelled from a Victor Meyer apparatus when 0·05 g of ethanol ($C_2H_6O$) is vaporized in the apparatus?

8. Describe how you would determine the relative molecular mass of ethoxyethane by Victor Meyer's method.

1·0 g of the volatile chloride of a metal M displaced 74·2 cm$^3$ of air, measured at 17°C and a pressure of 101 300 Pa (corrected for the vapour pressure of water). The chloride contained 34·46 per cent of M and the heat capacity of M was 0·481 J g$^{-1}$ K$^{-1}$. Find the exact relative atomic mass of M and the formula of its chloride. Cl = 35·46; 1 mole of a gas occupies 22·4 dm$^3$ at s.t.p. (C.L.)

## Graham's Law

9. (Part question.) 141·4 cm$^3$ of an inert gas diffused through a porous plug in the same time as it took 50 cm$^3$ of oxygen to diffuse through the same plug under identical conditions. Calculate the relative atomic mass of the inert gas and state the law involved in your calculation.

Explain why diffusion in solution is very much slower than in gases. Indicate *very briefly* how you would demonstrate diffusion in solution by means of a simple experiment. (W.J.E.C.)

10. (Part question.) Into the ends of a horizontal wide glass tube there are inserted, simultaneously, plugs of cotton wool soaked in concentrated ammonia solution and concentrated hydrochloric acid, respectively. After a short time a white disc of solid ammonium chloride forms across the tube. If the distance between the inner surfaces of the cotton wool plugs is 50 cm, how far from the ammonia plug would you expect the disc to be formed? (C.L.)

11. Describe ONE method for comparing the rates of diffusion of gases.

A gaseous compound X contained 46·1 per cent of carbon and 53·9 per cent of nitrogen. In 20 second 50 cm$^3$ of X diffused through a porous plug and the same volume of oxygen diffused in 15·7 second. Deduce the molecular formula of X. What volume of carbon dioxide would diffuse in 20 second under the same conditions? C = 12; N = 14; O = 16. (C.L.)

12. (Part question.) A certain volume of ethanoic acid vapour was found to diffuse in 580·8 sec while the same volume of oxygen, with the same experimental conditions, took 300 sec to diffuse. What deductions can be made as a result of this experiment? (J.M.B.)

## Formulæ of Gases

13. A gaseous hydride $A_xH_y$ decomposes on heating as follows: 4 volumes of $A_xH_y$ = 1 volume of gaseous A + 6 volumes of hydrogen.

From this result, and assuming Avogadro's hypothesis *only*, deduce the simplest possible formula for $A_xH_y$. Show your reasoning in full. (J.M.B.)

**14.** (Part question.) In an experiment to determine the molecular formula of ammonia 24 cm³ of ammonia were sparked in a eudiometer until no further change in volume occurred. The product was then exploded with 24 cm³ of oxygen. After cooling to the original temperature the volume of residual gas was 18 cm³ of which 6 cm³ were absorbed by pyrogallol. All measurements were made at the same pressure. Deduce the formula for ammonia, giving your reasoning in full.                                                                  (J.M.B.)

**15.** 20 cm³ of a gaseous hydrocarbon X were exploded with 100 cm³ (excess) of oxygen. After the explosion the volume was 90 cm³, and this on treatment with alkali decreased to 50 cm³. What was the molecular formula of X?

**16.** 15 cm³ of a gaseous hydrocarbon Y were exploded with 115 cm³ of oxygen (excess). After the explosion the volume was 85 cm³, and on treatment with alkali there was a further contraction of 45 cm³. What was the molecular formula of Y?

### Thermal Dissociation

**17.** Summarize the uses which have been made of Avogadro's law, giving examples of each use.

Ammonium carbamate dissociates according to the equation:

$$NH_4CO_2NH_2 \rightleftharpoons 2NH_3 + CO_2$$

5 g of ammonium carbamate were completely vaporized at 200°C. Calculate the total volume of gas produced at a pressure of 98 700 Pa, if the carbamate is completely dissociated at that temperature.

Describe an experiment by which you could show that dissociation had occurred (other than by measurement of the density of the vapour).                    (S.U.)

**18.** What do you understand by 'dissociation'? By what methods can the dissociation of a gaseous substance be recognized? The relative density of phosphorous pentachloride vapour at 250°C and under atmospheric pressure is 57·6. What is the percentage of dissociation of the compound under these conditions? P = 31; Cl = 35·5.                                                                          (O. and C.)

**19.** The density of partially dissociated dinitrogen tetraoxide at 105°C and standard pressure is 1·555 kg m⁻³. What is the percentage dissociation?

## MORE DIFFICULT QUESTIONS

**20.** Give a concise description of the essential features of Victor Meyer's method of determining the relative vapour density of a volatile liquid.

In such a determination 0·1784 g of an oxychloride Y of sulphur displaced 36·8 cm³ of air measured over water at 17°C and a total pressure of 100 300 Pa. The hydrochloric acid formed by the complete hydrolysis of 0·2380 g of Y yielded a precipitate of 0·5734 g of silver chloride. Calculate the relative molecular mass of the vapour of Y, and deduce its molecular formula. Write a structural formula for Y and an equation for its hydrolysis. State the gas laws which are involved in your calculation.

The vapour pressure of water at 17°C = 2 000 Pa. The density of hydrogen at s.t.p. = 0·0899 kg m⁻³. Ag = 107·9; Cl = 35·46; S = 32·07; O = 16·00.
                                                                          (W.J.E.C.)

**21.** The relative density of partially dissociated dinitrogen tetraoxide vapour at 65°C is 29. Calculate (i) the degree of dissociation, (ii) the volumes of $N_2O_4$ and $NO_2$ respectively contained in 1 dm³ of the vapour at this temperature.

**22.** A mixture of propane, $C_3H_8$, with a gaseous hydrocarbon of the alkene series occupied 24 cm³. To burn the mixture completely 114 cm³ of oxygen were required, and after combustion 72 cm³ of carbon dioxide were left. Calculate (*a*) the formula of the alkene, (*b*) the composition of the mixture by volume. All volumes were measured at the same temperature and pressure.    (J.M.B.)

23. (Part question.)  A mixture of methane, ethene and ethyne has a relative density of 11·3.  When 10·0 cm³ of this mixture and 30·0 cm³ of oxygen are sparked together over aqueous potassium(I) hydroxide the volume contracts to 5·5 cm³ and then disappears when benzene-1, 2, 3-triol (pyrogallol) is introduced. All volumes are measured under the same conditions of temperature, pressure, and humidity.  Calculate the composition of the original mixture.    (J.M.B.)

# CHAPTER 4

# Relative molecular masses in solution

When a solid dissolves in a liquid the freezing point of the liquid is lowered. Subject to certain limitations, the freezing point depression is in accordance with the following laws:

**Blagden's Law.** *The depression of the freezing point of the solvent is proportional to the mass of solute dissolved in a given mass of solvent.*
**Raoult's Law.** *The depression of the freezing point of the solvent is inversely proportional to the relative molecular mass of the solute.*

Thus, if a mass $m$ of a solute dissolved in 100 g of a solvent produces a freezing point depression of $t$ (where $t$ is in degC), we have

$$t \propto \frac{m}{M_r}$$

where $M_r$ is the relative molecular mass of the solute. Now $m/M_r$ measures the number of molecules of dissolved substance. It follows that the depression of freezing point is proportional to the number of dissolved molecules. Hence, if equimolecular proportions of two different solutes dissolve in the same mass of the same solvent the freezing point depressions are the same.

Closely associated with depression of freezing point by a dissolved substance are three other properties of solutions. These are elevation of boiling point, lowering of vapour pressure, and osmotic pressure. The magnitude of all four effects depends on the number of dissolved molecules in a given amount of solvent and is independent of the nature of the dissolved substance. Any of the four can be used to determine the relative molecular mass of the dissolved substance. Since the four properties are closely connected in this way they are described as *colligative properties* (Latin *colligare*, to bind).

## Method 1. Cryoscopic Method

The *freezing point constant*, or *cryoscopic constant*, of a solvent is the amount by which its freezing point is lowered in a *molal* solution—that is, when 1 mole of a solute dissolves in 1000 g of the solvent—assuming that Blagden's law and Raoult's law hold at this concentration. In practice this assumption is often untrue, and we have to calculate cryoscopic constants from results obtained with more dilute solutions. Thus the freezing point depression of water when $\frac{1}{10}$ mole (34·2 g) of sucrose dissolves in 1000 g of water is 0·186 degC. Assuming that Blagden's law holds, we find that the cryoscopic constant of water is 1·86 degC mol$^{-1}$ kg$^{-1}$. The cryoscopic constants of some other commonly used solvents are as follows:

|  | $t/°C$ mol$^{-1}$ kg$^{-1}$ |
|---|---|
| Benzene | 5·12 |
| Ethanoic acid | 3·90 |
| Camphor | 40·0 |

(Cryoscopic constants are sometimes given for 100 g of solvent. For this amount of water the depression is 18·6 degC mol$^{-1}$.)

If we know the cryoscopic constant for a solvent we can use it to find the relative molecular mass of a dissolved substance. Thus, let $m$ = mass of solute in 1000 g of solvent, $M_r$ = relative molecular mass of solute, $t$ = observed depression of freezing point (degC), and $K$ = cryoscopic constant of solvent (degC). Then, since a mass $m$ of solute produces a depression of $t$ and $M_r$ g a depression of $K$, we have the simple proportion

$$m : M_r \, g = t : K$$

If we know any three of these quantities we can calculate the fourth.

This is one of the most frequently used methods of measuring relative molecular masses. It is particularly useful for organic compounds like sucrose which decompose before their boiling points are reached. Clearly we cannot find the relative vapour densities of such compounds by the Dumas or Victor Meyer methods.

**Beckmann's Freezing Point Apparatus.** As the depression of freezing point measured is small (usually less than 1 degC), a very sensitive thermometer is required to record it. The Beckmann thermometer (Fig. 4–1(a)) is used. Since we require only the difference between the freezing point of the pure solvent and that of the solution, the thermometer need not register the actual temperature. The thermometer scale covers only about 6 degC, and is graduated in hundredths of a degree. At the top of the thermometer there is a reservoir for mercury so that we can set the thermometer at the beginning of the experiment for the range of temperature expected. The glass tube A (Fig. 4–1) contains a known mass of solvent, in which the bulb of the Beckmann thermometer is immersed. This tube is fitted with a wire stirrer, $S_1$, and has a side arm, M, through which the solute can be introduced. The tube A is surrounded by a wider tube, B, which provides an air jacket to ensure uniform cooling. The freezing agent contained in the vessel C consists of a mixture of ice and salt and is kept stirred by the stirrer $S_2$.

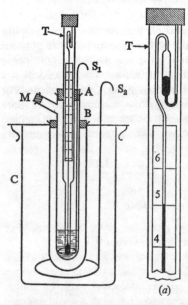

First, we determine the freezing point of the pure solvent. We place a weighed amount of solvent in A and keep both solvent and freezing agent well stirred. Almost invariably supercooling occurs, the temperature of the solvent falling below the true freezing point before solidification starts. When the solid solvent begins to separate the temperature rises to the true freezing point, owing to the latent heat of fusion evolved. When the temperature has risen to its highest point we note the reading on the Beckmann thermometer.

*Fig.* 4–1. Beckmann's freezing point apparatus

We now remove the inner tube and allow the solvent to melt. We introduce a weighed amount of the substance under investigation (in pellet form) through the side arm M and stir until it is completely dissolved. We then replace the tube in the apparatus and find the freezing point of the solution as for the solvent. Note that when the solution freezes the pure solvent alone separates out.

The depression of freezing point for a known mass of solute in a known mass of solvent has now been found. We can make further determinations by adding further weighed quantities of solute, thus obtaining the depression of freezing point at different concentrations. In each case we calculate the relative molecular mass and take the average of the results.

**Example.** *0·55 g of nitrobenzene in 22 g of ethanoic acid depressed the freezing point of the latter by 0·78 degC. Calculate the relative molecular mass of nitrobenzene. (The cryoscopic constant for 1000 g of ethanoic acid is 3·90 degC mol⁻¹.)*

0·55 g of nitrobenzene dissolved in 22 g of ethanoic acid is equivalent to $0.55 \times \dfrac{1000}{22}$ g of nitrobenzene in 1000 g of ethanoic acid.

$$m : M_r \, g = t : K$$

$$0 \cdot 55 \times \frac{1000}{22} \, g : M_r \, g = 0 \cdot 87°C : 3 \cdot 90°C$$

$$\therefore \quad M_r = 0 \cdot 55 \times \frac{1000}{22} \times \frac{3 \cdot 90}{0 \cdot 78}$$

$$= 125$$

**Rast's Method.** This method is based on the use of melted camphor as a solvent. Camphor has an unusually high cryoscopic constant; hence relatively large depressions of freezing point (or melting point) occur even with small concentrations of solute. This allows us to use an ordinary (360-degree) thermometer instead of a Beckmann thermometer. The procedure consists essentially of measuring the melting point of pure camphor and that of a mixture of the solute with camphor. We can use the ordinary melting point apparatus of organic chemistry or perform the experiment on a somewhat larger scale, as now described for the determination of the relative molecular mass of naphthalene.

**Experiment.** First find the melting point of a mixture of naphthalene and camphor. Weigh by difference (to three decimal places) about 2 g of powdered camphor in a small dry ignition tube. Add to the tube about 0·1 g of naphthalene, and find the mass of the latter. Holding the tube in a pair of tongs, warm it above a very small Bunsen flame until the mixture melts. Insert the bulb of a 360-degree thermometer into the liquid, using a short length of rubber tubing round the thermometer to fix the latter in the tube (Fig. 4–2). A slit cut in the tubing allows for expansion. Let the liquid cool and solidify, and then clamp the thermometer so that most of the ignition tube is immersed in a bath of liquid paraffin. With a mixture of compounds there is a melting point range, and the melting point of the mixture is the temperature at which the last trace of solid disappears. Place the oil bath on a tripod and gauze, and heat using a medium flame, meanwhile keeping the liquid well stirred with a glass rod. Above 130°C heat more gently. The mixture melts first to a cloudy liquid. Read the temperature at which the liquid becomes clear; this is the melting point of the mixture.

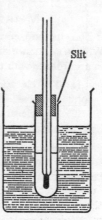

Slit

*Fig. 4–2.*
Apparatus used for Rast's method

If possible repeat the experiment, using different proportions of camphor and naphthalene until three consistent values are obtained for the depression produced by 1 g of naphthalene in 1000 g of camphor. Use the average of these values in the calculation.

With the same thermometer, but a different tube, find the melting point of pure camphor. This should be approximately 176°C. Again three determinations should be made to yield an average value.

**Limitations of the Cryoscopic Method.** A correct value for relative molecular mass (that is, a value corresponding to the usual molecular formula) is given only under the following conditions:

(1) When the solution freezes the pure solvent alone must separate out. (In some cases—*e.g.*, naphthalen-2-ol dissolved in naphthalene—the solute forms a 'solid solution' with the solvent as it crystallizes.)

(2) There must be no chemical reaction between solute and solvent— for example, phosphorus pentachloride cannot be used as solute with water as solvent because the two react together.

(3) The dissolved substance must not dissociate appreciably into ions. Thus the method is unsuitable for strong acids, alkalis, or salts, which are largely dissociated in aqueous solution. It can, however, be applied to organic acids, like ethanoic acid, that dissociate only slightly when dissolved in water.

For some organic compounds in organic solvents the relative molecular mass obtained by the cryoscopic method is higher than that determined by relative vapour density methods. This is due to association; if two molecules associate to give one, the relative molecular mass is doubled. Thus ethanoic acid dissolved in benzene exists as $(CH_3COOH)_2$ molecules.

(4) Solutions must be dilute because the Blagden and Raoult laws hold only for dilute solutions.

## Method 2.  Elevation of Boiling Point

This method corresponds in many ways to the cryoscopic method. When a substance dissolves in a liquid the boiling point of the liquid rises in accordance with the following laws:

(1) The elevation of boiling point is proportional to the mass of solute dissolved in a given mass of solvent.

(2) The elevation of boiling point is inversely proportional to the relative molecular mass of the solute.

The elevation of boiling point shown by a molal solution is called the *boiling point constant*, or *ebullioscopic constant*, of the solvent. The boiling point constants of some common solvents are as follows:

|  | $t/°C\ mol^{-1}\ kg^{-1}$ |
|---|---|
| Water | 0·51 |
| Ethanoic acid | 3·08 |
| Benzene | 2·67 |
| Trichloromethane | 3·66 |

The calculation of the relative molecular mass is similar to that used in the freezing point method.

The chief difficulty in finding relative molecular masses by this method arises from superheating the liquid above its true boiling point. To overcome the difficulty Beckmann's original apparatus has been modified. Two forms of apparatus are now commonly used.

**Cottrell's Method.**  We heat the solvent in a wide tube, A (Fig. 4–3); this is fitted with a side arm leading to a reflux condenser, which condenses the vapour of the boiling solvent and returns it to the boiling tube. A is closed by a stopper, through which passes a second tube, B, open at the lower end. B contains a Beckmann thermometer. The bulb of the thermometer is above the liquid and is protected by the inner tube from drops of condensed liquid falling on to it from the side arm. Partially immersed in the liquid is the Cottrell 'pump', C, consisting of a small inverted glass funnel with a stem which divides into two narrow tubes. These are bent round at the top, so that, when the liquid boils, the vapour

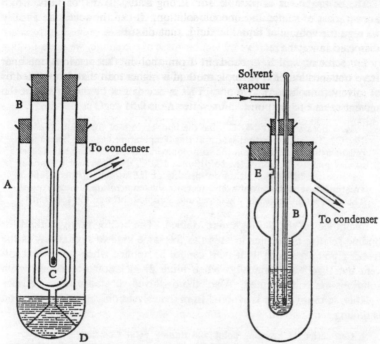

*Fig.* 4–3. Cottrell's apparatus          *Fig.* 4–4. Landsberger's apparatus

forces liquid to rise up the narrow tubes and spray over the bulb of the thermometer. This prevents superheating, and as a further precaution a platinum wire is sealed through the bottom of the tube A at D. Heating is by a microburner, which is shielded from draughts.

First, we read the Beckmann thermometer for boiling solvent. We use a known volume of solvent and calculate its mass from its density. We then introduce a weighed amount of solute (in pellet form) through the side arm, and read the temperature for the boiling solution.

We can use this method for measuring the relative molecular mass of benzoic acid dissolved in acetone or that of naphthalene dissolved in chloroform.

**Landsberger's Method.** In this method (Fig. 4–4) the solvent is boiled in a separate flask. The vapour passes into the solution, which is contained in a graduated tube surrounded by a second tube. The solution is raised to its boiling point by the latent heat of condensation of the solvent vapour. When the solution boils the solvent vapour passes through the solution without condensing. It then passes through a hole, E, in the boiling tube so that the latter becomes jacketed with the vapour of the boiling solvent. Radiation is thus reduced to a minimum. The bulb B on the inner tube prevents liquid from being splashed through E.

The boiling point of pure solvent is first found. We then introduce a known mass of solute and find the boiling point of the solution. Finally we read the volume of liquid in the graduated tube.

In calculating the relative molecular mass of the solute, we use a boiling point constant which is different from that in the previous method. Here the constant ($K'$) is for 1 mole of substance dissolved in 1000 cm$^3$ of solvent (or solution). We obtain $K'$ by dividing $K$ by the density of the solvent at its boiling point. For water $K'$ is 0·54 degC mol$^{-1}$ dm$^{-3}$.

Why does steam at 100°C raise the temperature of the solution *above* 100°C? The reason is that steam at this temperature has a higher vapour pressure than the solution at the same temperature. Thus if we had a layer of steam at 100°C above the solution at 100°C, the difference of vapour pressure would lead to steam condensing at the surface of the solution, and the latent heat evolved would increase the temperature of the solution. The same situation exists when steam is bubbled through the solution.

**Limitations of the Boiling Point Method.** The boiling point method for finding relative molecular masses has the same kind of limitations as the freezing point method; that is, it cannot be applied when solute and solvent act together chemically, when ionic dissociation occurs, or when solutions are concentrated. Also, the dissolved substance must be non-volatile, so that vapour is evolved from the solvent only when the solution is boiling.

**Calculation of Freezing point and Boiling point Constants.** It can be shown by thermodynamics that if $K$ is the cryoscopic constant of the solvent (per 1000 g), $T$ the freezing point of the solvent on the Kelvin scale, and $L$ the latent heat of fusion per gram of the solvent,

$$K = \frac{RT^2}{1000 L}$$

where $R$ is the gas constant. Now the value of $R$ is approximately 8·31 J mol$^{-1}$ K$^{-1}$, and therefore we have

$$K = \frac{0 \cdot 00831 \text{ J mol}^{-1} \text{ K}^{-1} \times T^2}{L}$$

For water the freezing point is 273 K, and the latent heat of fusion is 333 J g$^{-1}$. Thus

$$K = \frac{0 \cdot 00831 \times 273^2}{333} = 1 \cdot 86 \text{ K mol}^{-1} \text{ kg}^{-1}$$

Similarly it can be shown that, if $K$ is the boiling point constant (per 1000 g of solvent), $T$ the boiling point of the solvent on the Kelvin scale, and $L$ the latent heat of vaporization per gram of the solvent,

$$K = \frac{0 \cdot 00831 \text{ J mol}^{-1} \text{ K}^{-1} \times T^2}{L}$$

Thus for water $T = 373$ K, and $L = 2\,250$ J g$^{-1}$. By substitution in this expression we obtain $K = 0 \cdot 51$ K mol$^{-1}$ kg$^{-1}$.

## Method 3. Lowering of Vapour Pressure

If we fill a barometer tube with mercury and invert it in a trough of mercury, the mercury in the tube falls to a level A, as shown in Fig. 4–5a, leaving a vacuum above A. The height of the mercury column AB represents the atmospheric pressure.

If we treat a second barometer tube similarly and introduce a drop of ethanol into the vacuum, the ethanol evaporates completely and the level of the mercury falls. This is because of the pressure exerted by the molecules of ethanol vapour. At this stage the vapour is *unsaturated*. If we introduce further single drops of ethanol the mercury level drops farther until the space above the mercury becomes *saturated* with vapour. Some liquid now remains on the surface of the mercury, and there is equilibrium between the liquid and vapour (Fig. 4–5b). AB − CD measures the saturated vapour pressure of the ethanol at the prevailing temperature. The saturated vapour pressure is usually referred to simply as the 'vapour pressure.'

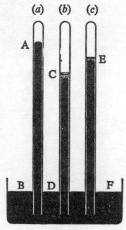

Fig. 4–5

*The* **vapour pressure** *of a liquid at a given temperature is the pressure exerted by the vapour when it is in equilibrium with the liquid.*

The vapour pressure of a solution of a solid in a liquid is less than that of the pure liquid. Thus if we use a solution of ethanamide in ethanol instead of ethanol in the last experiment, there is a smaller decrease in the height of the mercury to E (Fig. 4–5c). EF − CD is the lowering of the vapour pressure of the ethanol.

A more convenient method of measuring the difference in vapour pressure of a solvent and a solution is to use a differential tensimeter as described below. Propanone is used instead of ethanol because it has a higher vapour pressure at a given temperature.

**Experiment.** Set up the apparatus shown in Fig. 4–6. The two 250-cm³ conical flasks, A and B, are connected to a pressure gauge containing phenylamine (which has a negligible vapour pressure at room temperature) and to a strong suction pump. Poly(chloroethene) (PVC) tubing is employed for all connections except between the apparatus and the pump, where rubber pressure tubing is used. Screw clips are attached to the connections at $S_1$ and $S_2$. Test the apparatus for airtightness before commencing the experiment.

Put 100 cm³ of propanone into A, and a solution of 6 g of dinitrobenzene in 100 cm³ of propanone into B. Add a few fragments of porous pot to both flasks to promote steady boiling. Place each flask in a beaker

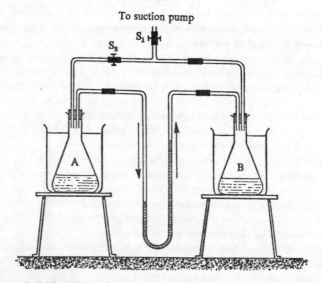

*Fig.* 4–6. Differential tensimeter used to measure lowering of vapour pressure

of water standing on a tripod. With screw clips $S_1$ and $S_2$ open, start the suction pump. Solvent alone boils first, but the solution boils soon afterwards. If boiling becomes violent, partially close $S_1$.

Allow the liquids to continue boiling for ½ minute to remove air from the apparatus. Screw up $S_1$ tightly, and then $S_2$. The level of the phenylamine in the pressure gauge falls on the solvent side and rises on the solution side. Leave the flasks to warm up to room temperature. When equilibrium is reached note the difference in the heights of phenylamine and convert the pressure difference into Pa(Nm⁻²). (Take the density of phenylamine to be 1·02 g cm⁻³.) This is the lowering of vapour pressure for the solution at room temperature. To allow for solvent evaporated from the solution, measure the volume of solution left and take this as the volume of propanone in which the given mass of dinitrobenzene is dissolved.

Determinations of lowering of vapour pressure can be made at other temperatures by placing the flasks in a thermostat at the required temperature. Vapour pressures of pure propanone are given on p. 93.

*Notes.* (1) When the experiment is finished first open $S_2$. Then disconnect the pump and shake out any water in the tube on the pump side of $S_1$. Finally open $S_1$ gradually.

(2) The actual vapour pressure of a pure solvent at a given temperature is measured by means of the apparatus described on p. 92.

According to kinetic theory, the vapour pressure of a solution is lower than that of the solvent because part of the liquid surface of the solution is occupied by solute molecules. This makes it more difficult for solvent molecules to escape. No corresponding barrier, however, exists for the return of the molecules. Therefore an equilibrium is established between escaping and returning molecules at the surface of a solution when the number of solvent molecules in the space above the liquid is less than in the case of the pure solvent.

**Laws for Vapour Pressure of Solutions.** As with freezing point depression and boiling point elevation, the lowering of the vapour pressure of a solvent by a solute, at a given temperature, is directly proportional to the concentration of the solute, and inversely proportional to its relative molecular mass. More important than the actual lowering of the vapour pressure is the *relative lowering of the vapour pressure*. If $p$ is the vapour pressure of the pure solvent, and $p'$ the vapour pressure of the solution at the same temperature, the relative lowering of the vapour pressure is $(p - p')/p$. The relative lowering of the vapour pressure is (i) independent of the temperature, (ii) proportional to the concentration of solute, and (iii) constant when equimolecular proportions of different solutes are dissolved in the same mass of the same solvent.

Furthermore, from measurements of vapour pressure made with solutions in different solvents, Raoult established that, if $n_1$ and $n_2$ are the number of mole of solute and solvent

$$\frac{p - p'}{p} = \frac{n_1}{n_1 + n_2} = \frac{n_1}{n_2} \text{ (if the solution is dilute)} \qquad (1)$$

This is known as *Raoult's vapour pressure law*.

The numbers of mole of solute and solvent are obtained by dividing the masses in gram by the relative molecular masses. If $m_1 =$ mass of solute, $m_2 =$ mass of solvent, $M_1 =$ relative molecular mass of solute, and $M_2 =$ relative molecular mass of solvent, then

$$\frac{p - p'}{p} = \frac{m_1/M_1}{m_2/M_2} \qquad (2)$$

The vapour pressure laws of solutions hold only under the conditions described for the boiling point method of finding relative molecular masses. The solute must be non-volatile, must not react with the solvent, and must not dissociate appreciably into ions. Also, solutions must be dilute.

**Example.** *The vapour pressure of water at* $50°C$ *is* $12\ 333\ Pa$. *At this temperature a solution of* $9·14\ g$ *of carbamide (urea) in* $150·0\ g$ *of water has a vapour pressure of* $12\ 106\ Pa$. *Find the relative molecular mass of carbamide.*

From equation (2) we have

$$\frac{12\ 333 - 12\ 106}{12\ 333} = \frac{9·14/x}{150·0/18}$$

which gives $x = 59·7$

**Measurement of Vapour Pressure of a Liquid by Isotensimeter.** Several methods exist for finding the vapour pressure of a pure liquid or solution at a given temperature. Raoult's vapour pressure law can be verified

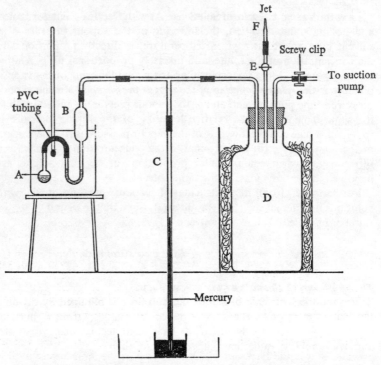

*Fig.* 4–7. Measurement of vapour pressure of a liquid by isotensimeter

from the difference between the vapour pressure of a dilute solution of known concentration of dinitrobenzene in propanone, measured with a differential tensimeter, and the vapour pressure of pure propanone at the same temperature, determined by means of an isotensimeter (Fig. 4–7). An isotensimeter is described on the next page.

**Experiment.** A (Fig. 4–7) is a bulb of about 3 cm diameter blown on the end of a short length of glass tubing. The bulb is half filled with the liquid and a few pieces of porous pot are added. B is a 20-cm³ pipette, the top part of which has been bent through an angle of 180°. The bend is also filled with the liquid. After the liquids have been introduced A and B are connected by PVC tubing, which is also used for other connections. A, the lower part of B, and the connection are immersed in water in a litre beaker, which contains a thermometer. The top of B is connected to a mercury manometer, C, and to a large glass bottle, D. The purpose of the latter is to increase the volume of vapour in the apparatus and prevent fluctuations in the manometer readings due to draughts, and so on. As a safety precaution in case it collapses under reduced pressure, D is placed in a box or tin and surrounded with cloths. The tube, E, is fitted with a tap and attached to a short-length of glass tubing, F, which has been drawn out to form a fine jet at the end. D is also connected to a suction pump by glass tubes and a short length of PVC tubing, which carries a screw clip, S.

With the tap on E closed and S open, start the suction pump. The liquid in A soon begins to boil under the reduced pressure and vapour displaces air from A and B. Allow the boiling to continue for about a minute and then close the screw clip S, tightly. Wait for a few minutes to allow the liquid to warm up to the temperature of the water in the beaker. Open the tap on E *very gradually*, so that air is admitted slowly through the jet on F. Close the tap when the levels of the liquid in the bend of B become the same. Read the height of the mercury in the manometer and the barometric pressure. From the difference the vapour pressure of the liquid at the temperature of the water bath can be calculated.

Table 4–1 shows the vapour pressures of some common liquids at various temperatures.

**Table 4–1.** VAPOUR PRESSURES OF LIQUIDS ($p$/Pa)

| | \multicolumn{8}{c}{Temperature/°C} | | | | | | | |
|---|---|---|---|---|---|---|---|---|
| | 0 | 10 | 20 | 30 | 40 | 60 | 80 | 100 |
| Water | 613 | 1 227 | 2 333 | 4 240 | 7 373 | 20 870 | 47 330 | 101 325 |
| Ethanol | 1 600 | 3 200 | 5 866 | 10 530 | 18 000 | 47 060 | — | — |
| Propanone | 8 933 | 15 470 | 24 670 | 37 330 | 56 130 | — | — | — |
| Benzene | 3 467 | 6 000 | 10 000 | 16 000 | 24 400 | 52 260 | 101 100 | — |

**Vapour Pressure, Boiling Point, and Freezing Point.** The vapour pressure of a liquid increases with an increase in the temperature until it becomes equal to the pressure of the atmosphere above, when the liquid boils. When a non-volatile solute is dissolved in the liquid the vapour pressure at any temperature is lowered. Hence the liquid must be heated to a higher temperature to bring its vapour pressure up to the pressure of the atmosphere. Again, at the freezing point of the pure liquid the liquid is in equilibrium with the solid; both have the same vapour pressure. Since a solution has a lower vapour pressure than the pure solvent, the temperature at which the solution is in equilibrium with the solid solvent is lower than for the pure solvent; that is, the solution has a lower freezing point than the solvent.

Taking water as the solvent, we can illustrate the effects of decrease in vapour pressure on boiling point and freezing point as in Fig. 4–8. AB is the vapour pressure curve of water, and BC that of ice. DG and HM are the vapour pressure curves of two solutions of different concentration.

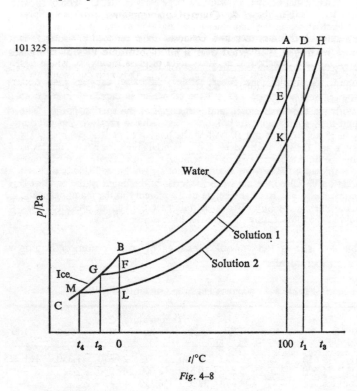

*Fig.* 4–8

The first, more dilute, solution boils at $t_1$ and freezes at $t_2$, while the second boils at $t_3$ and freezes at $t_4$.

Let us first consider the boiling points of the two solutions. AD and AH represent the elevations of boiling point of the solutions, and AE and AK the decreases in vapour pressure. If the solutions are dilute A, D, and H are close to each other, and ED and KH are approximately parallel straight lines. Then the triangles ADE and AHK are similar triangles, in which AD : AH = AE : AK. That is, the elevations of boiling point are proportional to the decreases in vapour pressure.

Similarly we can show that the depressions of freezing point, FG and LM, are related to the decreases in vapour pressure, BF and BL. If the solutions are dilute, G and M are close to B, and FG and LM are approximately straight lines. Since there is little change of vapour pressure over a small range of temperature, FG and LM are also approximately parallel with the temperature axis and with each other. Then BFG and BLM are similar triangles, in which FG : LM = BF : BL. That is, the depressions of freezing point are proportional to the decreases in vapour pressure.

## Method 4. Osmotic Pressure

**Semipermeable Membranes and Osmosis.** We have already seen that if a jar of hydrogen is inverted over a jar of bromine vapour the gases diffuse together because the molecules of both are in a state of rapid movement. A similar, but much slower, diffusion process also occurs with liquids and solutions. If a layer of water is carefully run on to a layer of purple potassium(I) manganate(VII) solution so as to avoid mechanical mixing, the ions of the potassium(I) manganate(VII) slowly diffuse into the upper layer until the colour of the liquid is the same throughout. At the same time molecules of water move in both directions between the upper and lower layers.

We have also seen that two gases intermingle if separated by a porous pot, although a lighter gas diffuses more rapidly than a heavier one. Similarly, if we separate water and aqueous potassium(I) manganate(VII) by a cellophane membrane, solute and solvent still diffuse, but at different rates. This can be illustrated by the experiment described below.

**Experiment.** Fasten a cellophane membrane tightly over one end of a glass tube, which is open at both ends. Fill the membrane and tube with a concentrated solution of potassium(I) manganate(VII). Insert a rubber stopper with a long glass tube into the upper end of the tube, and then place the membrane in a beaker of water (Fig. 4–9).

After a time the liquid is seen to be rising in the long tube. Also, the colour of the manganate(VII) ions is noticed in the water in the beaker.

We must conclude that the membrane is permeable to both water molecules and ions of the solute. If the apparatus is left long enough the liquid ceases to rise in the tube and then falls until the level is the same inside and outside. A dynamic equilibrium has then been reached. The solutions on both sides of the membrane are equal in concentration, and water molecules (and ions) are passing through the membrane at equal rates in both directions.

If this experiment is modified by using sucrose instead of potassium(I) manganate(VII) as solute, we obtain a different result. Thus water transfers as before from the solvent to the solution side of the membrane, but the solute does not move in the opposite direction. When a membrane permits the passage of the solvent particles, but not those of solute, it is described as *semipermeable*. Since the liquids on opposite sides of the membrane can never become equal in concentration of solute, we might expect the transference of solvent to continue indefinitely. In practice it is stopped by the hydrostatic pressure developed on the solution side of the membrane; this hydrostatic pressure tends to force water through the membrane in the opposite direction.

Suppose that solvent and solution are contained in the limbs of a U-tube and that they are separated by a semipermeable membrane (Fig. 4–10). Further, imagine that above the solution there is a movable piston. With this arrangement, water would pass through the membrane to the solution, and the piston would rise. This could be prevented, however,

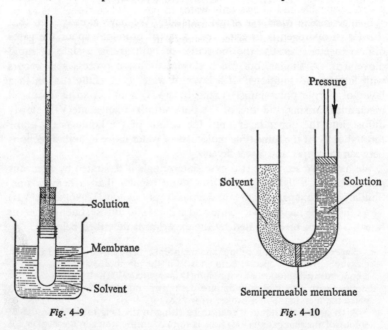

Fig. 4–9                                                Fig. 4–10

by applying just the right external pressure to the piston. If the external pressure were too large water would be forced through the membrane from solution to solvent.

> The **osmotic pressure** *of a solution is the pressure which must be applied to the solution to balance the tendency of solvent to flow from the solvent side to the solution side of a semipermeable membrane.*

Note that a solution in itself has no osmotic pressure. 'Osmotic pressure of a solution' only possesses meaning if it is understood that the phrase refers to a given set of conditions, namely, when the solution is separated from the solvent by a semipermeable membrane. Again, the pressure which is applied does not stop the diffusion of water molecules into the solution. It merely causes water molecules to diffuse more rapidly through the membrane in the opposite direction. When the pressure is correctly adjusted a dynamic equilibrium is established, and water molecules diffuse in both directions at equal rates.

**Natural and Artificial Semipermeable Membranes.** Semipermeable membranes commonly occur in plants and animals. A plant cell (Fig. 4–11) has a cellulose wall, inside which is a lining of protoplasm enclosing the cell sap, a dilute solution of salts and other substances. The cellulose wall is permeable to both water and dissolved material, but the protoplasm is semipermeable and allows only water to pass. When the cell is in its normal condition the layer of protoplasm is pressed tightly against the cell wall by the solution inside. Osmosis in plant cells is most readily

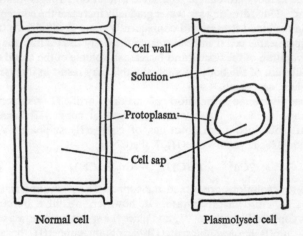

Cell wall

Solution

Protoplasm

Cell sap

Normal cell                    Plasmolysed cell

*Fig.* 4–11. Plasmolysis of plant cell

demonstrated by means of a thin section of beetroot, in which the cell sap is red. If we place the section in a concentrated solution of sucrose, and examine it under the microscope, we see the protoplasm withdraw from the cell wall and collect as a more or less spherical red blob in the middle of the cell (Fig. 4–11). The cell is now *plasmolysed*, because of diffusion of water from the more dilute solution inside the membrane to the more concentrated solution outside.

If we repeat the experiment with increasingly dilute solutions of sucrose, we find that plasmolysis fails to occur when the concentration of the sugar solution has fallen to about 6 per cent. More dilute solutions also fail to cause plasmolysis. Hence we conclude that a 6 per cent solution of the sucrose has the same osmotic pressure as the cell sap. Solutions which have the same osmotic pressure are said to be *isotonic*.

Another example of a semipermeable membrane is the membrane outside the white of an egg. If we dissolve the shells from two eggs of similar size by dilute hydrochloric acid and place one egg in water and the other in concentrated brine, there is a large difference in size after a few hours. The egg in the water swells owing to the diffusion of water through the membrane to the solution (of salts, etc.) inside. The egg in brine shrinks because water is lost from the dilute solution inside to the concentrated solution outside. By interchanging the eggs we can reverse the process and restore the eggs to their original size.

The last experiment illustrates the danger which arises when people who are shipwrecked and without a fresh water supply start drinking sea water. The concentration of salts in sea water is about 3·5 per cent, while that of dissolved substances in the body fluids is only about 1 per cent. The kidneys are unable to excrete a solution stronger than about 2 per cent. Thus, drinking sea water gradually increases the concentration of salts in the blood stream, and consequently water is withdrawn through the semipermeable cell membranes from the body tissues into the blood. The dehydration of the tissues and increase in volume of the blood disturb the equilibrium of the body. Eventually this may result in intense thirst, heart failure, and death.

Traube discovered a method of making artificial semipermeable membranes in 1867. These were composed of copper(II) hexacyanoferrate(II), obtained from solutions of copper(II) sulphate (VI) and potassium(I) hexacyanoferrate(II), $K_4Fe(CN)_6$.

$$2Cu^{2+} + Fe(CN)_6^{4-} \rightarrow Cu_2Fe(CN)_6 \downarrow$$

If the two solutions are mixed the copper(II) hexacyanoferrate(II) is obtained as a brown precipitate. If, however, we fill a burette with aqueous copper(II) sulphate(VI) and place the spout just below a solution of potassium(I) hexacyanoferrate(II) we obtain copper(II) hexacyanoferrate(II) as a thin membrane by allowing a small amount of the copper(II) salt solution to ooze from the spout. The membrane forms a small transparent bag. If the potassium(I) salt solution is dilute and the copper(II) salt solution is concentrated, the bag swells as water passes from the dilute solution to the concentrated one (Fig. 4–12).

Copper(II) hexacyanoferrate(II) membranes have been used in most of the work carried out in connection with osmotic pressure. As prepared above, however, they are very weak and incapable of withstanding the large pressures sometimes encountered—osmotic pressures up to 250 atm have been recorded. For most experiments, therefore, a method of strengthening the membrane is employed. This was first used by Pfeffer in 1877 and consists of depositing the membrane in the walls of a porous earthenware pot. The pot is steeped in water to remove air from the pores, filled with copper(II) sulphate(VI) solution and placed in a solution of potassium(I) hexacyanoferrate(II). The solutions diffuse towards each other and react in the walls of the pot. The membrane obtained in this way will withstand high pressures.

Other examples of artificial membranes are found in 'chemical gardens'. These are plant-like growths of insoluble silicates produced by dropping crystals of salts of certain metals (such as iron, cobalt, nickel, and cadmium) into sodium(I) silicate(IV) solution, *e.g.*,

$$Co^{2+} + SiO_3^{2-} \rightarrow CoSiO_3 \downarrow$$

The precipitate formed on a crystal is a semipermeable membrane, and water flows from the dilute solution outside to the concentrated one inside. The internal pressure developed bursts the membrane, and the metal ions freed react with more silicate(IV) ions and extend the precipitate. By repeated bursting and 'self-healing' the precipitate takes the form of a hollow tube.

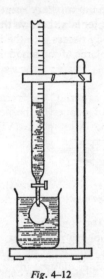

*Fig.* 4–12

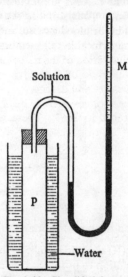

*Fig.* 4–13. Simplified form of Pfeffer's osmometer

## Measurement of Osmotic Pressure

**Pfeffer's Method.** Pfeffer determined the osmotic pressure of a large number of solutions. The simplified form of apparatus in Fig. 4–13 illustrates the principle of his method. A copper(II) hexacyanoferrate(II) membrane was deposited in the walls of a porous pot, P. The solution was placed in the pot, which was connected to a closed manometer, M, containing mercury and nitrogen. The solution filled the pot and the whole of the apparatus between the pot and the mercury. The pot stood in water in the usual way. The osmotic pressure of the solution was obtained from the extent to which the nitrogen was compressed in the closed limb of the manometer.

The weakness of Pfeffer's membranes limited his work to dilute solutions, giving osmotic pressures up to 50 atm. With improved apparatus and stronger membranes, however, Pfeffer's method was used by the Americans Morse and Fraser about 1910 to measure osmotic pressures up to 250 atm.

**Berkeley and Hartley's Method.** The apparatus given in a simplified form in Fig. 4–14 was used about 1907 by the Earl of Berkeley and E. G. J. Hartley to measure osmotic pressures. The semipermeable membrane of copper(II) hexacyanoferrate(II) was formed in a horizontal porous pot, A, which, at each end, carried glass capillary tubes, $T_1$ and $T_2$, bent at right angles. The porous pot was enclosed in a gunmetal casing, B. Water filled A completely and rose part of the way up the capillary tubes. The solution of which the osmotic pressure was to be found occupied the space

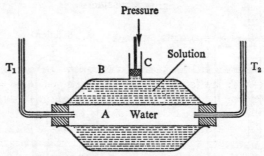

*Fig.* 4–14. Berkeley and Hartley's osmometer

between the porous pot and the metal casing. The tendency for water to pass from A to the solution was balanced by applying pressure externally through the tube C and the pressure applied was measured by a pressure gauge. If the external pressure was too small, water diffused from A and the water levels in the capillary tubes fell. If too great a pressure was applied, water diffused from the solution into A and the water levels rose. The osmotic pressure of the solution therefore equalled the pressure recorded when the water levels remained stationary. In practice it was difficult to adjust the pressure so as to obtain no movement in either direction and an average was taken of the readings when the water columns were slowly rising and slowly falling. With this apparatus it was possible to measure osmotic pressures up to 150 atm.

## Laws of Osmotic Pressure

The laws of osmotic pressure were established by the Dutchman van't Hoff about 1885 from the experimental results of Pfeffer. Van't Hoff found that there was a close resemblance between the osmotic pressure laws and the gas laws.

**Osmotic Pressure and Boyle's Law.** According to Boyle's law for gases

$$pv = k \quad \text{or} \quad p = k \times \frac{1}{v}$$

That is, the pressure of a gas is proportional to its concentration. For substances in *dilute* solution it is also found that *the osmotic pressure is proportional to the concentration.*

**Osmotic Pressure and Charles's Law.** For a given volume of a gas the pressure is proportional to the thermodynamic, or kelvin, temperature. *The osmotic pressure of a solution of given concentration is also proportional to the kelvin temperature;* that is,

$$\frac{\Pi}{T} = \text{a constant}$$

where $\Pi$ (capital pi) is osmotic pressure.

**Osmotic Pressure and the Ideal Gas Equation.** By combining Boyle's law and Charles's law for gases, we obtain the equation $pv = RT$, which shows the relation between pressure, volume, and temperature for 1 mole of a gas. Since analogous laws apply to osmotic pressure, a similar expression $\Pi v = R'T$ can be deduced for dilute solutions, where $\Pi$ is the osmotic pressure and $v$ the dilution—that is, the reciprocal of the concentration in mole per m³. Van't Hoff showed that the constant $R'$ for osmotic pressure agrees very closely with the gas constant $R$. The value of the latter is $8 \cdot 31 \ \text{J mol}^{-1} \text{K}^{-1}$. The value of $R'$ can be calculated from the equation $\Pi v = R'T$. Thus the osmotic pressure of a $0 \cdot 1$ molar solution of sucrose ($C_{12}H_{22}O_{11}$) at 0°C is $228 \times 10^3$ Pa. Then

$$R' = \frac{\Pi v}{T} = \frac{228 \times 10^3 \times 0 \cdot 01}{273} = 8 \cdot 35 \ \text{N m mol}^{-1} \text{K}^{-1}$$

$$= 8 \cdot 35 \ \text{J mol}^{-1} \text{K}^{-1}$$

**Osmotic Pressure and Avogadro's Hypothesis.** The close analogy between the laws of osmotic pressure and those of gas pressure led van't Hoff to suggest a gaseous theory of solution, in which he compared the dissolved molecules in a solution to molecules of a gas moving in a confined space. He imagined that all the solvent was removed from a solution, so that the solute molecules were left moving about like a gas in the same volume previously occupied by the solution. Then, according to van't Hoff, the osmotic pressure of the original solution should be equal to the pressure exerted by the 'gas'. This means that osmotic pressure, like gas pressure, should be independent of the nature of the particles and the nature of the solvent, depending only on the concentration of dissolved molecules and on the temperature.

Van't Hoff's conclusions are in fact true. Thus equal volumes of all gases at the same temperature and pressure contain the same number of molecules. By comparing the concentrations of solutions which exert the same osmotic pressure at the same temperature we find that *equal volumes of these solutions contain the same number of dissolved molecules.*

One mole of a gas at 273 K and 101 325 Pa (1 atm) occupies 22·4 dm³.

One mole of a solute dissolved in 22·4 dm³ of solution at 273 K has an osmotic pressure of 101 325 Pa.

The correspondence between osmotic pressure and the calculated 'gas' pressure holds only for dilute solutions; the divergence increases with

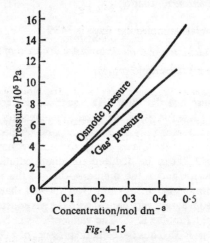

*Fig.* 4–15

concentration. This is shown by the graphs of the osmotic pressure and calculated 'gas' pressure of sucrose solutions, at 0°C, that are plotted in Fig. 4–15.

**Calculations on Osmotic Pressure.** In practice we seldom determine relative molecular masses from osmotic pressures, because accurate measurement of the osmotic pressure is difficult, and easier methods are usually available (*e.g.*, the cryoscopic method). The osmotic pressure method is chiefly used for finding the relative molecular masses of plastics, which are often extremely high.

In calculating the relative molecular mass of a solute from the osmotic pressure of its solution we use van't Hoff's gaseous theory of solution. We mentally remove the solvent from the solution and then perform the calculation as for a gas. Similarly we can calculate the osmotic pressure of a solution of known concentration at a given temperature if we know the relative molecular mass of the solute.

**Example 1.** *An aqueous solution of sucrose containing* 19·15 g *of the sugar per dm³ has an osmotic pressure of* 136 300 Pa *at* 20°C. *Find the relative molecular mass of sucrose.*

Imagine the sugar to be a gas in the volume occupied by the solution. 1 dm³ of 'gas' at 20°C and 136 300 Pa pressure has a volume at s.t.p.

of $1 \times \dfrac{273}{293} \times \dfrac{136\ 300}{101\ 300}$ dm³ $= 1·254$ dm³

At s.t.p. 1·254 dm³ of 'gas' contain 19·15 g of sucrose.

$\therefore$      22·4 dm³ at s.t.p. contain $\dfrac{19·15 \times 22·4}{1·254}$ g

$$= 342·3 \text{ g}$$

Therefore the relative molecular mass is 342·3.

**Example 2.** *Calculate the osmotic pressure in pascal of a* 2 *per cent solution of glucose* ($C_6H_{12}O_6$) *at* 18°C. *(C = 12, H = 1, O = 16.)*

The relative molecular mass of glucose (by adding relative atomic masses) = 180.

A 2 per cent solution of glucose contains 2 g of glucose in 100 g of solution, but, since the solution is dilute, 100 g of solution = 100 cm³ approximately.

The most straightforward way of solving this type of problem is to put the answer term at the end of the line and correct for one factor at a time. Thus,

180 g of glucose in 22 400 cm³ of solution at 0°C have an osmotic pressure of approx. 101 300 Pa.

2 g of glucose in 22 400 cm³ of solution at 0°C have an osmotic pressure

of $101\ 300 \times \dfrac{2}{180}$ Pa

2 g of glucose in 100 cm³ of solution at 0°C have an osmotic pressure of

$101\ 300 \times \dfrac{2}{180} \times \dfrac{22\ 400}{100}$ Pa

2 g of glucose in 100 cm³ of solution at 18°C have an osmotic pressure

of $101\ 300 \times \dfrac{2}{180} \times \dfrac{22\ 400}{100} \times \dfrac{291}{273}$

$$= 268\ 000 \text{ Pa}$$

**Exceptions to the Osmotic Pressure Laws.** The laws of osmotic pressure hold only for dilute solutions of non-electrolytes or weak electrolytes. In this respect osmotic pressure is similar to the other colligative properties of solutions.

*Concentrated Solutions.* With solutions containing less than 1 mole of solute per dm³ the osmotic pressure is approximately proportional to concentration. With more concentrated solutions osmotic pressure increases rapidly with increase in concentration. Fig. 4–16 shows the variation of osmotic pressure with concentration for sucrose solutions, at 30°C, up to a concentration of 2·5 mole per dm³ of solution. The reason for the values being larger than the theoretical ones is not known with certainty. It is probable that 'solvation' of the solute occurs—that is, some of the solvent molecules become attached to solute molecules, thus increasing the effective concentration. With sucrose and water, solvation could take place by 'hydrogen bonding' (Chapter 7).

*Strong Electrolytes.* The osmotic pressure of a solution of a strong electrolyte like sodium(I) chloride is always more than the value calculated

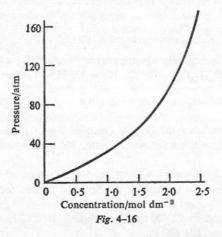

*Fig.* 4–16

on the assumption that the solute exists in solution as molecules corresponding to the chemical formula. This is explained by dissociation of the solute into ions, which increases the effective number of particles (*e.g.*, $NaCl \rightleftharpoons Na^+ + Cl^-$). Van't Hoff employed a factor $g$ (until recently, $i$) to represent the relation between the observed osmotic pressure and that calculated on the assumption that the solute was a non-electrolyte.

$$g = \frac{\text{observed osmotic pressure}}{\text{calculated osmotic pressure}}$$

The value of $g$ depends on the solute, the solvent, and the concentration. The 'abnormally' high osmotic pressures of strong electrolytes in solution are paralleled by similar 'abnormalities' in freezing point depression, boiling point elevation, and lowering of vapour pressure. For a solution of given concentration, the value of $g$ is the same in all four cases, showing clearly the colligative character of these properties of solutions. Any of the four can be used to measure the degree of dissociation of a strong electrolyte in solution.

*Association.* If the molecules of a solute are in association, the osmotic pressure of the solution is less than calculated from the simple formula. Thus the osmotic pressure of a solution of ethanoic acid in benzene is half the calculated value, corresponding to association into $(CH_3COOH)_2$ molecules. A similar effect has already been noted for freezing point depression.

### Cause of Osmosis.

*Callendar's Vapour Pressure Theory.* This theory regards a semi-permeable membrane as containing a large number of very fine pores. These are too small to permit the passage of a liquid, but allow diffusion of the vapour of the liquid. Suppose that such a membrane is used as a partition between a solution and pure solvent. Since the vapour pressure of the solution is less than that of the solvent, vapour will diffuse through the pores from the solvent side of the membrane to the solution side, thus tending to increase the volume of the solution. The vapour pressure of a solution can be increased, however, by applying an external pressure to the solution. If the vapour pressure is increased sufficiently it becomes equal to that of the solvent, and the vapour then diffuses through the pores at equal rates in both directions. The osmotic pressure is therefore the external pressure which must be applied to the solution to make its vapour pressure equal to that of the solvent.

*Solvent Bombardment Theory.* According to this theory osmosis is the result of unequal bombardment pressure by the solvent molecules on the two sides of a semipermeable membrane. On the solvent side impacts are made by solvent molecules only. On the solution side, part of the liquid surface is occupied by solute molecules, and so a given area of the membrane is bombarded by fewer solvent molecules than on the solvent side (Fig. 4–17). Hence the solvent molecules diffuse into, and through the membrane more slowly on the solution side than on the solvent side. (We have already used a similar kinetic explanation for the lower vapour pressure of a solution as compared with that of the pure solvent.) External pressure on the solution increases the velocity of the solvent molecules on this side and thus their rate of diffusion. The osmotic pressure is the external pressure required to equalize the rates of diffusion of solvent molecules in both directions.

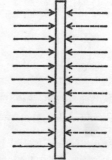

*Fig.* 4–17. (Dotted lines represent impacts by solute molecules)

Coupled with the solvent bombardment theory of osmosis is the selective solubility theory of semipermeability. On the solution side the membrane is bombarded by molecules of solute as well as by those of solvent. The membrane is able to dissolve solvent molecules, but not those of solute. Since the two bombardment pressures act independently, like the partial pressures of two gases mixed together, the pressure due to the solute molecules cannot affect the rate of diffusion of solvent molecules.

The solvent bombardment theory of osmosis and the selective solubility theory of semipermeability are both illustrated by the experiment now described. In this experiment we use hydrogen sulphide to represent the solvent, hydrogen the solute, and water the semipermeable membrane. (Carbon dioxide can be used instead of hydrogen sulphide, and air instead of hydrogen.)

**Experiment.** Fit two litre flasks, A and B (Fig. 4–18), with tightly fitting rubber stoppers, through which pass a longer glass tube and a short right-angled tube. Attach short lengths of rubber tubing and

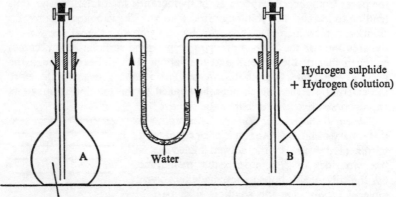

*Fig.* 4–18. Apparatus used to illustrate the solvent bombardment theory of osmosis

screw clips to the tops of the longer tubes. Fill one flask, A, with hydrogen sulphide and the other flask, B, with a mixture of hydrogen sulphide and hydrogen. Close the screw clips after filling the flasks and connect the latter through a small pressure gauge half filled with coloured water. Open the clips for a moment to make the pressures in both flasks equal to that of the atmosphere and therefore equal to each other. Hydrogen sulphide dissolves readily in water, while hydrogen is insoluble. Thus the hydrogen sulphide can be regarded as the 'solvent,' the mixture of hydrogen sulphide and hydrogen as the 'solution,' and the water in the pressure gauge as the 'semi permeable membrane.'

Leave the apparatus standing for 2 hours. (If the water in the pressure gauge is first saturated with hydrogen sulphide a pressure difference will be observed in 10—15 minutes.) It will be observed, from the movement of the water in the pressure gauge, the pressure in A decreases, while that in B increases. This can only be because hydrogen sulphide is being transferred from the 'solvent' side of the semipermeable membrane to the 'solution' side. In this experiment the larger pressure of hydrogen sulphide in A as compared with the partial pressure of the gas in B corresponds to the greater bombardment pressure of a pure solvent as compared with that of the solvent in a solution.

**Relation between Osmotic Pressure and Vapour Pressure.** A solution of a non-volatile solute has a lower vapour pressure than the pure solvent, and at a given temperature the lowering of the vapour pressure is proportional to the molar concentration of the solute. Since the osmotic pressure of a (dilute) solution is also proportional to the molar concentration of the solute, the osmotic pressure must be proportional to the lowering of the vapour pressure. This conclusion can be deduced theoretically as shown below.

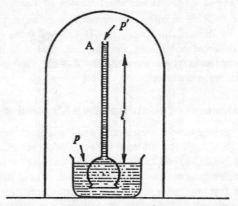

*Fig.* 4–19. Osmotic pressure and vapour pressure

Imagine the apparatus in Fig. 4–19. Suppose that the solution is in a thistle funnel, and is separated from the solvent by a semipermeable membrane tied over the mouth of the funnel. The whole apparatus is in a bell jar which has been exhausted of air. When equilibrium has been attained the liquid in the funnel will have risen to A, at a height $l$ in m above the level of the solvent, and the bell jar will be saturated with the vapour of solvent.

The osmotic pressure, $\Pi$ (in Pa), of the solution is represented by the hydrostatic pressure of the column, of length $l$, of solution. If the solution is dilute the density (in kg m$^{-3}$) of the solution is equal to the density, $\rho$, of the solvent. Hence, if $g$ is the acceleration of free fall in m s$^{-2}$,

$$\Pi = l\rho g \tag{1}$$

Let $p =$ the vapour pressure of the pure solvent at the given temperature and $p' =$ the vapour pressure of the solution. Since the system is in equilibrium, the vapour pressure $(p')$ at A must be the same inside and outside the funnel.

The difference between the vapour pressure of the solvent at its snrface and at a height $l$ above the surface is given by $p - p'$, and

$$p - p' = l\rho' g \qquad (2)$$

where $\rho'$ is the density of the solvent vapour at a pressure $p$.

Dividing (1) by (2), we obtain

$$\frac{\Pi}{p - p'} = \frac{l\rho g}{l\rho' g} = \frac{\rho}{\rho'} \qquad (3)$$

At a given temperature $p$ (and therefore $\rho'$) is constant; also $\rho$ is constant. We therefore see that

$$\Pi \propto p - p'$$

That is, for a dilute solution the osmotic pressure is proportional to the lowering of the vapour pressure. It follows from (3) that dilute solutions of different solutes in the same solvent will have the same osmotic pressure at a given temperature if their vapour pressures are equal. The converse is also true.

## Determination of High Relative Molecular Masses

Relative molecular masses vary over a wide range; those of plastics and proteins may be as high as a million. The classical methods of determination described in this chapter cannot normally be used for substances with relative molecular masses above a few hundred. For a solution of given concentration, freezing point depression, osmotic pressure, and so on are inversely proportional to the relative molecular mass of the solute. Thus if a 1 per cent solution of a substance of relative molecular mass 100 has a freezing point depression of 0·1 degC, the depression for a substance of relative molecular mass 10 000 would be only 0·001 degC. Such small differences of temperature cannot be measured with ordinary thermometers.

In modern times several methods have been devised for finding very high relative molecular masses. Some of these are now described.

*Adaptations of Classical Methods.* Most of the adaptations make use of the newer techniques of measurement employed in physics. Thus we can now measure accurately boiling point elevations as small as 0·0001 degC by thermocouples and amplifying devices which magnify minute differences of current strength. This has considerably extended the range of relative molecular masses which can be found by the boiling point method.

We can use osmometers of the type shown in Fig. 4–20 to determine relative molecular masses above 10 000. A plastic is dissolved in an organic solvent and the solution placed in a glass cell fitted with a capillary tube. The wide open end of the cell is ground flat and over it is stretched a cellophane membrane. The latter is supported by a perforated steel plate, which is bolted tightly to an upper steel plate. The cell is immersed in the solvent and the rise in level of the liquid in the capillary tube is measured. This may be a few millimetres or several centimetres. From the increase in height we calculate the osmotic pressure, and hence the relative molecular mass of the solute.

*The Ultracentrifuge.* When particles of a solid are suspended in a liquid they gradually settle under the action of gravity. The rate of settling depends partly on the masses of the individual particles. The tendency to settle is opposed by the thermal movement of the molecules of the liquid and, if the particles are small individual molecules or ions (as in ordinary solutions) they never settle. The macromolecules of proteins and plastics in solution have only a slight tendency to sediment, but this can be increased by means of an ultracentrifuge, which replaces the force of gravity by a much larger centrifugal force.

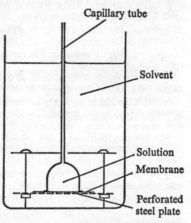

The ultracentrifuge was developed in the 1920s by the Swedish chemist Svedberg. It differs from the laboratory centrifuge (used for sedimenting fine precipitates in semimicro analysis) in having a much higher speed of rotation and in being fitted with an optical system so that the rate of

*Fig.* 4-20. Osmometer used for measuring relative molecular mass of a plastic

sedimentation can be measured. In modern forms of the instrument the speed of rotation may reach 50 000 rev/min, giving a centrifugal force of 150 000 times gravity (for the laboratory centrifuge the figures are about 2000 rev/min and 600). The rate of sedimentation is usually followed from changes in the refractive index of the solution as the large molecules are driven to the periphery. From this rate, and additional data such as the speed of rotation, we can calculate the relative molecular mass of the macromolecules. Thus it has been found that egg albumin has a relative molecular mass of about 42 000, while that of human hæmoglobin is about 65 000. The linear molecules of ethenyl plastics vary in relative molecular mass according to the length of the chain. For poly-(chloroethene) the values vary between 20 000 and 160 000.

*Radioactive End Groups.* With plastics we can often introduce a group containing a single radioactive atom (*e.g.*, $^{14}C$) at the end of a long molecular chain. We then find the number of such atoms in a known mass of the plastic by measuring the radioactivity of the sample. Hence we can calculate the relative molecular mass of the plastic.

## EXERCISE 4

*(Relative atomic masses are given at the end of the book)*

**Freezing-point Depression** ($K$ for water per 1 000 g $= 1\cdot86$ degC mol$^{-1}$).

**1.** A solution holding $1\cdot2$ g of ethanoic acid in 80 g of water freezes at $-0\cdot46°$C. What is the relative molecular mass of the acid?

**2.** Describe the freezing point method for the determination of relative molecular mass. $0\cdot48$ g of a substance X dissolved in 50 g benzene caused a freezing point depression of $0\cdot440$ degC. Calculate the relative molecular mass of X from these observations. (Depression of freezing point for 1 000 g of benzene is $5\cdot5$ degC mol$^{-1}$.) (Lond.)

**3.** (Part question.) If the freezing point constant for 1 000 g of water is $1\cdot86°$C mol$^{-1}$, at what temperature (theoretically) will a solution of $3\cdot33$ g of ethane-1, 2-diol, $C_2H_4(OH)_2$, in 14 g of water begin to freeze? (O.L.)

**4.** Liquid camphor freezes at $175°$C. A solution of $1\cdot54$ g of naphthalene ($C_{10}H_8$) in 18 g of camphor freezes at $148\cdot3°$C. What is the freezing constant (per 1000 g) of camphor?

**5.** In how much water should 10 g of glucose ($C_6H_{12}O_6$) be dissolved to obtain a solution freezing at $-0\cdot35°$C?

**6.** With the aid of a clearly labelled diagram outline the essential features of a suitable experimental method for the measurement of the freezing point depression of a solvent (*e.g.*, benzene) by a dissolved solute.

The freezing point of a sample of pure benzene was found to be $5\cdot481°$C. A solution of $0\cdot321$ g of the hydrocarbon naphthalene ($C_{10}H_8$) in 25 g of this benzene began to freeze at $4\cdot971°$C. A solution of $0\cdot305$ g of benzoic acid in 25 g of the same solvent began to freeze at $5\cdot226°$C. Calculate the molar freezing point depression constant for 1 000 g of benzene and hence calculate the relative molecular mass of benzoic acid in benzene solution. $C = 12$; $H = 1$.

(W.J.E.C.)

**Boiling-point Elevation**

**7.** (Part question.) $2\cdot00$ g of phosphorus raise the boiling point of $37\cdot4$ g of carbon disulphide by $1\cdot003°$C. What is the molecular formula of phosphorus in carbon disulphide? What reasons can you suggest for this result? (Boiling point elevation constant for carbon disulphide is $2\cdot35°$C for 1 mole in 1 000 g.)

(S.U.)

**8.** The boiling point of ethanol is $78°$C. Calculate the boiling point of a solution containing $2\cdot7$ g of ethanamide ($CH_3CONH_2$) in 75 g of ethanol. (Boiling constant for 1 000 g of ethanol $= 1\cdot15°$C mol$^{-1}$.)

**9.** Explain what is meant by the boiling point elevation constant of a liquid.

How would you measure the relative molecular mass of a compound by the method of the elevation of boiling point of a liquid? Give concise experimental details.

State, with reasons, why this method could be used with aqueous solutions to obtain the relative molecular mass of urea, but not that of ethanoic acid.

A solution of $2\cdot8$ g of cadmium(II) iodide in 20 g of water boiled at $100\cdot2°$C at standard pressure. Calculate the relative molecular mass of the solute and comment on the result. (The boiling point elevation constant for water is $0\cdot52°$C mol$^{-1}$ per 1 000 g.) (J.M.B.)

**Lowering of Vapour pressure**

**10.** (Part question.) The vapour pressure of pure water at $25°$C is 3 167 Pa. The vapour pressure of a solution of 4 g of a sugar in 100 g of water at the same temperature is 3 154$\cdot$5 Pa. What is the relative molecular mass of the sugar?

(O.L.)

**11.** The vapour pressure of carbon disulphide at a certain temperature is 5 333 Pa. At the same temperature a solution of 5 g of sulphur in 63 cm$^3$ of carbon disulphide has a vapour pressure of 52 230 Pa. The density of carbon disulphide is 1·27 g cm$^{-3}$. Find (i) the relative molecular mass, (ii) the molecular formula of sulphur in carbon disulphide.

**12.** What is the vapour pressure of a 3 per cent solution of camphor, $C_{10}H_{16}O$, in ethoxyethane, $C_4H_{10}O$, if the vapour pressure of pure ethoxyethane at the same temperature is 32 670 Pa?

### Osmotic Pressure

**13.** Define the terms: osmosis, osmotic pressure, isotonic solutions. Describe a method for the *accurate* determination of the osmotic pressure of an aqueous solution of sucrose.

A solution of 42·0 g of mannitol in 1 dm$^3$ of water has an osmotic pressure of 5·624 × 10$^5$ Pa at 20°C. Calculate the relative molecular mass of mannitol. Molar volume at s.t.p. = 22·4 dm$^3$. (Lond.)

**14.** (Part question.) What pressure would prevent the passage of water through a semipermeable membrane from water into a 2 per cent solution of sucrose (relative molecular mass = 342) at 12°C? (O.L.)

**15.** (Part question.) Write an equation showing the quantitative relationship between osmotic pressure, concentration of solution and temperature. State the limitations on the use of this equation.

At 25°C, the osmotic pressure of a solution containing 1·35 g of a protein per 100cm$^3$ of solution was found to be 1 216 Pa. Calculate the relative molecular mass of the protein. The molar volume at s.t.p. is 22·4 dm$^3$. (S.U.)

**16.** At what temperature would an aqueous solution containing 10 g of glucose, $C_6H_{12}O_6$, in 500 cm$^3$ have an osmotic pressure of 264 700 Pa?

**17.** Outline ONE experimental method by which the osmotic pressure of a solution has been measured. State the laws which apply to the osmotic pressure of dilute solutions.

An aqueous solution containing 5·2 g of ethanamide ($C_2H_5NO$) per dm$^3$ froze at −0·164°C. Calculate (*a*) the freezing point, (*b*) the osmotic pressure at 20°C, of a 1 per cent solution of glucose ($C_6H_{12}O_6$). H = 1; C = 12; N = 14; O = 16; 1 mole of a gas occupies 22·4 dm$^3$ at s.t.p. (C.L.)

**18.** The osmotic pressure of an aqueous solution of a non-electrolyte containing 8·15 g in 1·5 dm$^3$ of solution is 70 930 Pa at 25°C. What would be the freezing point of the solution?

## MORE DIFFICULT QUESTIONS

**19.** (*a*) Describe how you would determine the relative molecular mass of benzenecarboxylic acid in benzene by elevation of boiling point of the solvent. (Assume that the elevation constant of benzene is already known.)

(*b*) The following results were obtained in an investigation into the molecular state of ethanoic acid in benzene:

Freezing point of pure benzene = 5·533°C.

Freezing point of solution of 0·289 g ethanoic acid in 100 g benzene = 5·386°C.

Freezing point of a solution of 0·784 g tetrachloromethane in 43·0 g benzene = 4·930°C.

Tetrachloromethane does not associate or dissociate in benzene. Assuming that ethanoic acid partly associates to double molecules in benzene, calculate the ratio of associated to non-associated molecules in the solution. (S.U.)

**20.** (Part question.) The pressure ($p$) of water vapour in equilibrium with ice varies with the absolute temperature ($T$) according to the equation

$$\log p \, (+2 \cdot 125) = \text{const.} - 2\,650/T$$

A solution of 0·04 mole of sucrose in 100 g of water has a vapour presure of 574·7 Pa at its freezing point, compared with that of 610·7 Pa for pure water at 273·2 K (0°C). Calculate (*a*) the freezing point of the sugar solution, (*b*) the molar depression constant for water.     (C.L.)

**21.** State Raoult's law and use a clearly labelled vapour pressure/temperature diagram to explain how the change in the vapour pressure of a solvent (caused by the solution of non-volatile solute) is related to the simultaneous changes in its freezing point and boiling point. Indicate briefly how these changes depend on the concentration of the solute.

The vapour pressure of an organic liquid X at 20°C is 58 670 Pa, and that of a solution of 9·00 g of methyl octadecanoate (non-volatile) in 100 g of the same liquid is 57 400 Pa at 20°C. Calculate the relative molecular mass of X, given that the relative molecular mass of methyl octadecanoate is 298.     (W.J.E.C.)

**22.** At 20°C the vapour pressure of an aqueous solution of a non-electrolyte containing 3·266 g of solute per 100 cm$^3$ of solution is 2 334·4 Pa. At the same temperature the vapour pressure of water is 2 338·7 Pa. What would be the osmotic pressure in pascal of this solution at 20°C?

# CHAPTER 5

# Electrical structure of atoms and nuclear reactions

## Electrical Structure of Atoms

The atomic theory of matter formed the basis of theoretical chemistry during the nineteenth century. It was believed that atoms were indivisible and unchangeable, so that the efforts of the alchemists to convert base metals into gold were foredoomed to failure by the individuality of the atoms themselves. Both of these beliefs, the atom as the smallest unit of matter and the impossibility of transmuting the elements, had to be revised in the light of new discoveries made towards the end of the century. These discoveries were concerned with the discharge of electricity through gases at low pressure and the phenomenon of radioactivity. They showed that atoms are made up of still smaller particles, some of which are electrically charged.

Scientists became aware of the close connection between matter and electricity quite early in the last century from experiments on electrolysis. The laws of electrolysis were established by Faraday about 1833. From these it can be shown that during electrolysis electricity is transferred between the electrodes and the ions in solution in small definite amounts, In other words, electricity itself is 'atomic' in character—it consists of small fixed units. The unit charge is simply the quantity of electricity required to liberate one atom of hydrogen, silver, or any other monovalent element at an electrode. Its value can be easily calculated and shown to be $1.60 \times 10^{-19}$ coulomb. In 1871 Stoney suggested that the term *electron*, meaning 'atom of electricity,' should be used for this unit. A few years later the unit made a dramatic appearance in a totally different field.

**Electrons.** At ordinary pressures gases are very poor conductors of electricity. At low pressures, however, they become quite good conductors. If a gas is enclosed in a glass tube containing two electrodes and the pressure of the gas is reduced to about 700 Pa, a bright luminous discharge takes place when a sufficiently high voltage is applied. Discharge tubes of this kind are commonly used in fluorescent lights and neon advertising signs.

If the pressure of the gas in the tube is reduced to less than 1 Pa the luminous discharge is replaced by only very faintly luminous rays. These proceed from the cathode (Fig. 5–1) and hence are called *cathode*

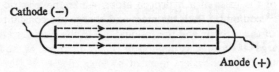

Cathode (−)

Anode (+)

*Fig.* 5–1. Production of cathode rays

*rays.* A discharge tube used under these conditions is called a *cathode-ray tube.* The cathode rays are emitted at right angles to the surface of the cathode and do not necessarily travel to the anode, for this may be placed at the side of the tube (Fig. 5–2). The rays have the following properties:

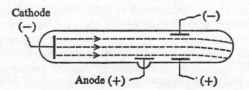

Cathode (−)

(−)

Anode (+)

(+)

*Fig.* 5–2. Deflection of cathode rays by an electrical field

- When they fall on the glass at the far end of the tube they cause the glass to fluoresce with a green light.
- If an obstacle is placed in their path they cast a sharply defined shadow on the end of the tube *remote from the cathode.*
- If the rays are allowed to strike the upper vanes of a small paddle wheel mounted in the tube, the wheel rotates in a direction *away from the cathode.*
- If the rays are passed between two oppositely charged plates they are deflected towards the positive plate and away from the negative plate (Fig. 5–2).
- They are deflected by a magnetic field in the direction which would be expected for negatively charged particles.

• They pass through a thin sheet of aluminium foil, which even the smallest gaseous atoms (helium) or molecules (hydrogen) cannot penetrate.

These properties can be explained only by assuming that cathode rays consist of a stream of negatively charged particles travelling away from the cathode in straight lines. The fact that the particles can pass through thin sheets of metal which are impervious to helium shows that they are smaller than atoms.

The mass of the particles and their electric charge were first determined by J. J. Thomson about 1897. The mass of a single particle is $9 \cdot 1 \times 10^{-28}$ g or 1/1837 of the mass of a hydrogen atom, while the electric charge is $1 \cdot 60 \times 10^{-19}$ coulomb. Both the mass and charge are constant whatever the nature of the cathode and the residual gas in the discharge tube. The particle charge is the same as the unit of charge involved in the liberation of one atom of a monovalent element by electrolysis. It was therefore concluded that the particles were identical with Stoney's 'electrons,' and this name was adopted for them.

Electrons can be produced by other means besides a discharge tube. They are given off when metal filaments are heated, as in radio valves, or when certain metals are exposed to light, as in photoelectric cells. Electrons are often transferred from one kind of material to another when the two are rubbed together—for example, sealing wax and wool, or Perspex and poly(ethene). The production of electrons from so many different kinds of substances leaves no doubt that these particles are one of the basic constituents of atoms in general.

**Protons.** Atoms are electrically neutral. It follows that if atoms contain electrons, they must also contain an equal amount of positive electricity in some form or other. We can demonstrate the existence of positively charged particles in a cathode-ray tube by boring small holes

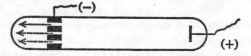

*Fig.* 5–3. Production of positive rays

in the cathode. When an electric discharge is passed we find that not only are cathode rays formed, but also rays are passing through the cathode in the opposite direction (Fig. 5–3). These rays are deflected by electric and magnetic fields in the opposite manner to electrons, showing that they consist of positively charged particles. They are called *positive rays*.

The mass and charge of the particles in positive rays are *not* constant.

The mass depends on the nature of the residual gas in the tube. For example, if the gas is hydrogen the mass of the particles is almost identical with that of a hydrogen atom; if the gas is oxygen we obtain particles corresponding in mass to an oxygen atom. It is therefore concluded that positive rays are produced by collisions between the molecules of residual gas and the electrons streaming away from the cathode. Although the electrons are very small their velocity is extremely high; it depends on the applied voltage but may be up to one third of the velocity of light. The electrons thus have a high energy. On collision, they break up some of the gas molecules into atoms and strip electrons from the atoms. Now if an atom loses an electron it is converted into a positive ion (*e.g.*, $H - e \rightarrow H^+$). The ion is attracted towards the cathode and, if the latter is perforated, may pass through it. Just as the unit of negative charge is the electron, so the unit of positive charge is that associated with a hydrogen ion.

The lightest positive ions are produced in a discharge tube when the residual gas is hydrogen. As we shall see later, a hydrogen atom possesses only one electron. Since the mass of the electron is only 1/1837 of that of a hydrogen atom the mass of the $H^+$ ion is almost the same as that of the neutral atom. The $H^+$ ion is called a *proton*. The ions formed from other residual gases may have one or more positive charges according to whether the neutral atoms have lost one or more electrons. Their masses, however, are very nearly whole-number multiples of that of a proton, indicating that protons, as well as electrons, are fundamental particles in the composition of atoms.

**Neutrons.** For a long time it was thought that atoms consisted only of electrons and protons. However, in 1932 it was found that when beryllium was bombarded with α-rays (see below) particles were given off which had properties quite different from those of electrons and protons. They were not deflected at all by electric or magnetic fields, and therefore possessed no charge. The new particles were investigated by Chadwick, who showed that they had almost the same mass as a proton. As they were neutral particles they were called *neutrons*. Later experiments have shown that neutrons are a constituent of all atoms except that of hydrogen. Neutrons therefore represent the third of the fundamental particles which make up atoms.

**Natural Radioactivity.** The phenomenon of radioactivity was discovered in 1896 by Becquerel, a French scientist, who found that a crystal of a uranium salt blackened a photographic plate even in complete darkness. Becquerel showed that the effect was due to the giving-off of very active rays from the uranium compound. It was then discovered that the uranium ore pitchblende was, mass for mass, more active than uranium, suggesting that the ore contained a still more powerful substance. As a result of experiments carried out by Marie Curie and her husband, Pierre Curie, two such substances were found in trace amounts. These were two new elements, to which the names polonium and radium were given. Radium proved to be two million times more radioactive than uranium itself. Radium, however, occurs naturally in very low concentrations; 1 g of radium is contained in $500 \times 10^3$ kg of pitchblende.

The rays given off by radium were investigated by the Curies and by Rutherford, who established that they were of three kinds. These were labelled $\alpha$, $\beta$, and $\gamma$ according to their penetrating power.

*α-rays.* These have the least penetrating power, being stopped by a sheet of paper or several cm of air. When a trace of a radium salt was put at the bottom of a hole bored in a lead block and the rays were passed through a magnetic field F (Fig. 5-4), the α-rays were deflected in a direction which showed that they consisted of positively charged particles. They proved to be helium ions ($He^{2+}$) with a velocity about one tenth that of light.

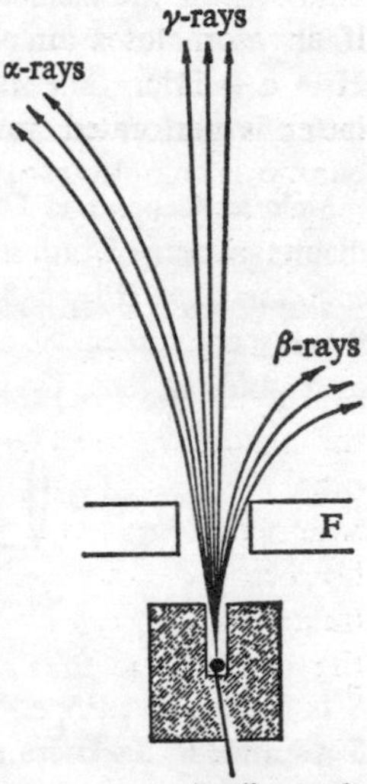

*β-rays.* These could pass through thin sheets of aluminium. A magnetic field deflected them much more readily than α-particles and in the opposite direction. They were found to be identical with electrons produced in cathode-ray tubes and to have a velocity approaching that of light.

*γ-rays.* These rays required several cm of lead or over 1 metre of concrete to absorb them completely. They were not deflected by a magnetic field. They were found not to consist of particles at all, but to be waves

Fig. 5-4. Deflection of rays from radium by a magnetic field

similar to X-rays, although of smaller wavelength. It was later discovered that γ-rays were extremely dangerous to living creatures, causing profound damage to the tissues.

In 1903 Rutherford put forward the theory that radioactivity was caused by the *disintegration*, or *decay*, of the large heavy atoms of radium, uranium, etc., into simpler atoms of other elements. This proved to be correct. Thus, when a radium atom of relative atomic mass 226 breaks up, it undergoes a series of changes, in which it loses altogether five helium ions, each of mass 4, and the surplus electrons. Eventually it is converted into an atom of lead of relative mass 206. The change takes place spontaneously and its rate cannot be altered in any way (*e.g.*, by changing the temperature). The precise reason for the disintegration is still unknown.

All elements of higher relative atomic mass than bismuth (209) are radioactive, and decay ultimately to lead. Radioactive forms of lighter elements also occur naturally to a minute (but significant) extent. Thus a small proportion of carbon in the carbon dioxide of the atmosphere consists of a radioactive form ($^{14}C$) of the element, and traces of radioactive potassium ($^{40}K$) are found in granite. Altogether some 30 elements exist naturally in radioactive forms.

Each radioactive form of an element has its own rate of disintegration. The different rates are denoted by the *half-life* period, which is the time required for half the element to change. This is the same whatever the original mass may be. The half-life of radium is about 1600 years. Thus 1 g of the element becomes $\frac{1}{2}$ g in 1600 years, a $\frac{1}{4}$ g in a further 1600 years, and so on. The half-life periods of radioactive elements vary from a fraction of a second to millions of years. One of the longest is that of ordinary uranium ($4.5 \times 10^9$ years).

**Spinthariscope and Determination of the Avogadro Constant ($N_A$).** The disintegration of radium atoms can be observed directly by means of a *spinthariscope* (Fig. 5–5). This consists of a tube closed at one end by a fluorescent screen, S, coated with zinc(II) sulphide and at the other by a magnifying lens, L. Screwed into the tube and a few millimetres from

L                                                    S

W

*Fig.* 5–5. Spinthariscope

S is a wire, W, tipped with a trace of a radium salt. After waiting for 5 minutes in darkness to allow the eye to become adapted, the observer views the screen through the lens. From 50 to 200 flashes of light, or scintillations, per second can be seen, each one caused by a separate α-particle striking the screen. When α-particles are brought to rest they acquire electrons from their surroundings, and become atoms of helium gas.

This experiment formed the basis of Rutherford and Geiger's method of finding the Avogadro constant ($N_A$), the number of molecules in 1 mole of a gas. The amount of radium salt was made sufficiently small to allow the observer to count the number of flashes obtained in a given time on a known area of screen. From this the number of helium atoms produced in this time could be calculated. The volume of helium gas collected over a (much longer) period of time was measured. By comparing the number of atoms and the volume of gas obtained in the same time it was possible to calculate the number of atoms in 22·4 dm³ at s.t.p. Since helium is a monatomic gas this was the same as the number of molecules in the molar volume of other gases. The value obtained for the Avogadro constant was $6.05 \times 10^{23}$ mol⁻¹, which agrees well with the modern value ($6.023 \times 10^{23}$ mol⁻¹).

**Detection of Radiation.** Two methods of detecting the products of radioactive change have already been mentioned, namely, the blackening of a photographic plate and the flashes produced on a fluorescent screen coated with zinc(II) sulphide. For more detailed study of radiation two special instruments are used.

*Geiger–Müller Counter.* This depends on the fact that radiation of any of the three types, $\alpha$, $\beta$, or $\gamma$, makes a gas into a conductor of electricity by ionizing it. There are various forms of Geiger–Müller tube, each adapted to the type and strength of the radiation to be investigated; essentially, however, the apparatus consists of a glass tube (Fig. 5–6), into which two metal electrodes are sealed, and which contains a mixture of 94 per cent argon and 6 per cent of ethanol vapour at a pressure of 1300 — 1600 Pa. The anode, A, is a fine tungsten wire running along the axis of the tube. The cathode, B, is a thin aluminium cylinder around

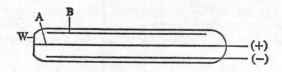

*Fig.* 5–6. Geiger–Müller counter

the inside of the tube. A potential difference of about 500 volt is maintained between the electrodes. Radiation enters the tube through a thin mica window at the end of the tube.

When an $\alpha$-particle, a $\beta$-particle, or a $\gamma$-ray enters the tube and strikes the argon atoms, it releases electrons from them, producing positively charged argon ions ($Ar \rightarrow Ar^+ + e$). Electrons and ions are attracted to the electrodes of opposite charge. Under the high voltage applied, the electrons, being very small and mobile, accelerate rapidly towards the wire anode. As they do so they collide with other argon atoms, causing these to ionize in turn and liberate more electrons. Thus a single $\beta$ or $\gamma$ ray entering the tube may produce a million or more electrons, which are discharged at the anode, while a similar number of argon ions are discharged at the cathode. The discharge causes a pulse of electricity to flow round the circuit. The latter includes a counting device which records the separate pulses and hence the number of particles or rays entering the tube. Alternatively, if the instrument is being used merely for detection of radioactivity, the pulses may be amplified and fed into earphones or a loudspeaker, where they produce the well-known 'clucking hen' sound. The purpose of the ethanol vapour in the tube is to 'quench' the discharge rapidly, so that the instrument is ready to record the arrival of the next particle or ray.

*Wilson Cloud Chamber*. With this apparatus, devised by C. T. R. Wilson in 1911, the path of an α-particle or β-particle is rendered visible as it travels through air or through some other gas. The principle of the apparatus is as follows. If the pressure of an enclosed volume of air or other gas is suddenly reduced—for example, by expansion—its temperature falls. If the air is saturated with water vapour it will be supersaturated with water vapour at the lower temperature, and hence some of the vapour will condense, forming a cloud. For condensation to occur it is necessary to have nuclei present, on which the water droplets can form. These nuclei are usually dust particles in the air. If dust is carefully excluded no cloud is produced, and the air remains supersaturated with

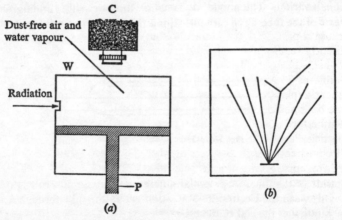

*Fig.* 5–7. (*a*) Simplified form of cloud chamber; (*b*) Type of tracks produced by α-particles

water vapour. In the Wilson apparatus α-particles and β-particles generate ionized particles as they travel through the supersaturated air and condensation takes place on these particles.

The cloud chamber (Fig. 5–7*a*) consists of a large glass cylinder, containing dust-free air and a little water. The cylinder is closed at one end by a window, W, and contains a piston, P. The air is expanded and cooled by dropping the piston and at the same time radiation particles are admitted from the side. As the charged particles travel through the supersaturated air they remove electrons from the molecules of oxygen and nitrogen in their path. Condensation of water vapour occurs on the remaining positive ions, and thus the paths of charged particles are shown by cloud tracks. These are photographed by a camera, C. Fig 5–7*b* shows the kind of tracks produced by α-particles. A forked track results when an α-particle collides with the nucleus of a nitrogen atom and forms a proton and a charged oxygen atom (see p. 131).

In recent years more sensitive forms of apparatus have been developed. One of these is the *bubble chamber*, which contains a superheated liquid instead of a supersaturated vapour. The liquid may be pentane or liquid hydrogen. In this apparatus the passage of a charged particle through the liquid is shown by a trail of small vapour bubbles formed from the liquid.

**Rutherford's Nuclear Model of the Atom (1911).** In 1909 Geiger and Marsden, working under Rutherford at Manchester University, found that when α-particles from radium were directed on to a very thin sheet of gold or platinum the great majority of the particles passed through the metal without change of direction. A few, however, were scattered through various angles, which in some cases were as high as 90° or more (Fig. 5–8). This effect was startling because α-particles were known to be extremely energetic. As Rutherford commented later, "It was about as credible as if you had fired a 15-inch shell at a piece of tissue paper and it came back and hit you."

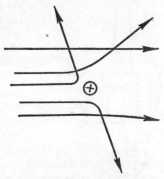

Rutherford concluded that, since α-particles are positively charged, the deflections could only be caused by the particles approaching closely to some concentrated form of positive charge. As only a small fraction (about 1 in 20 000) of the α-particles was scattered, it was clear that the body responsible for

*Fig. 5–8.* Deflection of α-particles by a charged nucleus

the scattering occupied a relatively small part of the metal atom. To explain the facts Rutherford suggested an atomic model in which all the protons of the atom were collected into a small central *nucleus*. (When neutrons were discovered in 1932 it became necessary to incorporate these in the nucleus as well. Subsequently a number of other sub-atomic particles (*e.g.*, *positrons* and *mesons*) were found to play a part in the structure of nuclei.)

The electrons required to make the atom electrically neutral were pictured as rotating round the nucleus in much the same way that the planets rotate round the sun, the attraction between electrons and nucleus being balanced by centrifugal force.

Although Rutherford's notion of planetary electrons has long been discarded, the existence of the positive nucleus has been supported by many later experiments. From the angles of scattering of α-particles we can calculate that the size of a nucleus is of the order of $10^{-13}$ cm, whereas that of an atom is of the order of $10^{-8}$ cm. Thus a nucleus is about 1/100 000 of the size of an atom. If we think of an atom as represented (in two dimensions) by a circular cricket field 100 metres in diameter, the nucleus would be represented by a pin-head placed at the centre of the field. Since an electron has roughly the same size as a nucleus and there are only a few dozen electrons at most in an atom, it follows that by far the largest part of an atom consists of empty space.

**Moseley's Experiments and Atomic Number.** In 1913 an experimental method of finding the number of electrons in an atom was discovered by Moseley, one of Rutherford's assistants. The method involves measuring the difference in wavelength of X-rays given off by different elements when they are bombarded with high-speed electrons.

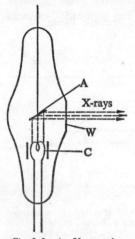

*Fig.* 5-9. An X-ray tube

We have already seen that X-rays are electromagnetic waves of very short wavelength. They are produced in a modified form of cathode-ray tube called an *X-ray tube* (Fig. 5-9). This is made of lead glass and is under high vacuum. The cathode, C, is a tungsten spiral strongly heated by a current. Electrons are given off by the tungsten and are accelerated to a high velocity by maintaining a large voltage (80,000 volt) across the tube. The electrons fall on to a metal anode, A, inclined at an angle so that the X-rays evolved pass out of the tube through a window, W, at the side.

In Moseley's experiments a number of different solid elements were used as the anode to generate X-rays. The X-rays were passed through a slit to obtain a narrow beam and diffracted by a crystal of potassium(I) hexacyanoferrate(II), which acts as a finely ruled diffraction grating. When the diffracted rays were allowed to fall on a photographic plate an X-ray spectrum was produced. This consisted of groups of lines (known as the K, L, M, etc. series), each line representing a definite wavelength in the rays (Fig. 5-10a). Moseley found that the wavelength of any one characteristic line from any one of the series varied with the element producing the X-rays. He showed that when the different wavelengths for this line were compared with the order of the elements in the Periodic Table there was a regular variation in the wavelength from element to element (Fig. 5-10b). This indicated that in an atom there was some fundamental quantity which increased by regular steps from one element to the next. The fundamental quantity could not be relative atomic mass because this increases unevenly. Moseley concluded that it could only be the number of protons in the nucleus or the number of electrons in the atom (these being equal). He therefore suggested that for successive elements in the Periodic Table the number of protons and electrons in the atom increased by one. Thus the hydrogen atom would have one proton and one electron. The helium atom would have two of each, that of lithium three of each, and so on through the Periodic Table. Moseley's theory was confirmed in 1920 by Chadwick, who succeeded in measuring the number of unit positive charges on different atomic nuclei from the scattering of α-particles.

In this way the concept of *atomic number* was derived. The atomic number of an element is the most important feature of its individuality. The number represents:

• The number of protons in the nucleus, and hence the number of unit positive charges on the latter.
• The number of external electrons in the atom.
• The order in which the element appears in the Periodic Table.

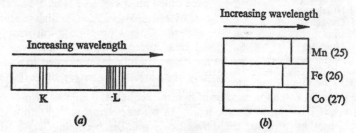

*Fig.* 5-10. (a) Part of an X-ray spectrum; (b) Changes in wavelength of a characteristic line for consecutive elements

**Composition of the Nucleus.** We have seen how the number of protons in an atomic nucleus can be found. How can the number of neutrons be determined? Sometimes this is derived from the relative atomic mass of the element. Protons and neutrons have about the same mass, while the mass of the external electrons is negligible in comparison. It follows that the mass of an atom is concentrated almost entirely in the nucleus and is approximately proportional to the number of protons and neutrons. Protons and neutrons together are called *nucleons*, and the combined total is known as the *mass number* (represented by the symbol $A$).

From our knowledge of relative atomic masses we know that hydrogen has the lightest atoms and that many other elements have relative atomic masses which are very nearly whole-number multiples of that of hydrogen. Thus it is fair to assume that the hydrogen nucleus contains only one of the mass units. This must be a proton, since the hydrogen nucleus has unit positive charge. Hence the hydrogen atom consists of a single proton with one external electron.

The second element, helium, has an atomic number of 2 and a relative atomic mass of 4—that is, a helium atom is approximately four times as heavy as a hydrogen atom. Hence the helium nucleus must contain 2 protons and 2 neutrons. In general, if an atom has a relative atomic mass $A_r$ and atomic number $Z$, the nucleus is composed of $Z$ protons and $A_r - Z$ neutrons, and there are $Z$ external electrons. This can be seen from the Table 5–1. The neutron number is denoted by the symbol $N$.

**Table 5–1.** COMPOSITION OF SOME ATOMIC NUCLEI

| Element | Atomic number ($Z$) | Rounded relative atomic mass ($A_r$) | Number of protons | Number of neutrons ($N$) |
|---|---|---|---|---|
| Hydrogen | 1 | 1 | 1 | 0 |
| Helium | 2 | 4 | 2 | 2 |
| Lithium | 3 | 7 | 3 | 4 |
| Carbon | 6 | 12 | 6 | 6 |
| Nitrogen | 7 | 14 | 7 | 7 |
| Oxygen | 8 | 16 | 8 | 8 |
| Iron | 26 | 56 | 26 | 30 |
| Uranium | 92 | 238 | 92 | 146 |

This table also indicates that in the atoms of the lighter elements of the Periodic Table the number of neutrons equals the number of protons, or is half the relative atomic mass. In heavier elements there are more neutrons than protons, the difference in number becoming larger with increase in atomic mass.

The relative atomic masses given in Table 5–1 are taken to the nearest whole number. These make it appear that relative atomic masses are exact multiples of that of the hydrogen atom. In fact this is not so. There are three reasons for the discrepancies.

• The hydrogen nucleus consists of a single proton, while the nuclei of other atoms contain protons and neutrons. These particles have slightly different masses.

On the atomic mass scale ($^{12}C = 12$)

| | |
|---|---|
| Mass of a proton | $= 1 \cdot 007\ 4$ units |
| Mass of a neutron | $= 1 \cdot 008\ 9$ units |
| Mass of an electron | $= 0 \cdot 000\ 55$ units |

• The mass of a nucleus containing protons and neutrons is not equal to the sum of the masses of the separate protons and neutrons.

Thus a helium nucleus containing 2 protons and 2 neutrons might be expected to have a mass of

$$(2 \times 1 \cdot 0074) + (2 \times 1 \cdot 0089) = 4 \cdot 0326 \text{ units}$$

The actual mass of the helium nucleus found by mass spectrometer is 4·0015 units. Thus in the combination of the 2 protons and 2 neutrons there is a loss in mass of 0·0311 atomic mass units. The loss of mass is explained by its conversion into energy in accordance with Einstein's law, $E = mc^2$. The energy evolved in formation of a nucleus from free protons and neutrons is called the *binding energy*. This varies for different nuclei. The evolution of energy stabilizes the nucleus since an equal amount of energy would be required to break it down again into free protons and neutrons.

● Most of the elements are mixtures of two or more kinds of atom known as *isotopes*. These have different masses because they contain different numbers of neutrons in the nucleus.

The actual relative atomic mass of an element is an average value, which depends on the separate masses of the atoms and their proportion. The number of neutrons is determined from the relative atomic masses found with a mass spectrometer.

**Determination of Relative Atomic Masses by Mass Spectrometer.** The relative atomic mass of an atom is its mass on a scale on which an atom of the $^{12}C$ isotope of carbon has a mass of exactly 12 units. This isotope of carbon is the one which has six protons and six neutrons in the nucleus; that is, its mass number is 12. The reasons for selecting this standard are given later.

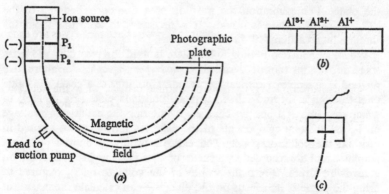

*Fig.* 5–11. (*a*) Simple form of mass spectrograph; (*b*) Mass spectrum of aluminium; (*c*) Collector plate used in electrical recording

The mass spectrometer, which was invented by Aston in 1919, has various forms according to the type of material being examined and the particular purpose for which the instrument is being used. Fig. 5–11*a* shows one of the simpler forms. In this a mixture of positive ions, all travelling with the same velocity, is passed through a magnetic field. The individual ions are deflected to an extent proportional to $\sqrt{(z/m)}$, where

$z$ is the charge of the ion and $m$ is its mass. Thus a lighter ion is deflected more than a heavier one with the same charge, and a doubly charged ion (*e.g.*, $Al^{2+}$) more than a singly charged ion (*e.g.* $Al^+$) of the same mass. The masses of the ions may be taken as equal to those of the neutral atoms.

Positive ions are produced in an ionizing chamber in various ways, which depend on the nature of the material. A trace of a salt (which contains positive ions) may be vaporized on an electrically heated tungsten filament. A gas (or vapour) at low pressure may be bombarded with electrons, or a spark discharge may be passed between two samples of a metal. Usually a mixture of ions results. Thus spark discharge of aluminium produces $Al^+$, $Al^{2+}$, and $Al^{3+}$ ions according to the number of electrons lost.

The mixture of positive ions is attracted towards a charged plate, $P_1$, maintained at a negative potential, and passes through a slit in the plate. A second negatively charged plate, $P_2$, maintained at a high voltage (500–2000 volt), accelerates the ions to a high and constant velocity. The beam emerging from the slit in $P_2$ passes through a strong magnetic field, which deflects the ions into a curved path. The degree of curvature depends on the mass and ionic charge, as explained above. At approximately 180° to the original direction of the beam the ions fall on a photographic plate and produce a mass spectrum consisting of a series of lines at different places (Fig. 5–11*b*). If the apparatus has been calibrated, the masses of the particles can be deduced from the positions of the lines on the plate. For calibration the position of a line corresponding to the $^{12}C$ isotope of carbon is found. It is necessary to keep the apparatus under a high vacuum ($10^{-5}$ Pa).

When, as above, a photographic plate is used for recording the mass spectrum the instrument is called a *mass spectrograph*. An alternative method is to employ electrical recording. Here ions of a given $z/m$ value pass through a slit and fall on a metal collector plate (Fig. 5–11*c*), to which they impart a positive charge. By altering the accelerating voltage on $P_2$ the different components of the deflected ion beam are focused in turn on the collector plate. The charges produced on the plate are amplified and are recorded by a meter or by a pen moving over paper on a revolving drum. From the values of the voltages on $P_2$ required to bring the particles into focus on the plate, we can calculate the masses of the particles (again after calibration). With the latest form of mass spectrometer we can analyse as little as a thousand-millionth of a gram of material.

**Isotopes.** When the various elements are examined by mass spectrometer about three-quarters of them are found to possess atoms with different atomic masses. (Of the better known elements F, Na, Al, P, Mn, Co, and As do not consist of isotopes.) Thus chlorine is composed of atoms of relative masses 35 and 37. Since the atomic number is 17 for both forms these must have 17 protons in the nucleus. The difference in atomic mass must therefore be due to the first kind of atom having 18 neutrons, while the second contains 20 neutrons. Each kind of atom is called a *nuclide*, this being defined as an atomic species in which all atoms have the same number of protons (or atomic number) and the same number of nucleons (or mass number). The two nuclides of chlorine are represented by $^{35}$Cl and $^{37}$Cl, or more fully $^{35}_{17}$Cl and $^{37}_{17}$Cl, the upper and lower numbers being the mass number and atomic number respectively.

Similarly, we can explain the different atomic forms of other elements by different numbers of neutrons in the nucleus. The different forms are called isotopes (Greek *isos*, same; *topos*, place) because they occupy the same position in the Periodic Table of the elements. Usually one particular isotope is more abundant than the others. Thus ordinary hydrogen contains 99·98 per cent of $^1_1$H atoms to 0·02 per cent of $^2_1$H atoms. In the first the nucleus is a single proton, while in the second it consists of a proton and a neutron and has a mass of 2. The two isotopes are distinguished by the names *protium* (symbol H) and *deuterium* (symbol D). Water also contains traces of a radioactive isotope called *tritium* (symbol T). This has one proton and two neutrons in the nucleus.

The chemical properties of isotopes of the same element are identical except that compounds containing a heavier isotope react more slowly (the effect is usually only noticeable with lighter elements). There are appreciable differences in physical properties, however, and this extends to compounds containing different isotopes. Ordinary water is a mixture of diprotium oxide, $^1$H$_2$O, with about 1 part in 6 400 of dideuterium oxide, 'heavy' water, $^2$H$_2$O. The latter can be separated from ordinary water by prolonged electrolysis of sodium(I) hydroxide solution, diprotium oxide being electrolysed more rapidly than dideuterium oxide. Table 5–2 shows some of the differences in physical properties of the two oxides.

**Table 5–2.** SOME PHYSICAL PROPERTIES OF DIPROTIUM OXIDE AND DIDEUTERIUM OXIDE

|  | H$_2$O | D$_2$O |
| --- | --- | --- |
| Density/g cm$^{-3}$ at 298 K | 1·000 | 1·105 |
| Freezing point/K | 273·15 | 276·97 |
| Boiling point/K | 373·15 | 374·55 |
| Heat of vaporization/J mol$^{-1}$ | 40 500 | 41 600 |

The higher freezing point, boiling point, and heat of vaporization of dideuterium oxide show that more energy is needed to melt and boil this oxide as compared with water. This reflects the greater mass of the D$_2$O molecule and parallels the increase in melting point and boiling point which occurs in an homologous series of organic compounds. The heavier the molecules the larger is the energy required to increase the molecular velocity sufficiently to make the substance melt or boil.

We can use the different atomic masses of isotopes to separate them. The electrolytic method, which has already been mentioned in connection with deuterium, has also been applied to lithium, the fused chloride being electrolysed with a mercury cathode. On the industrial scale, the different rates of diffusion of uranium hexafluoride, $UF_6$, permit partial separation of $^{235}U$ from $^{238}U$. The most efficient method of separation on a small scale (and the one most frequently used) is by means of an electromagnet as in the mass spectrometer. Instead of bringing the different ions to a focus on the same collector plate, we collect them separately in small receiving vessels suitably spaced.

Many isotopes different from those found in Nature have now been made artificially. Some of these are described shortly.

**Determination of Relative Atomic and Molecular Masses by Mass Spectrometer.** The modern definition of the relative atomic mass of an element on the $^{12}C = 12$ scale has already been given (p. 37). A mass spectrometer or mass spectrograph gives the relative atomic masses of separate isotopes on the carbon scale. When an element consists of only one kind of atom the relative isotopic mass is the relative atomic mass of the element. If, however, an element is made up of several isotopes, its relative atomic mass is an average value depending on the separate relative masses and proportions of the isotopes. In nature the proportions, or *abundancies*, are usually constant. We can determine abundancies from the relative intensities of the lines produced by the isotopes on the photographic plate of a mass spectrograph. Alternatively we can deduce them from the relative charges imparted to the collector plate of a mass spectrometer when the ions of different mass are focused in turn on the plate. For example, the abundancies of $^{35}Cl$ and $^{37}Cl$ are 75·4 per cent and 24·6 per cent, giving an average relative atomic mass of 35·453.

This method of finding relative atomic masses surpasses in accuracy any of the chemical methods described in Chapter 1. It also has the advantage of requiring only minute amounts of material.

The abundancies of elements are not always constant, and the variations result in different values being obtained for relative atomic masses according to the source of the element. Thus lead produced by decay of certain radioactive elements like radium has a lower relative atomic mass (206·50) than that of lead from non-radioactive sources (207·19). Usually the differences are much smaller and do not affect the value of the relative atomic mass unless this is being given with great precision.

The mass spectrometer can also be used for very accurate determination of relative molecular masses of gases and vapours. Thus by bombarding hydrogen chloride at low pressure with electrons one (or more) electrons may be lost by a molecule with the formation of a positive ion.

$$HCl - e \rightarrow HCl^+$$

Since both hydrogen and chlorine consist of isotopes, hydrogen chloride contains different kinds of molecules such as $^1H^{35}Cl$ and $^1H^{37}Cl$. These give rise to positive ions of different mass which are sorted out by the mass spectrometer. As with relative atomic masses, we find the average relative molecular mass from the masses and abundancies of the separate species of molecule or ion.

$^{12}C$ = 12 Standard of Relative Atomic Masses. The oxygen standard of relative atomic masses was replaced by the carbon standard in 1962, chiefly because of the occurrence of oxygen in different isotopic forms. Oxygen is a mixture of three isotopes, $^{16}O$, $^{17}O$, and $^{18}O$, the abundancies of these being 99·76 per cent, 0·04 per cent, and 0·20 per cent respectively. Prior to 1962 chemists and physicists were using two different 'oxygen standards.' The chemist's standard O = 16 was based on the element as it occurs naturally, that is, as a mixture of the three isotopes. The physicist, in calibrating a mass spectrograph, was using the line corresponding to the $^{16}O$ isotope. 'Chemical' relative atomic masses thus differed slightly (but appreciably) from 'physical' relative atomic masses, and to convert the latter into the former a conversion factor had to be used. The double standard led to much confusion. Calculation of various quantities (*e.g.*, the Avogadro constant, $N_A$, and the gas constant, $R$) depends on using precise' values of relative atomic masses, but some scientists were using the chemical scale, and others the physical scale, of relative atomic masses.

It might appear that the simplest solution to the problem was for all scientists to adopt one or other of the two oxygen standards. Chemists, however, were reluctant to accept the physicist's standard of $^{16}O$ = 16 because it involved changing relative atomic masses (and the dependent quantities) by about 1 part in 4000, a substantial difference. Physicists could not use the chemical standard because they require a single line on a photographic plate for calibrating a mass spectrograph. It was therefore agreed to adopt $^{12}C$ = 12 as a new standard for relative atomic masses because the $^{12}C$ isotope of carbon is particularly suitable for calibrating a mass spectrograph or mass spectrometer. It is the chief isotope of carbon and a large number of positive ions containing the isotope are available (*e.g.*, $CH_4^+$, $C_2H_6^+$, etc.). In adopting the new standard, chemists have reduced their relative atomic masses expressed on the O = 16 scale by only 43 parts in a million—a negligible change from the practical point of view. Thus the relative atomic mass of oxygen, previously 16 exactly, has now become 15·9994.

Summary of Developments in the Particle Theory of Matter. At the end of the nineteenth century the atomic theory underwent a dramatic change.

Discharge of electricity through rarefied gases showed that atoms contain negatively charged particles (electrons) and positively charged particles (protons). These must be present in equal numbers for an atom to be electrically neutral. The mass of an electron was found to be 1/1837, and that of a proton 1836/1837, of the mass of a hydrogen atom. From the size of their charges the electron was identified with the unit of electricity transferred to a monovalent positive ion at the cathode during electrolysis. Later (in 1932) evidence was obtained by Chadwick that all atoms except that of hydrogen also contain neutral particles (neutrons), which have a mass approximately equal to that of protons.

The first 'nuclear' model of the atom was that of Rutherford (1911). From the way that α-particles are deflected by atoms in a thin sheet of metal Rutherford concluded that the protons formed a small central nucleus (neutrons were added later), around which the electrons revolved. In 1913 Moseley devised an experimental method of measuring the number of electrons outside the nucleus (and hence the number of protons in the nucleus). This number was called the 'atomic number' of the element. The invention of the mass spectrometer by Aston in 1919 allowed measurement of the masses of individual atoms. Most elements were found to consist of atoms (isotopes) with different masses. Subsequent improvements in the instrument enabled it to be used for the very accurate determinations of relative atomic and molecular masses.

The discovery of natural radioactivity led to the overthrow of the belief that atoms of one element could not change into those of another. Thus radium atoms were found to decay spontaneously into helium and lead atoms. Radioactive changes also revealed the possibility of producing enormous amounts of energy from atoms. Later investigations have shown that transmutation of elements can be brought about artificially, as described in the next section.

## Nuclear Reactions

**Transmutation of Elements.** As we shall see in Chapter 7, ordinary chemical reactions consist simply of rearrangement of electrons in atoms and molecules. In these changes the atomic nuclei are not affected. In the disintegration of radium and other radioactive elements both nuclei and electrons are involved. A chemical change of this type is called a *nuclear reaction*. Nuclear chemical changes occur when the nuclei of atoms are bombarded with various kinds of high-speed particles.

*Bombardment with α-particles.* The possibility of artificial transmutation of the elements was first suggested by the discovery that different kinds of atoms are composed of the same fundamental units (protons, neutrons, and electrons). Essentially the problem consists of changing the number of protons and neutrons in the nucleus. To Rutherford a likely solution appeared to be collision between atomic nuclei and the fast-moving α-particles, or helium nuclei, shot out from radium atoms.

Although the targets occupied only a small fraction of the atomic volume a few helium nuclei might make direct hits and either combine with the atomic nuclei or break them up.

The theory proved correct. When Rutherford (1919) passed α-particles through nitrogen he found that protons (H+) and positive ions of oxygen (O+) were formed. These were identified from the lengths of the tracks which they produced in a Wilson cloud chamber (see Fig. 5–7*b*). The lighter proton had a longer track than the heavier oxygen ion, and the tracks diverged because of the similar charges on the ions. The oxygen was not ordinary oxygen of mass number 16, but the isotope of mass number 17. The nuclear reaction can be represented in either of the ways shown

$$\mathrm{^{14}_{7}N + {}^{4}_{2}He \rightarrow {}^{1}_{1}H + {}^{17}_{8}O}; \quad \mathrm{^{14}_{7}N(\alpha, p){}^{17}_{8}O}$$

In the first (equation) method it is usual to omit any charges on the particles and to show only the changes in mass number and atomic number. *These must balance* on the two sides of the equation. In the second (more concise) method we simply show in order the following:

Initial $\Bigg($ bombarding   expelled $\Bigg)$ nuclide
nuclide $\Bigg($ particle   , particle $\Bigg)$ formed

The particles concerned may be α-particles (α), protons (p), neutrons (n), electrons, or β–particles (e), photons (γ), etc.

Rutherford's achievement in transmuting elements was followed by others. Bombardment by α-particles brings about transmutation of most of the lighter elements. Heavy atoms are immune to attack by α-particles unless these are accelerated to still higher velocities. The reason is the strong repulsion between the positively charged nucleus and the positively charged α-particles. As the charge on the nucleus increases with atomic number it becomes more difficult for an α-particle to penetrate to the nucleus. The energy of the α-particles can be increased by means of an electrical field (using a cyclotron). With these more energetic particles nuclei of quite high atomic number undergo transmutation.

*Bombardment by Protons.* Since a proton (H+) has a smaller mass than an α-particle (He$^{2+}$) and carries only a single positive charge, there is a smaller force of repulsion between a given nucleus and a proton than between the nucleus and an α-particle. This led Cockcroft and Walton in 1932 to use protons for attacking atomic nuclei. The protons were obtained from a discharge tube and were accelerated to a high velocity by an electrical field. When they were directed on to a lithium target small amounts of helium were produced:

$$\mathrm{^{7}_{3}Li + H^{1}_{1} \rightarrow {}^{4}_{2}He + {}^{4}_{2}He}$$

In this reaction a lithium nucleus had evidently first captured a proton to give an unstable nucleus of relative mass 8. The latter had then broken up to yield two stable helium nuclei of relative mass 4. For the first time, scientists had split the atom by purely artificial means.

Since 1932 fast-moving protons have frequently been used for bringing about transmutation. Sometimes the proton captured by the nucleus is retained, so that a new nucleus of the next element is formed. Thus a fluorine nucleus becomes a neon nucleus:

$$^{19}_{9}F + ^{1}_{1}H \rightarrow ^{20}_{10}Ne$$

*Bombardment by Neutrons.* Today, nuclear changes are usually made by means of neutrons. These are produced in abundance in nuclear reactors, so that transmutations can be performed on a larger scale than formerly. Neutrons are more effective than $\alpha$-particles or protons for attacking nuclei because they have no electrical charge and therefore are not repelled by a nucleus. Slow neutrons (those with relatively low energies) combine with the nuclei of most elements to produce an isotope of the target element, the mass number being increased by one unit. This happens with phosphorus, excess energy being liberated as $\gamma$-rays.

$$^{31}_{15}P + ^{1}_{0}n \rightarrow ^{32}_{15}P + \gamma; \qquad ^{31}_{15}P(n, \gamma)^{32}_{15}P$$

With fast neutrons the additional energy is sufficient to cause a proton to be ejected from the 'compound nucleus' first formed. The relative atomic mass therefore remains the same, but a nucleus of the next lower element is formed. Thus nitrogen is changed into an isotope of carbon.

$$^{14}_{7}N + ^{1}_{0}n \rightarrow ^{14}_{6}C + ^{1}_{1}H; \qquad ^{14}_{7}N(n, p)^{14}_{6}C$$

**Radioisotopes.** As the name implies, these are isotopes which are radioactive. When atomic nuclei are bombarded with neutrons the resulting nuclei are often unstable. They decay spontaneously with emission of various kinds of particles and, possibly, $\gamma$-rays. The precise reason for the instability is unknown, but it appears to be associated with the ratio of the number of neutrons to that of protons. Both $^{32}P$ and $^{14}C$ are radioactive. They lose an electron, or $\beta$-particle, from the nucleus, one of the neutrons being converted into a proton. Consequently the mass number remains the same, while the atomic number increases by 1 to give a nucleus of the next higher element.

$$^{32}_{15}P \rightarrow ^{32}_{16}S + ^{0}_{-1}e$$

(Half-life    (Stable)
14 days)

$$^{14}_{6}C \rightarrow ^{14}_{7}N + ^{0}_{-1}e$$

(Half-life    (Stable)
5570 years)

Artificial radioisotopes are now manufactured on a considerable scale by exposing suitable elements or their compounds to neutron bombardment. They are widely used in industry, medicine, and scientific research.

**Production of New Elements.** Until 1940 the heaviest known atomic nucleus was that of uranium, element 92 in the Periodic Table. Four lighter elements were still unknown, these having atomic numbers of 43, 61, 85, and 87. Since 1940 scientists have not only made these elements artificially, but have also produced thirteen new elements with atomic nuclei heavier than the uranium nucleus and with higher atomic numbers. The latest of these transuranium elements (number 105) was obtained in the U.S.A. in 1970.

All the new elements have been prepared by nuclear transformations. The starting point for the transuranium elements is uranium itself ($^{238}_{92}U$). When this is bombarded with neutrons it becomes an unstable isotope ($^{239}_{92}U$), which decays by emission of a $\beta$-particle into the next higher element neptunium ($^{239}_{93}Np$). The process of neutron capture and $\beta$-decay can be repeated, giving in turn the new elements as far as fermium (100). Fermium has such a short half-life, however, that another method was needed to obtain the subsequent elements. This consisted of bombarding somewhat lower elements with relatively heavy positive ions. Thus lawrencium (103) was produced from californium (98) by bombardment with boron ions ($B^{4+}$).

$$^{250}_{98}Cf + {}^{11}_{5}B \rightarrow {}^{257}_{103}Lw + 4^{1}_{0}n$$

The transuranium elements are all radioactive metals with short half-lives. Although some have been obtained in only minute amounts their chemistry has been investigated and we know that they form part of a closely related family of elements, known as the *actinide series*.

**Nuclear Fission.** Naturally occurring uranium is a mixture of two radioactive isotopes having relative atomic masses of 235 and 238. The metal, which is extracted from the ore pitchblende, contains only about 0·7 per cent of the first isotope. This can be separated from the heavier isotope by means of the different rates of diffusion of the two forms of the hexafluoride, $UF_6$.

When uranium-235 is bombarded with slow neutrons a peculiar change takes place. The nucleus first captures a neutron and then breaks up into two nuclei of roughly equal size which fly apart at great speed. This kind of change is described as *nuclear fission*. ($^{238}U$ undergoes fission with fast neutrons. Thorium, proactinium, and plutonium are also fissile elements.) The nature of the fragments produced depends on how the neutron collides with the nucleus, but usually a mixture of particles results. Thus fission of one $^{235}U$ nucleus might give barium and krypton, while that of another might yield xenon and strontium. Usually the products are radioactive isotopes of the normal elements.

There are two important aspects of nuclear fission. Firstly, it is accompanied by the shooting-out of two or three fresh neutrons at great speed. These may escape from the metal or they may collide with other $^{235}U$

atoms, causing the latter to undergo fission with the production of more neutrons. Clearly, if we can arrange matters so that at least one neutron from each fission brings about another fission the reaction will be self-sustaining and in time all the $^{235}U$ atoms will be split up. A reaction of this kind, which consists of a series of repeated steps, each initiated by the previous one, is called a *chain reaction*. The fission process can be represented as shown in Fig. 5–12.

The second important aspect of the fission is that it is attended by the evolution of a very large amount of energy. When a $^{235}U$ nucleus captures a neutron, and then splits up, the combined masses of the particles formed

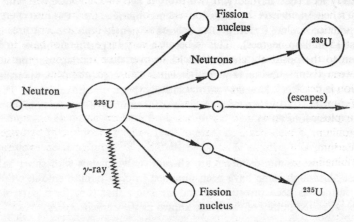

*Fig.* 5–12. Diagrammatic representation of fission of a $^{235}U$ nucleus resulting in a chain reaction

is less than the original mass. The balance is transformed into energy in accordance with Einstein's law, the energy appearing partly as kinetic energy of the particles formed and partly as $\gamma$-radiation.

We can easily show from the Einstein equation that a huge amount of energy corresponds to a small amount of mass. Since the velocity of light is $3 \times 10^8$ m s$^{-1}$ the energy which would be produced by complete conversion of 1 g of mass into energy would be

$$E = mc^2 = 10^{-3} \times (3 \times 10^8)^2 \text{ J}$$
$$= 9 \times 10^{13} \text{ J}$$
$$= 9 \times 10^{10} \text{ kJ}$$
$$= 90\ 000\ 000\ 000 \text{ kJ}$$

In contrast, the amount of energy liberated by the burning of 1 g of carbon is about 34 kJ. We can see that enormously greater quantities of energy are involved in nuclear changes than in ordinary chemical changes.

Energy obtained by nuclear fission has been applied in two ways—in atomic bombs and in atomic power plants. In the first the output of energy is uncontrolled; in the second it is regulated. The first atomic bomb (dropped on Japan in 1945) derived its energy from the extremely rapid chain reaction involved in fission of a few pounds of uranium-235. Great Britain was the first country to build a full-scale atomic power station. This was at Calder Hall in Cumberland, where production of electricity from uranium fission began in 1956.

**Nuclear Fusion.** This type of change is the opposite of nuclear fission. It consists of joining two light nuclei to form a heavier nucleus. We have already seen that, in theory, if two protons and two neutrons combine to give a helium nucleus there is a conversion of part of the mass into energy. In practice nuclear fusion occurs only at extremely high temperatures (at least 15 million degrees). This is because very large energies have to be given to the positively charged nuclei to overcome the strong repulsion between them. Owing to the high temperature requirement a nuclear fusion is described as a *thermonuclear* reaction.

Temperatures of the order needed for nuclear fusion can be obtained by explosion of an atomic bomb based on nuclear fission of uranium or plutonium. Thus when a mixture of the two heavy isotopes of hydrogen, deuterium, and tritium, is exposed to the heat from such an explosion combination occurs as follows:

$$_1^2D + _1^3T \rightarrow _2^4He + _0^1n + energy$$

This type of reaction is the basis of the hydrogen bomb. The lethal power of this bomb is partly due to the enormous amount of energy released and partly to the large quantity of $\gamma$-radiation produced by the detonating agent.

Attempts are being made at the present time to carry out thermonuclear reactions in the laboratory under controlled conditions, so far with only limited success. In these experiments the high temperatures necessary are obtained (for brief periods) by discharging extremely powerful pulses of electricity through the mixed gases at low pressure.

Thermonuclear reactions take place in the sun and other stars, producing temperatures of some 15 million degrees. Since the energy which we obtain from coal, oil, water power, etc., was all derived from the sun, we live indirectly on energy obtained by the nuclear fusion process.

### EXERCISE 5

**1.** Three kinds of fundamental particles are of great importance in chemical theory. Name them and give their relative masses and charges.

What is meant by (a) the atomic number ($Z$) and (b) the relative atomic mass ($A_r$) of an element? For sodium $Z = 11$ and $A_r = 23$; how must the nucleus of this atom be built up? How is the fractional relative atomic mass of chlorine ($Cl = 35 \cdot 5$) explained? How does the explanation conflict with Dalton's concept of the nature of *atoms* and which other of his postulates concerning elements are no longer acceptable?                          (J.M.B.)

**2.** (Part question.) Outline the principles of a *mass spectrometer*. Explain how it can be used to separate isotopes. (O. and C.)

**3.** At various times the following have been used as basic standards for relative atomic masses: $H = 1$, $O = 16$, $^{12}C = 12$. Discuss the advantages and disadvantages of these different standards.

**4.** Explain the difference between (*i*) atomic mass and relative atomic mass, (*ii*) atomic number and mass number. Why is the mass of a helium atom not exactly four times that of a hydrogen atom?

**5.** What are *isotopes*? How can the existence of isotopes be explained in the cases of nitrogen (14 and 15), oxygen (16, 17, and 18), and copper (63 and 65). The numbers in brackets refer to the mass numbers of the constitutent isotopes. The atomic numbers of nitrogen, oxygen, and copper are 7, 8, and 29 respectively.

**6.** (Part question.) Explain the meaning of the following equations which represent the net changes occurring in certain natural processes:

(i) $^{214}_{83}Bi = {}^{206}_{82}Pb + 3_{-1}^{0}e + 2^{4}_{2}He$

(ii) $^{214}_{82}Pb = {}^{207}_{82}Pb + 2_{-1}^{0}e + {}^{4}_{2}He$ (W.J.E.C.)

## MORE DIFFICULT QUESTIONS

**7.** Give a concise description of the characteristic features of natural radio-activity which are of chemical interest. Show how radioactive emission affects the relative positions of the emitter element and its immediate disintegration product in the periodic table.

A heavy element A disintegrates via a total of two *alpha* emissions and four *beta* emissions to yield an element B. What is the chemical relation between these two elements?

Describe briefly **one** piece of experimental evidence which points to the existence of an atomic *nucleus*. (W.J.E.C.)

**8.** (*a*) List the three main fundamental particles which are constituents of atoms. Give their relative charges and masses.

(*b*) Similarly name and differentiate between the radiations emitted by naturally occurring radioactive elements.

For the particle which is common to lists (*a*) and (*b*) name **two** methods by which it can be obtained from non-radioactive metals.

Complete the following equations for nuclear reactions by using the periodic table provided to identify the elements *X*, *Y*, *Z*, *A* and *B* and add the atomic and mass numbers where these are missing.

$$^{214}_{82}Pb \rightarrow {}_{83}X + {}_{-1}^{0}e$$

$$^{27}_{13}Al + {}_{0}^{1}n \rightarrow {}^{24}Y + {}_{2}^{4}Z$$

$$^{14}_{7}N + {}_{2}^{4}Z \rightarrow {}^{17}_{8}A + {}_{1}B$$

(J.M.B.)

**9.** What is meant by *atomic fission* and *atomic fusion*? Describe and explain one example of each of these processes.

# CHAPTER 6

# Electron distribution in atoms

So far we have given little attention to the electrons present in atoms. In Rutherford's 'planetary' model of the atom the electrons were simply pictured as rotating round the nucleus at various distances. It was supposed that the electrons were prevented from falling into the oppositely charged nucleus by centrifugal force. However, it was pointed out that a moving charged particle like an electron should radiate energy continuously. This would result in a decrease in velocity of the electron, which would eventually be drawn into the nucleus. For this reason the electron part of Rutherford's model was soon discarded, although the concept of the nucleus remained.

Modern ideas on electron distribution in atoms originated with the model of the hydrogen atom proposed by the Danish physicist Niels Bohr in 1913. Bohr's theory was based on certain discoveries about the spectrum of atomic hydrogen. We shall therefore describe these discoveries before explaining Bohr's theory.

**Emission Spectrum of Atomic Hydrogen.** When electricity passes through a discharge tube containing hydrogen at low pressure many of the molecules break up into single atoms. These atoms emit both visible and invisible radiations, some of the latter being in the infrared, and some in the ultraviolet part of the spectrum. If the radiations are analysed by spectrograph they yield a line spectrum, which can be photographed. Each line represents a definite wavelength of radiation. The series of lines in the visible part of the spectrum is the *Balmer series*. The three most prominent lines in this series are labelled $H_\alpha$, $H_\beta$, and $H_\gamma$ (Fig. 6-1). Another series of lines (the *Lyman series*) is found in the ultraviolet, and a further series occurs in the infrared.

When we measure the wavelengths corresponding to the lines in the various series we find a surprising regularity. The wavelengths are in accordance with the simple equation

$$\frac{1}{\lambda} = R_H \left( \frac{1}{n_1^2} - \frac{1}{n_2^2} \right)$$

where $\lambda$ is the wavelength, $R_H$ is a constant known as Rydberg's constant, and $n_1$ and $n_2$ are simple whole numbers. If $n_1$ is taken to be 1, $n_2$ can be

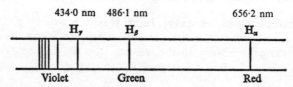

*Fig.* 6-1. Lines in the Balmer series in the emission spectrum of atomic hydrogen

2, 3, 4, etc. If $n_1$ equals 2, $n_2$ can be 3, 4, 5, etc. Thus, assuming $n_1$ to be 2 and $n_2$ to be 3, and inserting the value of $R_H$ in the equation, we have

$$\frac{1}{\lambda} = 109\ 678 \left( \frac{1}{2^2} - \frac{1}{3^2} \right) \text{cm}^{-1}$$

From this we obtain

$$\lambda = 0.00006565 \text{ cm} = 656.5 \text{ nm}$$

The value for the wavelength is that, *in vacuo*, of the red line ($H_\alpha$) in the Balmer series.

If the value 2 is kept for $n_1$ and $n_2$ is taken to be 4, the calculated value of $\lambda$ is 486·3 nm, which corresponds to the measured wavelength of the green line ($H_\beta$) in the Balmer series. If $n_1$ again equals 2 and $n_2$ is 5, the value of $\lambda$ is that of the violet line ($H_\gamma$) in the series. Note that in calculating the wavelength of the lines in the Balmer series we keep the value of $n_1$ as 2 throughout. In the same way, by putting $n_1$ equal to 1 and taking $n_2$ to be 2, 3, 4, etc., we can deduce correctly the wavelengths of the lines in the Lyman series in the ultraviolet region.

**Bohr's Theory of Electron Energy Levels.** In 1913 Bohr showed how the regularity in the wavelengths of the lines in the atomic spectrum of hydrogen could be explained. His explanation was revolutionary in that he based it on Planck's quantum theory instead of on classical electromagnetic theory. The quantum theory was first advanced in 1900 to explain certain findings in the radiation of energy. The fundamental postulate of the theory is that matter cannot absorb or emit energy in continuous amounts, but only in small discrete units (called *quanta*). In applying this theory to the hydrogen atom Bohr made two important assumptions as now described.

According to Bohr, the single hydrogen electron could travel round the nucleus in various possible orbits (assumed for simplicity to be circular), but only certain orbits were permissible. These orbits were ones in which the electron possessed a whole number of quanta of energy.

•   The first of Bohr's assumptions was that no energy was radiated by the electron while it was rotating in a permissible orbit.

The permissible orbits, or *energy levels*, were specified by giving them *quantum numbers* of 1, 2, 3, etc. Under normal conditions the electron occupied the orbit nearest to the nucleus—that is, the energy level of quantum number 1. It was then in its state of lowest energy, the *ground state*.

By means of an electric discharge or strong heat it was possible to 'excite' the atom, the electron absorbing energy which caused it to 'jump' to one of the higher energy levels. The excited state was, however, an unstable state, and there was a strong tendency for the electron to get rid of its extra energy and return to the ground state. The return might take place in one step or it might occur in stages, like a ball rolling down a flight of stairs. According to which level it fell the electron gave out some or all of its surplus energy in the form of radiation.

•   The second of Bohr's assumptions was that the wavelength of the radiation emitted was determined by the energy difference of the electron in the two levels.

Bohr deduced that, if $E_1$ was the energy of the electron in the higher level and $E_2$ that in the lower level, $1/\lambda$ was proportional to $E_1 - E_2$.[1] Now the energy of the electron in any level and the theoretical wavelength corresponding to $E_1 - E_2$ could be calculated. The calculated wavelengths agreed closely with those actually present in the spectrum of atomic hydrogen. Thus both of Bohr's assumptions were justified.

We can now see the significance of the numbers $n_1$ and $n_2$ in the equation

$$\frac{1}{\lambda} = R_{\mathrm{H}} \left( \frac{1}{n_1^2} - \frac{1}{n_2^2} \right)$$

given in the last section. In a discharge tube containing millions of hydrogen atoms the individual atoms are excited to different extents, and electron transitions of many kinds are taking place. All transitions from the same higher level to the same lower level are accompanied by the

---

[1] The frequency $\nu$ (nu), of the radiation emitted is given by: $h\nu = E_1 - E_2$, in which $h$ is a universal constant known as Planck's constant ($6 \cdot 63 \times 10^{-34}$ J s). The frequency and wavelength of the radiation are related by the equation: $\nu = c/\lambda$, where $c$ is the velocity of light.

emission of the same quantum of energy as radiation of specific wave-length. The radiation produces a line in the emission spectrum. Thus the return of the electron from the third, fourth, and fifth energy levels to the *second* (Fig. 6–2) results in the three lines labelled $H_\alpha$, $H_\beta$, and $H_\gamma$ in the Balmer series. Similarly each line in the Lyman series represents an electron transition from a higher energy level to the *first*. Evidently in this equation $n_1$ is the number of the energy level to which the electron returns, while $n_2$ is the number of the energy level from which the electron falls.

It is also significant that the Lyman series is in the ultraviolet part of the spectrum, while the Balmer series is in the visible region. Greater

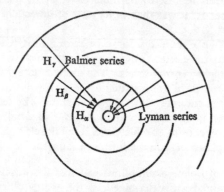

*Fig.* 6–2. Electron transitions producing lines in the Balmer series and Lyman series in the emission spectrum of atomic hydrogen (orbits not drawn to scale)

energy is associated with the shorter wavelength radiation in the ultra-violet. Thus larger energy quanta are concerned in electron transitions involving the first energy level than in those involving the second. This would be expected because in the first level the electron is closer to the attracting nucleus and considerably more energy is needed to raise it to a higher level. In what part of the spectrum ought we to find lines produced by electron transitions involving the third level?

Fig. 6–1 shows that the differences between the wavelengths of the radiations in the Balmer series decrease progressively towards the violet end of the spectrum. Eventually the lines become so close together that they form practically a continuous band of light. It is clear from the convergence of the lines that the differences in energy between successive energy levels becomes smaller with increasing distance of the levels from the nucleus. The convergence limit is reached when the electron is removed completely from the atom. This is represented for electron transitions in both the Balmer series and the Lyman series in Fig. 6–3. The horizontal lines depict the different energy levels of the electron in the hydrogen atom. The lowest of these lines (the horizontal axis) represents the ground state of the electron, while the shaded portion of the diagram represents complete removal of the electron. The energy required to remove the electron completely is called the *ionization energy* of the atom. The measurement of this important quantity is described below.

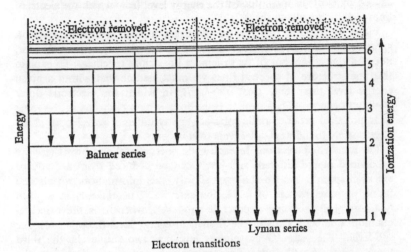

Fig. 6–3. Convergence of energy levels

**Energy Levels and Sub-levels in Higher Atoms.** Atoms containing more than one electron are not amenable to mathematical treatment by the Bohr method because the calculations involved are too complex. Nevertheless, Bohr's theory of the different possible energy levels of the hydrogen electron has provided a basis for explaining how electrons are distributed in larger atoms. The emission spectra of atoms of sodium and chlorine, for example, also consist of lines corresponding to definite wavelengths, and it is reasonable to suppose that these are again due to electron transitions between different energy levels. The energy of an electron in a given level varies from element to element because the charge on the atomic nucleus is different in each case. Thus each element has its own characteristic *emission spectrum*, or pattern of lines, by which the element can be identified. In many cases the radiations are in the visible part of the spectrum and produce distinctive colours of light. This enables us to identify a number of metal ions (*e.g.*, $Na^+$, $Sr^{2+}$) by their characteristic flame test colours.

We have seen how the total number of electrons in an atom can be found by Moseley's method. Emission spectra of atoms provide evidence that the electrons are distributed round the nucleus in various energy levels. Electrons occupying the same level are said to be in the same *quantum shell*. The different quantum shells are denoted by their *principal quantum numbers* 1, 2, 3, 4, etc., starting with the shell nearest to the nucleus. An alternative method of specifying them is by the letters *K, L, M, N*, etc. These letters correspond with those given to the different series of lines in X-ray spectra. The *K* lines in an X-ray spectrum are caused by electron transitions involving the *K* shell of electrons, the *M* lines to those involving the *M* shell, etc.

When we examine the lines in the emission spectra of atoms with a spectroscope of high resolving power we find that many of them consist, not of a single line, but of two or more lines close together. To explain this fine structure of spectral lines we must assume that within a given energy level there may exist *sub-levels*, in which the electrons differ slightly in energy. Thus similar transitions of two electrons in the same quantum shell produce radiations of slightly different wavelength if the electrons are in different sub-levels to start with.

The number of electrons in each main energy level and how these are subdivided at the different sub-levels can be deduced from a study of emission spectra. As we shall see shortly, this information can also be found by measurement of ionization energies. Theoretically in a given quantum shell there are *n* possible sub-levels, where *n* is the principal quantum number. Thus in the first main energy level there is only one sub-level. The second main energy level has two sub-levels, the third three, and the fourth four. The various sub-levels are indicated by the letters *s, p, d,* and *f*. Thus an electron in the first quantum shell is in the 1*s* sub-level. If it is in the second quantum shell it may be in either the 2*s* or 2*p* sub-level. In the third quantum shell it could occupy a 3*s*, 3*p*, or 3*d* sub-level, while in the fourth it might be in a 4*s*, 4*p*, 4*d*, or 4*f* sub-level.

**Determination of Ionization Energies.** As indicated in Fig. 6–3, if sufficient energy is imparted to an electron it can be removed completely from the atom. The latter is then left as a positive ion.

*The energy required for complete removal of an electron in its ground state from an atom is called the* **ionization energy** *of the atom.*

The value of this energy depends on the energy level and sub-level occupied by the electron in its lowest energy state and also on the nuclear charge.

We can deduce the ionization energy of an atom from its spectrum, or we can measure it directly by means of a valve containing gaseous atoms (*e.g.*, those of helium or argon) at 100 Pa pressure or less. The principle of the determination is illustrated by Fig. 6–4, and is as follows. The tungsten filament, F, gives off electrons when it is heated electrically.

We can charge the grid, G, positively to different voltages, which we read with a voltmeter. The plate, P, has a small negative charge. When the potential on G is zero no current flows between G and P. If we give G sufficient positive potential, however, the electrons emitted by the filament are accelerated towards the grid, pass through it, and ionize the atoms between the grid and the plate. The electron ejected from each atom is attracted to the grid, while the positive ion is attracted to the plate. A current thus passes between grid and plate, and we measure this with a sensitive ammeter. The minimum grid voltage, $V_1$ (Fig. 6–5) required to

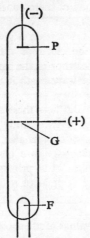

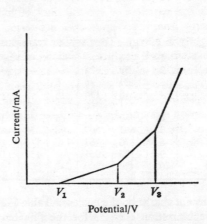

*Fig.* 6–4. Measurement of ionization potential

*Fig.* 6–5. Graph showing relation between current and grid potential

give the bombarding electrons sufficient energy to ionize the atoms is called the *ionization potential*. The *ionization energy* for one electron has the same numerical value, but is expressed in electron volt (eV). In this case the value in electron volt is $V_1$. (Ionization energies are often expressed in kilojoule per mole. To convert electron volt to kilojoule per mole the former must be multiplied by 96·5.)

Strictly speaking, the ionization energy found as just described is the *first ionization energy*; that is, it is the energy required to remove the most loosely bound electron in the atom. If we gradually increase the voltage on the grid beyond the first ionization potential there is a corresponding increase in current. This is because of the greater energy of the bombarding electrons. The latter not only release electrons, but now give them some kinetic energy as well. At a certain voltage, $V_2$ (Fig. 6–5), the value of the current increases sharply. At this stage a bombarding electron has sufficient energy to release a second electron from the atom.

The energy in electron volt required for this to occur is the *second ionization energy*. The value here is $V_2$.

Similarly there are values for the third, fourth, fifth, etc., ionization energies until the atom has been completely stripped of its electrons. Many of the values are not measured directly, but are obtained from spectroscopic data. The values increase with removal of each electron because the remaining electrons are attracted more strongly by the

**Table 6–1.** SUCCESSIVE IONIZATION ENERGIES FROM
HYDROGEN TO NEON

(*Note.* The values should be read from right to left)

| | At. no. | Ionization energy/eV | | | | | | | | | |
|---|---|---|---|---|---|---|---|---|---|---|---|
| | | 1s | | 2s | | 2p | | | | | |
| H | 1 | 14 | — | — | — | — | — | — | — | — | — |
| He | 2 | 54 | 25 | — | — | — | — | — | — | — | — |
| Li | 3 | 123 | 76 | 5 | — | — | — | — | — | — | — |
| Be | 4 | 218 | 154 | 18 | 9 | — | — | — | — | — | — |
| B | 5 | 341 | 260 | 38 | 25 | 8 | — | — | — | — | — |
| C | 6 | 489 | 393 | 65 | 48 | 24 | 11 | — | — | — | — |
| N | 7 | 665 | 507 | 98 | 78 | 48 | 30 | 15 | — | — | — |
| O | 8 | 870 | 737 | 138 | 114 | 77 | 55 | 35 | 14 | — | — |
| F | 9 | 1125 | 951 | 184 | 157 | 114 | 87 | 61 | 35 | 17 | — |
| Ne | 10 | 1470 | 1200 | 242 | 208 | 158 | 128 | 100 | 65 | 45 | 22 |

constant charge on the nucleus. Table 6–1 shows the values of the successive ionization energies for the first ten elements in order of atomic number. (The grounds on which the electrons are allocated to the various energy levels and sub-levels are explained in the next section.)

**Ionization Energies and Electron Distribution.** Ionization energies provide a great deal of information about the arrangement of electrons in atoms. The information is conveniently summarized under the headings given below.

*Total Number of Electrons in an Atom.* This number is equal to the number of separate ionization energies possessed by the atom. We thus have an alternative method to that of Moseley for finding the atomic number of an element.

*Number of Quantum Shells Occupied and the Number of Electrons in Each.* We can deduce these numbers for a given element by plotting the successive ionization energies against the order in which the electrons are removed from the atom. Fig. 6–6 shows the graph given by argon. To obtain a reasonable scale the logarithm of the ionization energy has been plotted instead of the ionization energy itself.

The ionization energies for successive electrons fall into three groups

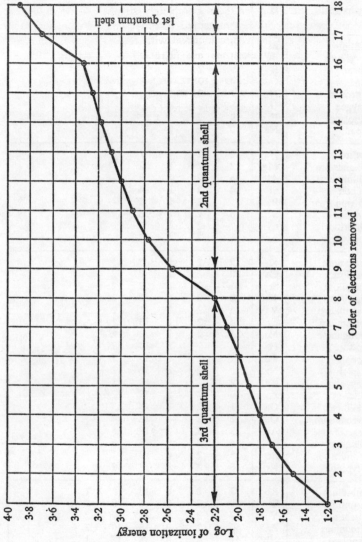

*Fig.* 6–6. Graph showing variation of successive ionization energies of argon

**Table 6–2.** ELECTRON DISTRIBUTIONS IN ATOMS OF THE ELEMENTS

| Element | | At. No. | Electron Distribution | | | | | | |
|---------|---|---------|---|---|---|---|---|---|---|
| | | | K | L | M | N | O | P | Q |
| Hydrogen | H | 1 | 1 | | | | | | |
| Helium | He | 2 | 2 | | | | | | |
| Lithium | Li | 3 | 2 | 1 | | | | | |
| Beryllium | Be | 4 | 2 | 2 | | | | | |
| Boron | B | 5 | 2 | 3 | | | | | |
| Carbon | C | 6 | 2 | 4 | | | | | |
| Nitrogen | N | 7 | 2 | 5 | | | | | |
| Oxygen | O | 8 | 2 | 6 | | | | | |
| Fluorine | F | 9 | 2 | 7 | | | | | |
| Neon | Ne | 10 | 2 | 8 | | | | | |
| Sodium | Na | 11 | 2 | 8 | 1 | | | | |
| Magnesium | Mg | 12 | 2 | 8 | 2 | | | | |
| Aluminium | Al | 13 | 2 | 8 | 3 | | | | |
| Silicon | Si | 14 | 2 | 8 | 4 | | | | |
| Phosphorus | P | 15 | 2 | 8 | 5 | | | | |
| Sulphur | S | 16 | 2 | 8 | 6 | | | | |
| Chlorine | Cl | 17 | 2 | 8 | 7 | | | | |
| Argon | Ar | 18 | 2 | 8 | 8 | | | | |
| Potassium | K | 19 | 2 | 8 | 8 | 1 | | | |
| Calcium | Ca | 20 | 2 | 8 | 8 | 2 | | | |
| Scandium | Sc | 21 | 2 | 8 | 9 | 2 | | | |
| Titanium | Ti | 22 | 2 | 8 | 10 | 2 | | | |
| Vanadium | V | 23 | 2 | 8 | 11 | 2 | | | |
| Chromium | Cr | 24 | 2 | 8 | 13 | 1 | | | |
| Manganese | Mn | 25 | 2 | 8 | 13 | 2 | | | |
| Iron | Fe | 26 | 2 | 8 | 14 | 2 | | | |
| Cobalt | Co | 27 | 2 | 8 | 15 | 2 | | | |
| Nickel | Ni | 28 | 2 | 8 | 16 | 2 | | | |
| Copper | Cu | 29 | 2 | 8 | 18 | 1 | | | |
| Zinc | Zn | 30 | 2 | 8 | 18 | 2 | | | |
| Gallium | Ga | 31 | 2 | 8 | 18 | 3 | | | |
| Germanium | Ge | 32 | 2 | 8 | 18 | 4 | | | |
| Arsenic | As | 33 | 2 | 8 | 18 | 5 | | | |
| Selenium | Se | 34 | 2 | 8 | 18 | 6 | | | |
| Bromine | Br | 35 | 2 | 8 | 18 | 7 | | | |
| Krypton | Kr | 36 | 2 | 8 | 18 | 8 | | | |
| Rubidium | Rb | 37 | 2 | 8 | 18 | 8 | 1 | | |
| Strontium | Sr | 38 | 2 | 8 | 18 | 8 | 2 | | |
| Yttrium | Y | 39 | 2 | 8 | 18 | 9 | 2 | | |
| Zirconium | Zr | 40 | 2 | 8 | 18 | 10 | 2 | | |
| Niobium | Nb | 41 | 2 | 8 | 18 | 12 | 1 | | |
| Molybdenum | Mo | 42 | 2 | 8 | 18 | 13 | 1 | | |
| Technetium | Tc | 43 | 2 | 8 | 18 | 14 | 1 | | |
| Ruthenium | Ru | 44 | 2 | 8 | 18 | 15 | 1 | | |
| Rhodium | Rh | 45 | 2 | 8 | 18 | 16 | 1 | | |
| Palladium | Pd | 46 | 2 | 8 | 18 | 18 | | | |
| Silver | Ag | 47 | 2 | 8 | 18 | 18 | 1 | | |
| Cadmium | Cd | 48 | 2 | 8 | 18 | 18 | 2 | | |
| Indium | In | 49 | 2 | 8 | 18 | 18 | 3 | | |
| Tin | Sn | 50 | 2 | 8 | 18 | 18 | 4 | | |
| Antimony | Sb | 51 | 2 | 8 | 18 | 18 | 5 | | |

| Element | | At. No. | Electron Distribution | | | | | | |
|---|---|---|---|---|---|---|---|---|---|
| | | | K | L | M | N | O | P | Q |
| Tellurium | Te | 52 | 2 | 8 | 18 | 18 | 6 | | |
| Iodine | I | 53 | 2 | 8 | 18 | 18 | 7 | | |
| Xenon | Xe | 54 | 2 | 8 | 18 | 18 | 8 | | |
| Caesium | Cs | 55 | 2 | 8 | 18 | 18 | 8 | 1 | |
| Barium | Ba | 56 | 2 | 8 | 18 | 18 | 8 | 2 | |
| Lanthanum | La | 57 | 2 | 8 | 18 | 18 | 9 | 2 | |
| Cerium | Ce | 58 | 2 | 8 | 18 | 19 | 9 | 2 | |
| Praseodymium | Pr | 59 | 2 | 8 | 18 | 20 | 9 | 2 | |
| Neodymium | Nd | 60 | 2 | 8 | 18 | 21 | 9 | 2 | |
| Promethium | Pm | 61 | 2 | 8 | 18 | 22 | 9 | 2 | |
| Samarium | Sm | 62 | 2 | 8 | 18 | 23 | 9 | 2 | |
| Europium | Eu | 63 | 2 | 8 | 18 | 24 | 9 | 2 | |
| Gadolinium | Gd | 64 | 2 | 8 | 18 | 25 | 9 | 2 | |
| Terbium | Tb | 65 | 2 | 8 | 18 | 26 | 9 | 2 | |
| Dysprosium | Dy | 66 | 2 | 8 | 18 | 27 | 9 | 2 | |
| Holmium | Ho | 67 | 2 | 8 | 18 | 28 | 9 | 2 | |
| Erbium | Er | 68 | 2 | 8 | 18 | 29 | 9 | 2 | |
| Thulium | Tm | 69 | 2 | 8 | 18 | 30 | 9 | 2 | |
| Ytterbium | Yb | 70 | 2 | 8 | 18 | 31 | 9 | 2 | |
| Lutetium | Lu | 71 | 2 | 8 | 18 | 32 | 9 | 2 | |
| Hafnium | Hf | 72 | 2 | 8 | 18 | 32 | 10 | 2 | |
| Tantalum | Ta | 73 | 2 | 8 | 18 | 32 | 11 | 2 | |
| Tungsten | W | 74 | 2 | 8 | 18 | 32 | 12 | 2 | |
| Rhenium | Re | 75 | 2 | 8 | 18 | 32 | 13 | 2 | |
| Osmium | Os | 76 | 2 | 8 | 18 | 32 | 14 | 2 | |
| Iridium | Ir | 77 | 2 | 8 | 18 | 32 | 17 | | |
| Platinum | Pt | 78 | 2 | 8 | 18 | 32 | 17 | 1 | |
| Gold | Au | 79 | 2 | 8 | 18 | 32 | 18 | 1 | |
| Mercury | Hg | 80 | 2 | 8 | 18 | 32 | 18 | 2 | |
| Thallium | Tl | 81 | 2 | 8 | 18 | 32 | 18 | 3 | |
| Lead | Pb | 82 | 2 | 8 | 18 | 32 | 18 | 4 | |
| Bismuth | Bi | 83 | 2 | 8 | 18 | 32 | 18 | 5 | |
| Polonium | Po | 84 | 2 | 8 | 18 | 32 | 18 | 6 | |
| Astatine | At | 85 | 2 | 8 | 18 | 32 | 18 | 7 | |
| Radon | Rn | 86 | 2 | 8 | 18 | 32 | 18 | 8 | |
| Francium | Fr | 87 | 2 | 8 | 18 | 32 | 18 | 8 | 1 |
| Radium | Ra | 88 | 2 | 8 | 18 | 32 | 18 | 8 | 2 |
| Actinium | Ac | 89 | 2 | 8 | 18 | 32 | 18 | 9 | 2 |
| Thorium | Th | 90 | 2 | 8 | 18 | 32 | 19 | 9 | 2 |
| Protactinium | Pa | 91 | 2 | 8 | 18 | 32 | 20 | 9 | 2 |
| Uranium | U | 92 | 2 | 8 | 18 | 32 | 21 | 9 | 2 |
| Neptunium | Np | 93 | 2 | 8 | 18 | 32 | 22 | 9 | 2 |
| Plutonium | Pu | 94 | 2 | 8 | 18 | 32 | 23 | 9 | 2 |
| Americium | Am | 95 | 2 | 8 | 18 | 32 | 24 | 9 | 2 |
| Curium | Cm | 96 | 2 | 8 | 18 | 32 | 25 | 9 | 2 |
| Berkelium | Bk | 97 | 2 | 8 | 18 | 32 | 26 | 9 | 2 |
| Californium | Cf | 98 | 2 | 8 | 18 | 32 | 27 | 9 | 2 |
| Einsteinium | Es | 99 | 2 | 8 | 18 | 32 | 28 | 9 | 2 |
| Fermium | Fm | 100 | 2 | 8 | 18 | 32 | 29 | 9 | 2 |
| Mendelevium | Md | 101 | 2 | 8 | 18 | 32 | 30 | 9 | 2 |
| Nobelium | No | 102 | 2 | 8 | 18 | 32 | 31 | 9 | 2 |
| Lawrencium | Lw | 103 | (2 | 8 | 18 | 32 | 32 | 9 | 2) |

The removal in turn of the first eight electrons is accompanied by a more or less steady rise in ionization energy. After the removal of the eighth electron, however, there is a big jump in the amount of energy required to detach the next electron. This is marked by a sharp rise on the graph. We can therefore deduce that the first eight electrons removed occupy the same quantum shell or energy level. This will be the third, or *M* shell, since electrons are removed first from the *outer* part of the atom.

Another large increase in ionization energy takes place after the removal of the sixteenth electron. Hence the second group of eight electrons forms

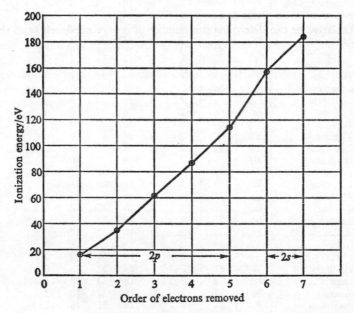

*Fig.* 6–7. Graph showing ionization energies in the second quantum shell of fluorine

the second quantum shell, the *L* shell. We are finally left with two electrons, and these constitute the third and innermost shell, the *K* shell.

Graphs similar to that given for argon can be constructed for the other elements, and in the same way they indicate the number of quantum shells in the various atoms and the number of electrons in each. Table 6–2 shows the arrangement of electrons for the different elements. There may be up to seven quantum shells in an atom. The capacity of the shells for holding electrons increases with the distance of the shells from the nucleus. Thus the maximum number of electrons in the first four shells is 2, 8, 18, and 32.

*Energy Sub-levels in each Main Energy Level.* By plotting the ionization energies of the individual electrons in a quantum shell against the order of removal of the electrons we obtain evidence of the small energy differences within a main energy level. The graph in Fig. 6–7 is for the second quantum shell of the fluorine atom (see Table 6–2). There are seven electrons in this shell.

We see that for successive removal of the first five electrons there is a steady rise in ionization energy. Then there is a sharp increase, indicating that the last two electrons are bound to the nucleus somewhat more firmly than the first five. We can conclude that in the second main energy level there are two sub-levels. These are labelled $2s$ and $2p$, and are occupied by two and five electrons respectively.

Similarly, we can determine the number of energy sub-levels (and the number of electrons in each) for other quantum shells. As stated earlier, the number of possible sub-levels is equal to $n$, the principal quantum number of the shell. We have here an indication that electron distribution in atoms follows a simple mathematical pattern. This is confirmed when we examine the maximum numbers of electrons which can occupy the various energy sub-levels. The numbers are as follows:

| Energy sub-level | Number of electrons |
|:---:|:---:|
| *s* | $2 \times 1 = 2$ |
| *p* | $2 \times 3 = 6$ |
| *d* | $2 \times 5 = 10$ |
| *f* | $2 \times 7 = 14$ |

Bearing in mind that the first quantum shell has one energy sub-level, the second two, etc., we see that the numbers of electrons required to fill the first four quantum shells are 2, 8, 18, and 32 respectively. These numbers are in accordance with the general formula $2n^2$, where $n$ is the principal quantum number of the shell. The fifth and subsequent quantum shells always contain an incomplete number of electrons.

**Electron Orbitals.** Why is there a difference of energy between, say, $s$ and $p$ electrons in the same quantum shell? To explain this we must look more closely into the movement of electrons round the nucleus. In Bohr's atomic model the electrons were pictured as travelling round the nucleus in circular (or elliptical) orbits. In the 1920s it was found, however, that electrons have a dual character. Sometimes they behave like particles, and sometimes like waves (electrons can be diffracted by crystals in the same way that light waves are diffracted by a grating). This led to the treatment of electron distribution in atoms by a new mathematical technique called *wave mechanics*. We are not concerned here with this technique, but merely with some of the conclusions which have followed its application.

Wave mechanics has confirmed the arrangement of electrons in groups and sub-groups of energy levels, but the electron orbits of the Bohr atomic model have disappeared. Indeed, the term 'orbit' is no longer used. Nowadays an electron is regarded, not as travelling round the nucleus in a fixed path, but as occupying a certain region in space round the nucleus. This region is called the *orbital* of the electron. Electron orbitals in a free atom are described as *atomic orbitals*.

The geometrical forms of electron orbitals vary with the energy level and sub-level in which the electrons occur. The sizes and shapes of the orbitals are calculated from spectral data by means of wave mechanics. Thus if $\Psi$ (psi) is the amplitude of the wave system which represents an electron, the electron orbital is the region in which $\Psi$ has an appreciable

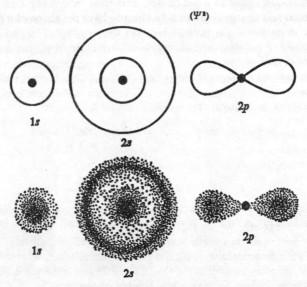

$(\Psi^2)$

1s

2s

2p

1s

2s

2p

*Fig.* 6–8. Forms of electron orbitals and probability distributions for 1s, 2s, and 2p electrons

magnitude. The probability of the electron being located at different points within the orbital at any instant is not the same at all points. The *probability distribution* is determined by magnitudes of $\Psi^2$. In the case of a 1s electron both the orbital and probability distribution have the form of a sphere surrounding the nucleus (Fig. 6–8).

We can represent the probability distribution in detail by dots, the density of which decreases with increasing distance from the nucleus. This is the picture which would be obtained if the dots represented the position of the electron averaged over a suitably large interval of time. Note, however, that the electron distribution is in three dimensions, and not in two as shown in the diagram. Another way of visualizing the probability distribution of an electron is to think of the negative charge of the electron as spread out in space, so that the density of the charge varies in the same manner as the density of the dots in Fig. 6–8. The total charge of the 'charge cloud' must, of course, be that of the single electron, but at different distances from the nucleus the effective charge will have only fractional values (0·8, 0·4, 0·2, 0·1, etc.) of the unit charge on the electron. Theoretically the charge cloud extends to infinity, so that the electron has no definite size.

The orbital and charge cloud of a $2s$ electron consist of two concentric spheres. The density of the charge cloud is again greatest at the centre. Then with increasing distance from the centre the density falls away to zero, increases again to a maximum, and then falls away once more. The orbital and charge cloud of a $2p$ electron have the shape of a figure 8 in three dimensions, the nucleus being in the middle. The geometrical forms of $d$ and $f$ orbitals and charge clouds are more complicated and we shall not describe them.

There are two reasons for the difference of energy between $s$ electrons and $p$ electrons in the same quantum shell. Firstly, the average distance of $s$ electrons from the nucleus is less than that of $p$ electrons, so that the former are more firmly held by the nucleus. Secondly, the $p$ electrons are partially screened from the full nuclear charge by the $s$ electrons. Similar considerations apply to $d$ and $f$ electrons. Thus the relative energies required for removal of electrons in the same quantum shell are in the order $s > p > d > f$.

**Pairing of Electrons.** An electron not only has a linear motion, but also spins on its axis like a top. It therefore behaves like a small magnet and has a magnetic field. According to *Pauli's exclusion principle* an orbital can be occupied by only two electrons at the same time. When two electrons occupy the same orbital their spins are always in opposite directions (clockwise and anti-clockwise), and the electrons are then said to be *paired*. Two paired electrons form a stable combination because their magnetic fields reinforce each other. To separate the electrons energy is required. There is no direct proof of Pauli's principle, but it forms the theoretical basis for the electronic structures of atoms, molecules, and ions. If it is assumed we find that there is a logical and consistent explanation for these structures.

When the energy sub-levels in a quantum shell are full all the orbitals are occupied by two paired electrons. We have already seen that the maximum numbers of electrons which can be present in $s$, $p$, $d$, and $f$ sub-levels are 2, 6, 10, and 14. This means that the maximum numbers of *pairs* of electrons in these orbitals are 1, 3, 5, and 7 respectively. Hence there can be only one orbital at the $s$ sub-level, while at the $p$, $d$, and $f$ sub-levels there can be three, five, and seven orbitals. This is illustrated by Table 6–3, which shows the electron distribution in the first three completed quantum shells. In this table an orbital is represented by a square and two paired electrons by two arrows pointing in opposite directions.

We can regard atoms of elements following hydrogen in atomic number as formed by adding in succession one further electron and one proton, the appropriate number of neutrons (in accordance with the relative atomic mass) being added at the same time to the nucleus. This procedure is known as the *Aufbau* ('building') principle.

**Table 6-3.** ELECTRON DISTRIBUTIONS IN THE FIRST THREE COMPLETED QUANTUM SHELLS

| Quantum shell | K | L | M |
|---|---|---|---|
| Main energy levels, sub-levels, and number of electrons in each | | | $d$ [↑↓] [↑↓] [↑↓] [↑↓] [↑↓] <br> $p$ [↑↓] [↑↓] [↑↓] <br> $s$ [↑↓] |
| | | $p$ [↑↓] [↑↓] [↑↓] <br> $s$ [↑↓] | |
| | $s$ [↑↓] | | |
| Total number of electrons | 2 | 8 | 18 |

The following rules hold for the addition of the extra electron at each stage:

* Not more than two electrons may occupy one orbital.
* The added electron goes into an orbital in the lowest available energy level and sub-level (thus giving minimum potential energy, or maximum stability, to the system).
* If two orbitals of equivalent energy are available the electron enters one which is not already occupied by another electron, and the spins of any unpaired electrons are parallel.

Table 6–4 gives the distribution of electrons in different energy levels and sub-levels for the first ten elements in order of atomic number. The single electron of the hydrogen atom in its ground state occupies the $1s$ orbital. The orbital can accommodate one more electron of opposite spin. This arrangement occurs in the helium atom, which has the electron configuration $1s^2$ (the superscript showing the number of electrons in the $s$ orbital). The first quantum shell is now complete, and we have an extremely stable arrangement of electrons, helium being devoid of chemical properties. The 'building' process is then continued in accordance with the principles given, the second quantum shell being completed in neon, which also has a very stable electronic structure ($1s^2\ 2s^2\ 2p^6$.) The electron configurations of the elements which follow neon are discussed later.

Table 6–4. ELECTRON DISTRIBUTIONS OF THE FIRST TEN ELEMENTS IN ORDER OF ATOMIC NUMBER

| Element | Atomic number | Electron distribution | | |
|---------|---------------|-----|-----|-----|
| | | $1s$ | $2s$ | $2p$ |
| H  | 1  | ↓  |    |          |
| He | 2  | ↑↓ |    |          |
| Li | 3  | ↑↓ | ↓  |          |
| Be | 4  | ↑↓ | ↑↓ |          |
| B  | 5  | ↑↓ | ↑↓ | ↓        |
| C  | 6  | ↑↓ | ↑↓ | ↓  ↓     |
| N  | 7  | ↑↓ | ↑↓ | ↓  ↓  ↓  |
| O  | 8  | ↑↓ | ↑↓ | ↑↓ ↓  ↓  |
| F  | 9  | ↑↓ | ↑↓ | ↑↓ ↑↓ ↓  |
| Ne | 10 | ↑↓ | ↑↓ | ↑↓ ↑↓ ↑↓ |

**Summary of Developments in the Particle Theory of Matter.** The developments described in this chapter are concerned with the arrangement of electrons in atoms. In Rutherford's 'planetary' model of the atom no particular distribution of the electrons was suggested. The first experimental evidence of an orderly arrangement of the electrons came from emission spectra of atoms, and more specifically from the emission spectrum of atomic hydrogen. Bohr showed how the latter spectrum could be explained by a theory of electron distribution based on Planck's quantum theory. He demonstrated that the wavelengths of the lines in the spectrum could be calculated if it were assumed that the hydrogen electron could exist in different energy levels and that radiations of different wavelength were caused by transitions of the electron between the various levels.

Bohr's theory was then extended to the electron systems of larger atoms. For elements with an atomic number greater than 2 it was necessary to assume the existence of energy sub-levels within the main energy levels (quantum shells) in order to explain the 'fine' structure of atomic spectra. The manner in which electrons are arranged in different energy levels and sub-levels can now be determined from the relative values of the ionization energies of the electrons.

The discovery that an electron can behave both as a particle and a wave system has led to a more fundamental mathematical treatment of electron distribution. Wave mechanics has confirmed Bohr's theory of energy levels, but does not allow electrons to have fixed orbits. It indicates that an electron may be anywhere within a region of space called an orbital. The shapes of orbitals vary with the energy level and sub-level in which the electrons occur. If we assume that only two electrons can occupy an orbital (in which case they have opposite spins) we can consistently explain the electron configurations of atoms.

We have now brought the particle theory of matter up to date in broad outline as regards atoms. The present-day atomic model has been built up from three main components: Rutherford's nuclear atom, Bohr's theory of electron energy levels, and the concept of electron orbitals derived from wave mechanics. In subsequent chapters we shall use this atomic model to explain the behaviour of atoms.

## EXERCISE 6

**1.** Explain how the emission spectrum of atomic hydrogen led to Bohr's theory of discrete energy levels for electrons in atoms.

**2.** What is meant by the *ionization energy* of an atom? In what units is it expressed? Explain (*i*) why the ionization energies of atoms of different elements are not the same, (*ii*) why the second ionization energy of an atom is always larger than the first.

**3.** What light is thrown by ionization energies on the manner in which electrons are distributed in atoms?

**4.** Plot a graph showing how successive ionization energies for the second quantum shell of neon vary with the number of electrons removed. (Use the values of ionization energies given at p. 144.) What deductions can be made from the graph?

**5.** Explain (*i*) atomic orbital, (*ii*) 'pairing' of electrons, (*iii*) Pauli's exclusion principle. Discuss the significance of the latter in regard to the manner in which electrons are distributed in atoms.

**6.** (Part question.) Draw up a table to show the names, symbols, atomic numbers and arrangements of electrons for the first ten elements of the Periodic Table.

An atom has 14 neutrons in the nucleus and an atomic number of 13. State the composition of the nucleus and describe the arrangement of the electrons, name the element of which the atom is an isotope, and state its valency.

(C.L.)

# CHAPTER 7

# The bonding of atoms

**Why Do Atoms Combine?** Only with the rare gases of the atmosphere (and mercury vapour) do elements exist in the form of individual atoms at ordinary temperatures. The atoms of all other elements occur in a combined form. The oxygen of the atmosphere is sometimes described as 'free' oxygen because it is not combined with any other element. Nevertheless, the atoms are combined with each other (in the form of $O_2$ molecules) just as much as the atoms in water ($H_2O$). Similarly in 'free' iron the metal atoms have joined together to give crystals of iron.

The reason why atoms usually exist in a combined form must be that they have an attraction for each other. As we shall see later, even rare-gas atoms have a mutual attraction, although this is so small that it becomes apparent only at low temperatures. The attraction between atoms is evident because whenever atoms are combined energy is required to separate them. It follows that energy must be evolved when the atoms combine. The emission of energy means that a more stable system is formed. Hence the basic reason for atoms combining is that the combination represents a more stable form of matter than the individual atoms.

**Electronic Theory of Valency.** In ordinary chemical reactions the nuclei of the atoms are unaffected. The reactions must therefore involve changes in the electron systems of the atoms. Now if we examine the properties of compounds we find that two classes of compounds can be distinguished, according to whether or not they conduct electricity when in the fused state. The first class, comprising salts like sodium(I) chloride and strong bases like sodium(I) hydroxide, are good conductors in the fused state and are decomposed by a current. Compounds of the second class, of which water and naphthalene ($C_{10}H_8$) are examples, are practically non-conductors in the fused state and are not electrolysed. It thus appears that there are two types of bonding between atoms. One type leads to electrical conductivity and electrolysis in the fused state, the other to absence of these properties. The difference can be illustrated practically as now described.

**Experiment.** As the melting point of common salt is rather high (804°C) anhydrous zinc(II) chloride (m.p. 262°C) is used to represent salts. Lead(II) chloride, lead(II) bromide, and tin(II) chloride can also be used, but their melting points are appreciably higher (400°–500°C).

Set up the apparatus shown in Fig. 7–1. The electrical circuit includes a switch, S, an ammeter, A, and a lamp resistance, L (one 250-watt lamp gives a current of about 0·6 amp with a d.c. voltage of 200 volt).

Connect one lead by means of a crocodile clip to the rim of a nickel crucible, which acts as the anode. Attach the other lead to a thick copper wire, which forms the cathode. Fill the crucible with anhydrous

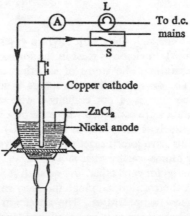

*Fig.* 7–1. Experiment to show the electrical conductivity of fused zinc(II) chloride

zinc(II) chloride, put it on a clean (preferably new) pipe-clay triangle, and melt the salt over a Bunsen flame. Then reduce the size of the flame, lower the copper cathode into the melt, and switch on the current. Test for chlorine evolved at the anode by damp litmus paper (which is bleached) or by damp starch-potassium(I) iodide paper (which turns blue-black). At the end of five minutes switch off the current and dissolve the film of zinc salt left on the copper cathode by dipping the latter in water. A shiny grey deposit of zinc will be found adhering to the copper. If the cathode is put into a test tube containing dilute sulphuric(VI) acid, bubbles of hydrogen will be evolved at the zinc surface.

Repeat the experiment with a second crucible containing melted naphthalene or paraffin wax. No current is produced in this case, and there is no sign of chemical reaction at the electrodes.

The electronic theory of valency, originated by Kossel and Lewis in 1916, was prompted by the remarkable stability of the rare-gas elements (although a number of compounds of them can now be prepared). This stability is associated with the presence in the atoms of a group of eight electrons (two in helium) in the outer quantum shell. Table 7–1 gives the electron configurations of the rare gases.

**Table 7–1.** ELECTRONIC STRUCTURES OF THE RARE GASES

| Element | Atomic number | Electrons in shells | | | | | |
|---------|---------------|-----|---|----|----|----|---|
| | | *K* | *L* | *M* | *N* | *O* | *P* |
| Helium | 2 | 2 | | | | | |
| Neon | 10 | 2 | 8 | | | | |
| Argon | 18 | 2 | 8 | 8 | | | |
| Krypton | 36 | 2 | 8 | 18 | 8 | | |
| Xenon | 54 | 2 | 8 | 18 | 18 | 8 | |
| Radon | 86 | 2 | 8 | 18 | 32 | 18 | 8 |

The atomic numbers of the rare gases provide another illustration of the simple mathematical basis of electron arrangement mentioned previously, the numbers being in accordance with the general formula

$$2(1^2 + 2^2 + 2^2 + 3^2 + 3^2 + 4^2)$$

Table 6–2 indicates that in every rare gas except helium (where only the $1s$ orbital contains electrons) the $s$ and $p$ orbitals of the outer quantum shell are filled, giving a total of eight electrons. Thus neon has an electron configuration $1s^2 2s^2 2p^6$. The atoms of all other elements contain less than eight electrons in their outermost quantum shell. This incompleteness appears to be one of the chief causes of the chemical activity of other elements. When atoms combine they often acquire the electron pattern of a rare gas; that is, they obtain an external 'octet' of electrons (or a 'duplet' in the case of the helium structure). In general this is accomplished either by the transfer of one or more electrons from one atom to the other, or by atoms sharing electrons, which thus become the property of both. The first method of combination is described as *electrovalency*, and the second as *covalency*.

Besides these two kinds of valency bond there are various weaker attractive forces which can cause atoms or molecules to hold together. These forces are described later in the chapter. Bonding of a special type occurs between the atoms in metals.

## Electrovalency

**Formation of Ions.** A bond which is formed by transfer of one or more electrons from one atom to another is called an *electrovalent, ionic,* or *heteropolar* bond. As an illustration of this type of linkage, consider the redistribution of electrons when sodium combines with chlorine. The atomic numbers of sodium and chlorine are 11 and 17 respectively. The 11 external electrons of the sodium atom are in shells containing 2 (*K*), 8 (*L*), and 1 (*M*) electrons, while those of chlorine are arranged 2 (*K*), 8 (*L*), and 7 (*M*). By giving an electron to the chlorine atom the sodium atom acquires a structure of the neon type. By gaining one electron the chlorine atom conforms to the argon pattern. Thus the atoms of both elements gain a stable rare-gas structure, although they differ from rare-gas atoms in that the sodium atom, having lost an electron, has now one positive charge, while the chlorine atom, having gained an electron, has now one negative charge. They are no longer atoms but ions bound together by electrostatic attraction.

In expressing the change we usually represent only the electrons in the shells affected, and omit the core consisting of the nucleus and the electrons in the remaining shells. Sometimes we use dots and crosses to distinguish between the electrons in the outer valency shells of the two atoms, but this is really unnecessary.

$$\text{Na} \cdot \ + \ \cdot \ddot{\underset{..}{\text{Cl}}} : \ \rightarrow \text{Na}^+ + \ \left[ : \ddot{\underset{..}{\text{Cl}}} : \right]^-$$

$$\quad\ 2,8,1 \qquad 2,8,7 \qquad 2,8 \qquad\ 2,8,8$$

The chlorine atom with an electron configuration $1s^2 2s^2 2p^6 3s^2 3p^5$ can accept the electron from the sodium atom because one of its three *p* orbitals contains only one electron. For the incoming electron to fit into the vacant place its spin must be opposite to that of the electron already present. Thus a single electrovalent bond is formed by the pairing of two electrons of opposite spin.

In formation of sodium(I) oxide ($Na_2O$) one oxygen atom gains two electrons from two sodium atoms, while in calcium(II) chloride ($CaCl_2$) two chlorine atoms each acquire one electron from a calcium atom.

$$2\text{Na} \cdot \ + \ \ddot{\underset{..}{\text{O}}} : \ \rightarrow 2\text{Na}^+ + \ \left[ : \ddot{\underset{..}{\text{O}}} : \right]^{2-}$$

$$\text{Ca} : \ + \ 2 \cdot \ddot{\underset{..}{\text{Cl}}} : \ \rightarrow \ \text{Ca}^{2+} + \ 2 \left[ : \ddot{\underset{..}{\text{Cl}}} : \right]^-$$

These equations merely illustrate the process of electron transfer. Ions are usually represented by their chemical symbols and charges, *e.g.*, $Na^+$, $Ca^{2+}$, $Cl^-$, and $O^{2-}$. Charged atoms are *simple* ions. If the charge is located on a group of atoms, as a whole, the ion is a *complex* ion. Examples of complex ions are $NH_4^+$, $NO_3^-$, $SO_4^{2-}$, and $CO_3^{2-}$. In electrovalent compounds the oppositely charged ions are held together by electrostatic attraction, *e.g.*, $Na^+Cl^-$, $[Na^+]_2O^{2-}$, and $Ca^{2+}[Cl^-]_2$. Unless we particularly desire to draw attention to the charges on the ions we usually omit them and write the formula of the electrovalent compound in the conventional manner, *e.g.*, $NaCl$, $Na_2O$, and $CaCl_2$.

**Properties of Electrovalent Compounds.** The general properties of electrovalent compounds may be summarized as follows:

- At ordinary temperatures they invariably exist as crystalline solids although individual particles such as $Na^+Cl^-$ may exist in the vapour state. The crystal consists of an 'infinite assembly' of ions, that is, a large and indefinite number of oppositely charged ions joined together in a regular manner.

  The force between oppositely charged ions is an extremely powerful one. If we could assemble a thimbleful of free sodium ions in London and an equal number of free chloride ions in Birmingham, a hundred miles away, the force of attraction would be over 10 000 N.

- Electrovalent compounds have high melting points and boiling points owing to the strong attraction between the oppositely charged ions.
- In the fused state they are good conductors of electricity and are electrolysed by a current. This is explained by the breakdown of the crystal into 'free' ions, which are attracted to, and discharged at, the electrode of opposite sign.
- They often dissolve readily in water, but are only sparingly soluble in organic solvents like ethoxyethane and benzene.
- When oppositely charged ions react together in aqueous solution they do so almost instantaneously.

An example of the last is the immediate precipitation of white silver(I) chloride when silver(I) nitrate(V) solution is added to a solution of sodium(I) chloride ($Ag^+ + Cl^- \rightarrow AgCl \downarrow$).

**Sizes of Ions and Atoms.** The term 'size' is used in rather a special sense in regard to ions and atoms. There is no method of measuring the size of an isolated ion or atom. Indeed, these particles cannot be said to have a definite size because electron charge clouds theoretically extend to infinity. However, in *combination* with other ions or atoms the particles behave as if they have a size. Thus, when oppositely charged ions are joined in a crystal there is a particular distance between the nuclei of the ions. If we measure the distances for a number of electrovalent compounds by X-ray diffraction, we find that a characteristic *ionic radius* can be assigned to each ion, so that the distance between the nuclei is the sum of the two ionic radii, $r_1$ and $r_2$ (Fig. 7–2). $r_1$ and $r_2$ therefore represent

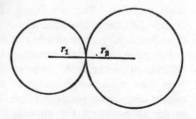

*Fig. 7–2*

the sizes of the ions. However, the radius of an ion varies to some extent according to the number of oppositely charged ions which surround it in the crystal.

The size of a neutral atom in combination is represented by its *atomic*, or *covalent*, *radius*. For a metal this is half the distance between adjacent atomic nuclei in the solid metal. For a non-metal it is half the distance between the nuclei of two similar atoms joined by a single covalent bond. Another important quantity for a neutral atom is the *van der Waals radius*, which is half the distance between the nearest atomic nuclei of neighbouring molecules in the solid. This radius is always larger than the covalent radius. Thus in solid chlorine the van der Waals radius is 0·180 nm.

When we compare the sizes (radii) of atoms with those of the corresponding ions we find a marked difference. Positively charged ions are smaller, and negatively charged ions are larger, than the neutral atoms. This can be seen from the comparison now given.

ATOMIC AND IONIC RADII/nm

| Na | 0·186 | Mg | 0·160 | Al | 0·143 | Cl | 0·099 | O | 0·073 |
|---|---|---|---|---|---|---|---|---|---|
| $Na^+$ | 0·095 | $Mg^{2+}$ | 0·065 | $Al^{3+}$ | 0·050 | $Cl^-$ | 0·181 | $O^{2-}$ | 0·140 |

Note that the radius of the singly charged sodium ion is about half that of the neutral atom. With the doubly charged magnesium ion the radius is reduced by considerably more than half, while the radius of the trebly charged aluminium ion is little more than one third of that of the aluminium atom. These differences in size between ion and atom are readily understood. In all three cases formation of the ion from the atom is accompanied by disappearance of the outer quantum shell of electrons. Furthermore, loss of electrons takes place without any alteration of the nuclear charge. The remaining electrons are therefore more strongly attracted by the nucleus and are drawn closer to it. As may be expected, this effect is larger the greater the number of electrons lost.

The *increase* in size attending the formation of a negative ion by addition of one or more electrons to the atom is not so easily explained. Here there is no disappearance of an electron shell; the added electrons merely complete a partially filled shell. The gaining of electrons results in the attraction of the nucleus for each electron being reduced, and hence the electron shells move farther away from the nucleus. Again the effect is larger the greater the number of electrons gained. The change in size is too large, however, to be accounted for entirely in this way (for $Cl^-$ the number of electrons only increases from 17 to 18). With non-metallic atoms the covalent radius probably does not truly represent the effective size of a single free atom. This size corresponds more closely with the van der Waals radius.

If we assume that the relative sizes of atoms and ions are given by their atomic or ionic radii, we can illustrate (Fig. 7–3) the changes of size which occur when $Na^+$ and $Cl^-$ ions are formed from the atoms.

The differences in size between atoms and their ions provide additional evidence that salts liks sodium(I) chloride are indeed composed of ions, and not of atoms. The differences agree with the theory of loss or gain of electrons.

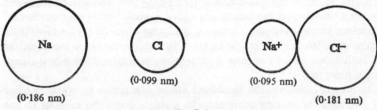

(0·186 nm)    (0·099 nm)    (0·095 nm)    (0·181 nm)

*Fig. 7–3*

**Isoelectronic Series.** The gaining of a rare-gas pattern of electrons by an ion does not mean that it has the chemical stability of a rare-gas atom. The ion may not only react with oppositely charged ions or with neutral molecules (*e.g.*, $H_2O$), but it can usually be reconverted to an atom (*e.g.*, by electrolysis). The ion differs from a rare-gas atom in having a different number of positive charges on the nucleus and therefore a different size. We can see this by comparing the radii of ions and rare-gas atoms which are *isoelectronic*, that is, which possess the same number of electrons. An example of an isoelectronic series is $O^{2-}$, $F^-$, Ne, $Na^+$, and $Mg^{2+}$. Each member of this series has ten electrons, but the charge on the nucleus

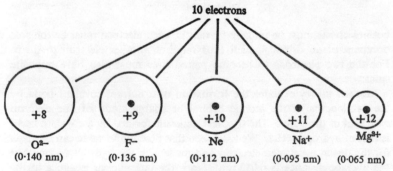

*Fig. 7–4.* Decrease in radius of the members of an isoelectronic series with increase in nuclear charge. (*Note.* The atomic radius given for neon is an estimated value.)

increases from $+8$ in $O^{2-}$ to $+12$ in $Mg^{2+}$. Fig. 7–4 portrays how the increase in nuclear charge is accompanied by a decrease in the size of the ion or atom.

Since an ion which is isoelectronic with a rare-gas atom differs in size from the latter, the compactness of the electrons round the nucleus is not the same in the two cases. This is probably one of the reasons (besides the difference in nuclear charge) for the difference in chemical reactivity.

## Normal Covalency

**Nature of the Covalent Bond.** Most compounds resemble naphthalene in being non-conductors of electricity in the fused state. Hence, if we are correct in ascribing conductivity to the presence of ions the majority of compounds do not contain electrovalent bonds. It is supposed that the linkages in these compounds are formed by the sharing of electrons between atoms, and the bonds are called *covalent* or *homopolar* bonds. Covalent bonds may join atoms of the same element (as in H—H) or atoms of different elements (as in H—Cl), but in either case the electrons of both atoms usually acquire a rare-gas pattern (there are many exceptions, however).

In *normal covalency* the combined atoms contribute an equal number of electrons for sharing purposes. In a single bond the number is one from each. For two electrons from different atoms to become shared

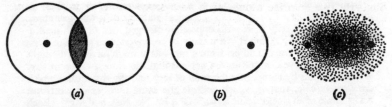

       *(a)*                     *(b)*                 *(c)*

*Fig.* 7–5. (*a*) Overlapping of atomic orbitals of hydrogen electrons; (*b*) Resulting molecular orbital; (*c*) Charge cloud of the two electrons in the molecule

both electrons must be unpaired; that is, each electron must be the sole occupant of an orbital, which thus requires another electron to fill it. For the two electrons to become paired they must also have opposite spins.

Modern theory explains the formation of a normal covalent bond by the overlapping of the atomic orbital or charge cloud of one electron with that of the other. The simplest molecule formed by a covalent bond is that of hydrogen ($H_2$). We have seen that no definite value can be given for the distance between electron and nucleus in the hydrogen atom, but the average distance is 0·053 nm. Since the internuclear distance in the hydrogen molecule is found to be 0·074 nm, there must be considerable overlapping of the atomic orbitals or charge clouds of the two electrons. This is represented in Fig. 7–5*a*. When two atomic orbitals overlap they coalesce and produce a single *molecular* orbital. The molecular orbital of the two electrons of the hydrogen molecule is represented (in two dimensions only) in Fig. 7–5*b*. This is the space in which the electrons will be found at any instant.

Fig. 7–5*c* shows the charge cloud of the two electrons in the hydrogen molecule. This charge cloud, which is formed by combination of the two charge clouds of the separate electrons, shows the electron density in and around the molecule by means of dots. There is a concentration of charge in the central region between the two nuclei. Consequently, over a period the two electrons spend more of their time in the region between the nuclei than we would expect from the probability distributions of the electrons in the isolated atoms. The concentration of negative charge in the region between the positively charged nuclei is responsible for holding the atoms together against the mutual repulsion of the nuclei, *the attractive force thus set up constituting the covalent bond*. We see that the covalent bond (like the electrovalent bond) is basically electrostatic in character, depending on attraction between electrons and atomic nuclei.

By the pairing of their single electrons both hydrogen atoms in the $H_2$ molecule acquire the electron pattern of helium.

$$H \cdot + \cdot H \rightarrow H{:}H$$

Similarly two chlorine atoms, each with seven electrons in their outer valency shell, attain the electron pattern of argon by sharing two electrons, which revolve round both nuclei.

$$:\overset{..}{C}l\cdot + \cdot\overset{..}{C}l: \rightarrow :\overset{..}{C}l:\overset{..}{C}l:$$

Covalent bonds can also form between atoms of different elements. Thus the oxygen, nitrogen, and carbon atoms, with six, five, and four electrons in their outer shells, combine with two, three, and four hydrogen atoms respectively, gaining in each case an external octet of electrons. The hydrides have the following electronic formulæ:

$$H{:}\overset{..}{O}{:} \qquad H{:}\overset{H}{\underset{..}{N}}{:} \qquad H{:}\overset{H}{\underset{..}{C}}{:}H$$
$$\dot{H} \qquad\qquad H \qquad\qquad H$$

Double and triple covalent bonds between two atoms are produced by the sharing of four and six electrons, half of which are contributed by each atom. Examples are seen in the molecules of ethene, ethyne, and nitrogen. These molecules are represented below, firstly by the classical method of valency bonds and secondly by the electronic formulæ.

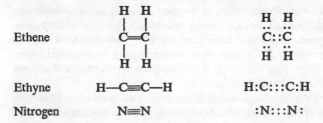

| | | |
|---|---|---|
| Ethene | H H<br>│ │<br>C═C<br>│ │<br>H H | H H<br>.. ..<br>C::C<br>.. ..<br>H H |
| Ethyne | H—C≡C—H | H:C:::C:H |
| Nitrogen | N≡N | :N:::N: |

From these examples we see that each unit of normal covalency (represented by a line in the classical formulæ) consists of one pair of shared electrons. For convenience in writing electronic structures we often retain the single line to indicate a pair of shared electrons.

Normal covalency does not always result in atoms gaining an external octet (or duplet) of electrons. Thus the boron atom in boron trichloride, $BCl_3$, has only six electrons in its outer shell. On the other hand the phosphorus atom in phosphorus pentachloride vapour, $PCl_5$, has ten electrons in the outer shell, while the sulphur atom in sulphur hexafluoride, $SF_6$, has twelve.

Boron trichloride    Phosphorus pentachloride    Sulphur hexafluoride

Since a single covalent bond is the pairing of an electron from one atom with an electron from another we should expect the number of bonds formed by an atom to correspond with the number of unpaired electrons in the atom. Table 6–4 indicates that the carbon atom in its ground state has only two unpaired electrons. Carbon, however, normally shows a valency of four. The reason for this is that when the carbon atom combines with other atoms it absorbs energy which causes the two paired electrons in the $2s$ sub-level to become unpaired, so that four electrons are available for bond formation. Similar unpairing of paired electrons occurs in many other atoms through absorption of energy. The atoms are then said to be in an excited state. This is why boron can have a valency of three (in $BCl_3$), phosphorus of five (in $PCl_5$), and sulphur of six (in $SF_6$). A fuller explanation of the use of the newly unpaired electrons in bonding is given at p. 179.

**Properties of Normal Covalent Substances.** (1) Whereas electrovalent bonds exist only in compounds, substances containing normal covalent bonds may be elements (non-metallic) or compounds. These usually exist as discrete molecules. The complexity of the latter may vary, however, from diatomic molecules to structures containing many thousands of atoms (as in proteins and plastics). In some cases there are no molecules of definite size. Instead we find infinite assemblies of atoms which can be regarded as giant molecules. An example is the diamond crystal, in which carbon atoms are joined by covalent bonds with a tetrahedral distribution (Fig. 7–6). Silicon, silicon(IV) oxide, $SiO_2$, and silicon carbide, SiC, have similar structures.

(2) The simpler covalent substances have low melting points and boiling points. Thus all compounds which are gases or liquids at ordinary

temperatures are covalent. Melting points and boiling points depend very much, however, on molecular size. Thus silicon(IV) oxide, which has a giant molecule, melts at over 1700°C. The melting points of proteins and many plastics are so high that the compounds decompose before their melting points are reached.

(3) In the fused state covalent compounds are non-conductors of electricity.

(4) Although some covalent compounds are soluble in water (for reasons given in Ch. 9) they usually dissolve more readily in organic solvents like benzene.

(5) As a rule covalent compounds react together slowly (organic chemistry provides many examples of this).

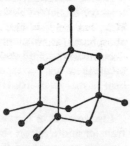

*Fig.* 7-6. Structure of carbon (diamond)

**Covalent Bond Energies and Bond Strengths.** The energy evolved when a covalent bond is formed between two free atoms in the gaseous state is called the *bond energy*, or, more correctly, the *bond energy of formation*. We can find its value for different kinds of bonds either spectroscopically or from thermochemical measurements. Some experimentally determined values for bond energies are as follows:

|  | Bond energy $E/\text{kJ mol}^{-1}$ |
|---|---|
| $H + H \rightarrow H_2$ | 431 |
| $O + O \rightarrow O_2$ | 493 |
| $N + N \rightarrow N_2$ | 931 |
| $Cl + Cl \rightarrow Cl_2$ | 238 |
| $H + Cl \rightarrow HCl$ | 430 |

As we might expect, the bond energy of formation of a double bond between two atoms is larger than that for a single bond, while the bond energy of a triple bond is larger again. This is seen from the bond energies (in kJ mol$^{-1}$) now given for the single, double, and triple bonds between two carbon atoms.

$$\underset{334}{C\text{—}C} \qquad \underset{606}{C\text{=}C} \qquad \underset{796}{C\text{≡}C}$$

Note, however, that the energy for the double bond is appreciably less than twice that for the single bond, and the energy of the triple bond is less than three times that of the single bond.

Bond energies are very important because all chemical reactions involve the making and breaking of bonds. The strength of a bond is represented by its *bond dissociation energy*. This is the energy (per mole) required to break a bond between two atoms. For a simple diatomic molecule like $H_2$ the bond dissociation energy is the same numerically as the bond energy of formation. In polyatomic molecules, however, the bond dissociation energy varies. This is because the attraction between two atoms is usually affected by neighbouring atoms in the molecule. The energy required to remove one hydrogen atom from a water molecule ($H_2O \rightarrow H + OH$) is 493 kJ mol$^{-1}$, while the energy needed to detach the hydrogen atom from the remaining O—H group is only 426 kJ mol$^{-1}$.

Normal covalent bonds are usually strong, and the energies required to break them are correspondingly high. This is shown by the stability to heat of molecules like $H_2O$, $CO_2$, HCl, and $CH_4$.

**Electrovalent or Covalent Bonding?** When atoms combine both electron transfer and electron sharing are possible. We know that in sodium(I) chloride the bonding is electrovalent, while in hydrogen chloride it is covalent. It is conceivable, however, that a sodium atom and a chlorine atom might share two electrons (one from each atom) and thus form a molecule Na—Cl. Similarly a hydrogen atom might transfer its one electron to a chlorine atom to give an ionic particle H$^+$Cl$^-$. We will investigate the alternatives from the energy point of view.

If there are two ways in which two atoms can combine we should expect that one to be favoured which produces the greater evolution of energy and hence the more stable system. We can see whether hydrogen and chlorine are more likely to form an electrovalent bond or a covalent bond if we prepare an energy 'profit and loss' account for the various steps involved in the two methods of combination. Starting with molecular hydrogen ($H_2$) and molecular chlorine ($Cl_2$), we first have to expend some energy to dissociate the molecules into free atoms. These energy items are the same whether the atoms subsequently combine to form H$^+$Cl$^-$ or H—Cl. We can therefore omit them from our energy account and start with free atoms.

To obtain H$^+$ ions from hydrogen atoms energy must again be expended. For 1 mole of hydrogen atoms this equals the ionization energy, which we here express in kJ mol$^{-1}$.

$$H \rightarrow H^+ + e \quad \text{(Energy absorbed} = 1350 \text{ kJ mol}^{-1})$$

Next we transfer the electron to a chlorine atom, which results in the liberation of energy equal to 385 kJ mol$^{-1}$ (this is called the *electron affinity* of chlorine).

$$Cl + e \rightarrow Cl^- \quad \text{(Energy evolved} = 385 \text{ kJ mol}^{-1})$$

Finally we allow the H$^+$ ion and the Cl$^-$ ion to approach each other from infinity until they are at the distance found experimentally for the two atoms in the HCl molecule. Energy is again liberated, the amount of which can be calculated.

$$H^+ + Cl^- \to H^+Cl^- \qquad (\text{Energy evolved} = 1087 \text{ kJ mol}^{-1})$$

The total energy evolved in forming an electrovalent bond between molar amounts of hydrogen and chlorine atoms would thus be

$$(-1350 + 385 + 1087) = 122 \text{ kJ mol}^{-1}$$

We take the energy produced in forming a covalent bond to be the average of the energies of the H—H bond and the Cl—Cl bond. The average is 334 kJ for the combination of moles of hydrogen and chlorine. Hence the formation of a covalent bond in preference to an ionic bond is favoured by an energy difference of $(334 - 122) \text{ kJ mol}^{-1}$ of hydrogen chloride. As we have seen earlier, the energy actually obtained in the combination of hydrogen and chlorine atoms is not 334 kJ, but 430 kJ mol$^{-1}$ of hydrogen chloride. The reason for the higher value is explained shortly.

Similarly we can compare the energy changes involved in electrovalent and covalent bonding between sodium and chlorine atoms. Here we find that the larger evolution of energy results from electrovalent bonding. The chief reason for this is the much lower ionization energy of sodium compared with that of hydrogen, the values being 493 and 1350 kJ mol$^{-1}$ respectively.

We see that whether electrovalent or covalent bonds are formed between atoms depends on a number of energy factors. One of the most important is the magnitude of the ionization energies required to remove the outer valency electrons. Alkali metals like sodium and alkaline-earth metals like calcium have relatively low ionization energies and form ionic bonds with chlorine. Trivalent aluminium and tetravalent carbon, having much higher ionization energies, form covalent chlorides.

**Intermediate Character of Covalent and Electrovalent Bonds.** When a covalent bond is formed between two similar atoms, *e.g.*, H—H or Cl—Cl, the pair of electrons forming the bond is shared equally between the two atoms. However, this is not so when the bond joins two different atoms. The chlorine atom in the HCl molecule has a stronger attraction for electrons than the hydrogen atom, and the two electrons forming the bond spend more time under the influence of the chlorine nucleus than under that of the hydrogen nucleus. In other words, there is an *electron displacement* towards the chlorine nucleus. As a result of the displacement the chlorine atom acquires a *partial* negative charge, represented by $\delta-$ (delta minus), and the hydrogen atom acquires a *partial* positive charge, represented by $\delta+$. Thus the covalent bond in hydrogen chloride has a certain amount of ionic character, and the conventional formula H—Cl does not adequately represent the structure of the molecule. A more accurate representation is to place an arrowhead at the end of the bond line, as in (i) below, to show the direction of electron displacement. Alternatively, each atom can be labelled with its appropriate partial charge, as in (ii).

$$H \rightarrow Cl \qquad \overset{\delta+\ \ \delta-}{H-Cl}$$
(i)                (ii)

When different parts of a molecule have opposite partial charges the molecule is said to possess an *electric dipole*. Practical evidence of the existence of electric dipoles in molecules can be obtained in several ways. Thus hydrogen chloride molecules tend to set their axes in the direction of an electric field (Fig. 7-7). This reduces the strength of the field. We can therefore detect the presence of the dipoles by their effect on the capaci-

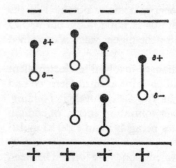

tance of a condenser when hydrogen chloride is used as the dielectric between the plates. Using the change in capacitance of the condenser, we can measure the amount of ionic character in the H—Cl bond. This is about 17 per cent. Other methods of detecting electric dipoles in molecules are by measuring bond lengths and by infrared spectra.

The formation of an intermediate type of bond, rather than a 'pure' covalent bond, results in evolution of greater energy and a stronger bond.

*Fig.* 7-7. Orientation of hydrogen chloride molecules in an electric field

Thus the theoretical energy of formation of a 'pure' covalent bond H—Cl is 334 kJ mol$^{-1}$, while the measured value for the actual bond is 430 kJ mol$^{-1}$.

Oxygen and nitrogen atoms have a greater attraction for electrons than hydrogen atoms. Thus in the molecules of water and ammonia there are electron displacements in the O—H bonds and N—H bonds towards the oxygen and nitrogen atoms. These displacements produce partial charges on the atoms as now shown.

$$\overset{\delta-}{O} \qquad\qquad \overset{\delta-}{N}$$
$$\delta+H \diagup \quad \diagdown H^{\delta+} \qquad \delta+H \diagup \ \big| \ \diagdown H^{\delta+}$$
$$H^{\delta+}$$

The O—H bonds in water have about 30 per cent of ionic character, and the N—H bonds in ammonia about 17 per cent. As we shall see later, this ionic character in the bonds has a profound effect on the properties of water and ammonia.

Intermediate character of bonding also occurs in electrovalent compounds. If an atom A has a much smaller attraction for a bond pair of electrons than an atom B the two electrons are displaced towards B to such an extent that they become the property of B. Instead of being covalent, the bond is now electrovalent $(A^+B^-)$. Electrovalent bonding is thus merely a more extreme form of electron displacement.

The bond formed, however, is not 100 per cent ionic in character. Owing to its charge the $A^+$ ion attracts the outer electrons of the $B^-$ ion and causes deformation of the outer electron orbitals towards itself (Fig. 7–8). The $A^+$ ion is said to *polarize* the $B^-$ ion because the electrical centre of gravity of the negative charges in the $B^-$ ion no longer coincides with that of the positive charges. Since some of the electrons of the anion are brought closer to the nucleus of the cation the positive charge on the cation is somewhat reduced. Also, since some of the electrons of the anion are removed to a greater distance from their own nucleus, the negative charge on the anion is decreased. Thus, instead of the charges on the ions being full charges, they are actually only partial charges. This means that the electrovalent bond has a certain amount of covalent character. We can estimate the amounts of covalent character in different electrovalent bonds from the differences between the observed and theoretical

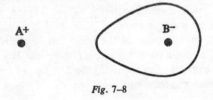

$A^+$  $B^-$

*Fig.* 7–8

bond energies. Thus the bond between two oppositely charged ions in the sodium(I) chloride crystal is found to have about 5 per cent of covalent character.

Evidently the division of chemical bonds into electrovalent and covalent is an oversimplification. 'Pure' electrovalency and 'pure' covalency represent extremes of bonding. When unlike atoms combine the bonds are always intermediate between the two extremes. A bond is often described as 'essentially ionic' or 'essentially covalent,' which means that it tends to one or other of the two extremes. The amounts of ionic and covalent character present in bonds vary, however, over a wide range.

**Pauling's Electronegativity Scale.** The relative attractions of two atoms for the electrons of a single covalent bond between them are called their *electronegativities*. The American scientist Linus Pauling deduced the electronegativities of atoms of different elements from the differences between the observed bond energies and the theoretical ones, the latter being for equal sharing of the bond electrons. There is excellent agreement between the electronegativities deduced in this way and the positions of the elements in the Periodic Table. This can be seen from Table 7–2.

**Table 7-2.** ELECTRONEGATIVITIES OF ELEMENTS (after Pauling)

(*Note.* Since electronegativity is a ratio of similar quantities it has no units.)

| H 2·1 | | | | | | |
|---|---|---|---|---|---|---|
| Li 1·0 | Be 1·5 | B 2·0 | C 2·5 | N 3·0 | O 3·5 | F 4·0 |
| Na 0·9 | Mg 1·2 | Al 1·5 | Si 1·8 | P 2·1 | S 2·5 | Cl 3·0 |
| K 0·8 | Ca 1·0 | | | | | Br 2·8 |
| Rb 0·8 | Sr 1·0 | | | | | I 2·5 |
| Cs 0·7 | Ba 0·9 | | | | | |

Any atom in this table behaves as an electron-attracting agent towards an atom with a lower electronegativity value. Hence, if two atoms A and B are joined by a single covalent bond and B has a higher electronegativity than A, the electrons of the bond are displaced towards B. It follows that A will have a partial positive charge and B a partial negative charge. We can calculate the approximate amount of ionic character in the bond A—B from the following empirical rule devised by Hannay and Smyth:

$$\text{Percentage of ionic character} = 16(x_A - x_B) + 3 \cdot 5(x_A - x_B)^2$$

where $x_A$ and $x_B$ are the electronegativities of A and B. If we calculate the amount of ionic character in the H—Cl bond from this rule the value obtained (17·2 per cent) agrees with the value (17 per cent) found experimentally.

**Covalent Bond Lengths.** The distance between the nuclei of two atoms joined by a covalent bond is called the *bond length*. Bond lengths can be measured by diffraction of X-rays by crystals (see later), diffraction of electrons by gases, and by spectroscopic methods. We may regard a single covalent bond between two similar atoms, A—A or B—B, as a pure covalent bond (that is, without any ionic character). We may then assume that the length of a pure covalent bond A—B between the different atoms would be the average of the lengths of the bonds A—A and B—B. In this way we can assign a *single-bond covalent radius* to each atom, so that when the radii for any two atoms are added together the result is the length of a pure covalent bond between the atoms. Some values for these radii are given in Table 7-3.

Table 7–3. SINGLE-BOND COVALENT RADII/nm

| H | C | N | O | Cl | Br | Na |
|---|---|---|---|----|----|----|
| 0·037 | 0·077 | 0·074 | 0·073 | 0·099 | 0·114 | 0·186 |

When we compare the lengths of single covalent bonds found in practice with those obtained by adding together the two radii we find that the additive principle holds only when the bond joins two atoms of about the same electronegativity. Thus nitrogen and chlorine have the same electronegativity. In nitrogen trichloride ($NCl_3$) the theoretical length of the N—Cl bond is 0·173 nm, while the observed length is 0·174 nm.

If atoms of different electronegativity are joined by a single covalent bond the bond length is less than that obtained by adding the single-bond covalent radii. Thus for H—Cl the calculated bond length is 0·136 nm, whereas the observed value is 0·128 nm. Similarly the calculated length of the O—H bond in ice is 0·110 nm, while the actual length is 0·099 nm. In both cases the decrease in bond length is due to the presence of ionic character in the bond and the consequent strengthening of the bond. *Decrease in bond length means increase in bond strength.* Comparison of observed bond lengths with those calculated for 'pure' single covalent bonds may thus provide evidence of ionic character in the bonds.

Double and triple covalent bonds are stronger than single bonds, and they are also shorter in length. Thus for the carbon–carbon bonds in ethane, ethene, and ethyne we have the lengths now shown.

| C—C | C=C | C≡C |
|-----|-----|-----|
| 0·154 nm | 0·133 nm | 0·120 nm |

## Co-ordinate (Dative) Covalency

When atoms have united by means of ionic or normal covalent bonds their combining powers are not necessarily exhausted. If one of the atoms or ions in the compound possesses an unshared, or *lone*, pair of electrons it may use them to establish a bond with another atom or ion which has an empty orbital in its outer quantum shell. An example is seen in the reaction which occurs between ammonia and anhydrous aluminium(III) chloride. The nitrogen atom of the ammonia molecule has an unshared pair of electrons. An aluminium atom has only three valency electrons, and when it has combined with three chlorine atoms by covalent bonds it still has only six electrons in its outer shell. It thus requires two electrons to complete its octet. Also, the nitrogen atom has a partial negative charge due to electron displacement in the N—H bonds, while the aluminium atom has a partial positive charge due to electron displacement towards the chlorine atoms in the Al—Cl bonds. Hence there is an attraction between the nitrogen atom and the aluminium atom. Combination between ammonia and anhydrous aluminium(III) chloride takes place with evolution of energy.

$$
\begin{array}{cc}
\text{H} & \text{Cl} \\
| & | \\
\text{H—N:} + \text{Al—Cl} \\
| & | \\
\text{H} & \text{Cl}
\end{array}
\;\rightarrow\;
\begin{array}{cc}
\text{H} & \text{Cl} \\
| & | \\
\text{H—N:Al—Cl} \\
| & | \\
\text{H} & \text{Cl}
\end{array}
$$

A bond produced by one atom (or ion) giving a share in a lone pair of electrons to another atom (or ion) is a *co-ordinate covalent*, or *dative*, bond. The atom which gives a share in two of its electrons is the *donor* atom, and the receiving atom the *acceptor* atom.

Other methods of representing the bond between the nitrogen atom and the aluminium atom in the addition compound of ammonia and anhydrous aluminium(III) chloride are as follows:

$$
\underset{\text{(I)}}{H_3N \rightarrow AlCl_3} \qquad \underset{\text{(II)}}{\overset{+}{H_3N}—\overset{-}{AlCl_3}}
$$

In the first formula we indicate the bond by an arrow pointing from the donor atom to the acceptor atom. The use of the arrow here is consistent with its use in the formula $H \rightarrow Cl$ (p. 168). In both cases an electron displacement has occurred. In the second formula the pair of shared electrons is represented by a line, but the charges produced on the two atoms are shown. The reason for the charges is as follows. The two electrons forming the new bond may be regarded as belonging to the nitrogen atom for half their time and to the aluminium atom for the other half. Since the nitrogen atom provides both electrons it loses two electrons for half their time, which is equivalent to complete loss of one electron, or a gain of one positive charge. Conversely we can regard the aluminium atom as acquiring one electron, which is equivalent to a gain of one negative charge. Actually the charges on the nitrogen and aluminium atoms do not stay as full charges. Re-adjustment occurs in the electron displacements in the two halves of the new molecule, so that the full charges are reduced to partial charges. For this reason we describe the full charges represented in the formula as *formal* charges.

There is no difference in character between a co-ordinate covalent bond and a single normal covalent bond. Both are formed by two electrons with opposite spins. The only difference lies in the method by which the bond is produced. In normal covalency the bonding electrons are derived from different atoms, while in co-ordinate covalency they are contributed by the same atom. 'Co-ordination' thus describes a method of forming a bond, and not the nature of the bond itself.

Co-ordination between two covalent molecules often results in the formation of ions. Thus when ammonia is mixed with hydrogen chloride an ionic compound, ammonium chloride, forms. The hydrogen atom of the hydrogen chloride carries a partial positive charge, and the attraction between the hydrogen atom and the nitrogen atom (with its partial negative charge) is so strong that the hydrogen atom is detached from the hydrogen chloride molecule. Both electrons in the H—Cl bond are left with the chlorine atom, which thus becomes a negative ion.

$$
\begin{array}{ccc}
\text{H} & & \text{H} \\
| & & | \\
\text{H—N:} + \text{H—Cl} & \rightarrow & \text{H—N}^+\text{—H} + \text{Cl}^- \\
| & & | \\
\text{H} & & \text{H}
\end{array}
$$

Here again the positive charge acquired by the nitrogen atom is only a formal charge. We usually represent it as belonging to the ammonium radical as a whole—thus, $NH_4^+$.

We have seen previously that the two hydrogen atoms of the water molecule carry partial positive charges, while the oxygen atom has a partial negative charge. The oxygen atom also has two lone pairs of electrons. The water molecule can take part in co-ordination reactions either through its hydrogen atoms or its oxygen atom. A hydrogen atom is involved when ammonia is dissolved in water, some ammonium ions and hydroxyl ions being produced (the solution is feebly alkaline).

$$
\begin{array}{cc}
\text{H} & \\
| & \\
\text{H—N:} + \text{H—}\overset{..}{\text{O}}: & \rightleftharpoons NH_4^+ + OH^- \\
| & | \\
\text{H} & \text{H}
\end{array}
$$

When acids are dissolved in water the oxygen atoms of the water molecules behave as electron donors. Thus with hydrogen chloride *oxonium* ions and chloride ions are formed. To make the change clearer in this case we shall include in the equation the partial charges on the atoms of the reacting molecules.

$$
\begin{array}{c}
\overset{\delta+}{\text{H}}\text{—}\overset{\overset{..}{\delta-}}{\text{O}}: + \overset{\delta+}{\text{H}}\text{—}\overset{\delta-}{\text{Cl}} \rightleftharpoons \left[ \text{H—}\overset{..}{\text{O}}\text{—H} \right]^+ + \text{Cl}^- \\
| \qquad\qquad\qquad\qquad | \\
\overset{}{\text{H}}{}^{\delta+} \qquad\qquad\qquad\quad \text{H}
\end{array}
$$

Or,     $$H_2O + HCl \rightleftharpoons H_3O^+ + Cl^-$$

For simplicity we usually write the 'ionization' of an acid in water as a straightforward dissociation of the acid molecule (*e.g.*, $HCl \rightleftharpoons H^+ + Cl^-$). Simple $H^+$ ions are far too reactive, however, to exist alone in water. They are invariably present as hydrated hydrogen ions, or as oxonium ions.

A number of other essentially covalent chlorides besides hydrogen chloride give rise to ions as a result of co-ordination with water molecules. If anhydrous aluminium(III) chloride is dissolved in water the solution gives the reactions of aluminium ions and chloride ions. This is because the aluminium atom has been attracted away as a positive ion ($Al^{3+}$) by the water molecules. Co-ordination occurs with six water molecules to give a 'hydrated' aluminium ion.

$$AlCl_3 + 6H_2O \rightarrow Al(H_2O)_6^{3+} + 3Cl^-$$

Many positive metal ions exert an attraction for the lone-pair electrons of water and ammonia molecules. A metal ion increases the electron displacements in the bonds of the latter by induction, the 'polarizing' effect being greater the larger the charge on the ion. Thus co-ordination occurs very readily between water, or ammonia, molecules and metal ions with double or treble charges. Very often in these reactions four or six molecules of water or ammonia are added on to give a complex ion. Further examples of these complex ions are the yellow hydrated iron(III) ion, $Fe(H_2O)_6^{3+}$, the blue hydrated copper(II) ion, $Cu(H_2O)_4^{2+}$, and the deep blue tetraamminecopper(II) ion, $Cu(NH_3)_4^{2+}$. The structures of the last two complex ions are shown below.

$$\left[ \begin{array}{c} H_2O \\ \downarrow \\ H_2O \rightarrow Cu \leftarrow OH_2 \\ \uparrow \\ H_2O \end{array} \right]^{2+} \qquad \left[ \begin{array}{c} NH_3 \\ \downarrow \\ H_3N \rightarrow Cu \rightarrow NH_3 \\ \uparrow \\ NH_3 \end{array} \right]^{2+}$$

Hydrated copper(II) ion            Tetraamminecopper(II) ion

## Weaker Binding Forces

**Hydrogen Bond.** This is a dipole–dipole attraction which occurs between a hydrogen atom attached to a strongly electronegative atom and a second strongly electronegative atom with a lone pair of electrons. The electronegative atoms are usually fluorine, oxygen, or nitrogen. A hydrogen bond is always a weak one (its energy is only 8–40 kJ per mole), and it is doubtful whether we should regard it as a 'true' valency bond or not.

The strongest hydrogen bonds are in hydrogen fluoride, where they are responsible for 'association' of the simple HF molecules into larger aggregates. Associated molecules usually break down readily under the influence of heat into simple molecules, but in hydrogen fluoride the hydrogen bonds persist even in the vapour state. Gaseous hydrogen fluoride consists mainly of a mixture of $H_2F_2$ and $H_3F_3$ molecules.

In hydrogen fluoride the hydrogen bond is due to the large electron displacement $H \rightarrow F$, the small size of the hydrogen atom, and the presence of lone pairs of electrons in the fluorine atom. The small size of the hydrogen atom enables a lone pair of electrons to approach closely to the hydrogen nucleus and produce an electrostatic attraction. The hydrogen bond is usually represented by a broken line as follows:

$$\overset{\delta+}{H}—\overset{\delta-}{F}----\overset{\delta+}{H}—\overset{\delta-}{F}$$

In ice crystals water molecules are held together by hydrogen bonds of the type O—H----O—H (this type of hydrogen bond is often called a 'hydroxyl' bond). X-ray analysis of ice crystals shows that the oxygen atoms have a tetrahedral configuration as represented in Fig. 7-9, each oxygen atom being attached to two hydrogen atoms by normal covalent bonds and to two others by longer hydrogen bonds. The network arrangement of atoms gives ice an extremely 'open' structure, which causes it to have a low density.

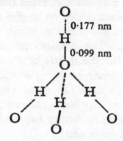

Fig. 7-9. Structure of ice

In liquid water hydrogen bonding results in the molecules being associated into larger aggregates, but these are not definite polymers as in hydrogen fluoride, where the bonds are stronger. Owing to the thermal movement of the molecules the hydrogen bonds in water are constantly being broken and re-formed. The structure of liquid water is discussed later.

Alcohols, amines, and alkanoic acids are all associated in liquid form owing to hydrogen bonding. For the first two we can represent this as in (i) and (ii) below.

$$
\begin{array}{cc}
\quad\overset{\displaystyle R}{\underset{\displaystyle |}{}} \quad \overset{\displaystyle R}{\underset{\displaystyle |}{}} & \quad\overset{\displaystyle R}{\underset{\displaystyle |}{}} \quad \overset{\displaystyle R}{\underset{\displaystyle |}{}} \\
\text{----H—O---H—O---} & \text{----H—N---H—N----} \\
& \underset{\displaystyle |}{H} \quad \underset{\displaystyle |}{H} \\
\text{(i)} & \text{(ii)}
\end{array}
$$

Hydrogen bonds of a special type occur in the lower alkanoic acids. If methanoic acid or ethanoic acid is dissolved in benzene, and the relative molecular mass of the acid is found by the cryoscopic method, the value we obtain is twice that of the simple molecules. In the liquid state or in solution in benzene the acids exist in a dimeric form containing *two* hydrogen bonds. This is illustrated below for methanoic acid.

$$
\begin{array}{ccc}
 & \text{O—H···O} & \\
 \diagup & & \diagdown \\
\text{H—C} & & \text{C—H} \\
 \diagdown & & \diagup \\
 & \text{O···H—O} &
\end{array}
$$

The very small energy (about $8 \text{ kJ mol}^{-1}$) of the N—H---N type of hydrogen bond is thought to be significant in connection with biochemical processes. The building-up of complex protein molecules from amino-acid residues probably depends on this kind of hydrogen bond. Owing to the small change of energy involved, the bond is easily established and just as easily broken. This explains why many biochemical reactions take place at ordinary temperatures.

**Van der Waals Forces.** The strength of dipole–dipole attraction between molecules varies with the nature of the molecules. In hydrogen bonding it is sufficiently strong to bring about association of simple molecules into larger aggregates. In some cases the electric dipoles present are too weak to cause any lasting association at ordinary temperatures, and there is merely an attraction between the molecules. This is true for HCl, HBr, and HI molecules.

Attraction also exists, however, between non-polar molecules such as $H_2$, $N_2$, and $O_2$. It may be explained by reference to the hydrogen molecule. On average the latter is electrically neutral because over a period of time the two bonding electrons are shared equally by the two nuclei. At any instant, however, the two electrons are not midway between the two nuclei, but are more under the influence of one nucleus than the other. Hence one part of the molecule is negatively charged with respect to the other part; that is, the molecule contains a temporary electric dipole. This can induce a similar dipole in an adjacent molecule and so the two molecules attract each other.

The attractive force due to this cause is usually very weak (its energy is only $2–20 \text{ kJ mol}^{-1}$), and it functions only when the molecules are very close together. The size of the force increases, however, with the number of electrons in the molecules, so that it may bring about cohesion of larger molecules into liquids or solids even at ordinary temperatures. Examples are liquid bromine ($Br_2$) and solid iodine ($I_2$). With gases like hydrogen and oxygen, cohesion at ordinary temperatures is prevented by the high thermal energy of the molecules. These gases liquefy and solidify only at low temperatures, when the thermal energy is much reduced. The fact that the monatomic rare gases can be liquefied and solidified shows that an attractive force exists even with single atoms.

The attraction between molecules because of temporary dipoles operates *in addition* to any dipole–dipole attraction which the molecules may have for each other. For convenience we group the two forces together and call them *van der Waals forces* after the Dutchman van der Waals, who first showed that intermolecular attraction was one of the factors responsible for the gas laws not holding at high pressures or low temperatures.

## Shapes of Molecules and Ions

**Directed Covalent Bonds.** Usually a chemical 'bond' is thought of as an attractive force between two particular atoms or ions, the attraction being strong enough to hold the two together. However, in a sodium(I) chloride crystal each sodium ion is attracted by six equidistant chloride ions as well as by the more distant ones, while it is repelled by other sodium ions in its vicinity. We therefore cannot say that a valency bond exists between any two particular ions. In other words an electrovalent bond has no directional character. In contrast a covalent bond is formed

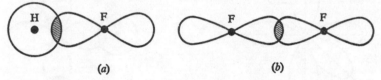

*Fig.* 7–10.   (*a*) The *s-p* bond in the HF molecule; (*b*) The *p-p* bond in the F₂ molecule

between two specific atoms, and we usually represent the direction of the bond by the line joining the two atomic nuclei. A covalent bond thus possesses directional character.

The covalent bond in a hydrogen molecule is formed by the overlapping of the 1*s* atomic orbitals or charge clouds of the two combining atoms. The orbital or charge cloud of an *s* electron is spherical and extends equally in all directions from the centre. It is said to have *spherical symmetry*. Hence overlapping of the atomic orbitals or charge clouds can occur with equal facility in any direction. When the bonding electrons consist of one *s* electron and one *p* electron (as in the hydrogen fluoride molecule) or of two *p* electrons (as in the fluorine molecule) the situation is different. The orbital or charge cloud of a *p* electron has the shape of a figure 8 in three dimensions, and is concentrated in a particular direction (see p. 150). With both the hydrogen fluoride and fluorine molecules overlapping of charge clouds occurs in the direction of maximum electron density—that is, along the axes of the charge clouds of the *p* electrons (Fig. 7–10). This is because the greater the overlapping of the charge clouds (from the density point of view) the greater is the evolution of energy and the stronger is the bond produced.

All quantum shells after the first possess three $p$ orbitals. These are equivalent in energy, but are orientated in space so that their axes are mutually at right angles to each other (Fig. 7–11). Thus the orbitals or charge clouds of $p$ electrons, unlike those of $s$ electrons, have directional character. In the oxygen atom one $p$ orbital is filled by a lone pair of electrons, while each of the other two contains a single electron. In the combination of hydrogen and oxygen two O—H bonds are formed by overlapping of the charge clouds of two $s$ electrons belonging to two hydrogen atoms with the charge clouds of the two single $p$ electrons of the oxygen atom (Fig. 7–12). We see that the shape of the water molecule is triangular and, if no other factors were involved, the angle between the two bonds would be 90°. The angle found experimentally is 105°. The reason for the discrepancy is explained later.

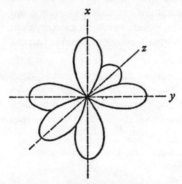

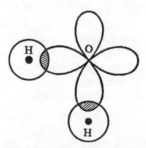

*Fig.* 7–11. Orientation of orbitals of $p$ electrons

*Fig.* 7–12. The two $s$–$p$ bonds in the water molecule

In the nitrogen atom all three $p$ orbitals of the second quantum shell are occupied by single electrons. Combination takes place with three hydrogen atoms by means of three $s$–$p$ bonds, which we should expect to be at right angles to each other. Here again, however, the bond angles found practically (107°) are considerably larger.

**Hybridization of Atomic Orbitals.** We have seen that when an atom is in an 'excited' state through absorption of energy, unpairing of its electrons may occur. If there is an unoccupied orbital in a higher energy sub-level one of the unpaired electrons is 'promoted' to the vacancy and both electrons become available for bond formation.

The *beryllium* atom in its ground state has an electron configuration $1s^2\,2s^2$. When beryllium combines with chlorine the $2s$ electrons become unpaired and one is 'promoted' to one of the three empty orbitals in the $2p$ sub-level. One might expect that the beryllium atom would combine with two chlorine atoms by means of an $s$–$p$ bond and a $p$–$p$ bond. This would result in two bonds of different strengths, a $p$–$p$ bond being stronger than an $s$-$p$ bond because of greater overlapping of the charge clouds. Actually the bonds in beryllium(II) chloride ($BeCl_2$) are equal in strength. This is because of a process known as *hybridization* of atomic orbitals. The wave systems of the two beryllium electrons interact to give two similar orbitals, with axes at 180° to each other. A beryllium(II) chloride molecule is thus linear (Cl–Be–Cl). Since the new orbitals are derived from one $s$ orbital and one $p$ orbital we describe them as '$sp$ hybrid' orbitals.

The advantage of hybridization of atomic orbitals is that the hybrid orbitals are concentrated in particular directions, so that there is greater overlapping of charge clouds, producing stronger bonds. Similar $sp$ hybridization of orbitals occurs in the mercury atom when it combines with two chlorine atoms. Mercury(II) chloride also has a linear molecule (Cl—Hg—Cl).

The *boron* atom has one more electron than the beryllium atom, its electron configuration being $1s^2\,2s^2\,2p$. Two of the three $p$ orbitals in the second quantum shell are unoccupied. In the formation of boron tri-chloride one of the $2s$ electrons is 'promoted' to one of the empty $p$ orbitals, giving two $p$ electrons. Hybridization occurs between the remaining $s$ orbital and the two $p$ orbitals, and three equivalent $sp^2$ hybrid orbitals are produced. The axes of these lie in the same plane at 120° to each other. Thus the boron trichloride molecule is flat and has the shape now shown.

$$Cl$$
$$/$$
$$Cl—B \rangle 120°$$
$$\backslash$$
$$Cl$$

Boron trichloride

The *carbon* atom again has one more electron than the boron atom. Its electron configuration is $1s^2\,2s^2\,2p_x\,2p_y$, the $2p_x$ and $2p_y$ signifying that two different $p$ orbitals in the second quantum shell are occupied by single electrons. 'Excitation' of the carbon atom results in one of the $2s$ electrons being 'promoted' to the empty $2p_z$ orbital, so that now all three $p$ orbitals contain one electron. Hybridization takes place between the $s$ orbital and the three $p$ orbitals, giving four equal $sp^3$ hybrid orbitals. These are directed outwards from the carbon atom so that they form the tetrahedral angle of 109° 28′ with each other. Thus a molecule of methane or tetrachloromethane has the shape of a regular tetrahedron, the carbon atom being at the centre and the hydrogen or chlorine atoms at the four corners—*e.g.*,

$$Cl$$
$$|$$
$$C$$
$$/\ |\ \backslash$$
$$Cl\ \ Cl\ \ Cl$$
$$Cl$$

Tetrachloromethane

In the formation of the ethene molecule ($H_2C{=}CH_2$) which contains a 'double' bond, hybridization of the atomic orbitals of the 'excited' carbon atom takes place differently. After 'promotion' of one of the $2s$ electrons to the $2p$ sub-level the remaining $s$ orbital combines with only two of the $p$ orbitals ($sp^2$ hybridization), so that each carbon atom has three $sp^2$ hybrid orbitals and one unchanged $p$ orbital. As with the boron atom, the axes of the $sp^2$ hybrid orbitals lie in the same plane and are at 120° to each other. Two of the hybrid orbitals are used in bonding with two hydrogen atoms, and the third in establishing a single bond with the other carbon atom. The axis of the remaining $p$ orbital is at right angles to the axes of the hybrid orbitals, and the second bond between the carbon atoms is formed by overlapping of the two $p$ orbitals or charge clouds as shown in Fig. 7–13. Maximum overlapping will occur if the orbitals have the same symmetry with respect to the C—C axis, that is, if all six atomic nuclei lie in the same plane. Overlapping then takes place both above and below the plane.

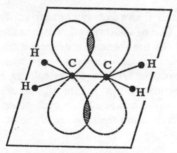

*Fig.* 7–13.  Formation of a $\pi$ bond in the ethene molecule

We see that the $p$—$p$ bond in ethene is different from the $p$—$p$ bond in the fluorine molecule. In the latter overlapping of orbitals occurs in a direct line between the atomic nuclei ('collinear' overlapping). A covalent bond formed by collinear overlapping of atomic orbitals is called a $\sigma$ (*sigma*) *bond*. The H—H bond, the O—H bond, and the four C—H bonds in methane are all $\sigma$ bonds. So also are the bonds formed by the $sp^2$ hybrid orbitals in ethene. The $p$—$p$ bond between the carbon atoms in ethene is not in the line of the atomic nuclei, but is parallel to this line. A bond of this type (formed by 'collateral' overlapping of $p$ orbitals) is known as a $\pi$ *bond*.

The overlapping of charge clouds in a $\pi$ bond is less than in a $\sigma$ bond. As a result less energy is evolved in its formation, and the bond is weaker. Thus the double bond in ethene is not composed of two similar single bonds, but of a relatively strong $\sigma$ bond and a relatively weak $\pi$ bond. The 'bond energy' of the former is 347 kJ mol$^{-1}$, while that of the latter is only 260 kJ mol$^{-1}$. The presence of the weaker bond in the double bond of alkenes is responsible for the high reactivity of these hydrocarbons.

In ethyne ($HC{\equiv}CH$) hybridization of the atomic orbitals of the two carbon atoms is of the $sp$ type. After 'promotion' of one of the $2s$ electrons to the $2p$ sub-level the orbital of the other $s$ electron undergoes hybridization with only one of the three $p$ orbitals. This results in two $sp$ hybrid orbitals with axes at 180° to each other as with the beryllium atom. One of the hybrid orbitals is used for bonding with a hydrogen atom, the other for bonding with the second carbon atom. Thus all four atomic nuclei are in a straight line and the bonds mentioned are $\sigma$ bonds.

The cases of the beryllium atom and carbon atoms in ethyne differ in that the carbon atoms still have two $p$ electrons for forming bonds. The axes of the orbitals of these electrons are at right angles to each other and to the axis of the molecule. Overlapping of the orbitals or charge clouds of the $p$ electrons takes place as in ethene, giving in this case *two* $\pi$ bonds. For clarity only the axes of the overlapping $p$ orbitals are shown in Fig. 7–14, and the directions of the resulting $\pi$ bonds are represented by dotted lines. The triple bond in ethyne thus consists of one $\sigma$ bond and two weaker $\pi$ bonds.

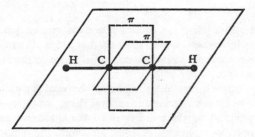

*Fig.* 7–14. The two $\pi$ bonds in the ethyne molecule

**Explanation of Molecular Shape by Repulsion of Orbitals.** In the last two sections we have seen how the directions of the covalent bonds from a central atom can be explained by the *types* of orbitals used in bonding. An alternative way of explaining the bond directions is based on the *number* of orbitals occupied by electron pairs in the outer valency shell of the central atom. The bond directions (and hence the shape of a molecule or ion) are in accordance with three basic principles, the importance of these decreasing in the order now given.

- *The orbitals of the electron pairs tend to become as widely separated as possible.*

The electrical field of one electron pair exerts a strong repulsion on the electrical field of another electron pair, so that the axes of the orbitals tend to form the maximum possible angles with each other. In applying the principle, however, we have to take into account the two kinds of electron pairs—those which constitute a covalent bond between two atoms, and those which are simply lone pairs. This leads us to the second principle.

● *Orbitals of lone-pair electrons exert a bigger repulsion than those of bond-pair electrons.*

This is because the orbital of a lone pair is concentrated closer to the nucleus of the central atom than that of a bonding pair, the orbital of the latter being drawn out to a greater distance by the second nucleus. We can think of the lone pair as occupying a 'fatter' orbital than a bonding pair. Thus the angles between the axes of the orbitals decrease in the following order: lone pair–lone pair > lone pair–bond pair > bond pair–bond pair.

● *Repulsion between orbitals is increased by increase of electronegativity of the central atom.*

The more electronegative the central atom the greater are the electron displacements in covalent bonds towards that atom. As in the previous principle, the more concentrated electrical field near to the nucleus results in bigger repulsion.

The first principle determines the general shape of the molecule or ion. The last two, while not altering the general shape, cause deviations in its regularity by modifying the bond angles. We shall now see how these principles apply to molecules or ions, in which the central atom has different numbers of occupied orbitals in its outer valency shell. We shall first consider cases in which only single covalent bonds are present.

*Two Occupied Orbitals.* The maximum angle for the axes of only two occupied orbitals is 180°. Hence the molecule is linear, as in beryllium(II) chloride and mercury(II) chloride.

*Three Occupied Orbitals.* The maximum angle between three bonds is 120°, for which the axes of the three orbitals must lie in the same plane. The boron atom in boron trichloride ($BCl_3$) forms single bonds with three chlorine atoms and there are no lone pairs of electrons in its valency shell. Hence the molecule is 'flat,' and has the shape shown at p. 179.

*Four Occupied Orbitals.* This is a common case since four occupied orbitals give a rare-gas pattern of electrons for the outer shell. In the molecules of methane ($CH_4$) and silicon tetrachloride ($SiCl_4$) and in the ammonium ion ($NH_4^+$) the central atom is joined to four similar atoms and has no external lone pairs of electrons. The molecule therefore has the shape of a regular tetrahedron with the four similar atoms at the corners, the inter-bond angles having the tetrahedral value of 109° 28'.

In the ammonia molecule ($NH_3$) the nitrogen atom has three bonding pairs of electrons and one lone pair. The axes of the orbitals again have a tetrahedral distribution, but owing to the greater space occupied by the lone pair the bonding pairs are forced closer together than in methane, the H—N—H angle being reduced to 107°. As shown below, the molecule

has the shape of a triangular pyramid with the nitrogen atom at the apex
and the three hydrogen atoms at the corners.

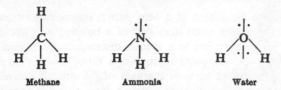

| Methane | Ammonia | Water |

In the water molecule the oxygen atom has two bonding pairs of
electrons and *two* lone pairs. The two bonding pairs are brought still
closer together, the H—O—H angle being only 105°. The molecule is
V-shaped, as shown above.

The effect of difference in electronegativity of the central atom can be
seen by comparing the bond angles in the series of similar molecules
$NH_3$, $PH_3$, and $AsH_3$, or in $H_2O$ and $H_2S$. In the first three molecules the
central atom has three bond pairs and one lone pair of electrons, while
in the last two there are two bond pairs and two lone pairs. In both
cases the electronegativity of the central atom decreases in the order
shown, and we find a corresponding decrease in the bond angles, as
indicated below.

| $NH_3$ | $PH_3$ | $AsH_3$ |
|--------|--------|---------|
| 107° | 94° | 92° |

| $H_2O$ | $H_2S$ |
|--------|--------|
| 105° | 92° |

An exception to the tetrahedral distribution of the four orbital axes of
the central atom or ion is found in the tetraamminecopper(II) ion,
$Cu(NH_3)_4^{2+}$. The bonds from the ion to the four nitrogen atoms lie in the
same plane, as represented in the adjoining diagram. This is the result of a

$$H_3N \begin{array}{|c|} \hline \\ Cu^{2+} \\ \\ \hline \end{array} NH_3$$

special type of hybridization of orbitals, the square planar structure giving
stronger bonds than the tetrahedral structure. (Hybridization occurs between one $d$, one $s$, and two $p$ orbitals, producing four $dsp^2$ hybrid orbitals.)

*Five or Six Occupied Orbitals.* In the vapour of phosphorus pentachloride ($PCl_5$) the phosphorus atom has five bonding pairs of electrons.

The five chlorine atoms are at the five corners of a trigonal pyramid with the phosphorus atom at the centre. A molecule of pentacarbonyliron(0), Fe(CO)₅, has a similar shape.

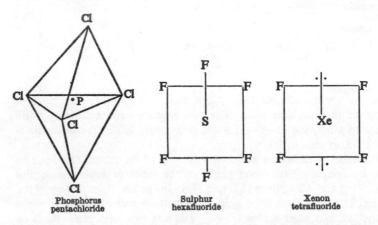

Phosphorus
pentachloride

Sulphur
hexafluoride

Xenon
tetrafluoride

The sulphur atom of sulphur hexafluoride (SF₆) has six orbitals occupied by bonding pairs of electrons. The shape of the molecule is octahedral, the sulphur atom being at the centre and the fluorine atoms at the six corners. For simplicity the octahedral structure is usually represented as shown above. This type of structure is again very common, the twelve external electrons forming a stable group. Other examples are seen in the hexacyanoferrate(II) ion, $Fe(CN)_6^{4-}$, the hexaaquaaluminium(III) ion, $Al(H_2O)_6^{3+}$, and the hexaamminecobalt(III) ion, $Co(NH_3)_6^{3+}$.

In xenon tetrafluoride (XeF₄) the xenon atom has four bonding pairs of electrons and two lone pairs. The shape of the molecule is square planar, the four fluorine atoms being at the corners of a square and the xenon atom in the centre. The axes of the lone pair orbitals are at right angles to the square.

**Molecules and Ions with Multiple Bonds.** The general principles already given apply to molecules and ions containing double and triple bonds, providing we regard these as equivalent to single bonds. There is a great deal of uncertainty concerning the multiple character of the bonds in many cases. The multiplicity of bonds is closely connected with their length, and frequently the lengths do not show clearly whether a bond is single, double, or triple. For example, the lengths of the sulphur–oxygen bonds in SO₂, SO₃, and SO₄²⁻ are all about the same. Many of these doubtful bonds are explained by the theory of *resonance*, according to which bonds can be intermediate in character between single and double or between double and triple.

The multiplicity of the bond does not affect the shape of the molecule or ion. This depends only on the number of atoms attached to the central atom and on any lone pairs of electrons possessed by the latter. This can be seen from the examples now given, in which classical structural formulæ are used.

*Linear*

H—C≡N
Hydrogen
cyanide

O=C=O
Carbon
dioxide

H—C≡C—H
Ethyne

*V-shaped*

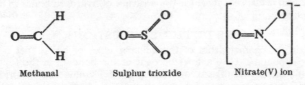

Sulphur dioxide

Ethene

(In ethylene the two halves of the molecule are V-shaped.)

*Triangular and planar*

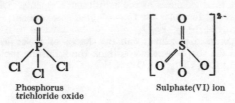

Methanal

Sulphur trioxide

Nitrate(V) ion

The carbonate ion ($CO_3^{2-}$) also comes in this group.

*Tetrahedral*

Phosphorus
trichloride oxide

Sulphate(VI) ion

## EXERCISE 7

1. The atomic nuclei of atoms X and Y contain the following: X, 7 neutrons and 7 protons; Y, 14 neutrons and 12 protons. Write down (*a*) the mass numbers of X and Y, (*b*) the atomic numbers of X and Y, (*c*) the electronic structure of (*i*) a hydride of X, (*ii*) an oxide of Y. Describe the reaction between the hydride of X and hydrochloric acid in terms of electronic structures.

How may the differences in physical properties of sodium(I) chloride and tetrachloromethane be explained in terms of their electronic structures? (C.L.)

2. Explain concisely what is meant by the terms: electrovalent bond, covalent bond, co-ordinate bond. Compare and contrast the physical properties usually displayed by compounds containing electrovalent and covalent bonds.

Write electronic formulae for the following compounds: (i) magnesium(II) chloride, (ii) ethene, (iii) ammonium chloride, and comment on the types of bonds involved.                                                                    (Lond.)

3. Name *two* general properties of covalent compounds and *two* general properties of electrovalent compounds. Explain carefully, using tetrachloromethane as an example, the nature of the covalent link.

Classify the following substances as electrovalent or covalent and give in each case their electronic structures: (*a*) hydrogen chloride, (*b*) calcium(II) sulphide, (*c*) magnesium(II) chloride, (*d*) ammonia.                                        (J.M.B.)

4. Write electronic formulae for any *six* of the following compounds and state the type (or types) of valence encountered in each compound you select. (N.B. Valence electrons only need to be shown and distinctive symbols should be used to identify electrons supplied by different elements appearing in the same formula): (*a*) chloroethane; (*b*) methanoic acid; (*c*) hydrazine ($N_2H_4$); (*d*) methylammonium chloride; (*e*) sulphuric(VI) acid; (*f*) sodium(I) sulphate(IV); (*g*) phosphorus trichloride oxide; (*h*) nitric(V) acid; (*i*) sodium(I) chlorate(V); (*j*) aluminium(III) chloride.                                                          (W.J.E.C.)

5. Why are the bonds between dissimilar atoms never 100 per cent electrovalent or covalent? Explain with reasons the kind of bonds you would expect to be formed between atoms of (i) nitrogen and iodine, (ii) aluminium and fluorine, (iii) boron and chlorine.

6. What is meant by *hydrogen bonding*? Explain how this type of bonding arises in (i) hydrogen fluoride, (ii) ice, (iii) methylamine, (iv) ethanoic acid. What practical evidence is there for the existence of hydrogen bonds in ice and ethanoic acid?

## MORE DIFFICULT QUESTIONS

7. Explain the characteristics of the different types of bonds that occur in chemical compounds. Give a brief account of the bonding in the following: (*a*) tetrachloromethane, (*b*) carbon (diamond), (*c*) ammonium chloride, (*d*) potassium(I) hexacyanoferrate(II), (*e*) ethanoic acid dissolved in benzene.                                                                                 (C.L.)

8. The following compounds have formulae which do not appear to be consistent with the normal valancies of the constituent elements. Discuss possible electronic or structural formulae for four of these compounds: $H_2O_2$; $KHF_2$; $KI_3$; $I_2O_5$; CO; NO; $NO_2$; $Fe_3O_4$.                                (O. and C.)

9. By what theoretical principles can the shapes of molecules and ions be deduced? Sketch the shapes which you think the following molecules or ions should have: $PCl_3$, $CS_2$, $ClO_3^-$, $H_3O^+$, $C_6H_6$.

# CHAPTER 8

# The periodic table

**History.** The earliest attempt to show a connection between the atoms of different elements was made by an Englishman called Prout in 1815, only seven years after the publication of the atomic theory. Prout noticed that many of the 'atomic weights' determined by Dalton, Berzelius, and others were approximately whole-number multiples of that of hydrogen, and this led him to suggest that atoms of different elements were composed of hydrogen atoms in various numbers. The exceptions to Prout's 'whole-number rule' soon proved so numerous, however, that this attractive hypothesis had to be abandoned. Today we realize that it had a substantial basis of truth. Since the mass of an atom is approximately proportional to the number of protons and neutrons in its nucleus and since a hydrogen nucleus consists of one proton, the rule holds approximately for particular isotopes.

Following the introduction in 1858 of Cannizzaro's method of finding relative atomic masses, chemists soon observed a periodic relationship between the properties of elements and their relative atomic masses. In 1864 Newlands showed that if the elements were arranged in order of their 'atomic weights', elements with similar properties appeared at regular intervals. Part of Newlands' table is shown below.

| H | Li | Be | B | C | N | O |
|---|----|----|----|-----|----|----|
| F | Na | Mg | Al | Si | P | S |
| Cl | K | Ca | Cr | Ti | Mn | Fe |

Chemically similar elements such as lithium, sodium, and potassium usually occurred in every eighth position. This led Newlands to put forward the following *periodic law*: *The properties of the elements are a periodic function of their 'atomic weights'*. Newlands' 'law' was badly received. Some elements had been assigned incorrect relative atomic masses and were wrongly placed in the table, and also no allowance was made for the possibility of undiscovered elements. Hence the periodicity of properties was not sufficiently comprehensive to convince the majority of chemists of the truth of the law.

In 1869 Newlands' basic idea was incorporated in a new form of table devised by a Russian, Mendeléeff. He arranged all the known elements in order of their relative atomic masses in such a way as to show the relationships between the elements more clearly. He did this by improving on Newlands' classification in two ways. He left spaces to be filled by undiscovered elements, and he placed triads of elements in a special group (Group VIII) after Newlands' third, fifth, seventh, and ninth rows. His table thus consisted of eight vertical groups divided horizontally into two 'short periods' of 7 elements and 'long periods' containing $7 + 3 + 7$ ($= 17$) elements. After the discovery of the rare-gas elements towards the end of the century another group (Group 0) was added to Mendeléeff's table to accommodate these elements. This brought the number of elements in the short periods to 8 and the number in the long periods to 18. Subsequently the third and fourth long periods were combined to give one 'very long' period of 32 elements by including all the 'lanthanide,' or rare-earth, elements in one position. Mendeléeff's form of the Periodic Table (with the later modifications) is shown in Table 8–1.

Table 8–1. SHORT FORM OF THE PERIODIC TABLE (after Mendeléeff)

| Group 0 | Group I | Group II | Group III | Group IV | Group V | Group VI | Group VII | Group VIII |
|---|---|---|---|---|---|---|---|---|
| | H | | | | | | | |
| He | Li | Be | B | C | N | O | (H) | |
| Ne | Na | Mg | Al | Si | P | S | F | |
| Ar | K | Ca | Sc | Ti | V | Cr | Cl | Fe Co Ni |
| | Cu | Zn | Ga | Ge | As | Se | Mn | |
| Kr | Rb | Sr | Y | Zr | Nb | Mo | Br | Ru Rh Pd |
| | Ag | Cd | In | Sn | Sb | Te | Tc | |
| Xe | Cs | Ba | La, etc. | Hf | Ta | W | I | Os Ir Pt |
| | Au | Hg | Tl | Pb | Bi | Po | Re | |
| Rn | Fr | Ra | Ac | Th | Pa | U | At | |

The usefulness of Mendeléeff's classification soon became apparent. First, it summarized in concise form a considerable amount of information about the elements. Chemically similar elements such as the alkali metals or the halogens appeared in corresponding groups, and there was a gradation in the properties of the elements and their compounds both in the periods and in the groups. Thus across the periods there was a decrease in electropositive, or metallic, character of the elements and an increase in electronegative, or non-metallic, character. Down the groups the elements changed in the opposite manner, becoming more electropositive or less electronegative with increasing relative atomic mass.

One immediate use of the Periodic Table was in checking relative atomic masses. Although many elements had more than one valency, particular valencies were associated with particular groups. If the valencies of the elements in Groups I to VII were deduced from the highest oxides the values varied from one to seven, as now shown.

$$Na_2O \quad CaO \quad Al_2O_3 \quad SiO_2 \quad P_2O_5 \quad SO_3 \quad Cl_2O_7$$

If, however, the valency were defined with respect to hydrogen atoms (or equivalent organic radicals) the valencies varied from one to four, as illustrated by the following formulæ:

$$NaH \quad CaH_2 \quad Al(CH_3)_3 \quad SiH_4 \quad PH_3 \quad H_2S \quad HCl$$

By inspection it was possible to see where an element would fit most suitably into the Periodic Table. The valency and correct multiple of the equivalent mass to adopt as relative atomic mass could then be deduced. This method was used to correct several doubtful relative atomic masses, including that of beryllium (an exception to Dulong and Petit's law).

Another use of the Periodic Table was to predict the existence of undiscovered elements, for which gaps had been left in the Table. Many of the properties of the missing elements could be inferred from their positions in the Table. Thus Mendeléeff predicted the properties of gallium, scandium, and germanium before these elements were found. They were discovered a few years later, and the close agreement between the predicted and the observed properties provided dramatic confirmation of the periodic law. Again, after the discovery of helium and argon in the 1890s it was thought probable that other rare gases existed which would fall in Group 0. By examining the constituents of liquid air, Ramsay revealed the presence of neon, krypton, and xenon.

An outstanding defect of the Mendeléeff classification was that in three cases pairs of elements had to be included in inverse order of their relative atomic masses so as to maintain correct relationships between the elements. These pairs were argon (39·9) and potassium (39·1), cobalt (58·9) and nickel (58·7), and tellurium (127·5) and iodine (126·9). This difficulty was resolved only in modern times, when the basis of classification was changed from relative atomic masses to atomic numbers.

**Classification of Elements by Atomic Number.** The periodicity in the properties of the elements when arranged in order of their relative atomic masses was too striking to be explained by chance. It indicated the existence of some fundamental pattern, or orderly arrangement, in the atoms of the elements, but the nature of the pattern remained a mystery until the electrical structures of atoms were discovered. It was then clear that the properties of the elements vary regularly, not with their relative atomic *masses,* but with their atomic *numbers,* which for the most part run parallel with relative atomic masses. The same element, however, can have different relative atomic masses in different isotopes, whereas the atomic number is characteristic for each element. As we saw in the last chapter, the properties of an element depend both on the number and arrangement of the electrons in its atoms, so that periodicity of properties must be associated with periodicity of electronic structure. Put in its modern form, the periodic law reads as follows:

*The properties of the elements are a periodic function of their atomic numbers.*

The change in the basis of classification not only provided a theoretical background for the Periodic Table, but also clarified several doubtful points. Atomic numbers justified the placing of argon and potassium, cobalt and nickel, and tellurium and iodine in inverse order of their relative atomic masses. The prediction of undiscovered elements was placed on a surer footing because the atomic numbers of missing elements were known beforehand. Thus it could be seen that four elements between hydrogen and uranium remained to be discovered. In the 1930s all four elements were prepared artificially (in very small amounts) from other elements. The reason why they had never been isolated from natural sources then became clear. They were all radioactive and had very short half-lives. These elements were: technetium (43), Tc; promethium (61), Pm; astatine (85), At; and francium (87), Fr.

The relationship of the elements in the middle of the long periods was also clarified. In these elements the relationships are horizontal, rather than vertical. Also, the inclusion of the lanthanides in one place in the Periodic Table could now be justified on theoretical, as well as on practical, grounds. These points are discussed at pp. 204 and 207.

**Modern 'Long' Form of the Periodic Table.** The 'short' form of the Periodic Table given in Table 8–1 has now been superseded by the 'long' form shown in Table 8–2. This is derived directly from the electronic configurations of the atoms, and thus shows the relationships between the elements in the clearest possible manner.

We have seen how the electronic structures of the first ten elements in the Table could be built up theoretically by starting with the hydrogen atom and 'feeding in' electrons one by one. In this process we used certain principles, the most important being that only two electrons can occupy one orbital and that the electron added each time goes into the lowest available energy level and sub-level. Thus in hydrogen and helium the $1s$ level is filled, and from lithium to neon the $2s$ and $2p$ sub-levels are completed in turn by addition of a further eight electrons.

The electron configurations of the elements which follow neon can be derived by continuing the process of adding electrons in accordance with the principles used previously. In the third quantum shell there are $s$, $p$, and $d$ sub-levels. The $3s$ sub-level is first filled in (Na and Mg) and then the $3p$ sub-level is completed (Al–Ar). Note that in argon (electron configuration $1s^2 2s^2 2p^6 3s^2 3p^6$) only the $s$ and $p$ sub-levels of the third quantum shell are full, and the special stability of the rare-gas pattern of

## Table 8–2. PERIODIC TABLE OF THE ELEMENTS (SHOWING ATOMIC NUMBERS)

GROUPS

←————— NON-METALS —————→

| | IA | IIA | IIIA | IVA | VA | VIA | VIIA | VIII | | | IB | IIB | IIIB | IVB | VB | VIB | VIIB | O |
|---|---|---|---|---|---|---|---|---|---|---|---|---|---|---|---|---|---|---|
| Period 1 | H 1 | | | | | | | | | | | | | | | | (H 1) | He 2 |
| Period 2 | Li 3 | Be 4 | | | | | | | | | | | B 5 | C 6 | N 7 | O 8 | F 9 | Ne 10 |
| Period 3 | Na 11 | Mg 12 | | | | | | | | | | | Al 13 | Si 14 | P 15 | S 16 | Cl 17 | Ar 18 |
| Period 4 | K 19 | Ca 20 | Sc 21 | Ti 22 | V 23 | Cr 24 | Mn 25 | Fe 26 | Co 27 | Ni 28 | Cu 29 | Zn 30 | Ga 31 | Ge 32 | As 33 | Se 34 | Br 35 | Kr 36 |
| Period 5 | Rb 37 | Sr 38 | Y 39 | Zr 40 | Nb 41 | Mo 42 | Tc 43 | Ru 44 | Rh 45 | Pd 46 | Ag 47 | Cd 48 | In 49 | Sn 50 | Sb 51 | Te 52 | I 53 | Xe 54 |
| Period 6 | Cs 55 | Ba 56 | La* 57 | Hf 72 | Ta 73 | W 74 | Re 75 | Os 76 | Ir 77 | Pt 78 | Au 79 | Hg 80 | Tl 81 | Pb 82 | Bi 83 | Po 84 | At 85 | Rn 86 |
| Period 7 | Fr 87 | Ra 88 | Ac† 89 | | | | | | | | | | | | | | | |

←—— TRANSITION ELEMENTS ——→

INNER TRANSITION ELEMENTS

| * LANTHANIDES | Ce 58 | Pr 59 | Nd 60 | Pm 61 | Sm 62 | Eu 63 | Gd 64 | Tb 65 | Dy 66 | Ho 67 | Er 68 | Tm 69 | Yb 70 | Lu 71 |
|---|---|---|---|---|---|---|---|---|---|---|---|---|---|---|
| † ACTINIDES | Th 90 | Pa 91 | U 92 | Np 93 | Pu 94 | Am 95 | Cm 96 | Bk 97 | Cf 98 | Es 99 | Fm 100 | Md 101 | No 102 | Lw 103 |

electrons does not coincide with the completion of a quantum shell. Indeed, there is no rare gas corresponding to the completion of quantum shells after the first two.

The fourth period of the Table presents an important departure from regularity in the order of the energy sub-levels: the sub-levels overlap to some extent. This is due to the increase in energy of $s$, $p$, $d$, and $f$ sub-levels in the same quantum shell. In the third and fourth shells the sub-levels increase in energy in the following order: $3s < 3p < 4s < 3d < 4p$,

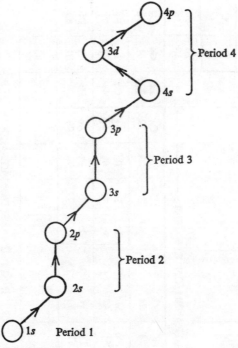

*Fig.* 8–1. Order of filling of energy sub-levels in the first four periods of the periodic table

etc. Hence, after the filling-in of the $3p$ sub-level in argon (Fig. 8–1), the next two electrons go into the $4s$ sub-level (in K and Ca), and not into the $3d$ sub-level. However, with the completion of the $4s$ sub-level subsequent additions are made in the $3d$ sub-level. Table 6–2 reveals that the filling-in of the five orbitals of the $3d$ sub-level starts with scandium (atomic number 21) and finishes with copper (atomic number 29). After completion of the $4s$ and $3d$ sub-levels further electrons fill the $4p$ sub-level.

The elements from scandium to copper inclusive, in which the electron added each time goes into the penultimate shell, are called *transition elements*. They are discussed more fully later (p. 204). Further series of transition elements occur in the second, third, and fourth long periods. Here again there is overlapping of energy sub-levels of different quantum shells, so that the sub-levels are filled in irregularly by electrons. The order of filling-in is shown in Table 6–2. Each of the last two series of transition elements contains an inner group in which the extra electron is added, not to the penultimate shell, but to the one below this. These elements (the *lanthanides* and *actinides*) are called *inner* transition elements.

## Periodicity of some Fundamental Properties

**Atomic Radii.** When the atomic radii of elements in successive periods of the Periodic Table are plotted against the atomic numbers (which represent the nuclear charges) the series of curves shown in Fig. 8–2 is obtained. One significant fact at once emerges from consideration of these curves. An atom containing only a few electrons is not necessarily smaller than one containing a large number. Compare the atomic radius of lithium (atomic number 3) with that of iodine (atomic number 53): clearly, in the iodine atom the electrons must be more closely packed than in the lithium atom.

The radius of an atom is determined chiefly by two factors. One is the attraction of the positively charged nucleus for the electrons and the other is the 'screening' of outer electrons from the nucleus by those in the inner shells. We have seen that an atom can be regarded as formed from the preceding one by adding an extra proton (and possibly neutrons) to the nucleus and an extra electron to the quantum shells. The increased nuclear charge resulting from addition of the proton tends to draw *all* the electrons closer to the nucleus and thus decrease the size of the atom. The screening effect arises from the mutual repulsion between the electrons in inner shells and those in outer shells. The latter are removed to a greater distance from the nucleus, so that they are not attracted as strongly by its positive charge. The screening effect thus tends to increase the atomic radius.

In any given *period* of the Periodic Table after the first there is a decrease in atomic radius from the alkali metal in Group I to the halogen element in Group VII. This contraction is explained by the increase in nuclear charge, the addition of the extra electron to the outer quantum shell making no appreciable difference to the screening effect of the inner shells. In Period 4 and subsequent periods the steady decrease in the size of the atom with increase in nuclear charge is interrupted by the transition elements. For a given period these have about the same atomic radius. This is because the effect of the increased nuclear charge is roughly balanced by the greater screening effect produced by adding an extra electron to the *penultimate* shell.

In a family of related elements in the same *group* there is a progressive increase in atomic radius. We see this clearly with the alkali metals, the alkaline-earth metals, and the halogens. Successive members of these

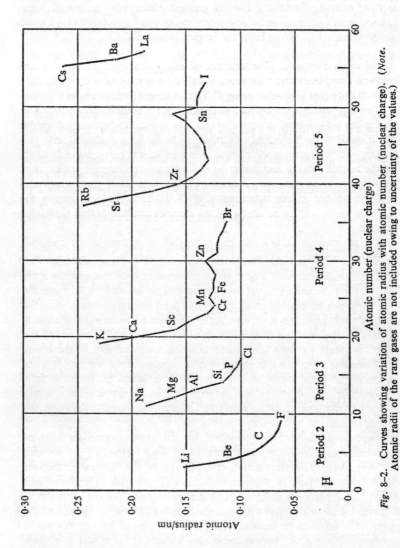

*Fig.* 8–2. Curves showing variation of atomic radius with atomic number (nuclear charge). (*Note.* Atomic radii of the rare gases are not included owing to uncertainty of the values.)

families contain an extra inner shell of electrons, and the additional screening effect due to the extra shell outweighs the effect of the increased nuclear charge. Screening has its greatest effect with the alkali metal atoms, which contain only one electron in their outer quantum shell. These elements therefore have the largest atomic radii.

**Ionization Energies.** The amount of energy required to remove one electron completely from an atom is called its *ionization energy* or, more correctly, its *first ionization energy*. The manner in which the first ionization energy varies with nuclear charge is shown graphically in Fig. 8–3. In a given *period* there is a general increase in ionization energy from the alkali metal at the beginning of the period to the rare gas at the end. We can explain this increase as follows. The ionization energy depends partly on the nuclear charge and partly on the distance between the nucleus and the electron to be removed. Across a period the nuclear charge increases and the atomic radius decreases (Fig. 8–2), both factors causing the electron to be held more firmly by the nucleus and increasing ionization energy.

The increase in ionization energy across a period is not uniform, however. In Period 2 the general trend is reversed in passing from beryllium to boron and from nitrogen to oxygen. Similar reversals occur for corresponding elements in the later periods, so that even the irregularities are periodic. The reason for the reversals is the unexpectedly high ionization energy of the first element, and not the low ionization energy of the second. A completed *s* sub-level of electrons (as in beryllium and magnesium) has rather special stability and more energy is required to remove one of the two paired *s* electrons than a single *p* electron (present in boron and aluminium). The extra stability of an *s* pair is also reflected in the relatively high ionization energies of zinc, cadmium, and mercury. Similar additional stability is associated with the half-completed *p* sub-level in nitrogen and phosphorus, so that the latter have higher ionization energies than the elements which follow them.

In the fourth period (and in later ones) the general increase in ionization energies is again interrupted by the transition elements. Although the nuclear charge in these elements increases progressively, the effect of this increase is roughly balanced by the extra screening effect caused by adding an electron to the inner shell. Hence the ionization energies of the transition elements are about the same.

For a family of elements in the same *group* the ionization energies decrease with increasing nuclear charge. This again is illustrated by the alkali metals, the alkaline-earth metals, and the halogens, as well as by the rare gases. In successive members of these families the electron to be removed is separated from the nucleus by an additional shell of electrons, and with greater distance of the electron from the nucleus less energy is required to detach it in spite of the higher charge on the nucleus.

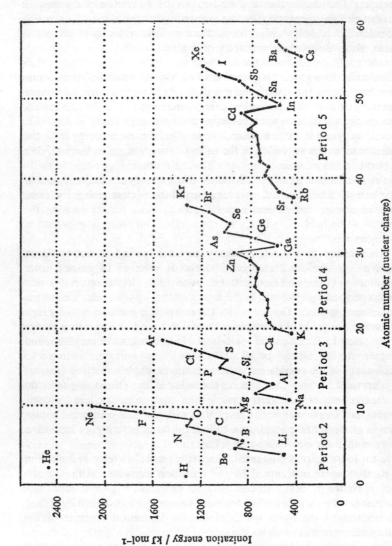

*Fig.* 8–3. Curves showing variation of ionization energy with atomic number (nuclear charge)

The properties of the elements are closely linked to their ionization energies. The magnitudes of these energies are a measure of the metallic character of the elements. The first ionization energies of metals are nearly all below 800 kJ mol$^{-1}$, while those of non-metals are nearly all above this value. According to modern theory of the metallic state, metals owe their characteristic properties to the readiness with which they contribute their valency electrons to form a general electron 'cloud,' which holds the atoms together. Down the groups the members of the various families become more metallic as the ionization energy decreases. Thus in the middle groups there is a change in character of the elements from non-metal to metal. Across the periods from left to right the elements become less metallic with increase in ionization energy. The result of these two trends is that non-metals (numbering only about 20) are confined mostly to the top right-hand corner of the Periodic Table. As noted previously, ionization energies are one of the most important factors in determining whether ionic or covalent bonds exist between atoms.

**Electronegativities.** In Chapter 7 we saw that the electronegativity of an element expresses on an arbitrary scale the relative attraction of one of its atoms for the electrons of a single covalent bond formed with an atom of another element. Table 7–2 reveals how electronegativities vary in the periods and the groups. As with ionization energies, the electronegativities increase from left to right across the periods, but decrease down the groups. The variations in ionization energy and electronegativity show similar trends because they depend essentially on the same factors, namely, nuclear charge and atomic radius. The higher the charge on a nucleus the more strongly is the electron pair of a covalent bond attracted by that nucleus. Also, the smaller the radius of the atom the greater the attraction because the electron pair is then closer to the nucleus. Electronegativities increase across a period owing to the increase in nuclear charge and simultaneous decrease in atomic radius. In a group electronegativities decrease because the effect of the larger atomic radius is greater than that of the increased nuclear charge.

As we have seen earlier, differences in the electronegativities of atoms give ionic character to covalent bonds, or produce electric dipoles in molecules. Physical and chemical properties are profoundly affected. Thus electric dipoles in water molecules are responsible for most of the outstanding properties of water, *e.g.*, its relatively high boiling point and surface tension, its action as a solvent, and its readiness to undergo co-ordination.

## Variation of Properties in the Periods

In studying the way in which the properties of the elements vary in the periods it is best to think of the transition elements as completely removed from the Periodic Table. As we saw in the last section, these elements interrupt the general trends in a period owing to the peculiar constitution of their atoms. We shall consider transition elements separately. Zinc, cadmium, and mercury (Group IIB) are also a special case. They are not transition elements, but they are more closely related to the latter than to the alkaline-earth metals in Group IIA. If we also disregard these three elements the remaining elements in successive periods are electronically similar as regards their outer quantum shells, and we have logical grounds for expecting general trends in one period to be repeated in the other periods. We shall illustrate the general trends by means of the third period, which can be regarded as typical.

As will be shown, gradation in properties is not confined to the elements themselves, but is found also in their compounds. Note, however, that the Periodic Table is not a rigid classification, and exceptions to the general trends in properties occur both in the elements and their compounds. Discovering the reasons for these exceptions is one of the chief points of interest in studying the Table.

### Variation of Properties in the Third Period (Na–Ar).

*Melting Point.* The melting points of the elements are a measure of the amount of energy which must be supplied to break down the regular arrangement of atoms or molecules in the crystal. They therefore indicate the strength of the forces holding the atoms or molecules together in the crystal. As the strength of these forces varies according to the type of crystal formed the melting points of the elements in a period do not change uniformly.

The third period starts with the alkali metal sodium, the atoms of which contain one loosely bound electron in the outer quantum shell. The electron is readily contributed to the electron cloud of a metallic crystal. Since, however, each atom can provide only one valency electron, the binding force in the crystal is weak, so that sodium has a relatively low melting point. The next element, magnesium, can contribute two electrons per atom to an electron cloud, thereby increasing the strength of the metallic bond. Thus we find a large increase in melting point from sodium to magnesium. The sharp rise in melting point is not maintained, however, in the third element, aluminium. Although the latter has three electrons in its outer shell its melting point is little above that of magnesium. This may be because only two of the three electrons are used in forming an electron cloud.

Silicon, which follows aluminium, is usually described as a non-metal. Nevertheless, it has certain metallic properties, such as its lustrous grey appearance, its appreciable electrical conductivity, and its ability to alloy with metals. These properties have not yet been satisfactorily explained. Silicon uses its four valency electrons to form an infinite three-dimensional assembly of atoms joined by single covalent bonds. These have a tetrahedral distribution similar to that of the bonds between carbon atoms in

Table 8-3. VARIATION IN PROPERTIES OF THIRD-PERIOD ELEMENTS AND THEIR COMPOUNDS

| | Na | Mg | Al | Si | P | S | Cl |
|---|---|---|---|---|---|---|---|
| Quantum shells | 2, 8, 1 | 2, 8, 2 | 2, 8, 3 | 2, 8, 4 | 2, 8, 5 | 2, 8, 6 | 2, 8, 7 |
| Atomic radius/nm | 0·186 | 0·160 | 0·143 | 0·117 | 0·110 | 0·104 | 0·100 |
| Melting point/°C | 98 | 650 | 660 | 1423 | 44 (yellow) | 120 | −101 |
| Density of solid/g cm$^3$ | 0·97 | 1·74 | 2·70 | 2·40 | 1·82 | 2·07 | 1·56 |
| Atomic volume/cm$^3$ | 23·7 | 14·0 | 10·0 | 11·6 | 17·0 | 15·5 | 22·6 |
| *Hydride* | | | | | | | |
| Bonding | $NaH$ ionic | $MgH_2$ covalent | $AlH_3$ covalent | $SiH_4$ covalent | $PH_3$ covalent | $H_2S$ covalent | $HCl$ covalent |
| Reaction with water | $H_2$ | $H_2$ | $H_2$ | $H_2O(OH^-$ catalyst) | no action | weak acid | strong acid |
| *Chloride* | | | | | | | |
| Bonding | $NaCl$ ionic | $MgCl_2$ ionic | $AlCl_3$ covalent | $SiCl_4$ covalent | $PCl_5$ covalent (vapour) | $(S_2Cl_2)$ (covalent) | — |
| Melting point (°C) | 808 | 714 | 192 | −68 | 160 | (−76) | — |
| Reaction with water | none | none | slight hydrolysis | complete hydrolysis | complete hydrolysis | (complete hydrolysis) | — |
| *Oxide* | | | | | | | |
| Bonding | $Na_2O$ ionic | $MgO$ ionic | $Al_2O_3$ ionic | $SiO_2$ covalent | $P_2O_6$ covalent | $SO_3$ covalent | $Cl_2O_7$ covalent |
| Melting point (°C) | 1193 | 3075 | 2300 | 1728 | 563 | 30 | −91 |
| Character | basic | basic | amphoteric | acidic | acidic | acidic | acidic |
| *Hydroxide* | | | | | | | |
| Character | $NaOH$ strong base | $Mg(OH)_2$ weak base | $Al(OH)_3$ amphoteric | $H_2SiO_3$ weak acid | $H_3PO_4$ weak acid | $H_2SO_4$ strong acid | $HClO_4$ strong acid |

a diamond crystal. The melting point of silicon is very high because melting involves breaking the strong covalent bonds.

Phosphorus and sulphur exist as molecules ($P_4$ and $S_8$), and the only force holding the molecules together in the solid state is the weak van der Waals force of attraction. Thus both elements have relatively low melting points. The van der Waals force increases, however, with the number of atoms in the molecule, so that sulphur melts at a higher temperature than (yellow) phosphorus.

The last two elements of the period are the halogen chlorine and the rare gas argon. The combining capacity of chlorine atoms limits combination to the formation of diatomic molecules ($Cl_2$). As the latter have little attraction for each other the melting point (and boiling point) is low, and the element is a gas at ordinary temperatures. Atoms of the rare gases cannot combine, and the van der Waals attraction between them is even smaller than that between chlorine molecules. Argon melts at $-189°C$.

*Density and Atomic Volume.* The density of a solid element depends on the mass (or masses) of its atoms, their size, and the amount of space between them. We know that the masses of the atoms increase across a period from left to right, while the atomic radius decreases. Therefore, if the atoms of successive elements were packed together with equal compactness, we should expect to find a uniform increase in density. Actually, for elements in the third period the closeness of packing of the atoms varies and hence the densities do not change uniformly.

Instead of comparing the densities of the solid elements it is more instructive to compare their *atomic volumes*, where

$$\text{Atomic volume} = \frac{\text{one mole}}{\text{density in g cm}^{-3}}$$

The atomic volume is not the same as the actual size of the atoms, but includes any space between them in the same way that the volume of a gas is partly the volume of the molecules and partly the volume of the intervening space. Atomic volume represents the volume in cubic centimetres occupied by one mole of an element. This is easily seen from the following:

$$\text{Density} = \text{mass in gram of } 1 \text{ cm}^3$$

$$\frac{1}{\text{Density}} = \text{volume in cm}^3 \text{ of } 1 \text{ g}$$

$$\therefore \quad \frac{\text{One mole}}{\text{Density}} = \text{volume in cm}^3 \text{ of one mole}$$

We have seen previously that one mole of different elements contains the same number of atoms. It follows that the atomic volumes of different elements are volumes occupied by the same number of atoms of the elements. This number is the Avogadro constant ($N_A$), which is equal to $6 \cdot 023 \times 10^{23} \, \text{mol}^{-1}$.

When we compare the atomic volumes of the third period elements we find a decrease with smaller atomic radius as far as aluminium. This is because the atoms of sodium, magnesium, and aluminium are tightly packed, although the method of packing is not quite identical. If we assume that the atoms of the metals are spherical and touching, the atomic volumes should be approximately proportional to the cube of the atomic radius. If this is checked it is found to be true. Thus the atomic volumes of sodium, magnesium, and aluminium are in the ratios $2 \cdot 37 : 1 \cdot 40 : 1$, while the ratios of the cubes of the atomic radii are $2 \cdot 20 : 1 \cdot 40 : 1$.

From aluminium to silicon there is an increase in atomic volume in spite of smaller atomic radius. We can explain this by the more open structure of the silicon crystal, each atom being in contact with only four other atoms instead of with twelve as in aluminium. The atomic volumes of phosphorus, sulphur, and chlorine are all higher than that of silicon. This is because all three elements consist of discrete molecules and the distances between the molecules are relatively large compared with those between the atoms in the molecules. As we should expect, the atomic volume varies in accordance with the number of atoms in the molecule, being largest for chlorine ($Cl_2$) and least for sulphur ($S_8$).

*Hydrides.* The formulæ of the hydrides vary with the number of electrons used by the different atoms in bonding with hydrogen atoms, the number increasing from one to four and then decreasing to one again. There is a striking change in the properties of the hydrides as the elements become more electronegative across the period. Hydrogen has a greater electronegativity than the earlier elements in the period, but a smaller one than the later elements. Hence the electron pair forming a bond X—H shifts progressively away from the hydrogen atom towards X. In sodium(I) hydride the electron pair actually belongs to the hydrogen atom, the hydride being composed of $Na^+$ and $H^-$ ions. The bonds in magnesium(II) hydride are largely ionic in character, but are mainly covalent. Aluminium(III) hydride, $AlH_3$, exists only in ethoxyethane solution, which yields polymers, $(AlH_3)_n$, on evaporation. With water the hydrides of sodium, magnesium, and aluminium yield hydrogen and the metal hydroxide. In general these reactions can be represented as follows:

$$\overset{\delta+}{X}—\overset{\delta-}{H} + \overset{\delta+}{H}(\overset{\delta-}{OH}) \rightarrow H_2 + X^+ + OH^-$$

In silane, $SiH_4$, the partial charges on the atoms are relatively small. Hydrogen is still evolved with water, but only if $OH^-$ ions are present to act as a catalyst. The phosphorus and hydrogen atoms in phosphine, $PH_3$, have the same electronegativity, and therefore partial charges are absent. Phosphine does not react with water.

In the two remaining hydrides, $H_2S$ and HCl, the bonds are again partially ionic, but the partial negative charge has moved from the hydrogen atoms to the sulphur or chlorine atom. With water these two hydrides form acids of increasing strength.

$$\overset{\delta-}{X}\!-\!\overset{\delta+}{H} + H_2O \rightleftharpoons X^- + H_3O^+$$

*Chlorides.* Chlorine is more electronegative than any of the other elements in the third period. Therefore in the chlorides of these elements the chlorine atoms always carry a negative, or partial negative, charge. However, this decreases as the elements become more electronegative across the period.

Sodium(I) chloride consists of $Na^+$ and $Cl^-$ ions, and magnesium(II) chloride (anhydrous) is sufficiently ionic to give ions and undergo electrolysis when fused. Anhydrous aluminium(III) chloride, however, is essentially covalent. When fused it does not conduct a current and is not electrolysed.

All three chlorides consist of infinite assemblies of ions or atoms, but there is a progressive fall in melting point from sodium(I) chloride to aluminium(III) chloride. This reflects the increasing weakness of bonding in the crystal.

From aluminium(III) chloride to silicon tetrachloride the melting point falls further. Solid silicon tetrachloride consists of $SiCl_4$ molecules, and relatively little energy is required to overcome the van der Waals force of attraction between them. In this case the chloride is a liquid at ordinary temperatures. Phosphorus pentachloride has a peculiar constitution. It exists in the vapour state as $PCl_5$ molecules, but at ordinary temperatures it is a white solid consisting of $PCl_4^+$ ions and $PCl_6^-$ ions. The ionic structure is responsible for the melting point being higher than one would expect. Sulphur hexachloride is too unstable to exist, although there is a stable hexafluoride, $SF_6$. The only stable chloride of sulphur at ordinary temperatures is disulphur dichloride, $S_2Cl_2$.

The extent of hydrolysis of the chlorides varies across the period. Sodium(I) chloride is not hydrolysed at all by water. The same is also true of magnesium(II) chloride, although the hydrated crystals undergo hydrolysis when heated, giving off hydrogen chloride and leaving a basic salt. A solution of (hydrated) aluminium(III) chloride is appreciably hydrolysed and turns blue litmus paper red. The chlorides of silicon, phosphorus, and sulphur hydrolyse completely in water.

*Oxides.* Oxygen is more electronegative again than chlorine. We should therefore expect more ionic character to be present in the bonds of the oxides than in those of the corresponding chlorides. Thus the oxides of sodium, magnesium, and aluminium are all essentially ionic compounds, and their very high melting points show the strength of the ionic bonding in the crystals.

In silicon(IV) oxide the bonds are mainly covalent, but contain a large amount (about 37 per cent) of ionic character. The formula $SiO_2$ is misleading. The crystal is not composed of molecules, but is an infinite three-dimensional assembly of silicon and oxygen atoms. Each silicon atom is joined to four oxygen atoms, and each oxygen atom to two silicon atoms, by single bonds, the bonds from the silicon atoms having a tetrahedral distribution (Fig. 8–4). The ratio of silicon atoms to oxygen atoms gives the empirical formula $SiO_2$. Again the strength of the bonding is revealed by the high melting point of the oxide.

*Fig. 8–4*

Phosphorus(V) oxide is a white solid consisting of $P_4O_{10}$ molecules. These contain electric dipoles owing to electron displacements in the bonds and are quite strongly attracted to each other. As a result the melting point is fairly high. In sulphur(VI) oxide and chlorine(VII) oxide the intermolecular attraction is progressively weaker and the melting points lower.

Across the period there is a gradual change in character of the oxides from strongly basic sodium(I) oxide to strongly acidic chlorine(VII) oxide. Aluminium(III) oxide, which is amphoteric, acts as a connecting link between the earlier basic oxides and the later acidic oxides.

*Hydroxides.* From the typical valencies of the elements in the third period we might expect a series of hydroxides of formulæ NaOH, $Mg(OH)_2$, $Al(OH)_3$, $Si(OH)_4$, $P(OH)_5$, $S(OH)_6$, and $Cl(OH)_7$. The hydroxides of the non-metals, however, split off water. Thus we have

$$Si(OH)_4 - H_2O \rightarrow H_2SiO_3 \text{ or } OSi(OH)_2$$
$$P(OH)_5 - H_2O \rightarrow H_3PO_4 \text{ or } OP(OH)_3$$
$$S(OH)_6 - 2H_2O \rightarrow H_2SO_4 \text{ or } O_2S(OH)_2$$
$$Cl(OH)_7 - 3H_2O \rightarrow HClO_4 \text{ or } O_3Cl(OH)$$

The basic character of the metal hydroxides NaOH and $Mg(OH)_2$ depends on their ability to split off the hydroxyl group as a whole in the form of an $OH^-$ ion. The acidic character of the oxy-acids of the non-metals is due to their losing $H^+$ ions from the hydroxyl groups. It is worth enquiring why the hydroxyl groups behave differently in the two cases. Owing to the high electronegativity of oxygen there always tends to be an electron displacement in the hydroxyl group towards the oxygen atom $(O \xleftarrow{\delta-} H \text{ or } O{-}\overset{\delta+}{H})$. This displacement is responsible for the tendency of the hydrogen atom to be removed as $H^+$ by a base. If, however, we imagine sodium(I) hydroxide to consist of NaOH molecules, we see that there would be a bigger electron displacement from the sodium atom to the oxygen atom (Na $\rightarrow$ O—H), sodium being much less electronegative than either oxygen or hydrogen. Thus there would be a greater tendency for the sodium atom to split off as a positive ion. This in fact is what happens, even the solid hydroxide being composed of $Na^+$ and $OH^-$ ions.

In sulphuric(VI) acid, $O_2S(OH)_2$, the two singly attached oxygen atoms exert a big attraction for the electrons of their bonds with the sulphur atom. The effect of the attraction does not stop here, but is transmitted to the two O—H groups, increasing the electron displacements in the bonds towards the oxygen atoms. Thus the two hydrogen atoms are more readily detached as protons ($H^+$) by a base. The molecule of phosphoric(V) acid, $OP(OH)_3$, has only one separately attached oxygen atom, and its 'inductive effect' has to be shared by three hydroxyl groups. Hence this acid is weaker than sulphuric(VI) acid. On the other hand, chloric(VII) acid, $O_3Cl(OH)$, has three oxygen atoms to one hydroxyl group and there is a correspondingly greater inductive effect. Accordingly chloric(VII) acid is stronger than sulphuric(VI) acid. It should be mentioned that the relative strengths of the oxy-acids are also governed by other factors, but that mentioned is one of the most important.

We now see why hydrated aluminium(III) hydroxide is amphoteric. Aluminium is more electronegative than sodium or magnesium, and therefore its hydroxide occupies an intermediate position. It is able to part with hydroxyl ions to a sufficiently strong acid or with hydrogen ions to a sufficiently strong base.

### Transition Elements.

*A transition element may be defined as an element which can use electrons from both the outer quantum shell and the penultimate shell for combination with other elements.*

As noted earlier, transition elements result from the overlapping of energy sub-levels of different quantum shells, so that when the atomic number is increased by one the extra electron goes into a lower shell than the one occupied by the outer electrons. Thus in the first long period the filling-in of the $3d$ sub-level, which can hold ten electrons, takes place as now shown.

|       | Sc | Ti | V | Cr | Mn | Fe | Co | Ni | Cu |
|-------|----|----|---|----|----|----|----|----|----|
| $4s$  | 2  | 2  | 2 | 1  | 2  | 2  | 2  | 2  | 1  |
| $3d$  | 1  | 2  | 3 | 5  | 5  | 6  | 7  | 8  | 10 |

In chromium and copper there are discontinuities in the uniform increase of number of electrons in the $3d$ sub-level. The second discontinuity results in there being only nine transition elements instead of ten, as might be expected. The reason for the discontinuities is the extra stability associated with the $3d$ sub-level when each of its five orbitals is occupied by a single electron (as in chromium) or when all five orbitals are occupied by paired electrons (as in copper). Thus when another electron is added to the $3d$ sub-level of a vanadium atom or nickel atom, one of the two $4s$ electrons is also drawn into this sub-level.

Whereas the properties of the other elements in a period show a more or less steady gradation, the properties of transition elements are generally similar. Thus the atomic radii are about the same, and so are the ionization energies and electronegativities. Some other properties characteristic of transition elements are given below.

*Metallic Character.* All the transition elements are metals, in which the bonds between the atoms are very strong, as shown by the high melting points. In the first long period the latter vary from 1083°C for copper to 1900°C for chromium. The basic reason for the strong bonding here is the small energy difference between the electrons in the $4s$ sub-level and those in the $3d$ sub-level. This makes it possible for electrons from *both* sub-levels to be contributed to the general electron cloud of the metal crystal. In potassium (m.p. 63°C) and calcium (m.p. 850°C) the contribution is limited to one and two electrons per atom respectively, and the metallic bonding is much weaker.

*Density.* This is one of the few properties of transition elements which shows gradation across the period. The densities of all these elements are high because the atoms are tightly packed together in the crystal. As the atoms of successive elements increase in mass while the atomic radius remains about the same, there is a progressive increase in density. Thus the densities/g cm$^{-3}$ of the transition elements of the first long period are as follows:

Sc 3·0    Ti 4·5    V 6·1    Cr 7·1    Mn 7·4    Fe 7·9

Co 8·9    Ni 8·9    Cu 9·0

*Magnetic Properties.* The presence of unpaired electrons in the atoms and ions of transition elements is responsible for the elements and many of their compounds being *paramagnetic*. A paramagnetic material tends to align its length with the lines of force of an applied magnetic field and to move from weaker to stronger parts of a field. Partly because of its orbital motion and partly because of its spin, an unpaired electron behaves like a small magnet with north and south poles. Hence, when an external magnetic field is applied, the electron orientates its spin so that its magnetic axis is parallel with the direction of the field. The effect is temporary and disappears when the external field is removed. (*Ferromagnetism*, shown by iron, cobalt, and nickel, is a special kind of paramagnetism. In this case the effect is much stronger and may be permanent.) The magnitude of the paramagnetic effect depends on the number of unpaired electrons in the atom or ion, and we can use it to determine this number. Substances containing only paired electrons are *diamagnetic*. They tend to set their length at right angles to the direction of an applied magnetic field and to move from stronger to weaker parts of the field.

*Variable Valency and Catalytic Power.* Transition elements show variable valency because they are able to lose electrons from both the outer quantum shell and the penultimate shell to give ions. Thus in the transition elements of the first long period the electrons of the $4s$ and $3d$ sub-levels are little different in energy and both can be used for valency purposes. In forming an iron(II) ion the iron atom loses its two $4s$ electrons, while the loss of these two electrons plus one $3d$ electron produces the iron(III) ion. The most frequent valency is two, but a valency of three is also common. Manganese is particularly versatile in its ability to exist in different oxidation states. Thus it forms oxides of formulæ $MnO$, $Mn_2O_3$, $MnO_2$, $Mn_3O_4$, and $Mn_2O_7$.

Ions of transition elements do not usually have the electron configuration of a rare-gas atom. For example, in $Fe^{2+}$ the electron arrangement is 2, 8, 14 and in $Cu^{2+}$ it is 2, 8, 17. These ions do not possess the stability of ions like $K^+$ and $Ca^{2+}$. They lose or gain electrons fairly readily, and so undergo oxidation or reduction; *e.g.*,

$$Fe^{2+} \underset{+e}{\overset{-e}{\rightleftharpoons}} Fe^{3+}$$

The ease with which ions of transition metals change their oxidation state enables them to act as catalysts. Thus both iron(II) and iron(III) ions catalyse the decomposition of hydrogen peroxide.

*Formation of Complex Ions.* Because of their high charge and small size, ions of transition elements are outstanding in their ability to form complex ions by co-ordination with molecules or negative ions which can act as electron donors. As we have seen earlier, the oxygen atoms of water molecules and the nitrogen atoms of ammonia molecules possess lone pairs of electrons which can be donated to metal ions. Some typical complex ions formed with ions of transition elements are the following:

| | |
|---|---|
| $[Cu(H_2O)_4]^{2+}$ | tetraaquacopper(II) ion |
| $[Cu(NH_3)_4]^{2+}$ | tetraamminecopper(II) ion (formerly called the cuprammonium ion) |
| $[Fe(H_2O)_6]^{3+}$ | hexaaquairon(III) ion |
| $[Co(NH_3)_6]^{3+}$ | hexaamminecobalt(III) ion |

The ammonia complexes are *ammines*, and the oxidation state of the metal ion is indicated by a roman numeral.

In the hexacyanoferrate(II) and hexacyanoferrate(III) ions the lone pairs of electrons are donated by cyanide ions.

$$Fe^{2+} + 6CN^- \rightarrow [Fe(CN)_6]^{4-}$$
$$Fe^{3+} + 6CN^- \rightarrow [Fe(CN)_6]^{3-}$$

In these reactions co-ordination is brought about by lone-pair electrons belonging to the carbon atom, and not the nitrogen atom. The cyanide ion has the electronic formula $[:C:::N:]^-$.

Molecules or ions which become attached to a central atom or ion by donating a share in a lone pair of electrons are *ligands*. The number of ligands joined to the central atom or ion in this way is the *co-ordination number* of the latter. This may vary from one to six, but is most commonly four or six.

*Coloured Ions.* It is often said that transition elements have coloured ions. The colours, however, are characteristic of the hydrated ions, not of the simple ions. Copper(II) sulphate(VI)-5-water and its solution are blue, while the anhydrous salt is white (strictly speaking, it is colourless). The blue colour is due to the ion $Cu(H_2O)_4^{2+}$. Similarly the yellow colour of iron(III) salts in aqueous solution is brought about by $Fe(H_2O)_6^{3+}$ ions, and the red colour of cobalt(II) salts by $Co(H_2O)_6^{2+}$ ions. As we saw in the last section, the water molecules in these complex ions are joined to the metal ions by co-ordinate covalent bonds. For the production of colour, however, the co-ordinated groups need not be water molecules. Thus an aqueous solution of tetraamminecopper(II) ions, $Cu(NH_3)_4^{2+}$, has an intense royal blue colour, and the $Fe(CN)_6^{4-}$ ions of potassium(I) hexacyanoferrate(II) solution are yellow.

The reasons for the colours of the complex ions of transition elements are complicated, and we shall not explain them in detail. It may be mentioned, however, that the colours are due to the absorption of some of the frequencies of visible light, the other frequencies being transmitted or reflected. Ions (*e.g.*, $Na^+$ and $Ca^{2+}$) with a completed outer quantum shell of eight electrons are colourless because visible radiation is not sufficiently energetic to bring about transitions of the outer electrons to a higher energy level. The same applies to $Cu^+$, $Ag^+$, and $Zn^{2+}$ ions, which have a moderately stable outer shell of 18 electrons. In the case of the complex ions of transition elements electrons can 'jump' into sub-levels which are only a little higher in energy, and the necessary energy can be absorbed from visible light. In the first series of transition elements the electrons chiefly affected are those of the $3d$ sub-level.

The *lanthanides* and *actinides* are called *inner* transition elements. In both cases they consist of atoms in which the *third* outermost quantum shell is progressively built up from 18 to 32 electrons. The lanthanides (see Table 8–2) include the fourteen elements from cerium to lutetium. Lanthanum has in its quantum shells 2, 8, 18, 18, 9, and 2 electrons. In the following elements the extra electron is added to the fourth quantum shell, the extension of this shell being completed in lutetium (2, 8, 18, 32, 9, 2). The two outermost shells, which provide the valency electrons, are identical for all fourteen elements, and hence the chemical properties correspond very closely.

The series of actinides was completed with the synthesis of lawrencium (Lw) in 1961. All these elements are radioactive, but many of their properties still await investigation. Present knowledge, however, indicates that they are very similar, as in the lanthanides.

## Variation of Properties in the Groups

**Numbering and Lettering of the Groups.** It is customary to number the various groups of the Periodic Table with roman numerals and (except in Groups 0 and VIII) to identify the groups further by adding the letters A or B. Thus we have Groups I A and I B, II A and II B, etc. The purpose of the numbers and letters is to show the relationships between the elements more clearly. All elements with the same group number have a common valency (or oxidation state) and form compounds of similar type. For example, sodium (I A) and copper (I B) give oxides $Na_2O$ and $Cu_2O$, while chromium (VI A) and sulphur (VI B) have oxides $CrO_3$ and $SO_3$. In the 'short' form of the Periodic Table (Table 8–1) a relationship between sodium and copper or between chromium and sulphur was implied from their presence in the same vertical column, but in the modern 'long' form of the Table the related elements are widely separated. Numbering brings them together mentally, if not physically.

Elements with the same group number and letter have the same type of electron configuration and constitute a natural 'family' of elements. This is not true, however, for elements with the same group number, but a different letter. Thus all the alkali metals (I A) have atoms containing one electron more than a rare-gas atom, and have very similar properties. The atoms of copper, silver, and gold (I B) also have one electron in their outer quantum shell, but the shell below is an 18-shell, and not an 8-shell. This is a fundamental difference. Not only are the properties of the coinage metals different from those of the alkali metals, but so also are those of the monovalent compounds. For example, copper(I) oxide and copper(I) chloride have little in common with potassium(I) oxide and potassium(I) chloride. There are even bigger differences between the elements in other A and B sub-groups. Chromium and sulphur are totally different in electron configuration, so that any similarity of the oxides $CrO_3$ and $SO_3$ is largely incidental. Attempts to relate elements in A and B sub-groups are often unjustified, and frequently the numbering and lettering of the groups are omitted altogether.

Leaving out the transition elements and the zinc sub-group, we can conveniently sub-divide the groups of the Periodic Table as now shown.

**Groups I and II.** These consist of elements with atoms containing one or two electrons more than a rare-gas atom. The elements are all metallic because of the ease with which these outer valency electrons form an electron cloud between the atoms. Whereas the atomic radius becomes smaller with increasing nuclear charge across a period, it becomes larger with increasing nuclear charge down a group. We have already noted that the latter is due to the screening effect of the completed quantum shells outweighing the effect of the increased nuclear charge.

Most of the gradations in the properties of the alkali metals are related to the increase of atomic radius down the group. Thus as the atomic radius becomes larger the positively charged nuclei are farther away from the electron cloud and the binding force in the crystal is weaker. This is reflected in the decrease in melting point. All the alkali metals crystallize in the body-centred cubic system (see later) and hence there is a similar degree of compactness in the packing of the atoms. Therefore, as the atomic radius increases, so does the atomic volume. The ratios of the cubes of the atomic radii from lithium to caesium are $1 : 1.8 : 3.5 : 4.1 : 5.1$, while the ratios of the atomic volumes are $1 : 1.8 : 3.5 : 4.3 : 5.3$. The good agreement between the two sets of values is evidence of similar tightness of packing of the atoms.

All the alkali metals form positive ions very readily by loss of their single valency electron. The ionization energies decrease with increase in

Table 8–4. SOME PHYSICAL PROPERTIES OF THE ALKALI METALS

| | Li | Na | K | Rb | Cs |
|---|---|---|---|---|---|
| Quantum shells | 2, 1 | 2, 8, 1 | 2, 8, 8, 1 | 2, 8, 18, 8, 1 | 2, 8, 18, 18, 8, 1 |
| Atomic radius/nm | 0·152 | 0·186 | 0·231 | 0·244 | 0·262 |
| Ionic radius/nm | 0·06 | 0·095 | 0·133 | 0·148 | 0·169 |
| Melting point/°C | 180 | 98 | 63 | 39 | 29 |
| Density/g cm$^{-3}$ | 0·53 | 0·97 | 0·86 | 1·53 | 1·90 |
| Atomic volume/cm$^3$ | 13·1 | 23·7 | 45·3 | 55·9 | 70·0 |
| Ionization energy/ kJ mol$^{-1}$ | 519 | 494 | 418 | 402 | 377 |

size of the atoms, which means that the metals become more electropositive from lithium to caesium. This is illustrated by the increasing violence with which the metals react with water.

Parallel with the increase in atomic radii is the increase in ionic radii. The latter leads to variations in the properties of the ionic compounds. An example is the melting points of the chlorides, which are as follows:

| | LiCl | NaCl | KCl | RbCl | CsCl |
|---|---|---|---|---|---|
| m.p./°C | 613 | 801 | 776 | 715 | 646 |

The decrease in melting point from sodium(I) chloride to caesium(I) chloride is readily explained by the increase in size of the metal ions. The melting point of lithium(I) chloride, however, is relatively low. This is due to the small size of the Li$^+$ ion compared with that of the Cl$^-$ ion (radius 0·181 nm). In the closely packed crystals the small lithium ions cannot shield the large chloride ions completely from each other. The binding force between the ions is therefore reduced and the melting point is lowered.

The tendency for salts of the alkali metals to combine with water of crystallization decreases from lithium to caesium. Thus, while about 75 per cent of lithium salts form hydrates, scarcely any rubidium or caesium salts do so. This reflects the decreasing attraction of the metal ions for water molecules as the ions become larger.

A feature of alkali metals is the great stability to heat of their carbonates, sulphates(VI) and nitrates(V). The temperatures needed to decompose the salts increase from lithium to caesium. The relative stabilities of the salts vary with the abilities of the cations to polarize the anions, that is, to attract the electrons of the latter. Distortion of the outer electron shells of the anions results in decreased stability and decomposition at a lower temperature. The polarizing power of the alkali metal cations is greatest for the small $Li^+$ ion and least for the large $Cs^+$ ion. Thus only moderate heat will decompose lithium(I) nitrate(V), and unlike other alkali-metal nitrates(V), which leave the nitrate(III), lithium(I) nitrate(V) yields the metal oxide, nitrogen dioxide, and oxygen.

The alkaline-earth metals of Group II are less electropositive than the alkali metals. Not only are the first ionization energies larger, but the formation of an ion $M^{2+}$ from an atom M involves the giving-up of *two* electrons. The latter results in a greater decrease in the size of the ions as compared with the size of the atoms. The combination of small size and double charge enables the cations of the Group II metals to polarize anions more effectively than the cations of the alkali metals. The hydroxides are weaker bases, and the oxy-salts decompose at lower temperatures. Within the Group the alkaline-earth metals show similar variations to the alkali metals, although the variations are less regular. Beryllium(II) carbonate is so unstable to heat that it decomposes at ordinary temperatures. Magnesium(II) carbonate gives carbon dioxide when heated gently, and the carbonates of calcium, strontium, and barium require increasingly high temperatures to decompose them.

**Groups III, IV, and V.** In each of these groups the atomic radius again increases down the group, so that the attraction between the nucleus and the outer electrons decreases. In the early members the attraction is too large for electrons to be contributed to an electron cloud and the elements are non-metals. Having high ionization energies, they combine with other elements by covalent bonds. In the later members, which have lower ionization energies, some of the outer electrons can be used for formation of an electron cloud or to produce positive ions. Thus the later members are metals. In Groups IV and V, the non-metals are linked to the metals by 'metalloids,' which partake of the character of both. Accompanying the change in properties of the elements are gradations in the properties of the compounds. For example, the hydrides become less stable to heat, and the oxides and hydroxides become less acidic and more basic.

An important feature of the middle group elements (with the exception of the first member of each group) is their ability to expand their outer quantum shells to twelve, or ten, electrons. Thus cryolite, $Na_3AlF_6$, contains the $[AlF_6]^{3-}$ ion, hexafluorosilicates(IV) the $[SiF_6]^{2-}$ ion, and phosphorus pentachloride vapour $PCl_5$ molecules. In expansion of the outer shell use is made of $d$ orbitals for bonding. This is impossible for the first elements in the groups because their atoms do not possess $d$ orbitals. The reason why silicon tetrachloride, but not tetrachloromethane, is hydrolysed by water is that a molecule of the former is able to attract, and co-ordinate with, two molecules of water (the silicon atom has a partial positive charge and the oxygen atoms partial negative charges). However, six atoms or groups cannot be comfortably accommodated round the small silicon atom unless (like fluorine atoms) they are very small. The addition compound is therefore unstable, and two molecules of hydrogen chloride are eliminated. By repetition of water addition and hydrogen chloride elimination silicic(IV) acid is produced, as now shown.

$$
\begin{array}{c}
\mathrm{H}\\
\quad\ \mathrm{O} \rightarrow \mathrm{Si} \leftarrow \mathrm{O}\\
\mathrm{H}\quad\quad\quad
\end{array}
\ \xrightarrow{\ -2\mathrm{HCl}\ }\ 
\begin{array}{c}
\mathrm{HO}\quad\ \mathrm{Cl}\\
\quad\ \mathrm{Si}\\
\mathrm{Cl}\quad\ \mathrm{OH}
\end{array}
$$

$$
\begin{array}{c}
\mathrm{H}\ \ \mathrm{HO}\quad\ \mathrm{Cl}\ \ \mathrm{H}\\
\quad\ \mathrm{O} \rightarrow \mathrm{Si} \rightarrow \mathrm{O}\\
\mathrm{H}\ \ \ \mathrm{Cl}\quad\ \mathrm{OH}\ \ \mathrm{H}
\end{array}
\ \xrightarrow{\ -2\mathrm{HCl}\ }\ 
\begin{array}{c}
\mathrm{HO}\quad\ \mathrm{OH}\\
\quad\ \mathrm{Si}\\
\mathrm{HO}\quad\ \mathrm{OH}
\end{array}
\ \xrightarrow{\ -\mathrm{H_2O}\ }\ \mathrm{H_2SiO_3}
$$

**Groups VI and VII.** The members of these groups have six electrons and seven electrons respectively in their outer quantum shells. With decrease in atomic radius across the periods electronegativities of the elements reach a maximum for each period in Group VII. With the exception of polonium the elements of both Groups are non-metallic because of the strong hold exerted by the atomic nuclei on the outer electrons. With increase in atomic radius down the Groups electronegativities are highest in the first members and lowest in the last. Thus fluorine is the most electronegative of all the elements. Polonium, the last element in Group VI is able to use some of its outer electrons to form an electron cloud and is a metal.

We can regard the halogen family as representative of the two Groups. Some physical properties of the halogens are listed in Table 8–5. All the

**Table 8–5.** SOME PHYSICAL PROPERTIES OF THE HALOGENS

| | F | Cl | Br | I |
|---|---|---|---|---|
| Quantum shells | 2, 7 | 2, 8, 7 | 2, 8, 18, 7 | 2, 8, 18, 18, 7 |
| Atomic radius/nm | 0·072 | 0·099 | 0·114 | 0·133 |
| Ionic radius/nm | 0·136 | 0·181 | 0·195 | 0·216 |
| Physical state | Yellow gas | Green gas | Red liquid | Black crystals |
| Melting point/°C | −220 | −101 | −7 | 114 |
| Boiling point/°C | −188 | −34 | 58 | 183 |
| Bond energy (X—X) /kJ mol$^{-1}$ | 268 | 238 | 192 | 151 |
| Electronegativity | 4·0 | 3·0 | 2·8 | 2·5 |

elements exist as diatomic molecules ($F_2$, etc.), but as these are only weakly attracted by the van der Waals force melting points and boiling points are low. With increase in size of the molecules, however, the van der Waals force becomes larger and there is a progressive rise in melting point and boiling point. The strength of the covalent bond between the two atoms in the molecule decreases from fluorine to iodine. This is to be expected from the increase in atomic radius, the bonding pair of electrons becoming further from the two attracting nuclei. Corresponding with the decrease in bond energy is a decrease in thermal stability of the molecules. Iodine dissociates appreciably into single atoms at 700°C, bromine at 1000°C, and chlorine at 1300–1400°C. Fluorine molecules do not dissociate even at 2000°C.

The halogens combine with other elements both by ionic bonds and by covalent bonds, the amount of ionic character in the bonds depending on the relative attractions of the combining atoms for the bond electrons. Thus in the hydrogen halides the percentages of ionic character in the bonds decrease from HF to HI. Corresponding with this decrease we

Table 8–6

|  | H—F | H—Cl | H—Br | H—I |
|---|---|---|---|---|
| Percentage of ionic character in bond | 43 | 17 | 11 | 5 |
| Bond length/nm | 0·092 | 0·128 | 0·141 | 0·160 |
| Bond energy/kJ mol$^{-1}$ | 614 | 430 | 364 | 297 |

find an increase in bond length and a decrease in bond energy. Three ways in which the decrease in bond energy affects the properties are the following:

• Thermal dissociation into free atoms takes place least readily with hydrogen fluoride and most readily with hydrogen iodide.
• The ease of oxidation of the hydrides increases from hydrogen fluoride to hydrogen iodide.
• The strengths of the acids in aqueous solution are in the order: HI > HBr > HCl > HF (see p. 463).

The readiness with which halogens are reduced to negative ions decreases from fluorine to iodine. Thus fluorine is the strongest oxidizing agent, and iodine the weakest. The decreasing attraction of the atoms for electrons is illustrated by the fact that fluorine displaces $Cl^-$ ions from metal chlorides, chlorine displaces $Br^-$ ions, and bromine displaces $I^-$ ions. Conversely the ease with which the anions can be oxidized to atoms decreases from $I^-$ to $F^-$. As we might expect, the strength of the bonds holding together metal ions and halide ions in a crystal is greatest for fluorides and least for iodides. This is illustrated by the decrease in the melting points of the sodium halides:

|  | NaF | NaCl | NaBr | NaI |
|---|---|---|---|---|
| m.p./°C | 992 | 801 | 755 | 651 |

Astatine, the last member of the halogen family, has not been discussed because little of its chemistry has so far been investigated. The element is radioactive and does not occur naturally. It has been prepared artificially only in minute amounts. It is a volatile solid soluble in tetrachloromethane, which indicates its non-metallic character. It forms an anion $At^-$, which is precipitated as AgAt by silver(I) nitrate(V) solution.

**Position of Hydrogen.** The allocation of hydrogen in the Periodic Table has given rise to much argument. Some authors have placed the element above lithium in Group I, others above fluorine in Group VII, while others again have given it an isolated central position to indicate that it is unique and has no genuine analogues. From the structural point of view hydrogen is related to both the alkali metals and to the halogens. Like lithium its outer quantum shell contains only a single $s$ electron, but like fluorine the atom is only one electron short of a rare-gas pattern of electrons. Hence hydrogen resembles the elements of both groups in being consistently monovalent in combination with other elements.

Hydrogen is similar to the alkali metals in being electropositive, and it is liberated at the cathode when aqueous solutions of acids are electrolysed. It is replaced by metals from acids, which have been described as hydrogen 'salts,' but which differ from true salts in being covalent, and not ionic, compounds.

The properties of hydrogen relate it much more closely to the halogens than to the alkali metals. It has a much higher ionization energy and electronegativity than any of the latter, and is therefore non-metallic in character. It resembles the halogens in existing as diatomic molecules, and the compounds which it forms with halogens also consist of gaseous diatomic molecules and are not salts. On the other hand, the hydrides of the alkali metals are similar to the halides in being salts, in which the hydrogen exists as a negative ion (*e.g.*, $Na^+H^-$). When the fused hydrides are electrolysed hydrogen is liberated at the anode.

**Group 0.** The atoms of all the rare gases except helium have an outer quantum shell of eight electrons. The existence of the gases as single atoms is accounted for by the unusual stability of the octet (or duplet) of electrons. As explained earlier, the theory of electrovalency and covalency is largely based on the tendency of atoms to acquire this stable electron grouping. The reasons for the stability of the octet are not fully understood, but one factor appears to be the symmetry of the electrical and magnetic fields associated with the four pairs of electrons (one $s$ pair and three $p$ pairs in each case). Ionization energies are high, making the production of positive ions difficult, and there is no tendency to form negative ions. At the same time the large energy required to unpair the $p$ electrons and promote one to a higher level acts as a barrier to the formation of covalent compounds.

At one time it was thought that the combination of rare-gas atoms with atoms of other elements could not take place. However, in 1962 xenon tetrafluoride, $XeF_4$, was obtained by direct combination of the elements. The new compound proved to be a white crystalline solid, which was quite stable even when heated. This was followed by the preparation of two further fluorides, $XeF_2$ and $XeF_6$, an oxide, $XeO_3$, and a compound $Xe^+[PtF_6]^-$, xenon hexafluoroplatinate(IV), containing xenon as a positive ion. In the formation of the three fluorides one, two, and three pairs of the $p$ electrons of the xenon atom become unpaired, and covalent bonds are established with fluorine atoms. Thus in $XeF_2$, $XeF_4$, and $XeF_6$ the xenon atom extends its outer quantum shell to 10, 12, and 14 electrons. Further research has resulted in the preparation of two fluorides of krypton, $KrF_2$ and $KrF_4$, and very small amounts of a fluoride of radon. So far no compounds of the lighter rare gases, helium, neon, and argon have been obtained.

## EXERCISE 8

1. (Part question.) Describe the modern theory of the structure of atoms, defining any particles you mention. Explain how this theory accounts for (a) the periodicity of elements, (b) the variations in valency within a short period, and (c) a characteristic valency in a group of the Periodic Table.        (S.U.)

2. What is meant by the terms *group, short period, long period, transition element* in the description of the periodic classification of the elements? Write out the periodic table of the elements from hydrogen to argon. Which of these elements form chlorides and what differences would you expect to find in the properties of these chlorides?

How can the position of an element in the periodic table be used, together with its equivalent, to estimate its relative atomic mass?        (C.L.)

3. Show how the electronic theory of valency provides an explanation of the regular variation of valency in passing from group to group of the periodic classification of the elements.

Tabulate as completely as the facts allow the following information about the elements Al, C, Ca, Cl, N, Na, Ne, S, rearranging the elements in the order of their groups:

(a) the group in the Periodic Table,

(b) the formula of the characteristic hydride,

(c) the action of the hydride on water,

(d) the formula of the oxide characteristic of the group to which the element belongs,

(e) the action of this oxide on water.        (J.M.B.)

4. State what you understand by the term 'transition metal'. Give *two* examples from the chemistry of each of the metals manganese, iron, and copper to illustrate their transitional character.        (O. and C.)

**5.** By means of a comparison of the properties of the elements and their compounds, discuss critically the inclusion of each of the following pairs of elements in the same group in the periodic table: (i) carbon and silicon, (ii) oxygen and sulphur. (Lond.)

**6.** By reference to the properties of the halogens and of their compounds with hydrogen, explain the features you would expect to find in a group in the Periodic Table. (O. and C.)

**7.** Discuss with reference to the Periodic Table the significance as regards the properties of the elements of (i) the electronegativities of the elements in any one period, (ii) the ionization potentials of the elements in any one group.

## MORE DIFFICULT QUESTIONS

**8.** Justify the statement: 'The atomic number of an element gives more information regarding its atomic structure and is of greater usefulness in its classification that its relative atomic mass.' (J.M.B.)

**9.** Discuss the variation of atomic radius with increase in atomic number and its relevance to the types of chemical bonding shown by the elements.
(W.J.E.C.)

**10.** Summarize, *clearly and concisely*, with illustrative examples, the main evidence to support or oppose the statement that 'the metallic character of the elements in the Periodic Classification decreases in progressing from left to right in the Table but increases in progressing from top to bottom.' (Lond.)

**11.** What is meant by the terms *basic oxide, acidic oxide,* and *amphoteric oxide*? Discuss the factors which influence the formation of these various types of oxides as illustrated by elements in the two short periods of the Periodic Table. (O. and C.)

**12.** The atomic numbers of the inert gases are 2, 10, 18, 36, 54, and 86. Make use of this information in deducing

(*a*) the properties of the elements having atomic numbers 15 and 56;

(*b*) the salient characteristics of the newly discovered elements astatine (atomic number 85) and francium (atomic number 87);

(*c*) the probable position in the periodic classification of a metallic element which has a heat capacity of $0.238 \text{ J g}^{-1} \text{ K}^{-1}$ and an atomic number between 47 and 52, and which forms two chlorides containing respectively 23.7 and 48.1 per cent of chlorine. (J.M.B.)

# CHAPTER 9

# The solid state

## Internal Structure of Solids

**Investigation of Crystal Structure by X-rays.** We have seen how evidence as to the particle nature of matter is provided by the diffraction patterns obtained when X-rays are passed through solids. The patterns show that nearly all solids are crystalline in composition and that the crystals consist of particles arranged in parallel planes. Nowadays it is more usual to obtain the diffraction patterns by *reflection* of X-rays at a crystal surface.

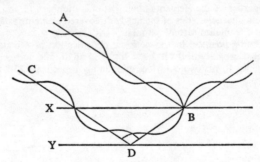

*Fig.* 9–1. Reinforcement of X-rays reflected from successive crystal planes

This technique (due to W. H. Bragg and W. L. Bragg) yields patterns from which the distance between the crystal planes can be calculated.

The principle of the method is as follows. If monochromatic X-rays (that is, rays of uniform wavelength) are rendered parallel by passing through slits, and are then allowed to fall on a crystal surface, the beam is partially reflected by particles in the surface layer and partially transmitted. At each successive layer reflection and transmission again take place. Fig. 9–1 shows a ray, CD, reflected from the second plane, Y, so that it follows the same path as a ray, AB, reflected from the first plane,

X. When the ray reflected at D emerges from the crystal it has travelled a greater distance than the one reflected from B. The difference depends on the distance between the planes and on the angle of incidence of the X-rays in the crystal. For certain values of the angle the difference in distance equals a whole number of wavelengths of the X-rays. The crests and troughs of both reflected waves then coincide and the waves reinforce each other. Consequently there is a reflection of maximum intensity, which produces a bright spot on a photographic plate. The conditions for these maxima are given by the *Bragg equation*

$$n\lambda = 2d \sin \theta,$$

where $n$ is a simple whole number, $\lambda$ is the wavelength of the X-rays, $d$ is the distance between the planes, and $\theta$ (theta) is the angle of incidence of the rays.

As $\theta$ is increased from zero strong reflections occur when $n = 1, 2, 3$, etc. In practice the intensity of the maxima decreases as $n$ increases, and it is usual to take the value of $\theta$ when $n = 1$. The wavelength of the X-rays can be measured by a very finely ruled grating of known spacing. Then, since $n$, $\lambda$, and $\theta$ are known, the distance, $d$, between the crystal planes can be calculated.

For complete analysis of a crystal by the Bragg method it is necessary to turn the crystal so that reflections are obtained in turn from different faces. If a sufficiently large crystal is not available the single crystal can be replaced by a mass of small crystals. As these have a random orientation it is not necessary to turn them. Some of the crystals will have the orientation required to give 'first-order' reflections ($n = 1$), while others will yield 'second-order' ($n = 2$) reflections. This *powder method* does not produce bright spots, but a series of concentric circles of varying width and intensity. A section of the circles is photographed on a strip of film as shown in Fig. 9-2. The resulting

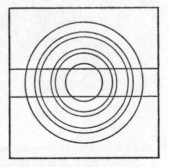

*Fig.* 9–2. Type of diffraction pattern from the powder method

diffraction pattern is characteristic for each substance, and can be used for its identification or to test its purity.

**Information Derived from X-ray Diffraction by Crystals.** Two kinds of basic information can be obtained from X-ray analysis of crystals. Firstly, we can find the distances between the various crystal planes and the angles at which the planes intersect. From these the arrangement of the particles and the distances between them can be ascertained. The particles may consist of atoms (as in carbon (diamond)), simple ions (as in NaCl), complex ions (as in $NH_4NO_3$), or molecules (as in iodine). Secondly, from X-ray diffraction patterns the electron density in different parts of a crystal can be deduced. Scattering of X-rays is caused, not by atomic or ionic nuclei, but by the electrons round the nuclei. Scattering is greater where the electron density is bigger, that is, in the immediate neighbourhood of the nuclei. Thus, if the intensities of the scattered rays are measured with a Geiger–Müller counter, the electron densities at various points in a crystal plane can be calculated and an electron density 'map' constructed.

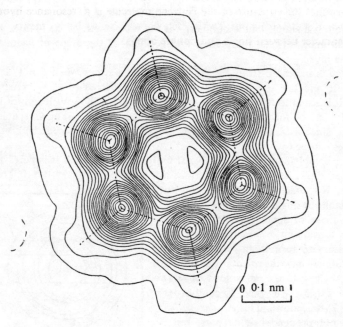

*Fig.* 9–3. Electron density map of the benzene molecule. (From "The Crystal Structure of Benzene at −3°C," by E. G. Cox, D. W. J. Cruickshank, and J. A. S. Smith, 1958, *Proc. Roy. Soc.* A, 247, p. 7.)

Fig. 9–3 shows an electron density map of the flat benzene molecule, $C_6H_6$, obtained by diffraction of X-rays by a benzene crystal at −3°C. In this diagram each contour line connects points of equal electron density in the plane of the molecule, and the positions of the six carbon atoms in the benzene ring are shown by electron density maxima. Hydrogen atoms are too small to scatter X-rays appreciably, and hence the positions of the six hydrogen atoms are not indicated by similar maxima. The positions are revealed, however, by curved projections in the contour lines extending outwards from the carbon atoms. The C—C and C—H bonds are represented by dotted lines.

From electron density maps of this kind we can find the shapes of molecules and ions, their structural formulæ, and the bond lengths (distances between the nuclei). Bond lengths in turn give information about the nature of the bonds, *e.g.*, the amount of ionic character present and whether the bonds are single or multiple. The lengths of the six carbon–carbon bonds in the benzene molecule are equal. This shows that the bonds are similar, and not alternately single and double bonds, as represented in the Kekulé structure for benzene. Furthermore, the observed bond length (0·139 nm) is intermediate between the lengths of the C—C bond (0·154 nm) and the C=C bond (0·133 nm). This supports the theory that the structure of the benzene molecule is a 'resonance hybrid' of two Kekulé structures, the carbon–carbon bonds being intermediate in character between single and double bonds.

**Crystal Structures.** X-ray analysis of crystals shows that these are built up from simple structural units composed of a few atoms, ions, or molecules. The crystal is merely a repetition of the basic unit, just as the pattern woven into a piece of cloth consists of the same design repeated over and over again. The crystal unit is called the *unit cell*, or *space lattice*.

There are seven types of unit cell, and these give rise to the seven crystal systems: cubic, rhombic, monoclinic, triclinic, tetragonal, rhombohedral, and hexagonal. In all except the hexagonal system the unit cell

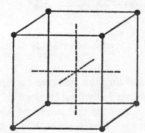

*Fig.* 9–4. Simple cubic lattice

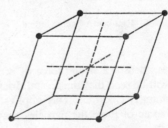

*Fig.* 9–5. Simple monoclinic lattice

is bounded by three pairs of opposite parallel faces. Three imaginary lines drawn between the centres of opposite faces represent three axes of symmetry. If a crystal is rotated through 360° round one of these axes it passes through a number of positions in which it has the same orientation in space. The system to which the crystal belongs is determined by the relative lengths of the axes of symmetry and the angles at which they meet.

In the simplest form of space lattice, the cubic type (Fig. 9–4), the three axes of symmetry are all equal in length and meet at 90°. In the monoclinic system (Fig. 9–5) the axes have different lengths and, while two of the axes are at right angles to each other, the third is inclined to the other two.

This is indicated in the name 'monoclinic.' Iron(II) sulphate(VI)-7-water and sucrose crystallize in the monoclinic system.

The cubic type of lattice commonly occurs in two forms, the *face-centred cubic* and the *body-centred cubic*. The first, which is illustrated in Fig. 9–6, is typical of a number of metals, including copper. The second, illustrated in Fig. 9–10a, is found in iron.

A sodium(I) chloride lattice (Fig. 9–7) consists of a face-centred cubic arrangement of sodium(I) ions (black circles) with chloride ions (plain circles) situated at the middle of each edge and in the centre of the cube. Alternatively we can regard the lattice as two interpenetrating face-centred cubes, one composed of sodium(I) ions, the other of chloride ions. Since the oppositely charged ions are closer together than the similarly charged

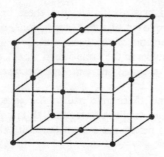

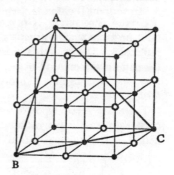

*Fig.* 9–6. Face-centred cubic lattice

*Fig.* 9–7. Space lattice of sodium(I) chloride

ones, there is a strong force of attraction holding the particles together. It will be seen that the crystal does not contain individual 'molecules' of sodium(I) chloride. This applies to other crystals composed of ions. In some of these the particles may be complex ions such as $NH_4^+$ and $CO_3^{2-}$.

It might be expected that the external form of a perfect crystal should always correspond to the form of the space lattice—that is, sodium(I) chloride having a cube as space lattice should invariably crystallize in cubes. Actually the same unit cell may give rise to more than one external shape. Common salt deposited from a solution containing a little carbamide (urea) may form octahedral crystals (Fig. 9–8). X-ray analysis shows that, while the planes in a cubic crystal consist of sodium(I) ions and chloride ions, successive planes of an octahedral crystal contain alternately only sodium(I) ions or chloride ions. This is illustrated in Fig. 9–7, where ABC is a plane containing only the sodium(I) ions. The alums, which also belong to the cubic system, normally form octahedral, and not cubic, crystals.

The method shown in Fig. 9–7 is convenient for drawing crystal lattices on paper, but it is misleading in one respect. It gives the appearance of the ions being separated by relatively large distances. The measurement of ionic radii shows that the ions are actually in contact. We should, of course, expect oppositely charged ions to pack together as closely as possible, and this is what occurs in practice. The closeness of packing depends on a number of factors, the chief of which are the relative numbers of positive and negative ions, their relative size, and, in the case of complex ions like $NO_3^-$ and $SO_4^{2-}$, their shape.

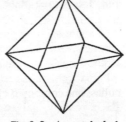

*Fig.* 9–8. An octahedral crystal

In sodium(I) chloride there is one positive ion to one negative ion, and the size of the negative ion is about twice that of the positive ion, the ionic radii being 0·181 nm for $Cl^-$ and 0·095 nm for $Na^+$. Here the most compact form of packing is for each metal ion to be in contact with six chloride ions and each chloride ion in contact with six metal ions; that is, each ion has a *co-ordination number* of six. A crystal layer containing both ions is shown in Fig. 9–9, where each ion is touching four of the opposite kind in the same plane. The other two are directly above and below in the adjacent planes.

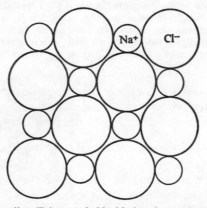

*Fig.* 9–9. Layer of sodium(I) ions and chloride ions in a sodium(I) chloride crystal

Most of the alkali metal halides have the same cubic type of space lattice as sodium(I) chloride. Exceptions are the chloride, bromide, and iodide of caesium. If francium is ignored, a caesium(I) ion is the largest of the alkali metal ions. The ratio of the ionic radius of $Cs^+$ (0·169 nm) to that of $Cl^-$ (0·181 nm) is 0·93. When the ions are simple and have equal or nearly equal size more compact packing is obtained by an arrangement resembling the *body-centred cubic* lattice found in iron (Fig. 9–10a). A caesium(I) chloride crystal has alternate layers of caesium(I) ions and chloride ions (Fig. 9–10b). The centre of the lattice is occupied by a caesium(I) ion (or chloride ion) in contact with eight oppositely charged ions, four in the plane above, and four in the plane below.

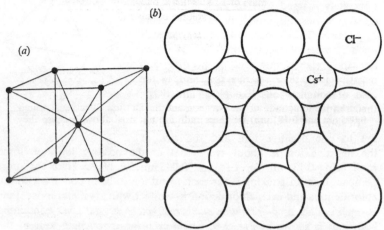

*Fig.* 9–10. (*a*) Body-centred cubic lattice of iron; (*b*) Alternate layers of caesium(I) ions and chloride ions in a caesium(I) chloride crystal

**Determination of the Avogadro Constant ($N_A$) from Unit Cell Dimensions.** The Avogadro constant can be calculated from the dimensions and other particulars of the unit cell of a crystal like that of sodium(I) chloride. Actually, in the case of the latter salt it is simpler to consider one of the eight similar sub-cells making up the unit cell represented

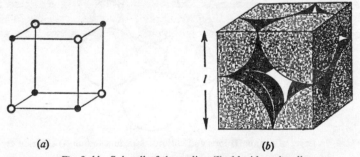

*Fig.* 9–11. Sub-cell of the sodium(I) chloride unit cell

in Fig. 9–7. One of these sub-cells is shown in Fig. 9–11a. The corners of the sub-cell are occupied by four $Na^+$ ions and four $Cl^-$ ions, but each ion is held in common by eight sub-cells, so that only one eighth of an ion belongs to any particular sub-cell (Fig. 9–11b). Thus in effect any sub-cell has in all half a sodium(I) ion and half a chloride ion to itself. We can regard these as making up half a 'molecule' of sodium(I) chloride.

Since the complete crystal is a large number of similar sub-cells, the density, $\rho$, of the crystal is the same as the density of a sub-cell. Also, if the 'relative molecular mass' of sodium(I) chloride is $M$ and Avogadro constant $N_A$, the mass of one 'molecule' of sodium(I) chloride is $M/N_A$ gram. Assuming the length of one side of the sub-cell to be $l$, we have

$$\text{Density} = \frac{\text{mass of 1 molecule} \times \text{number of molecules}}{\text{volume of sub-cell}}$$

That is,
$$\rho = \frac{\frac{1}{2}(M/N_A)}{l^3}$$

The density of a sodium(I) chloride crystal is $2 \cdot 163$ g cm$^{-3}$, the 'relative molecular mass' of sodium(I) chloride is $58 \cdot 45$, and the length found for a side of the sub-cell by X-ray diffraction is $0 \cdot 281$ nm. (The length, $l$, is the sum of the ionic radii of the Na$^+$ ion and the Cl$^-$ ion. The usual values $0 \cdot 095$ nm and $0 \cdot 181$ nm) for these radii are not used above because they are average values derived from different compounds.)

Substituting these values in the equation, we obtain

$$2 \cdot 163 = \frac{\frac{1}{2}(58 \cdot 45/N_A)}{(0 \cdot 281 \times 10^{-7})^3}$$
$$\text{and} \quad N_A = 6 \cdot 08 \times 10^{23} \text{ mol}^{-1}$$

This method of finding the Avogadro constant, $N_A$, can be applied to various kinds of crystal lattice formed both by compounds and elements (usually metals). These experiments yield an average value for $N_A$ of $6 \cdot 023 \times 10^{23}$. mol$^{-1}$.

Clearly, we can reverse our calculation and use the value of $N_A$ to find the length of side of a sub-cell or unit cell of sodium(I) chloride. Alternatively we can assume the values of both $N_A$ and $l$ and determine the wave length of X-rays used in diffraction from the Bragg equation. The sodium(I) chloride crystal is frequently used for this purpose.

**Layer Crystals.** These are crystals composed of layers of atoms or ions, the layers being held together either by the weak van der Waals force of attraction or by hydrogen bonds. The layers can have any

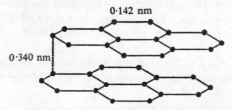

0·142 nm

0·340 nm

*Fig.* 9–12. Structure of carbon (graphite)

length and breadth, and in comparison with these the depth, or thickness, is negligible. Layer crystals are thus infinite two-dimensional assemblies of atoms or ions.

The simplest example of a layer crystal is found in carbon (graphite), which consists of sheets of carbon atoms joined together in the form of hexagonal rings by covalent bonds. In the crystal the sheets are super-imposed on each other rather like sheets of wire netting, and are held together by the van der Waals force. The carbon atoms in alternate layers are directly above each other, those in successive layers differing in relative position by one bond length. The distance (0·340 nm) between the layers is more than twice the distance (0·142 nm) between two adjacent atoms in a layer.

In other layer crystals each layer resembles a sandwich. Thus in the flat yellow hexagonal crystals of lead(II) iodide the layer consists of two sheets of iodine atoms enclosing a sheet of lead atoms. The bonds between the lead atoms and the iodine atoms are of the intermediate ionic–covalent type. Ignoring the spatial arrangement of the atoms, we can represent a section of one of the layers as now shown.

| I | I | I | I | I | I |
|---|---|---|---|---|---|
| Pb | Pb | Pb | Pb | Pb | Pb |
| I | I | I | I | I | I |

The layers are held together in the crystal by the van der Waals force between neighbouring sheets of iodine atoms.

Layer crystals are formed by talc (French chalk), mica, and many metal halides and hydroxides of general formula $MX_2$ or $MX_3$ (*e.g.*, $ZnCl_2$ and $Al(OH)_3$). In the hydroxides the cohesion is mainly due to hydrogen bonds between the —OH groups.

**Chain Crystals.** Silver(I) cyanide forms crystals in which the atoms are linked by normal covalent and co-ordinate covalent bonds in straight chains as follows:

$$—Ag—C≡N → Ag—C≡N → Ag—C≡N—$$

The chains of atoms lie side by side like cigarettes in a packet and are held together by the van der Waals force. As the length of the chains is indefinite the crystal consists of infinite one-dimensional assemblies of atoms.

Examples of chain, or linear, crystals also occur in metal halides. Thus in anhydrous copper(II) chloride the arrangement of atoms is:

Chain crystals are also formed by cellulose and some proteins. In these crystals very long molecules are held in parallel chains by hydrogen bonds between the chains.

**Water of Crystallization.** Many substances when crystallized from aqueous solution contain combined water, and are said to be 'hydrated.' The combined water is described as 'water of crystallization.' However, this is misleading because it implies that if the water is expelled the remaining anhydrous substance is amorphous, or non-crystalline, which is not so. The anhydrous substance also has a crystalline structure, which usually differs, however, from that of the hydrate. Thus hydrated aluminium(III) chloride forms rhombohedral crystals, while the anhydrous chloride has a layer structure. Water may enter into the composition of crystals in three ways, as now described.

(1) *Water Molecules Co-ordinated to Metal Ions.* Examples of hydrated metal ions have been given previously. The tendency of a cation to co-ordinate with water molecules varies with its size and charge. The ability of ions of similar charge to polarize water molecules decreases as the size of the ion increases. Thus small univalent ions like $Li^+$ undergo hydration and give rise to hydrated salts, while large univalent ions like $Cs^+$, $Ag^+$, and $NH_4^+$ are usually unhydrated and form salts without water of crystallization. As noted earlier, the polarizing power of cations on water molecules increases with the number of positive charges. Hence most divalent and trivalent metal ions are hydrated (usually by co-ordination with two, four, or six water molecules).

(2) *Water Molecules as Structural Linkages.* Water molecules often assist in the packing of atoms or ions in a crystal by acting as bridges between the particles—usually by forming hydrogen bonds of the type O—H----O. This explains the large number of molecules of water of crystallization present in such hydrated salts as $Na_2CO_3 . 10H_2O$. It also accounts for the presence of water of crystallization in organic hydroxy-compounds like ethanedioic acid-2-water, $(COOH)_2 . 2H_2O$.

In some crystals water fulfils both functions (1) and (2). Part is co-ordinated to cations, and part serves to link together cations and anions. An illustration is provided by $CuSO_4 . 5H_2O$. Four molecules of water are combined with the copper(II) ion, while the fifth forms a bridge between the hydrated ion and oxygen atoms of different sulphate(VI) ions. The simplified structure of copper(II) sulphate(VI)-5-water is:

Actually, the $Cu^{2+}$ ion is also co-ordinated to two oxygen atoms belonging to neighbouring $SO_4^{2-}$ ions, giving the copper (II) ion a co-ordination number of 6.

(3) *Clathrates.* By cooling aqueous solutions of various gases (often under pressure) we can obtain crystalline 'hydrates' of the gases. 'Hydrates' of argon, krypton, chlorine, phosphine, and several other gases have been prepared in this way. The crystals differ in form from ice crystals, but are usually unstable above 0°C. Their composition often approximates to one molecule of gas to six molecules of water, but is non-stoichiometric. Thus we have $Cl_2.6\cdot1H_2O$, $PH_3.5\cdot8H_2O$, and $Ar.5\cdot5H_2O$. These 'hydrates' are produced by the imprisoning of the gas molecules in cavities formed by the water molecules, but for this to happen the gas molecules must be a suitable size. Probably the water molecules, being themselves polar, polarize the gas molecules, setting up a weak dipole–dipole attraction. The crystals are not regarded as true chemical compounds. They are described as *clathrates*.

> Hydrated chromium (III) chloride, $CrCl_3.6H_2O$, provides an interesting example of inorganic isomerism. It exists in three forms, one violet and two green. If we add aqueous silver(I) nitrate(V) to a solution of the first, *all* the chlorine is precipitated as silver(I) chloride, but from solutions of the second and third only two thirds and one third of the chlorine are precipitated. This shows that in the violet form all the chlorine is present as chloride ion, but in the others only two thirds and one third of the chlorine occur as chloride ion. In the green varieties one and two chlorine atoms respectively must form part of the complex ion. We can represent the structures of the three chlorides as follows:
>
> $[Cr(H_2O)_6]^{3+}[Cl^-]_3$: all the chlorine precipitated by silver(I) nitrate(V)
> violet
>
> $[Cr(H_2O)_5Cl]^{2+}[Cl^-]_2.H_2O$: two thirds of the chlorine precipitated by silver(I) nitrate(V)
>
> $[Cr(H_2O)_4Cl_2]^+Cl^-.2H_2O$: one third of the chlorine precipitated by silver(I) nitrate(V)

## Correlation of Crystal Structure and Properties

**Hardness.** By 'hardness' we mean the resistance of a material to penetration. Penetration involves the forcing apart of the particles composing the material, and we should expect this to be more difficult the stronger the force holding the particles together. This is indeed the case. Leaving aside metals and alloys (which are considered later in this chapter), we find that the hardest materials are those which are infinite three-dimensional assemblies of atoms or ions joined by covalent or ionic bonds. In the first class are carbon (diamond) and quartz ($SiO_2$), and in the second varieties of aluminium(III) oxide (emery, rubies, etc.), fluorspar ($CaF_2$), and rock salt (NaCl).

In these hard materials strong bonds exist throughout the crystal. This is not so in layer crystals like carbon (graphite) and lead(II) iodide. In these the atoms or ions in a layer are held together by strong covalent or ionic–covalent bonds, but cohesion between the layers is due to the very weak van der Waals force. As a result cleavage between the layers takes place readily. The layers slide easily over each other and rub off or flake easily. This explains why carbon (graphite) feels greasy and why it is used as a lubricant and in pencils. Similarly talc, a naturally occurring form of magnesium(II) silicate(IV), is a good lubricant. It is also called 'soap-stone,' and is used in talcum powder because of its smoothness.

Some crystals consist of discrete molecules held together either by hydrogen bonds or by the van der Waals force. In both cases the cohesion between the molecules is weak, but is appreciably stronger when due to hydrogen bonding. Crystals of ice, sucrose, and ethanedioic acid-2-water, which contain hydrogen bonds, are hard to the touch, but are brittle. They fracture easily owing to the weakness of the hydrogen bonds. If cohesion is brought about by the van der Waals force, as in paraffin wax and naphthalene, the crystals are characteristically soft and greasy, as in carbon (graphite).

**Melting Point.** In a crystal the particles are in a constant state of vibration. When a crystal is heated energy is absorbed and the amplitude of vibration increases. A crystal melts when the thermal energy of its particles is increased sufficiently to overcome the cohesive force which holds them together. The process, however, is gradual as regards the crystal. When melting occurs the temperature remains constant even though heat is being supplied.

*The melting point is the temperature at which the solid and liquid are in equilibrium at a given external pressure* (usually 101 325 Pa).

The temperature at which a solid melts depends primarily on the type of crystal formed. In general, if a crystal is an infinite assembly of atoms or ions (whether in one, two, or three dimensions) it has a high melting point. This is because a large amount of energy has to be supplied to break the strong covalent or electrovalent bonds. If the crystal consists of finite molecules held together by hydrogen bonds or the van der Waals force the melting point is low. This is illustrated in the series of substances now given together with their melting points.

| | Infinite assemblies of atoms or ions | | Molecular crystals | | | |
|---|---|---|---|---|---|---|
| | NaCl | $SiO_2$ | $SiF_4$ | $H_2$ | $O_2$ | $H_2O$ |
| m.p./°C | 804 | 1728 | −127 | −259 | −219 | 0 |

Compounds containing bonds with a high percentage of electrovalent character usually have high melting points. Nevertheless, there is no direct relation between melting point and the amount of ionic character in the bonds. In $SiO_2$ there is only about 40 per cent of ionic character in the Si—O bonds, while the Si—F bonds of silicon tetrafluoride have over 50 per cent of ionic character. The wide difference between the melting point of sodium(I) chloride (or silicon(IV) oxide) and that of silicon tetrafluoride is due to the different types of crystal formed.

The van der Waals force of attraction between molecules increases with their size, that is, with their relative molecular mass. Thus solid oxygen has a higher melting point than solid hydrogen because it has a larger molecule. Oxygen, however, melts at a lower temperature than water in spite of its higher relative molecular mass. This is due to the hydrogen bonds between the water molecules in the ice crystal, additional energy being required to break these bonds.

**Conductivity in the Fused State.** If a crystalline compound is melted and the melt conducts a current and is electrolysed, the crystal must have broken down to give ions. We cannot assume, however, that the ions are present in the crystal. In compounds like sodium(I) chloride we have independent evidence that the crystal is composed of ions. Similarly we know that other compounds contain essentially covalent bonds and are composed of molecules. Between these extremes, however, are many compounds containing intermediate amounts of ionic and covalent character in their bonds.

The amounts of ionic and covalent character in the bonds of many halides like anhydrous zinc(II) chloride and anhydrous aluminium(III) chloride are not known with certainty. Zinc and aluminium have about the same electronegativity (1·5), and therefore according to Hannay and Smyth's rule the percentages of ionic character in Zn—Cl and Al—Cl bonds should be the same (32 per cent). In practice the fused zinc(II) chloride is an electrolyte, but fused aluminium(III) chloride is not. The fused oxide and tetrafluoride of silicon are non-electrolytes although the (calculated) percentages of ionic character in their bonds are 37 per cent and 52 per cent, both of which are higher than for Zn—Cl bonds. Thus there appears to be no direct correlation between the amount of ionic character in the bonds and the kinds of particles (ions or molecules) resulting from fusion.

**Solubility.** Whether a crystalline solid dissolves in a liquid largely depends on the strength of the forces holding the particles together in the crystal, but it also depends on the nature of the particles themselves and on the nature of the solvent.

*Ionic Crystals.* Ionic crystals often dissolve readily in water, but only slightly in organic solvents. The ions in sodium(I) chloride are held together by a strong electrostatic attraction and to separate them into free ions, either in the gaseous state or in solution, energy at least equal to the lattice energy (761 kJ $mol^{-1}$) must be supplied. (The *lattice energy* of an ionic compound is the energy required to dissociate 1 mole of the compound into free ions.) When sodium(I) chloride dissolves the necessary energy is derived mainly from hydration of the ions, as represented in Fig. 9–13.

Each sodium(I) ion gains an envelope of water molecules, which are orientated so that the oxygen atoms are directed (through their lone pairs of electrons) towards the metal ion and the hydrogen atoms away from it. The chloride ions also become surrounded by water molecules, but in this case the attraction is between the lone-pair electrons of the negatively charged ion and the hydrogen atoms with their partial positive charges. Since no definite numbers of water molecules become attached to the ions the 'combination' is usually attributed to simple ion–dipole attraction rather than to formation of definite valency bonds.

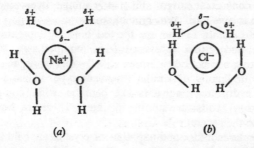

*Fig.* 9–13. Hydrated forms of (*a*) a sodium(I) ion, and (*b*) a chloride ion

Owing to the inductive effect of the ions on the attracted water molecules, polarization of the O—H bonds is increased, and so more water molecules are attracted by dipole–dipole attraction. Thus the envelope of water molecules round each ion may actually be two or three molecules thick. The energy given by complete hydration of the ions in sodium(I) chloride is 748 kJ mol$^{-1}$, made up of 481 kJ mol$^{-1}$ for the Na$^+$ ion and 267 kJ mol$^{-1}$ for the Cl$^-$ ion. This energy, together with a small amount of thermal energy taken from the solvent molecules (dissolving of the salt in water is slightly endothermic) is sufficient to counterbalance the lattice energy. The small solubility in water of salts like silver(I) chloride and barium(II) sulphate(VI) is due to the lattice energies greatly exceeding the energy evolved in hydration of the ions; that is, the attraction between the ions is much greater than that between the ions and water molecules.

Another factor involved in the dissolving of salts in water is the high relative permittivity of water. The larger the relative permittivity of the separating medium the smaller is the attraction between the particles. Consequently liquids of high relative permittivity are good insulators. As water has a relative permittivity of 80 (at 20°C) compared with 4·3 for ethoxyethane and 2·3 for benzene, it is a much better insulating medium than either of these liquids.

We can picture the dissolving of a sodium(I) chloride crystal in water as follows. First the water molecules attract the surface ions and draw them away from their neighbours sufficiently for other water molecules to penetrate between them. The insulating effect of the intruding molecules loosens the grip of the crystal on the displaced ions, so that the latter are readily surrounded ('dissolved') by water molecules.

The slight solubility in water of some metal hydroxides, *e.g.*, $Al(OH)_3$, is explained by these having infinite layer structures of the sandwich type, in which a sheet of metal ions is enclosed between two sheets of hydroxyl groups. The latter form many hydrogen, or hydroxyl, bonds (H-----O—H) with hydroxyl groups in adjacent sandwiches. The holding-together of the layers by these cross-linkages prevents the metal hydroxide from dissolving to any great extent. The freely soluble hydroxides of sodium and potassium have a different type of structure.

*Molecular Crystals.* These are formed both by inorganic substances (*e.g.*, iodine) and organic substances (*e.g.*, naphthalene). For the latter the general rule is that "Like dissolves like"—that is, organic compounds are soluble in liquids of similar chemical type. Thus hydrocarbons dissolve in hydrocarbon solvents like benzene, while compounds containing hydroxyl groups are soluble in water. There are, however, many exceptions to the general rule.

The molecules in iodine and naphthalene crystals are held together only by the weak van der Waals force. The same force operates between benzene molecules. In cases like this the thermal energy of the molecules is sufficient to bring about intermingling and solution occurs. Iodine and naphthalene, however, are only slightly soluble in water. This is due to the hydrogen bonding between water molecules. The latter tend to cohere, so that mixing with the solute molecules is largely prevented.

The presence of certain groups in organic molecules tends to increase the solubility in water. Such groups are described as *hydrophilic* ('water-loving'). They include —OH, —COOH, —NH$_2$, and —SO$_3$H. Thus ethanedioic (oxalic) acid and carbamide (urea) dissolve readily in water because solvent and solute molecules become linked by hydrogen bonds, as illustrated in (i) and (ii) below.

$$
\begin{array}{cc}
\underset{\text{(i)}}{\text{H}\!-\!\overset{\displaystyle\overset{X}{|}}{\text{O}}\cdots\text{H}\!-\!\overset{\displaystyle\overset{H}{|}}{\text{O}}}
&
\underset{\text{(ii)}}{\text{H}\!-\!\overset{\displaystyle\overset{Y}{|}}{\underset{\displaystyle\underset{H}{|}}{\text{N}}}\cdots\text{H}\!-\!\overset{\displaystyle\overset{H}{|}}{\text{O}}}
\end{array}
$$

Here the energy liberated in hydrogen bonding is sufficient to separate the organic molecules against the weak van der Waals force of attraction. As the molecules increase in size, however, so does the van der Waals force, and a stage is reached when the tendency of the organic molecules to cling together exceeds their tendency to bond with water molecules. This explains why the higher alcohols, carboxylic acids, etc., are insoluble in water. The possibilities of hydrogen bonding are increased if there is more than one hydroxyl group in the molecule, and this may result in compounds of quite high relative molecular mass dissolving in water. Thus sucrose ($C_{12}H_{22}O_{11}$) with eight —OH groups in its molecule is readily soluble.

## Metals

**Crystalline Structure of Metals.** The crystalline state of metals is easily shown by leaving a strip of zinc in lead(II) nitrate(V) solution. Lead is displaced by zinc and branching crystals of lead grow outwards from the surface of the zinc. Crystals of metallic elements differ from those of non-metallic elements in the ways now given.

* Metal crystals are infinite three-dimensional assemblies of atoms. Non-metal crystals are usually made up of discrete molecules (*e.g.*, $I_2$), although they may also consist of infinite assemblies of atoms in three dimensions, as in carbon (diamond), in two dimensions, as in carbon (graphite), or even in one dimension, as in selenium.
* The type of bond (the 'metallic' bond) holding the atoms together in a metal crystal is quite different.
* The bonding in metal crystals is non-directional in character.

The last represents an important difference between metal crystals and the infinite three-dimensional assembly of atoms in carbon (diamond). In the latter the packing of the atoms is determined by the tetrahedral distribution of the four valency bonds of the carbon atom, and each carbon atom has only four 'nearest neighbours.' Non-directional bonding also occurs in crystals of ionic compounds, but in this case the packing depends on the relative sizes and relative numbers of the different ions present. The atoms of a metal, however, are similar in size. Thus the restrictions on method of packing found in other infinite three-dimensional structures do not apply to metals. As a result metal atoms are able to pack together in a crystal with much greater compactness and the crystal has a high density.

In general metals crystallize either in the cubic system or in the hexagonal system. As we saw earlier, there are two common forms of cubic space lattice—the face-centred cubic and the body-centred cubic. Models of these two forms can be constructed from spheres of equal size as in Figs. 9–14 and 9–15.

The face-centred cubic lattice is called a *close-packed* arrangement because it represents one of the most compact methods of packing together spheres of equal size. In this arrangement each atom is touching twelve 'nearest neighbours,' that is, it has a co-ordination number of 12.

We can see this if we put together two of the face-centred cubes as in Fig. 9–16. Here dotted lines are drawn from one of the atoms to its twelve nearest neighbours (for clarity the remaining atoms in the lattice are omitted).

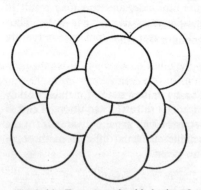

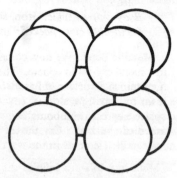

*Fig.* 9–14. Face-centred cubic lattice of copper (close-packed)

*Fig.* 9–15. Body-centred cubic lattice of iron (not close-packed)

In the body-centred cubic arrangement the atoms are not close-packed. Each atom has only eight nearest neighbours, although there are six other neighbours (the atoms in the centre of the six adjacent cubes) at slightly greater distances. This arrangement gives a high degree of compactness, although not quite as high as in the face-centred cubic lattice.

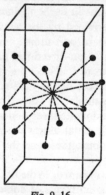

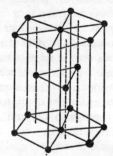

*Fig.* 9–16

*Fig.* 9–17. Unit cell of hexagonal close-packed structure

In the third common type of metal lattice the atoms again have a co-ordination number of 12. This type is known as the hexagonal close-packed (Fig. 9–17).

The structures of the various metals are divided fairly evenly between the three types, although a few metals have more complicated lattices.

We do not know why a metal adopts one type of structure in preference to another. Below are shown the kinds of lattice possessed by some of the well-known metals.

Face-centred cubic: aluminium, copper, silver, gold, platinum, and lead.

Body-centred cubic: iron, sodium, and potassium.

Hexagonal close-packed: magnesium and zinc.

**Metallic Bond.** We now come to the question of what causes the atoms in a metal crystal to cohere. The number of valency electrons in the outer quantum shell of a metal atom is usually only one, two, or three, and this is far too small for normal covalent bonds to be formed with the eight or twelve nearest neighbours in a metal crystal. Again, crystals of metals are distinguished from those of non-metals and compounds by their high electrical conductivity. To explain the latter we must assume that at least some of the electrons in a metal are able to move freely through the crystal, whereas the electrons in other solids are restricted to movement round particular atomic or ionic nuclei. In the second case the electrons are *localized*, while in the first some at least are *non-localized*. These considerations indicate a special type of binding force between the atoms in metals. The following is a simplified explanation of this force based on Pauling's theory of resonance.

Consider what happens when sodium vapour is cooled so that firstly liquid sodium, and then solid sodium, is formed. Sodium vapour consists mostly of single atoms, which have one electron in the $3s$ sub-level of their outer quantum shell. When the vapour liquefies the atoms are brought close together, and the charge clouds of the valency electrons of two adjacent atoms overlap. This results in combination of the two atoms, so that both complete their $3s$ orbitals and a molecule, $Na_2$, is formed (sodium vapour actually contains a small proportion of these molecules). In the molecule the bonding electrons have a combined charge cloud and occupy a common molecular orbital around both atomic nuclei (compare the hydrogen molecule in Fig. 7–5).

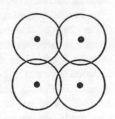

*Fig.* 9–18

However, the outer quantum shells of both sodium atoms are still far from full. In liquid or solid sodium interaction between the charge clouds of valency electrons is not limited to two atoms, but can occur with a large number of atoms. This is illustrated for only four atoms in Fig. 9–18. Here the valency electrons would form a combined charge cloud embracing all four atomic nuclei. They would not belong to any particular atom, but would 'resonate' between all four. As the electrons travel at high speeds, pairs of electrons could establish covalent bonds between different pairs of atoms many thousands of times per second, causing the atoms to cohere. Alternatively we can think of our system as composed of sodium ions held together by resonating electrons as represented in (i) and (ii) below.

$$Na^+ \quad {}_e^e \quad Na^+ \qquad Na^+ \; Na^+ \qquad Na^+ \quad {}_e^e \quad Na^+$$

$$e\;e \quad e\;e \qquad e\;e$$

$$Na^+ \quad {}_e^e \quad Na^+ \qquad Na^+ \; Na^+ \qquad Na^+ \qquad Na^+$$

(I)            (II)            (III)

For bonding to take place as described the resonance would have to be synchronized; that is, the transference of bonds between atoms would have to occur simultaneously. There is evidence, however, that the bonds resonate independently. This means that at a given instant a proportion of the atoms have two bonds, while others have none, as indicated in (iii) above. Pauli's exclusion principle limits the number of electrons in the $3s$ orbital to two, and for a sodium atom to form two bonds another orbital is required. The sodium atom, however, has three empty $p$ orbitals, so that the extra electrons can be accommodated in one of these. This presents no difficulty from the energy point of view because in metal crystals, where the atoms are tightly packed together, the energy differences between sub-levels in the same quantum shell are so small as almost to disappear completely. As we shall see in the next section, these empty orbitals in a partially filled quantum shell are an essential feature of metallic character. They are often referred to as *metallic* orbitals.

In a metal crystal resonance of the valency electrons extends throughout the crystal, so that the electrons form a composite electron cloud. The attraction between the metal ions and the electron cloud constitutes the *metallic bond*. Note that this bond is fundamentally covalent. Thus the distance between the nuclei in a sodium crystal is not twice the ionic radius (0·095 nm), but twice the single-bond covalent radius (0·186 nm). As mentioned earlier, metallic bonding differs from ordinary covalent bonding in being non-directional. In this respect it resembles electrovalent bonding.

**Electrical Conductivity of Metals.** Normally the electrons move at random in the electron cloud in a metal crystal. However, when a potential difference is applied to the two ends of a metal conductor there is a flow of electrons towards the positive pole, while fresh electrons are drawn into the conductor from the negative pole. This does not mean that the whole of the electron cloud moves bodily in the direction of the positive pole. To explain the electrical conductivity of metals the following analogy is sometimes used.

Imagine that we have a cardboard box partly filled with small steel balls arranged in layers. The balls represent the electrons in a metal crystal arranged in energy sub-levels, those in the bottom layer having the lowest energy. By means of a rod the balls in the upper layers can be moved about in random fashion, simulating the random movement of electrons in an electron cloud. Most of the balls remain in their own layer, but a few may climb on top of the others. This corresponds to some of the 3s electrons in the sodium crystal moving into the unoccupied 3p level.

Now suppose that a large and powerful magnet, M (Fig. 9–19a) is brought up to one end of the box. The steel balls in the top layers move as a whole towards the magnet, climbing on top of the lower layers to do so. In much the same way, when two parts of a metal have a different potential some of the electrons of the electron cloud resonate into an unoccupied orbital at a slightly higher energy level. They then move as a

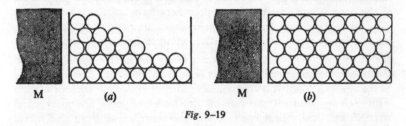

M  (a)  M  (b)

*Fig.* 9–19

body to the part of the metal deficient in electrons. Thus the empty ('metallic') orbital is of fundamental importance for the flow of electrons which constitutes a current.

Taking our analogy further, we can imagine that our cardboard box is completely full of layers of steel balls and that a lid is placed over the box (Fig. 9–19b). If the magnet is brought up as before there can be no movement of the balls towards it. This situation corresponds to trying to pass a current through an insulator such as diamond. In the latter the quantum shells occupied by electrons are full, and there is no orbital available to act as a conducting orbital for electrons. Conceivably, if the magnet were made increasingly strong the attraction on the balls would eventually become so great that the lid would burst, and the balls would then move towards the magnet as previously. Similarly, if a sufficiently high voltage is applied to diamond its insulation breaks down. This is because electrons jump from the outer full quantum shell into the empty one above it.

If the electrons had complete freedom of movement in flowing through the conduction level of a metal crystal there would be no electrical resistance. This arises from imperfections in the crystal lattice, the presence of impurities, and the thermal vibration of the ions. The conductivity of copper is greatly reduced by traces of other metals or of the oxide, and for electrical purposes it is necessary to purify the metal electrolytically. With metals of high purity the resistance is provided chiefly by the vibrations of the ions impeding the passage of the electrons. At higher temperatures the amplitude of vibration becomes larger, and the resistance increases. On the other hand at temperatures approaching absolute zero the resistance of many metals almost disappears, and the metals become *superconductors*.

**Mechanical Strength of Metals.** The strength of a metal is its resistance to deformation under stress. The latter may be applied in different ways. The piston of an internal combustion engine experiences a compressive stress, while a steel rope used for suspending a lift cage is subjected to a tensile, or pulling, stress. Hardness, malleability, ductility, and elasticity are different aspects of mechanical strength.

The strength of a metal depends on several factors—the cohesive force between the electron cloud and the ions, the size of crystals, imperfections in the crystal lattices, and the purity of the metal. The cohesive force is determined by the number of electrons per atom contributed to the electron cloud. The number may vary from one for the alkali metals to six for some of the transition metals (*e.g.*, manganese and iron). In so far as the strength depends on the cohesive force it shows a rough correlation with melting point. Thus sodium, tin, and lead have low mechanical strength and low melting points, while in manganese, iron, and nickel both are high.

Under small stresses metals are *elastic*. This is because the ions in two adjacent crystal planes can roll part way over each other without affecting the resonance of the electrons. When the stress is removed the ions spring back to their former positions. If the stress is increased beyond a certain value, the ions begin to slide completely over each other. Each ion acquires a new set of nearest neighbours, but it can form bonds with these just as readily as with the previous ones. This is responsible for malleability and ductility, the metal being able to undergo a certain amount of deformation without fracture.

Hardness is incompatible with malleability and ductility. A hard metal has a high resistance to deformation. Most pure metals are soft, but their hardness can be increased by mechanical working (*e.g.*, by rolling). This breaks up the larger crystals into small crystal grains, which have different orientations. The sliding of the ions over each other under stress then tends to be arrested at the grain boundaries. *Annealing* has the opposite effect. Annealing consists of heating the metal to some temperature below its melting point. Owing to their increased thermal energy the ions rearrange themselves into larger and more stable aggregates.

In practice the physical properties of metals are greatly affected by imperfections which are always present in the crystals. The imperfections may consist of 'holes' in the lattice where ions are missing, or they may be due to irregularities (*dislocations*) in the crystal planes. These imperfections reduce mechanical strength by facilitating slip. They allow ions to migrate from place to place, so that the imperfection is transferred to a different site. This may result in a dangerous fracture if the imperfection becomes located at a part of the metal which is under heavy stress.

**Alloys.** The hardness of a metal can often be increased by alloying it with another metal (or with carbon or silicon in the case of iron). As with other liquids, two molten metals may be immiscible (like oil and water), completely miscible (like ethanol and water), or partially miscible (like ethoxyethane and water). The latter means that the metals have a limited solubility in each other. If two metals are immiscible an alloy cannot be made. Thus when lead and aluminium are melted together they separate into two layers, the lighter aluminium being on top.

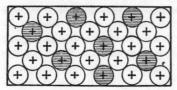

*Fig.* 9–20. Substitutional solid solution of silver in gold (or zinc in copper)

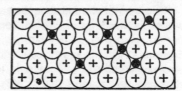

*Fig.* 9–21. Interstitial solid solution of carbon in iron above 906°C

Silver and gold are completely miscible. If a molten mixture of the two metals in any proportions is cooled and solidified a *substitutional solid solution* is formed. It is described in this way because an atom or ion of one metal can freely substitute an atom or ion of the other in the face-centred cubic lattice of either. A random distribution of 'solute' atoms in the 'solvent' metal is produced (Fig. 9–20). For a substitutional solid solution to be formed the atoms of the two metals must have about the same size and a similar electronic structure. This is true for silver and gold, the atomic radii actually being equal (0·144 nm). Two other metals which form substitutional solid solutions in all proportions are copper and nickel, which have atomic radii of 0·128 nm and 0·124 nm respectively.

As the difference in size between the atoms of the two metals increases the mutual solubility decreases. This is because the difference in size results in distortion of the lattice of the solvent metal and the extent to which the lattice can be distorted is limited. Thus in brass up to 38 per cent of copper atoms can be replaced by zinc atoms (radius 0·133 nm). If this proportion of zinc is exceeded a mixture of two kinds of crystals is produced on solidification. One kind is a solid solution of zinc in copper, the other a solid solution of copper in zinc.

Although most substitutional solid solutions consist of a random distribution of 'solute' atoms in 'solvent' atoms this is not always so.

Sometimes an orderly arrangement of 'solute' atoms is produced if the alloy is cooled sufficiently slowly. This happens with copper and gold.

A less common, but very important, type of solid solution is formed by carbon with iron. At ordinary temperatures iron has a body-centred cubic lattice, but when the metal is heated above 906°C the lattice changes to the face-centred cubic type. In the latter a 'hole' is left in the middle of the lattice, and this can be occupied by the much smaller carbon atom. In this way an *interstitial* solid solution of carbon in iron is formed (Fig. 9–21). However, in spite of their small size the carbon atoms produce some strain in the iron lattice, and not all the 'holes' in the crystal can be filled with carbon atoms. A maximum of 2 per cent by mass of carbon can be dissolved in the metal above 906°C. Iron containing up to 2 per cent of carbon is described as *steel*.

When iron containing dissolved carbon is cooled below 906°C the body-centred cubic lattice forms again. The distance between the metal atoms is now such that even the small carbon atoms cannot be accommodated between them without considerable strain. Hence most of the carbon is thrown out of solution. It does not, however, form particles of carbon, but combines with some of the iron to give crystals of a compound called cementite ($Fe_3C$). The latter hardens the metal and increases its strength, but reduces malleability. The forging and rolling of steel is usually carried out above 906°C, when the carbon is in solution and the metal is malleable.

Two metals which differ markedly in electropositive character may alloy together by compound formation. In some of these intermetallic compounds (*e.g.*, $Mg_2Sn$ and $Mg_2Pb$) the ordinary valencies of the metals are observed. In other cases (*e.g.*, $CuAl_2$ and $NaSn$) this is not so. The exact nature of many of these compounds is still unknown.

## EXERCISE 9

**1.** What evidence is there that a crystal of common salt is composed of ions, while an ice crystal consists of molecules? Explain the differences in physical properties which result from the different compositions of the crystals.

**2.** What factors determine the kind of crystal formed by an ionic compound? Describe the crystal structures of sodium(I) chloride and caesium(I) chloride, and explain why they differ.

**3.** Give a definition of 'melting point'. Explain, from the viewpoint of the kinetic theory, what changes take place when the temperature of a solid is raised to the melting point.

Describe concisely how you would measure the melting point of a solid, of which only a few milligram were available.

Explain carefully the differences between what takes place during the melting of pure and impure specimens of a substance, and show how the melting point may be used as a criterion of the purity and for the identification of substances.

(J.M.B.)

**4.** What is meant by 'water of crystallization'? Why is the term a misnomer? Explain with examples the different ways in which water of crystallization can enter into the composition of crystals.

**5.** Explain why: (i) sodium(I) chloride dissolves readily in water; (ii) hydrated aluminium(III) hydroxide is almost insoluble in water; (iii) iodine is only slightly soluble in water, but dissolves readily in ethanol; (iv) sucrose is only slightly soluble in ethanol, but dissolves readily in water.

**6.** What is a *metal*? Discuss from the point of view of their structure the following properties of metals: (i) high density, (ii) high melting point, (iii) good electrical conductivity, (iv) hardness, (v) malleability.

## MORE DIFFICULT QUESTIONS

**7.** (*a*) Distinguish in electronic terms between electrovalent (ionic), covalent, and dative covalent bonding.

(*b*) Show how the results of the study of the structures of solids illustrate these types of chemical bonding and demonstrate the reality of the molecule and ion as chemical entities. (J.M.B.)

**8.** Discuss the validity or otherwise of the term *molecule* when applied to the entities commonly formulated as follows: (*a*) $H_2O$, (*b*) $AlCl_3$, (*c*) $SiO_2$, (*d*) NaCl, (*e*) $NiSO_4 . 7H_2O$. (W.J.E.C.)

**9.** Suggest a method by which the Avogadro constant, $N_A$, could be found from measurements carried out on a copper crystal. (Copper crystallizes in the face-centred cubic system.)

**10.** What is an *alloy*? Describe the different ways in which two metals might form an alloy. Describe and explain (with examples) three ways in which the properties of metals can be improved by alloying.

# CHAPTER 10

# The gaseous and liquid states

## The Gaseous State

**Deviations from Boyle's Law.** In the simple model of a gas furnished by the kinetic theory (Ch. 2) a gas is pictured as an assembly of molecules travelling in straight lines with a high average velocity. The molecules are constantly colliding, but as they are supposed to be perfectly elastic no momentum is lost on impact. The simple kinetic theory makes two important assumptions:

* The volumes of the molecules themselves are negligible in comparison with the total volume of the gas.
* The molecules do not exert any attraction or repulsion towards each other.

However, no gas is a *perfect gas*, that is, one which obeys the gas laws at all temperatures and pressures. Thus Boyle's law does not hold for any gas at high pressures, the discrepancies being greater the lower the temperature. The deviations of gases from Boyle's law were investigated about 1870 by the French physicist Amagat, who used pressures up to 320 atm. Fig. 10–1 shows how the value of $pv$ varies with pressure for different gases.

In 1873 van der Waals pointed out two reasons why the value of $pv$ for a gas varies with the pressure at a fixed temperature. At high pressures the molecules are close together and the volume taken up by the molecules is no longer negligible. Increased pressure has no effect on the size of the molecules themselves. Thus, if $v$ is the volume of the gas and the space occupied by the molecules is $b$, the compressible part of the total volume is only $v - b$. Allowing for the volume of the molecules, we can amend Boyle's law to $p(v - b) =$ a constant.

Again, if the molecules of a gas attract each other, a molecule in the interior will be attracted equally on all sides by other molecules, but this will not be true for a molecule on the outside. The pressure of a gas is due to molecular impacts on the walls of the container. The velocity of a molecule just about to make an impact will be reduced by the attraction of molecules towards the interior of the gas. For a given mass of gas the number of molecules striking the walls in unit time is inversely proportional to the volume $v$. Also the number of molecules which exert an attraction on a molecule just before impact is again inversely proportional to $v$. We should therefore add to the observed pressure a term depending on the reciprocal of $v^2$. Hence, correcting for both the finite size of the

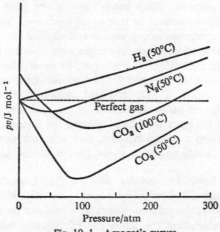

*Fig.* 10–1. Amagat's curves

molecules and their mutual attraction, we can write the law for gases in the form suggested by van der Waals, that is, for 1 mole of gas

$$\left(p + \frac{a}{v^2}\right)(v - b) = RT$$

where $a$ and $b$ are constants.

Van der Waals' equation gives a closer approximation to the behaviour of a gas than the simpler equation $pv = RT$. It will be noticed that the corrections applied affect the magnitude of $pv$ in opposite ways, the one tending to increase the product and the other to decrease it. Since the value of $pv$ for nitrogen and carbon dioxide decreases at first as the pressure is increased, the effect of molecular attraction is more important than the effect of molecular volume at lower pressures, although at higher pressures the position is reversed. With hydrogen (at 50°C) $pv$ increases from the beginning, indicating that the deviation due to the finite size of the molecules more than counterbalances the deviation due to molecular attraction.

Note that forces of both attraction and repulsion exist between molecules. Attraction arises from the van der Waals forces. Repulsion is due to interaction between the electron shells of the molecules. At relatively large distances attraction predominates between two molecules. If the molecules come very close together repulsion is the stronger force. Thus when two molecules approach each other directly their velocities increase until the forces of attraction and repulsion balance. After this, repulsion between the electron shells first brings the molecules to rest and then reverses the direction of the velocities.

**Avogadro's Hypothesis.** Since Boyle's law is not exact it follows that Avogadro's hypothesis cannot be true at all pressures. Suppose we have 1 dm$^3$ of a gas at atmospheric pressure and double the pressure while keeping the temperature constant. If the gas were a perfect gas the volume would decrease to 500 cm$^3$. For hydrogen, however, the volume would be slightly more than 500 cm$^3$, and for carbon dioxide it would be slightly less. If equal volumes of hydrogen and carbon dioxide at the same temperature contained an equal number of molecules at 1 atm pressure equal volumes of the gases would not contain the same number of molecules at 2 atm pressure. Actually, then, Avogadro's hypothesis will hold only when gases behave as perfect gases, that is, at very low pressures.

**Gay-Lussac's Law.** For similar reasons Gay-Lussac's law is only approximately true. If 200 cm$^3$ of hydrogen united with 100 cm$^3$ of oxygen for a certain value of the pressure this ratio would not be maintained at a different pressure because the alteration of pressure would affect the volumes of hydrogen and oxygen differently. At the standard pressure the ratio of combination by volume of hydrogen and oxygen is not exactly 2 : 1, but 2·0027 : 1 (Morley). Again, Gay-Lussac's law is valid only when gases are at very low pressures.

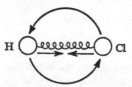

*Fig.* 10–2. Bond stretching and molecular rotation in the hydrogen chloride molecule

**Charles's Law.** The energy possessed by molecules because of their movement in straight lines is called *translational* energy. The temperature of a gas is determined by the magnitude of the translational energy of its molecules. In addition molecules normally have *rotational* energy and *vibrational* energy. Rotational energy is associated with rotation of the molecule as a whole (Fig. 10–2). Vibrational energy results from the valency bonds in the molecule behaving like spiral springs connecting small masses. The masses can vibrate in different ways. In diatomic molecules like $Cl_2$ and HCl vibration is limited to 'bond stretching,' the two atoms alternately approaching and receding from each other. In triatomic (and larger) molecules not only is there bond stretching, but the atoms also have a rocking vibration, in which they move from side to side, or up and down.

When molecules of a gas collide (particularly when the gas is at a high temperature) some of their translational energy can be converted into vibrational and rotational energy. If this happens the molecules are not behaving as perfectly elastic particles. Hence the principle of conservation of momentum does not hold, and Charles's law is not obeyed. Although the deviations due to this cause are relatively small they have considerable practical importance in connection with the behaviour of gases at high temperatures (as in the exhaust of a rocket motor).

The *internal* absorption of energy by gas molecules when heated is responsible for the ratio of the molar heat capacities ($C_p/C_V$) of polyatomic gases being less than the value $\frac{5}{3}$ (p. 58).

**Infrared Spectra.** Vibrations in the valency bonds of molecules are affected by absorption of energy from electromagnetic radiation. The vibrations respond, however, only to certain frequencies of radiation because the energy absorption occurs in definite quanta. Absorption of a quantum from the ultraviolet or more energetic visible regions of the spectrum may cause valency bonds to rupture. This happens with chlorine,

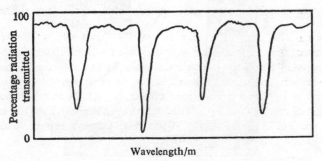

*Fig.* 10–3. General form of infrared absorption spectrum

which is dissociated into atoms. Less violent effects follow from absorption in the infrared, a quantum in this region having only about one twentieth of the energy of a quantum in the ultraviolet. Here absorption of a quantum merely causes the atoms to vibrate more vigorously.

Different quanta of energy are absorbed according to the atoms joined by a bond, the nature of the bond itself (whether single, double or triple), and the type of vibration (stretching or rocking). Thus when infrared radiation is passed through a thin film of substance (solid, liquid, or in solution) the latter gives rise to an infrared absorption spectrum of the form shown in Fig. 10–3, in which the percentage of radiation transmitted is plotted against wave length. Pronounced dips indicate wavelengths at which strong absorption occurs.

For absorption of infrared radiation by a molecule to occur the atoms must carry partial electric charges. This is because the oscillating electric field associated with the radiation has no effect on the neutral atoms of a molecule such as $Cl_2$. Thus an infrared absorption spectrum furnishes evidence of the polarization of covalent bonds. These spectra are characteristic for each compound and, like X-ray spectra, can be used in its identification. This is particularly useful in organic chemistry, where we often need to distinguish between different isomers. Infrared spectra can also be used for recognizing the presence of particular atomic groupings (such as $C - C$, $C = C$, or $C = O$), each grouping in a given class of compound absorbing radiation of specific wavelength.

## Liquefaction of Gases

**Critical State of a Gas.** Any gas can be liquefied at atmospheric pressure if it is cooled sufficiently. Many gases (but not all) can be liquefied at ordinary temperatures by compressing them. The conditions of temperature and pressure under which gases liquefy were first investigated by Andrews in 1869.

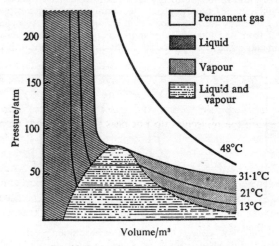

*Fig.* 10–4. Isothermals of carbon dioxide

Andrews subjected carbon dioxide to various pressures at different temperatures and plotted the effect of pressure on the volume (Fig. 10–4). The curve obtained at a given temperature is called an *isothermal*. At 48°C the volume of the gas decreased with increased pressure approximately in accordance with Boyle's law, giving a rectangular hyperbola. At 21°C, however, the volume first diminished in accordance with Boyle's law until the pressure was increased to about 60 atm. At this pressure there was a sudden break in the curve and liquid carbon dioxide appeared. The pressure remained constant until all the gas had been converted into liquid. Subsequent increase of pressure caused practically no change in volume, in accordance with the general rule that extremely high pressures are required to compress liquids appreciably.

Similar changes took place when the isothermals were constructed for temperatures below 21°C, except that the pressure required to liquefy the gas became smaller as the temperature decreased. Andrews found that liquefaction could be brought about at all temperatures below 31·1°C, but above this temperature no liquefaction occurred no matter how much the pressure was increased. This temperature was therefore called the *critical temperature* for carbon dioxide. The pressure (75 atm) required to liquefy the gas at the critical temperature was called the *critical pressure*, and the volume of 1 mole of the substance at the critical temperature and pressure the *critical volume*.

Every gas has its critical temperature, above which it cannot be liquefied by increase of pressure. The critical temperatures in degrees Celsius of some common gases are as follows:

| | | | |
|---|---|---|---|
| Sulphur dioxide | 157° | Nitrogen | −147° |
| Ammonia | 132° | Hydrogen | −240° |
| Oxygen | −119° | Helium | −268° |

Above its critical temperature a gas may be regarded as a 'permanent' gas, since it cannot be liquefied. Below the critical temperature it is more truly described as a vapour, since it can always be liquefied if the pressure is increased sufficiently.

Above the critical point liquid and gaseous states coincide. With increased pressure the molecules do not pass suddenly from a condition in which there is a considerable distance between them to one in which they are closely packed together. The change is gradual. We express this by saying that there is *continuity of state*. From the critical temperatures given for oxygen and nitrogen we see that liquid air cannot exist at ordinary temperature. If, however, air at ordinary temperatures is compressed to 3000 atm pressure its molecules become so tightly packed that its density is the same as that of water.

To understand the critical state of a gas we must know why gases liquefy. If liquefaction merely involved bringing the molecules together we should expect the change to be always continuous and gradual. In practice, below the critical temperature there is a sharp transition from the gaseous state to the liquid state. Two factors are involved in liquefaction of a gas. These are intermolecular attraction, which tends to bring the molecules together, and the thermal energy of the molecules, which tends to keep them apart. If a gas is at any temperature below its critical temperature and the pressure is gradually increased, the molecules come closer together and attract each other more strongly. Moreover, the molecular dipoles mutually reinforce each other, increasing the attraction still further. The effect of decreasing the distance thus builds up at an ever-increasing rate, and a stage is reached when the molecules 'fly together,' or liquefy. In much the same way iron filings suddenly jump to a magnet which is brought sufficiently close. Liquefaction of a gas occurs when intermolecular attraction becomes the dominant factor in comparison with the thermal energy of the molecules. Clearly we can achieve the same result by keeping the pressure constant and cooling the gas.

We can now see why a gas fails to liquefy above its critical temperature. Above this temperature the dominant factor is always the thermal energy of the molecules. The latter can be brought closer by increasing the pressure, but they cannot be made to 'fly together,' no matter how much the pressure is increased.

### Methods of Liquefying Gases.

(1) *Compression and External Cooling.* Any gas which has a critical temperature above room temperature can be liquefied by pressure alone. Such gases include sulphur dioxide, hydrogen sulphide, carbon dioxide, ammonia, chlorine, and hydrogen chloride. The pressure required for liquefaction is less the lower the temperature (see Fig. 10–4). Compression of a gas causes evolution of heat due to work done on the gas, a common

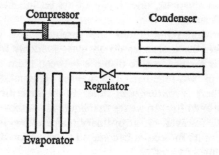

*Fig.* 10–5. Principle of the household refrigerator

illustration of this being the rise in temperature of air when compressed in a bicycle pump. For ease of liquefaction it is necessary to remove the heat of compression by external cooling.

A simple application of these principles is the ordinary household refrigerator (Fig. 10–5). The gas used is Arcton, difluoro-dichloro-methane ($CCl_2F_2$). This liquefies at $-30°C$ under atmospheric pressure. (Ammonia is still used as the refrigerant in many industrial refrigerating plants.) The gas is compressed by a pump and passes into the condenser coils, which are fitted with cooling fins. As heat is lost to the air the gas condenses to liquid. The liquid passes through a regulating valve, on the other side of which the pressure is low. Here evaporation occurs, latent heat of vaporization being taken from the surroundings with consequent cooling. After evaporation the Arcton returns to the compressor and is recycled.

A number of gases (*e.g.*, $Cl_2$, $SO_2$, $CO_2$) are liquefied on a large scale by compression and external cooling. Sometimes greater cooling is achieved by means of a gas which has already been liquefied. In the industrial preparation of liquid chlorine the gas is obtained as a by-product in making sodium(I) hydroxide from common salt by electrolysis. The chlorine is washed with water to remove dust and carbon dioxide, and dried with concentrated sulphuric(VI) acid. It is then compressed to about 4 atm and liquefied in condenser pipes surrounded by a jacket in which liquid Arcton is evaporating.

(2) *Cooling by Performance of External Work.* When a gas expands under adiabatic conditions (that is, so that heat is neither gained nor lost externally) it does work in increasing the volume against the external pressure. The work is performed at the expense of the molecular kinetic energy, which is reduced. The temperature of the gas therefore falls. This is the opposite of the heating effect obtained with the bicycle pump.

Cooling a gas by making it perform external work was first used in gas liquefaction by the Frenchman Cailletet in 1877. Among the gases which he succeeded in liquefying was oxygen. The oxygen was compressed and cooled in liquid sulphur dioxide. When the pressure was suddenly released the cooling effect of expansion was sufficient to reduce the temperature below the critical temperature of oxygen and liquefy the gas. This method of cooling is now often called the *Claude process* after the Frenchman Claude, who was the first to apply it to the liquefaction of air on a commercial scale.

(3) *Joule–Thomson Effect.* The value of $pv$ for nitrogen and most other gases decreases with increase of pressure up to moderately high pressures (Fig. 10-1). This indicates that the volume decreases to a bigger extent than with a perfect gas—that is, nitrogen is more compressible than a perfect gas at these pressures. This is due to the mutual attraction of the molecules. When nitrogen or air is expanded through a fine aperture work is done internally in separating the molecules against the attraction. The expenditure of energy in this way is done by the gas itself, which therefore becomes cooled. With hydrogen, which shows an increase in $pv$ with pressure, expansion through a fine aperture results in heat being evolved. However, if the temperature is reduced below $-79°C$ (the *inversion temperature*) the $pv$ curve becomes similar to that of nitrogen and cooling takes place on expansion. The consequent temperature change is called the *Joule–Thomson effect* after its discoverers J. P. Joule and Sir William Thomson (later Lord Kelvin).

The Joule–Thomson effect was used for the first liquefaction of hydrogen (b.p. −253°C). This was accomplished in 1898 by Dewar, whose invention of the vacuum flask did much to facilitate the storage of liquefied gases. Today liquid hydrogen is made on a large scale for use as a fuel in rockets. The method employed is essentially that of Dewar. Hydrogen is compressed to a pressure of 100 atm and cooled in liquid nitrogen. The gas then passes through an expansion valve and is cooled. By jacketing the tube containing hydrogen before expansion with the tube containing hydrogen after expansion the cooling in made cumulative. Ultimately, the gas reaching the valve is so cold that when it expands through the valve liquefaction occurs.

**Liquefaction of Air.** In the modern method of liquefying air cooling is achieved partly by the performance of external work by the gas and partly by the Joule–Thomson effect.

Filtered dust-free air is compressed to a pressure of 150 atm, the heat of compression being removed by cold water. Carbon dioxide is absorbed in sodium(I) hydroxide solution, and moisture by 'silica gel'. The air then travels down a spiral tube surrounded by a cooling jacket, through which is passing very cold nitrogen from the last stage of the process. The cooled compressed air then divides into two streams. One stream expands into a cylinder, where it drives an engine, which operates the compression pump of the earlier stage. The gas therefore does useful work, and its temperature is further reduced. The second stream expands through a fine nozzle. Here the fall in temperature due to the Joule–Thomson effect liquefies part of the air. The two streams then reunite and the partially liquefied air passes into a fractionating column, where liquid oxygen and liquid nitrogen separate. Much of the nitrogen remains as uncondensed cold gas. This passes out at the top of the column and is used, as described previously, for cooling the compressed air in the earlier stage.

The separation of liquid oxygen and liquid nitrogen is based on the fact that the mixture constitutes a pair of miscible liquids similar to a mixture containing methanol and water. The general principle of the separation is discussed in Chapter 12. At a pressure of 101 325 Pa liquid oxygen boils at −183°C and liquid nitrogen at −196°C. Liquid oxygen has a beautiful blue colour.

Argon (b.p. −186°C) is separated from liquid oxygen by further fractionation in an 'argon column.' Neon (b.p. −253°C) and helium (b.p. −269°C) are not liquefied in the process of liquefying air because of their extremely low boiling points. They can be obtained as liquids, however, by means of the Joule–Thomson effect if subjected to preliminary cooling by liquid hydrogen. It was in this way that the Dutchman Kammerlingh Onnes first prepared liquid helium in 1908. Some of the modern commercial forms of helium liquefier are based on the Joule–Thomson effect, while others utilize the Claude process. The peculiar behaviour of liquid helium below 2·19 K is described in Chapter 11.

## The Liquid State

**Relation of Liquids to Gases and Solids.** We have seen earlier that the tendency of the molecules in a gas to cohere owing to mutual attraction is outweighed by their tendency to remain apart as a result of their thermal energy. We have also seen that when a gas liquefies the situation is suddenly reversed. In a liquid intermolecular attraction is supreme, although the thermal energy of the molecules still prevents them from occupying fixed positions. In a solid attraction between the molecules is even more dominant, the movement of the molecules being restricted to vibration about a mean position. Thus, although the liquid state is intermediate between the gaseous and solid states, it appears to be closer to the latter than the former. This is confirmed in several ways.

(1) *Density.* The densities of most substances in the liquid state are little different from their densities in the solid state. This is because the average distance between the molecules in the liquid is nearly the same as that between the molecules in the solid. In the gas, however, the average distance between the molecules is greatly increased, and there is a corresponding reduction in density.

(2) *Energy Differences.* The specific latent heat of fusion represents the difference in energy between 1 g of a substance in the solid state at the melting point and 1 g of the substance in the liquid state at the same temperature. Specific latent heat of vaporization varies with temperature, but for a fixed temperature it similarly represents the energy difference per gram between the liquid and vapour states. If we multiply the latent heats of fusion and vaporization of different substances by their relative molecular masses, the products are the *molar heats of fusion* and the *molar heats of vaporization*. These represent the heat required to convert the same number of molecules of different substances from solid to liquid and from liquid to vapour respectively. Some values of molar heats of fusion ($L_f$) and molar heats of vaporization ($L_v$) are given below (molar heats of vaporization being at the boiling point of the liquid).

|  | $L_f/$J mol$^{-1}$ | $L_v/$J mol$^{-1}$ |
|---|---|---|
| Water | 5999 | 40 590 |
| Ammonia | 5614 | 23 280 |
| Ethanol | 4598 | 39 330 |
| Benzene | 9823 | 30 650 |

The conversion of the solid into liquid requires considerably less energy than the conversion of the liquid into vapour. The same is true of ionic compounds and metals. Thus in their energy content liquids are closer to solids than to gases.

(3) *Order and Disorder of Particles.* In earlier chapters we saw that

the characteristic feature of a solid is the orderly arrangement of its particles. In contrast a gas is characterized by the disorderly distribution of its particles. It was formerly thought that liquids resembled gases and had a random arrangement of particles, but we now know this is not true. The same forces which hold molecules together in a crystal still hold them together in the liquid, although less firmly. It may therefore be expected that the molecules in a liquid will tend to form structures similar to those in the crystal. Owing to the thermal movement of the molecules, however, such structures will be temporary, constantly breaking down and re-forming, and in many cases the structures will be only partially complete. This is illustrated by means of spheres in two dimensions in Fig. 10–6. In the 'solid' each sphere is in contact with six other spheres, while in the 'liquid' a given sphere may be in contact with six five, or even four spheres, as illustrated by the spheres numbered 1, 2, and 3 respectively.

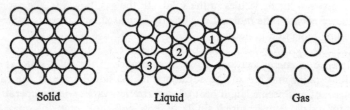

Solid                    Liquid                    Gas

*Fig.* 10–6. In a solid there is both 'short range' and 'long range' order of particles, in a liquid only 'short range' order. In a gas the particles are disordered.

Practical evidence of the presence of orderly structures in liquids is obtained from X-ray diffraction patterns of liquids. These patterns bear a general resemblance to those given by finely powdered crystals, except that the circles representing diffraction maxima are more spread out and blurred. If liquids were completely devoid of structure they would scatter X-rays continuously and there would be no diffraction maxima. From the X-ray patterns obtained we can often deduce approximate distances between the particles and hence the particle distribution. Thus we find that in liquid sodium each atom is in contact with an average of ten other atoms instead of twelve, as in the close-packed structure of the solid.

It was formerly thought that liquid water consisted of definite polymers such as $H_4O_2$ and $H_6O_3$, but this view is no longer held. When ice melts there is a general tendency for the tetrahedral structure (shown in Ch. 7) to persist. In the liquid there is a continual building-up and breaking-down of the tetrahedral structure resulting from hydrogen bonding, and this produces the effect of association into definite polymers. Ice differs from most other solids in melting with decrease in volume. In the collapsed tetrahedral structure of water at 0°C the molecules come together more closely than in ice, and this results in higher density. With increase of temperature still further collapse of the tetrahedral structure takes place, and the density increases. An opposing factor, however, is the normal tendency for expansion to occur with rise of temperature. Beyond 4°C the second factor becomes dominant, and the density of the liquid decreases. In ice each water molecule has four nearest neighbours. X-ray diffraction patterns given by water show that in the liquid just above 0°C the average number of nearest neighbours is 4·4, which is little more than in ice. This is evidence that the tetrahedral structure persists in the liquid. As the temperature rises, however, the average number of nearest neighbours increases, reaching a maximum of 4·9 at 63°C. This indicates that at higher temperatures it is increasingly difficult for a tetrahedral structure to be formed.

The picture we get of a liquid from the foregoing is more complicated than that of either a solid or a gas. The essential orderliness of the solid state enables us to treat solids mathematically (as in finding the dimensions of unit cells). Gases can also be treated mathematically because they have essentially a disorderly structure, which lends itself to a statistical approach (as in the deduction of average molecular velocities). The liquid state represents an ever-changing pattern of order and disorder, for which no satisfactory mathematical treatment has yet been evolved. From the order–disorder point of view, however, liquids appear to stand much closer to solids than to gases.

**Equilibrium between Solid, Liquid, and Vapour.** Each of the three homogeneous physical states (solid, liquid, and gas) in which a substance can exist is called a *phase*. Except under special conditions only two of these—solid and vapour or liquid and vapour—can be in equilibrium. If a little liquid is introduced into the vacuum above mercury in a barometer tube some of the liquid evaporates and an equilibrium is established between liquid and vapour. This is said to be a *one-component system* because only one chemical individual is involved in the equilibrium. An equilibrium of this kind can be affected in two ways; that is, there are two variable factors. The temperature can be changed, and so can the vapour pressure. However, these cannot be altered independently of each other. Thus, if the temperature is changed the vapour pressure also changes, and for any particular temperature there is always a fixed value for the vapour pressure.

Solids also exert a vapour pressure, although this usually becomes appreciable only at higher temperatures. Some solids (*e.g.*, iodine and naphthalene) have a measurable vapour pressure at ordinary temperatures, and hence they evaporate when exposed to the air. Ice evaporates below 0°C.

The relation between vapour pressure and temperature of a substance in the solid and liquid states can be represented graphically. The graph has the form shown in Fig. 10–7 for normal substances (*e.g.*, benzene), these being substances which melt with increase in volume. PQ shows the increase of vapour pressure of the solid with increase in temperature. The solid melts at Q and with further rise of temperature the vapour pressure curve QR of the liquid is obtained. QR represents the conditions of temperature and pressure under which the liquid and vapour phases are in equilibrium. The curve ceases at R, the critical point of the substance, and at higher temperatures liquid and vapour are indistinguishable.

At the melting point, Q, solid, liquid, and vapour are in equilibrium. Thus the melting point of the solid (or freezing point of the liquid) is the temperature at which the solid and liquid have the same vapour pressure. In the case being considered the substance is only under its own vapour

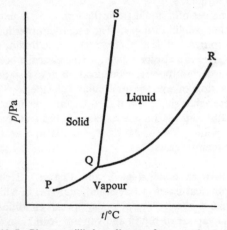

*Fig.* 10–7. Phase equilibrium diagram for normal substances

pressure. Therefore Q does not represent the normal melting point, or freezing point, which is measured at a pressure of 101 325 Pa. Q is called a *triple point*. Only at the temperature and pressure represented by Q can the three phases coexist in equilibrium.

The third curve, QS, shows the effect on the melting point, or freezing point, of increasing the pressure externally. The effect is in accordance with **Le Chatelier's principle**, which states:

*If a system in equilibrium is subjected to a constraint, that change takes place which tends to remove the constraint.*

This famous principle, which we shall meet frequently in considering physical and chemical equilibria, has been aptly called "the law of sheer cussedness." Thus, if the pressure on our system is increased the increase in pressure is opposed by the system reducing its volume. For normal substances decrease in volume occurs when liquid turns to solid. Hence at a higher pressure a higher temperature is required to melt the substance. This is indicated by QS sloping upwards to the right.

Fig. 10–8 is the phase equilibrium diagram for water. The lines P'Q', Q'R', and Q'S' represent the conditions of equilibrium between two phases as described in the case of normal substances. However, Q'S', representing the effect of pressure on melting point, or freezing point, slopes upwards to the *left*. Ice is exceptional in melting with decrease in volume, and in this case the melting point, or freezing point, is lowered by increased pressure. When water is under its own vapour pressure only it freezes at 0·0076°C, and its vapour pressure is then 612 Pa. At 101 325 Pa pressure it freezes at 0°C.

In both Fig. 10–7 and Fig. 10–8, the phase equilibrium diagram for a

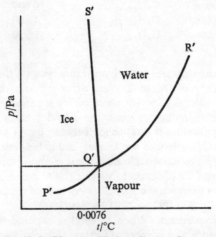

*Fig.* 10–8. Phase equilibrium diagram for water

substance below its critical temperature is divided into three *areas*. These show the conditions of temperature and pressure under which solid, liquid, and vapour respectively can exist. Each *line* on the diagram represents the conditions of equilibrium for two of the phases, and the *triple point* the conditions for all three phases to be in equilibrium.

### Physical Properties of Liquids.

(1) *Vapour Pressure*. The vapour pressure of a liquid depends on the temperature, and its molar heat of vaporization.[1]

---

[1] The following theoretical relationship can be deduced from thermodynamics:

$$\log_{10} p = -\frac{L_V}{2\cdot303\,RT} + C$$

where $p$ is the vapour pressure, $L_V$ is the molar heat of vaporization, $R$ is the gas constant, $T$ is the kelvin temperature, and $C$ is a constant.

At a given temperature the vapour pressure is governed by the relative tendencies of the molecules to disperse (because of their thermal movement) and to cohere (owing to their mutual attraction). The larger the molecules in the liquid the smaller is the molecular velocity, but the greater is the van der Waals attraction between the molecules. Hence in a series of similar liquids at the same temperature we should expect the vapour pressure to decrease with increasing relative molecular mass. This is found to be the case. Thus the vapour pressures of the first four liquid alkane hydrocarbons at 20 °C decrease in the order now shown.

|         | Formula      | Vapour pressure/Pa |
|---------|--------------|--------------------|
| Pentane | $C_5H_{12}$  | 56 250             |
| Hexane  | $C_6H_{11}$  | 16 000             |
| Heptane | $C_7H_{16}$  | 4 800              |
| Octane  | $C_8H_{18}$  | 1 330              |

Vapour pressure increases with temperature because the latter increases the thermal energy of the molecules, while the van der Waals forces decrease as the molecules become more widely separated. Thus at higher temperatures more molecules near the liquid surface acquire escape velocity, and equilibrium is established between liquid and vapour when a larger number of molecules are leaving and entering the liquid surface in a given time. Since the specific latent heat of vaporization is the energy required to convert 1 g of liquid to vapour this also decreases with rise of temperature. For example, the specific latent heat of vaporization of water falls from 2490 J $g^{-1}$ at 0°C to 2250 J $g^{-1}$ at 100°C.

(2) *Rate of Evaporation*. When a liquid is exposed to air its rate of evaporation depends on a number of factors, including its vapour pressure, molar heat of vaporization, molar heat capacity, and other factors. The most important of these is its vapour pressure. In general, those liquids evaporate most quickly which have the highest vapour pressure. This can be tested roughly by dropping $\frac{1}{2}$ cm$^3$ of ethoxyethane, benzene, water, and phenylamine on to filter-papers and noting how long the liquids take to disappear. The times will be in the order given, which is the same as the order of the vapour pressures.

(3) *Boiling Point*. *The boiling point of a liquid is the temperature at which its vapour pressure is equal to the pressure above it.* Just as melting point is a guide to the strength of the forces holding the particles together in the solid state, so boiling point is a measure of the cohesive forces in the liquid. If these forces are weak we find that not only is the boiling point low, but the difference between the melting point and boiling point is small. As the cohesive forces increase in strength the boiling point rises, and the interval between the melting point and boiling point becomes larger (Table 10–1).

Table 10-1

|  | m.p./°C | b.p./°C | Difference/°C |
|---|---|---|---|
| Hydrogen | −259 | −253 | 6 |
| Nitrogen | −210 | −195 | 15 |
| Benzene | 5·5 | 80 | 74·5 |
| Tetrachloromethane | −23 | 77 | 100 |
| Sodium(I) chloride | 804 | 1413 | 609 |
| Potassium(I) iodide | 723 | 1420 | 697 |
| Iron | 1539 | 2900 | 1361 |
| Tungsten | 3380 | 5700 | 2320 |

In liquid hydrogen and liquid nitrogen the small van der Waals attraction between the molecules is readily overcome by increase in temperature. With larger molecules (benzene and tetrachloromethane) the attraction is bigger and more energy is required to vaporize the liquid. In ionic compounds and in metals much of the strength of bonding in the solid state remains in the liquid state. We therefore find that boiling points are high, and there is a wide temperature gap between these and melting points.

The boiling points of liquids are also increased in those cases in which hydrogen bonding results in molecular association, the increase being due to the additional energy required to break the hydrogen bonds. Thus methane, which is unassociated, and has a relative molecular mass of 16, boils at −161°C, whereas water, with only a slightly higher relative molecular mass, boils at 100°C. The smaller degree of association in the lower aliphatic alcohols as compared with that in water is also reflected in the respective boiling points. Although the relative molecular mass (46) of ethanol is nearly three times that of water, ethanol boils at a lower temperature (78°C). In ethoxyethane both hydrogen atoms of the water molecule are replaced by alkyl groups, and hydrogen bonding is no longer possible. Ethoxyethane therefore consists of simple molecules, and although its relative molecular mass (74) is higher than that of ethanol it boils at a lower temperature (35°C).

A connection between the boiling point of a liquid, its relative molecular mass, and latent heat of vaporization was discovered by Trouton in 1884, and is called *Trouton's rule*. If $T$ is the boiling point on the Kelvin scale and $L_V$ is the *molar* heat of vaporization at the boiling point,

$$L_V/T = k$$

where $k$ is a constant. $k$ has a value of 88 J mol$^{-1}$ K$^{-1}$, approximately for most liquids. For liquids which obey Trouton's rule the boiling point on the kelvin scale is therefore proportional to the molar heat of vaporization. Table 10-2 shows how the rule applies in the case of some common liquids.

Table 10-2

| Liquid | $L_V$/J mol$^{-1}$ | Boiling point/K | $\dfrac{L_V}{T} = k$/J mol$^{-1}$ K$^{-1}$ |
|---|---|---|---|
| Ethoxyethane | 26 000 | 308 | 84·4 |
| Benzene | 30 650 | 353 | 86·8 |
| Tetrachloromethane | 29 850 | 350 | 85·3 |
| Propanone | 30 220 | 329 | 91·9 |
| Hexane | 29 480 | 342 | 86·1 |
| Water | 40 590 | 373 | 109 |
| Ethanol | 39 330 | 351 | 112 |

Trouton's rule does not hold for associated liquids like water and the lower alcohols. The reason for this is readily understood. The latent heat of vaporization of a liquid represents the energy expended in increasing the volume of the substance against the external pressure, and in overcoming the cohesion between the molecules. In unassociated liquids the cohesion is due only to the weak van der Waals force. In water and the lower alcohols there is also hydrogen bonding between the molecules, and more energy is required to break the hydrogen bonds. The latent heat of vaporization is thus larger, and the value of $k$ is increased. This provides strong evidence of association in these liquids.

(4) *Surface Tension.* Molecules in the interior of a liquid are attracted equally on all sides by the molecules around them. Molecules at the surface of a liquid are attracted only inwards and sideways. As a result the liquid surface is always under tension and tends to contract so as to reduce the surface area to a minimum. Thus in air drops of liquid tend to assume a spherical shape because for a given volume a sphere has the smallest surface area. The mutual attraction of the molecules in the liquid surface produces a resistance to penetration since work has to be done to force the molecules apart. The surface has a certain amount of 'hardness,' which is basically similar to the hardness of a solid metal surface. Thus we can float small metal objects such as a greased steel needle on water, and certain insects can move freely over a water surface without getting wet.

The **surface tension,** $\gamma$ (*gamma*), *of a liquid at a given temperature is defined as the force in newton acting parallel to the surface along a line one metre in length in the surface and at right angles to the line.*

If we imagine a cut one metre long to be made in the surface of the liquid the force tending to re-unite the two portions of the surface would be $\gamma$.

The magnitude of $\gamma$ varies with the strength of the attraction between the particles in the liquid. The surface tensions of fused ionic compounds are higher than those of molecular liquids, while those of fused metals are higher again. In each case the values decrease with rise of temperature and vary slightly with the nature of the gas in contact with the liquid. Some typical values for liquids in contact with air are given in Table 10–3, where boiling points are also included for comparison.

**Table 10–3**

|  | $\gamma/10^{-3}\,N\,m^{-1}$ | Boiling point/°C |
|---|---|---|
| *Molecular liquids* (at 20°C) | | |
| Ethoxyethane | 17·0 | 35 |
| Hexane | 18·4 | 69 |
| Ethanol | 22·3 | 78 |
| Propanone | 23·7 | 56 |
| Benzene | 28·9 | 80 |
| Water | 72·8 | 100 |
| *Fused salts* (at 1000°C) | | |
| Sodium(I) bromide | 88 | 1390 |
| Sodium(I) chloride | 98 | 1413 |
| *Fused metals* | | |
| Mercury (0°C) | 470 | 357 |
| Silver (970°C) | 800 | 2180 |
| Aluminium (700°C) | 840 | 2400 |

There is a rough correlation between surface tensions and boiling points. In general we can say that *liquids of high boiling point have a high surface tension, and vice versa.* For accurate comparison, however, we should have to measure the surface tensions at the boiling points of the liquids, and this is not possible.

### EXERCISE 10

**1.** Describe and interpret the results which are obtained when (i) a given mass of carbon dioxide is compressed isothermally at a series of temperatures between 0° and 50°C; (ii) a quantity of ethoxyethane is heated in a small tube attached to a pressure gauge and the pressure recorded is plotted against temperature, the experiment being repeated a number of times using the same tube but with a larger amount of ethoxyethane each time.

What explanation may be offered for the deviations in the behaviour of carbon dioxide from that required by the simple gas law, $pv = RT$, and how may this gas law be amended to take into account these deviations? (J.M.B.)

**2.** How do the volumes of (*a*) ideal (perfect) gases, (*b*) real gases vary with pressure and temperature? Illustrate your answer by referring (graphically or otherwise) to the real gases hydrogen, oxygen, and carbon dioxide, and indicate the two chief reasons for the failure of real gases to obey the ideal gas laws.

What do the terms 'critical temperature' and 'critical pressure' signify?

(J.M.B.)

**3.** What are the postulates of the kinetic theory of a perfect gas? From the kinetic equation, $pv = \frac{1}{3}Nmu_{r.m.s.}^2$, and any other necessary assumptions, deduce Boyle's law, Charles's law, Graham's law, and Avogadro's hypothesis. How may the observed deviations from Boyle's law in the behaviour of actual gases by explained? (O.L.)

**4.** Discuss the principles underlying the modern method of manufacturing oxygen from air.

**5.** Describe and explain how the physical properties of liquid water are affected by hydrogen bonding between the molecules. Why does an increase of pressure lower the freezing point of water, but raise the freezing point of benzene?

## MORE DIFFICULT QUESTIONS

**6.** Discuss the contributions made by Amagat, Andrews, Avogadro, and Boyle to our knowledge of the behaviour of gases.

A sample of nitrogen, known to contain traces of carbon dioxide and water vapour, occupied a volume of 200 dm³ at 18°C and 100 200 Pa pressure. When it was passed through concentrated sulphuric(VI) acid and a concentrated aqueous solution of potash, the respective increases in mass were 0·738 g and 14·52 g. What was the percentage composition by volume of the sample and the partial pressure of each constituent? (H = 1; C = 12; O = 16; molar volume at s.t.p. = 22·4 dm³.) (J.M.B.)

**7.** Show how the study of the deviations of the behaviour of gases from (a) Boyle's law and (b) the equation $pv = nRT$ led (i) to van der Waals modifications of the gas equation, (ii) to the development of methods for liquefying gases.

(J.M.B.)

**8.** What light is thrown on the constitution of liquids by (i) X-ray diffraction patterns, (ii) boiling points, (iii) latent heats of vaporization?

# CHAPTER 11

# Polymorphism and allotropy

A substance which can crystallize in more than one form is *polymorphic*. The same atoms or ions give different forms in different ways. With sodium(I) chloride, two crystal forms are built up from the same space lattice, or unit cell, but usually the different forms belong to separate crystal systems and have a different type of unit cell. When polymorphism occurs in elements the phenomenon is usually described as *allotropy*, but this term is also used to describe different liquid and gaseous forms of elements.

**Enantiotropic and Monotropic Polymorphism.** If we add potassium(I) iodide solution to mercury(II) chloride solution, we obtain mercury(II) iodide as a yellow precipitate which turns red on standing. X-ray analysis shows that both forms consist of layer crystals, in which the bonds are essentially covalent, but the arrangement of the atoms in the layers is different in the two cases. Mercury(II) iodide is therefore dimorphic. If we heat the red form slowly it passes into the yellow form at a temperature of 126°C. Below this temperature (the *transition temperature*) the red form is stable, while above it the yellow is the stable form.

$$\underset{\text{red}}{HgI_2} \underset{}{\overset{126°C}{\rightleftharpoons}} \underset{\text{yellow}}{HgI_2}$$

If we heat red mercury(II) iodide quickly it sublimes and condenses on a cool surface in the yellow form. This persists for a time, but is slowly converted back into the red variety, a change which can be accelerated by rubbing the substance with a glass rod. Below the transition temperature yellow mercury(II) iodide is *metastable*. This means that it is a form of intermediate stability, the Greek prefix *meta* signifying either 'between', or 'change'. This yellow solid is intermediate in stability because it has a lower energy than the vapour, but a higher energy than the red solid. Heat is evolved when the yellow form changes into the red. The appearance of the yellow polymorph before the red one is an illustration of **Ostwald's law of successive reactions**:

> *When it is possible for a stable and metastable form of a substance to be produced the metastable form is produced first.*

The rate at which a metastable form changes into a stable form at ordinary temperatures and pressures varies with the substance. With mercury(II) iodide the transformation occurs in a matter of minutes, but in other cases it may be too slow to be noticeable.

Two types of polymorphism are recognized according to whether the change from one solid form to the other is directly reversible, or not, at atmospheric pressure. If, as happens with mercury(II) iodide, there is a definite transition temperature, the polymorphism is *enantiotropic*. If the

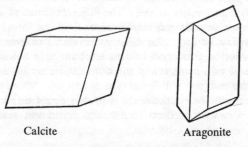

Calcite                              Aragonite

*Fig.* 11–1. Polymorphic forms of calcium(II) carbonate

change is not directly reversible (that is, there is no transition temperature) the polymorphism is *monotropic*. In this case only one form is stable over the whole temperature range at which the polymorphs exist, any other form being metastable.

Calcium(II) carbonate furnishes an interesting example of monotropic polymorphism. The stable form of this compound is the rhombohedral *calcite* (Fig. 11-1), which is found as chalk, limestone, marble, and Iceland spar. A metastable form is the rhombic *aragonite*, which occurs in coral. Aragonite is converted into calcite when it is heated to 400–500°C. Very small calcite crystals are produced from carbon dioxide and *cold* calcium(II) hydroxide solution, aragonite crystals from carbon dioxide and a *hot* solution. (We might expect from Ostwald's law that aragonite, and not calcite, would be precipitated from the hot solution. Actually, another metastable form is first deposited and then this changes rapidly to calcite.) The two forms behave differently towards a solution of a cobalt(II) salt.

**Experiment.** (i) Bubble carbon dioxide into cold calcium(II) hydroxide solution until a precipitate is formed. Add red cobalt(II) nitrate(V) solution to the tube. There is no change of colour.

(ii) Boil calcium(II) hydroxide solution and pass in carbon dioxide. Cool the precipitate under the tap and again add cobalt(II) nitrate(V) solution. The contents of the tube turn violet and if they are filtered a violet residue is left on the filter paper.

The different results in these experiments occur because aragonite has an appreciable solubility in water, whereas calcite is practically insoluble. A little aragonite dissolves and forms a violet precipitate of cobalt(II) carbonate (basic) with the cobalt(II) nitrate(V) solution.

## Allotropy

Allotropy is the property which some elements possess of existing in different forms with different properties. The chief differences are usually found in the physical properties, although there are sometimes differences in the chemical properties as well. The different forms of an element are called *allotropes*, or *allotropic modifications*.

*Enantiotropic allotropy* is the name given when one form of an element on being heated is converted into another form at a definite transition temperature—*e.g.*, $\alpha$-sulphur to $\beta$-sulphur. This is similar to enantiotropic polymorphism.

*Monotropic allotropy* is the name used when only one of the forms is stable at atmospheric pressure. There is no transition temperature, any other form being metastable whatever the temperature (providing both forms exist). Theoretically the metastable form is spontaneously converted to the stable form, but in practice the change may be too slow to be appreciable at ordinary temperatures. It usually takes place on heating, as in the conversion of phosphorus (white) to phosphorus (violet).

*Dynamic allotropy* describes the phenomenon when two forms coexist in equilibrium, the proportion of each present depending on the temperature—*e.g.*, dioxygen ($O_2$) and trioxygen ($O_3$).

**Allotropy of Sulphur.** At ordinary temperatures and pressures sulphur is dimorphic. It can exist as rhombic crystals ($S_\alpha$) and monoclinic crystals ($S_\beta$). Claims have been made for the existence of other crystalline forms, prepared under special conditions, but there is doubt about the authenticity of some of these forms and we shall ignore them. In plastic sulphur the element is obtained in an amorphous condition.

Sulphur exhibits enantiotropic allotropy. At ordinary temperatures the stable allotrope is the rhombic form ($S_\alpha$). If this is heated to just above 95·5°C (the transition temperature), it changes completely into the monoclinic form ($S_\beta$). The transition is marked by a change in direction of the vapour pressure–temperature curve (at A in Fig. 11–2). On further heating to 120°C, the monoclinic form melts (B). There are thus separate vapour pressure curves for $S_\alpha$, $S_\beta$, and liquid sulphur.

By heating sulphur (rhombic) rapidly we can raise it to its melting point,

C, without conversion into sulphur (monoclinic). Melting occurs at 113°C. Above the transition temperature $S_\alpha$ is unstable. When liquid sulphur is cooled needle-shaped crystals of $S_\beta$ are deposited. These can exist below the transition temperature (as indicated by the broken line AD), but are then metastable. The crystals change in a few hours into the

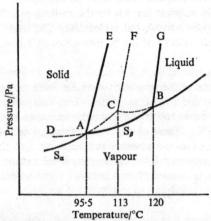

*Fig.* 11–2. Phase equilibrium diagram for sulphur

rhombic form. Notice that, in accordance with the general rule, the stable form has a lower vapour pressure than the unstable or metastable form at the same temperature.

The transition of $S_\alpha$ to $S_\beta$ is attended by increase in volume. So also are the melting of $S_\alpha$ and $S_\beta$. Hence, in accordance with Le Chatelier's principle, the transition temperature and both melting points are raised by increasing the pressure. This is shown by the lines AE, CF, and BG sloping upwards to the right.

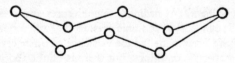

*Fig.* 11–3. The octasulphur molecule ($S_8$)

The structural units of sulphur (rhombic) have been shown by X-ray analysis to consist of $S_8$ molecules. These form a puckered ring (Fig. 11–3), alternate atoms lying in the same plane. The interbond angle is 105°, but it is uncertain whether the atoms are joined by single or double bonds. The structure of sulphur (monoclinic) is thought to be similar; that is, it probably consists of cyclic octasulphur molecules. The density of sulphur (rhombic) is 2·06 g cm$^{-3}$ while that of the monoclinic form is 1·96 g cm$^{-3}$.

Liquid sulphur provides an example of dynamic allotropy.[1] Just above the melting point (120°C) a mobile yellow liquid is formed. If we pour this into cold water it yields an 'amorphous' solid, which is completely soluble in carbon disulphide. The solid is probably microcrystalline sulphur (monoclinic). Determination of the freezing-point of the solution shows the dissolved sulphur consists of $S_8$ molecules. Hence it is concluded that liquid sulphur just above the melting point is composed of *cyclo*-sulphur (cyclic octasulphur molecules). The mobility of the liquid is explained by the ability of the $S_8$ molecules to move freely past each other.

If the liquid is heated to higher temperatures before it is poured into the cold water, part of the resulting solid is insoluble in carbon disulphide. There are also significant changes in the viscosity of the sulphur at higher temperatures. Above 160°C the viscosity increases rapidly, reaching a maximum at 187°C. These changes are explained as follows. At about 160°C the $S_8$ rings become unstable and open out into chains, which join by their end free valencies into very long molecular chains. The chainlike form of molecule (*catena*-sulphur) is insoluble in carbon disulphide.

$$Cyclo\text{-sulphur} \rightleftharpoons catena\text{-sulphur}$$
$$\text{(S}_8\text{ rings)} \qquad \text{(chains)}$$

The increase in viscosity above 160°C is caused by the long molecular chains (which may contain up to 100 000 atoms) becoming entangled with each other. After attaining its maximum at 187°C, the viscosity decreases because the long molecular chains begin to break up into shorter ones under the influence of heat. If liquid sulphur above 187°C is poured into cold water a rubberlike mass of 'plastic' sulphur is obtained. Like real rubber this can be stretched to several hundred times its length, and it gives a 'fibre' pattern on X-ray analysis. These facts confirm the presence of the long tangled chains.

**Allotropy of Phosphorus.** The two common forms of this element are now conventionally described as phosphorus (white) and phosphorus (violet). In the past the first form has often been known as yellow phosphorus, and the second as either red phosphorus or scarlet phosphorus. Phosphorus (white) is a metastable variety and changes slowly into its allotrope even at ordinary temperatures. The violet form cannot be converted directly into the white form, but must first be vaporized, phosphorus (white) being obtained on condensation of the vapour. The two forms are thus an example of monotropic allotropy. Phosphorus (white) has the higher vapour pressure at a given temperature, and is soluble in carbon disulphide, benzene, olive oil, etc. The violet form is insoluble in these liquids. In monotropic allotropy the metastable form usually has

---

[1] For a full discussion of the allotropy of liquid sulphur see *J. Amer. Chem. Soc.*, **65**, 639 and 648.

a greater solubility as well as a higher vapour pressure. Phosphorus (violet) has the greater density.

The two forms of phosphorus also differ chemically. Phosphorus (white) is poisonous, burns spontaneously in oxygen and chlorine, and gives phosphine with aqueous sodium(I) hydroxide. The violet form is non-poisonous, combines with oxygen and chlorine only when heated, and does not react with aqueous alkalis. Greater energy is associated with the white allotrope than with the violet one, as shown by the larger heat of combustion of the former.

$$4P + 5O_2 \rightarrow 2P_2O_5 \qquad \text{(Heat evolved} = 1200 \text{ kJ mol}^{-1})$$
(white)

$$4P + 5O_2 \rightarrow 2P_2O_5 \qquad \text{(Heat evolved} = 1460 \text{ kJ mol}^{-1})$$
(violet)

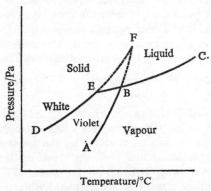

*Fig.* 11–4. Phase equilibrium diagram for phosphorus

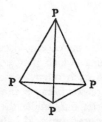

*Fig.* 11–5. The phosphorus molecule ($P_4$)

The phase equilibrium diagram of phosphorus (Fig. 11–4) is typical of elements showing monotropic allotropy. AB represents the variation with temperature of the vapour pressure of the stable form (in this case, the violet form) which melts at B. BC is the vapour pressure curve of the liquid. DE shows the vapour pressure of the metastable form (the white form) melting at E. The curve BE is a continuation of CB. By extrapolating the vapour pressure curves of the two solid forms until they meet at F, we can deduce the temperature and pressure at which both forms would be in equilibrium if melting did not occur. F represents a theoretical transition point. In monotropy the transition temperature is higher than the melting points of both forms, whereas in enantiotropic allotropy it is lower (see Fig. 11–2).

In phosphorus (white) four atoms are joined together by single covalent bonds to give a molecule $P_4$, which has the shape of a regular tetrahedron (Fig. 11–5). By forming three *p–p* bonds with three other atoms each phosphorus atom acquires four completed orbitals in its outer quantum shell. The angle (60°) between the bonds is much less than would be expected for four completed orbitals, and the bonds are probably in a condition of strain. This view is supported by the high chemical activity of phosphorus (white) and by its tendency to change spontaneously into the more stable violet form. The structure of phosphorus (violet) has not yet been established.

A third allotrope, phosphorus (black), is obtained by heating the white form at high pressures. Its chief interest is its resemblance to carbon (graphite). Thus it has a layer structure and conducts electricity.

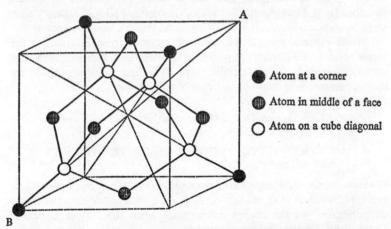

● Atom at a corner

▦ Atom in middle of a face

○ Atom on a cube diagonal

*Fig.* 11–6. Unit cell of the carbon (diamond) crystal. If the book is turned and the diagram viewed in the direction AB the tetrahedral distribution of the atoms will be seen to correspond to that given in *Fig.* 7–6.

**Allotropy of Carbon.** Carbon exists as carbon (diamond) and carbon (graphite). Carbon (charcoal) was once thought to be a third, amorphous, form, but X-ray analysis has shown that it is microcrystalline and has the same structure as carbon (graphite).

As seen in Chapter 7, carbon (diamond) consists of an infinite assembly of carbon atoms joined by single covalent bonds and having a tetrahedral distribution. The crystals belong to the cubic system. The unit cell is composed of a face-centred cube with one additional carbon atom situated on each of the four cube diagonals. The unit cell is illustrated in Fig. 11–6, but for clarity three of the corner atoms of the cube have been omitted. Each atom on a cube diagonal is attached to one corner atom and three atoms in the middle of a face.

Carbon (graphite) has the layer structure shown in Fig. 9–12. Some of the properties resulting from this structure have already been described. Each carbon atom in the interlocking hexagonal rings is joined to three other atoms by bonds which (as a result of resonance) are intermediate in character between single and double bonds. Evidence for this is found in the bond length (0·142 nm), which is intermediate between the length (0·154 nm) of the C—C bond in ethane and the length (0·133 nm) of the C=C bond in ethene. The bonds in carbon (graphite) resemble those between the carbon atoms in benzene, in which, however, the bond length is slightly less (0·139 nm). This allotrope is able to conduct electricity because one of the four valency electrons of each carbon atom is non-localized.[1] The non-localized electrons can travel not only round each hexagonal ring, but over the whole ring system composing a sheet of atoms. Thus carbon (graphite), like a metal, has a cloud of mobile electrons (it also has a metallic lustre). In contrast all four valency electrons in carbon (diamond) are localized and this is a non-conductor.

Carbon exhibits monotropic allotropy, carbon (graphite) being the stable form of the element at ordinary temperatures and pressures, and carbon (diamond) the metastable form. The first has a slightly lower energy content, as shown by the heats of combustion.

$$C + O_2 \rightarrow CO_2$$
carbon (graphite)    (Heat evolved $= 393\cdot3$ kJ mol$^{-1}$)

$$C + O_2 \rightarrow CO_2$$
carbon (diamond)    (Heat evolved $= 395\cdot1$ kJ mol$^{-1}$)

Owing to the small difference in energy content the transition of metastable carbon (diamond) to stable carbon (graphite) does not occur at ordinary temperatures, but the change takes place when the former is heated above 2000°C in the absence of air.

Owing to the relatively large distance between the sheets of atoms carbon (graphite) has a smaller density (2·3 g cm$^{-3}$) than carbon (diamond) (3·5 g cm$^{-3}$). Also, the transition of the first to the second is endothermic. Hence the transition should be favoured by high pressure and high temperature. This is actually the case. Diamonds are manufactured directly from carbon (graphite) in the U.S.A. by using pressures approaching 100 000 atm, a temperature of 2000°C, and various metal catalysts (*e.g.*, nickel).

### Further Examples of Allotropy.

*Tin.* Tin occurs in three forms: rhombic crystals, tetragonal crystals, and a form known as tin (grey), with the carbon (diamond) structure. The

---

[1] In carbon (graphite) (and benzene) hybridization of carbon atomic orbitals occurs in the same way as in ethene (p. 180). Promotion of one *s* electron to the *p* sub-level is followed by hybridization of the orbital of the remaining *s* electron with those of *two* of the three *p* electrons. This gives three *sp²* hybrid orbitals (which overlap with those of three adjacent atoms) and leaves one *p* electron, which is the non-localized electron.

different forms are connected by transition temperatures, so that the element is an example of enantiotropic allotropy.

$$\text{Tin (grey)} \underset{13°C}{\rightleftharpoons} \text{tin (tetragonal)} \underset{161°C}{\rightleftharpoons} \text{tin (rhombic)}$$

Ordinary white tin (tetragonal) keeps for long periods below 13°C without any apparent change, owing to the slowness of the transformation. Prolonged exposure of tin articles to temperatures below 0°C, however, produces grey spots on the metal ('tin plague'). Tin (grey) is a non-metallic form of the element. Its density is much lower than that of the white form and it is a very poor conductor of electricity.

*Dioxygen and Trioxygen.* These are the 'recommended' names for oxygen ($O_2$) and ozone ($O_3$) when the allotropy of the element is being discussed. Trioxygen is formed when a silent electric discharge is passed through dioxygen or when dioxygen is exposed to ultraviolet light. In both cases absorption of energy by the molecules brings about partial dissociation into free atoms, which then combine with other molecules:

$$O_2 \rightleftharpoons 2O \qquad O_2 + O \rightleftharpoons O_3$$

Dioxygen and trioxygen illustrate dynamic allotropy, the proportions of each gas present in the equilibrium mixture depending on the temperature. At room temperature the equilibrium proportion of trioxygen is too small to be measurable. The mixture of dioxygen and trioxygen from the silent electric discharge is not in equilibrium. If it is passed through a hot tube trioxygen changes back to dioxygen much faster, and almost no trioxygen remains. The overall conversion of dioxygen to trioxygen is endothermic. Therefore the formation of the latter is favoured by higher temperatures in accordance with Le Chatelier's principle. The equilibrium proportion of trioxygen in the mixture increases from about 0·1 per cent at 1600°C to about 1 per cent at 2300°C.

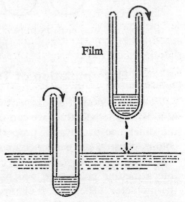

Fig. 11–7. Peculiar behaviour of liquid helium II

*Liquid Helium.* At atmospheric pressure helium condenses to a colourless liquid at 4 K. With further cooling the liquid does not freeze to a solid unless the pressure is increased to 25 atm. However, at 2·19 K and under its own vapour pressure a new kind of liquid helium is formed. Because of its peculiar properties this liquid is called helium II to distinguish it from the normal liquid, helium I.

The two liquids are immiscible and form separate layers. Helium II has a very high thermal conductivity, but practically no viscosity. If a vessel is lowered part-way into the liquid the helium climbs up the sides and into the vessel (Fig. 11–7). If the vessel is now raised the liquid climbs up the inside, flows down the outside, and drips from the lowest point. This 'siphoning' effect appears to be caused by a film of liquid (only about a hundred atoms thick) which spreads over the surface of the vessel. Liquid helium II has been described as a 'superfluid' and as a 'fourth state of matter.' Since there is a transition temperature the two forms of liquid are an example of enantiotropic allotropy.

*Hydrogen.* Hydrogen provides an illustration of dynamic allotropy. The element exists in two forms which differ in the relative directions of spin of the two atomic nuclei in the molecule. In *ortho-hydrogen* the two spins are aligned, while in *para-hydrogen* they are opposed (Fig. 11–8).

At ordinary temperatures hydrogen consists of about 75 per cent of the *ortho* form and 25 per cent of the *para* form. The proportion of the *para* form in equilibrium increases as the temperature is lowered and approaches 100 per cent near absolute zero. This change in composition is associated with a smaller energy content of the *para* form. The two forms differ slightly in physical properties (*e.g.*, b.p. and specific heat capacity), but they have identical chemical properties. Allotropy due to the same cause has been discovered in a few other diatomic molecules (*e.g.*, $N_2$ and $Cl_2$).

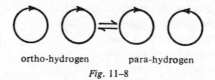

ortho-hydrogen        para-hydrogen

*Fig.* 11–8

## Determination of Transition Temperatures

When one form of an element or compound passes into another form at a definite temperature we can find the transition temperature by using the difference in physical properties of the two forms. Differences in colour, density, and energy content are frequently used.

METHOD 1. *Colour Change.* We introduce some red mercury(II) iodide into a melting point tube, which we then attach by a rubber band to a thermometer. The latter is clamped in a bath of liquid paraffin, which is then warmed slowly. We note the temperature at which the red colour changes to yellow.

METHOD 2. *By Dilatometer.* Since two forms of a substance have different densities there is a change in volume when one form passes into the other. Thus the conversion of sulphur (rhombic) into sulphur (monoclinic) of lower density is attended by increase in volume. We can detect the change by a dilatometer and use it to find the transition temperature of the sulphur.

**Experiment.** A simple form of dilatometer (Fig. 11-9) consists of a boiling tube fitted with a rubber stopper, through which passes a $\frac{1}{10}$-degree thermometer and a capillary tube of fairly wide bore (1-2 mm) and about 60 cm long. Behind the glass tube is fixed a millimetre scale.

Prepare 150 cm³ of medium sulphuric(VI) acid by diluting 50 cm³ of the concentrated acid to about three times its volume. This gives a solution boiling at about 117°C, which is sufficiently above the transition temperature under investigation. Drive off dissolved air from the acid by

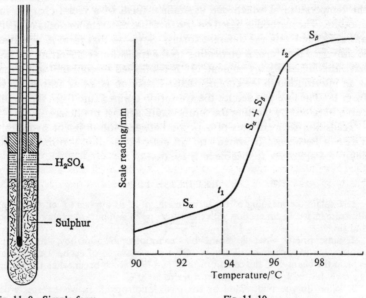

— H₂SO₄

— Sulphur

*Fig.* 11-9. Simple form
of dilatometer

*Fig.* 11-10

boiling it in a beaker for twenty minutes with a few pieces of porous pot. Meanwhile grind up some roll sulphur into pieces about half the size of a pea and remove any fine powder by sieving. Fill the boiling tube about three quarters full with the sulphur fragments. Allow the boiled acid to cool partially and cover the sulphur with the acid. Warm the tube to 40–50°C and stir the contents for several minutes until all entangled air has been removed from the sulphur. Then fill the tube nearly to the top with the air-free acid. Insert the stopper so that no air is trapped in the tube and the level of the liquid is near the lower end of the scale.

Clamp the boiling tube in a bath of liquid paraffin so that the tube is immersed to the level of the stopper. Heat the bath rapidly to about 90°C, stirring the paraffin continuously with an ordinary thermometer. Discontinue the heating temporarily to allow the thermometer inside the boiling tube to 'catch up'. Then resume the heating (and stirring), using a very small flame so that the temperature rises about one degree in two minutes. Note the readings on the scale from 90°C–100°C and plot them against the temperature as shown in Fig. 11–10.

Below the transition temperature the height of the liquid in the capillary tube rises slowly but steadily. About the transition temperature there is a more rapid increase in the height of the liquid column as the less dense allotrope is formed. Subsequently the height again increases slowly. Owing to thermometer 'lag' and the slow rate of transformation of one allotrope to the other, these changes do not occur sharply, but over a range of temperature readings. The transition temperature is taken as the mean of the two readings $t_1$ and $t_2$, at which the larger expansion begins and finishes.

METHOD 3. *Thermometric Method.* This is chiefly used to ascertain the temperature at which one crystalline form of a metal changes into another. The method is based on the fact that different amounts of energy are associated with the different forms. Suppose that form A of a metal is converted into form B on heating to the transition temperature. B must have the higher energy content. If B is allowed to cool uniformly there is an interruption in the cooling at the transition point as heat is evolved from B. Thus we can deduce the transition temperature from the cooling curve obtained by plotting the temperature against the time.

Transition temperatures also occur between the different hydrates of a salt or between a hydrated and an anhydrous salt. The methods just described are often applicable in these cases.

## EXERCISE 11

1. Explain the meaning of the term 'metastable' as applied to allotropy, and illustrate its use in connection with the allotropy of sulphur, phosphorus, carbon, and tin.

2. State briefly what is meant by 'enantiotropic' allotropy, 'monotropic' allotropy, and 'dynamic' allotropy. Give one example of each. Describe and explain the changes in the constitution of sulphur which occur when the element is slowly heated from room temperature to 2000°C.

3. What do you understand by the term 'allotropy'? In what respects does the allotropy of sulphur differ from that of phosphorus? Describe the preparation and properties of *two* allotropic forms of phosphorus, and explain how you would prove that the two are different forms of the same element. (O. and C.)

4. Give a general account of allotropy with particular emphasis on the distinctive characteristics and conditions for interconversion of the allotropes of (a) carbon, (b) phosphorus, (c) sulphur, and (d) tin. (J.M.B.)

5. Explain what is meant by a 'transition temperature.' Describe how you could find the temperature of transition of sulphur (rhombic) to sulphur (monoclinic).

# CHAPTER 12

# Solubility

We can define a *solution* as *any homogeneous mixture of two or more substances*. If we limit the substances to two we may say alternatively that a solution is a system in which one substance is uniformly distributed throughout the second substance. There is no restriction on the physical character of the substances, and these may be solid, liquid, or gaseous. Thus in principle we can have solutions consisting of any one of the nine possible combinations of solid, liquid, and gas, *e.g.*, gas–liquid, solid–solid, gas–solid, etc. A mixture of two gases is a solution of one gas in the other, while an alloy is an example of a solid–solid solution. In practice one of the substances is usually in excess. This is called the *solvent*, and the second substance is called the *solute*. In this chapter we shall confine our attention chiefly to solutions of solids, liquids, and gases in liquids.

## Solubility of Solids in Liquids

**Factors Affecting Solubility.** The amount of a solid which dissolves in a given amount of a liquid depends on the nature of the solvent and solute, the temperature, and the pressure. The effect of the latter is negligible unless the pressure is very high (see p. 273). The solubility of a solid is expressed by the *molality* of its saturated solution, that is

*the number of mole of solid which can be dissolved in 1000 g of the liquid at the given temperature in the presence of excess of the solid.*

The number of mole of solid is equal to the mass in gram divided by the sum of the relative atomic masses.

Some sparingly soluble substances are appreciably more soluble when they are in a very finely divided state. This is often important in qualitative and gravimetric analysis. Thus barium, strontium, and calcium are identified by precipitation as carbonates; sulphates(VI) in solution are estimated by precipitation as barium(II) sulphate(VI), which is filtered washed, dried, and weighed. In both cases precipitation is more complete if the precipitate is boiled, and then allowed to cool and stand. Boiling causes the finer particles to dissolve, but subsequently the solute is deposited on the larger particles and the precipitate is 'coarsened.' Not only is the solubility reduced, but the coarser precipitate is more readily filtered.

**Nature of Solvent and Solute.** Of all liquids water is the most powerful in the range of its solvent action, although solubilities vary greatly. Many substances which are commonly described as insoluble in water actually dissolve to a small extent. Thus at ordinary temperatures 1000 g of water dissolve $3 \times 10^{-7}$ mole of mercury, $1 \cdot 5 \times 10^{-3}$ mole of methane, and $8 \cdot 5 \times 10^{-6}$ mole of barium(II) sulphate(VI).

Water is the general solvent for inorganic substances, and liquids like ethoxyethane and benzene for organic compounds. Many of the latter, also dissolve in water if they contain —OH or —$NH_2$ groups. Water is described as an ionizing solvent because it brings about dissociation of ionic compounds into ions. We have seen that this property depends partly on the high dielectric constant of water and partly on its ability to hydrate the ions (with liberation of energy). Since these characteristics in turn are due to the O—H bonds being polarized, water is also described as a polar solvent. Ethoxyethane and benzene are non-ionizing, or non-polar solvents. The solvent action of these compounds has been discussed at p. 230.

Water has certain disadvantages as a solvent. Two of these are its reactive character and the limited range (0°–100°C) over which it can be used at ordinary pressures. A great deal of research has been carried out on solutions of inorganic substances in non-aqueous solvents, *e.g.*, liquid ammonia, liquid sulphur dioxide, liquid dinitrogen tetraoxide, and pure sulphuric(VI) acid. By means of these solvents we get systems which cannot exist in an aqueous medium. For example, sodium and potassium dissolve freely in liquid ammonia to give blue solutions. We can recover the metals by evaporation of the solutions. Several anhydrous metal nitrates(V) *e.g.*, copper(II) nitrate(V), were made for the first time by preparing them in liquid dinitrogen tetraoxide. Some of these have surprising properties. Thus, anhydrous copper(II) nitrate(V) is a blue crystalline compound, in which the bonding is *covalent*. It dissolves in certain organic solvents (*e.g.*, ethyl ethanoate), and it sublimes without decomposition below 200°C.

**Temperature.** The solubility of most solids increases with rise of temperature. This effect is in accordance with Le Chatelier's principle. Imagine a solute in equilibrium with its saturated solution. If the temperature is raised, more or less substance will dissolve according to which change is accompanied by absorption of heat. For most solids heat is absorbed when the substance dissolves in its *nearly saturated solution*. Therefore the solubility of these substances increases with rise of temperature. We cannot predict the effect of temperature on the solubility merely by the heat change which occurs when the solid is put into water. The heat evolved or absorbed then includes the heat of dilution, which may be large. Thus, although sodium(I) hydroxide dissolves in its nearly saturated solution with absorption of heat (and has a greater solubility at higher temperatures), there is evolution of considerable heat when the solid is added to water. Calcium(II) hydroxide is one of the few solids with a smaller solubility in water at higher temperatures, the solubility decreasing from 0·025 mole at 0°C to 0·01 mole at 100°C.

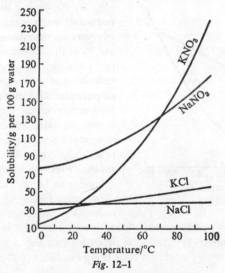

*Fig.* 12–1

In representing the variation of solubility with temperature graphically it is often convenient to express solubility in g of solute per 100 g of liquid. This method has been used in constructing the solubility curves shown in Fig. 12–1.

**Pressure.** The influence of pressure on the solubility of a solid is small because high pressures are required to affect the volumes of liquids, and extremely high pressures the volumes of solids. However, if we increase the pressure sufficiently an appreciable change in solubility may result. Thus at a pressure of 500 atm the solubility of sodium(I) chloride in water at ordinary temperatures is increased by about 2 per cent, while that of ammonium chloride is decreased by about 5 per cent. Both effects are in agreement with Le Chatelier's principle. According to the latter increased pressure should cause the equilibrium between solid and saturated solution to alter in the direction which is attended by decrease in volume (so as to relieve the applied pressure). Sodium(I) chloride dissolves in water with decrease in total volume, whereas the dissolving of ammonium chloride results in increase in total volume.

**Fractional Crystallization.** The different solubilities of substances are utilized in their purification by the process of fractional crystallization. When chlorine is passed into hot aqueous potassium(I) hydroxide both potassium(I) chloride and potassium(I) chlorate(V) are produced. If the solution is concentrated and cooled the salt which crystallizes out first is the one of smaller solubility at the lower temperature. The first crop of crystals thus contains potassium(I) chlorate(V) (Fig. 12–2), possibly

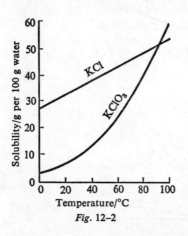

Fig. 12–2

mixed with a small amount of chloride. By dissolving these crystals in hot water and repeating the process of crystallization potassium(I) chlorate(V) is finally obtained free from the chloride. The mother liquor remaining after the first separation of potassium(I) chlorate(V) is relatively richer in the chloride, and by continuing the crystallization and separation of chlorate(V) a solution practically free from the latter results. From this solution the pure chloride can be crystallized.

In manufacturing potassium(I) nitrate(V) aqueous potassium(I) chloride and sodium(I) nitrate(V) are mixed. By concentrating at the boiling point sodium(I) chloride is deposited (see Fig. 12–1), the equilibrium in the following equation thus being displaced to the right:

$$K^+ + Cl^- + Na^+ + NO_3^- \rightleftharpoons K^+ + NO_3^- + NaCl \downarrow$$

The hot solution is filtered and cooled when potassium(I) nitrate(V), having the smallest solubility, crystallizes out first. It is purified by recrystallization.

$$K^+ + Cl^- + Na^+ + NO_3^- \rightleftharpoons KNO_3 \downarrow + Na^+ + Cl^-$$

**Supersaturated Solutions.**

*A supersaturated solution is one which contains more of the dissolved substance than is possible at the given temperature when the saturated solution is in contact with the solid substance.*

Supersaturation is observed chiefly with salts which contain water of crystallization—*e.g.*, $Na_2SO_4 . 10H_2O$ and $Na_2S_2O_3 . 5H_2O$. When we warm some crystals of the latter salt, sodium(I) thiosulphate(VI)-5-water, in a boiling tube we obtain a solution of sodium(I) thiosulphate(VI) in its own water of crystallization. If we close the warm tube with a plug of cotton wool to exclude dust particles, we can cool the solution to room temperature without crystallization taking place. Crystallization can be effected by adding to the solution a *nucleus* consisting of a small crystal of the solid or a crystal of an isomorphous substance. It also occurs if dust (which probably contains isomorphous particles) is allowed to enter the tube, and it is accelerated by shaking the tube. When crystallization takes place there is a considerable rise of temperature due to rehydration, and a solid mass of sodium(I) thiosulphate(VI)-5-water results.

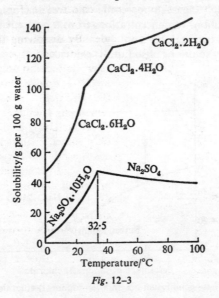

*Fig.* 12–3

**Discontinuous Solubility Curves.** If we find the solubility of sodium(I) sulphate(VI) at various temperatures from solutions made by dissolving the decahydrate, $Na_2SO_4.10H_2O$, in water, we find that it increases up to 32·5°C and appears to decrease above this temperature. The discontinuity occurs because the solid decahydrate loses all its water of crystallization when heated at 32·5°C (the *transition temperature*). Hence there are really *two* solubility curves. One, below 32·5°C, represents the solubility of $Na_2SO_4.10H_2O$; the other, above 32·5°C, is the curve for the anhydrous salt (see Fig. 12–3).

Discontinuous solubility curves are common with salts which contain water of crystallization. Different hydrates are often stable only over certain ranges of temperature, so that each hydrate has its own solubility curve. The curve for calcium(II) chloride consists of three separate parts with two transition temperatures (see Fig. 12–3).

**Eutectic Mixtures.** When a dilute solution of sodium(I) chloride is cooled from room temperature nothing happens until the temperature has fallen below 0°C because the freezing point of the solution is lower than that of the pure solvent. When the freezing point is reached pure ice separates. The cooling of a 10 per cent solution of the salt from 20°C is represented by PQ in Fig. 12–4, Q being the freezing point of the solution. If cooling goes on more ice separates and the solution becomes richer in the salt (QB). This continues until the solution contains 23·6 per cent of sodium(I) chloride, which occurs at about −21°C. At this point ice and the salt crystallize together, yielding a mixture which contains 23·6 per cent of the salt. Line AB represents the freezing points at standard pressure of solutions of concentrations from 0 per cent to 23·6 per cent of the salt. The fact that AB is not a straight line shows that Raoult's law is not obeyed by solutions of sodium(I) chloride.

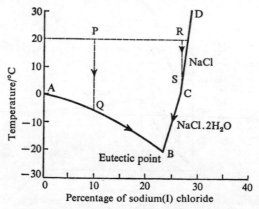

*Fig.* 12–4. Phase equilibrium diagram for sodium(I) chloride solutions

If a nearly saturated solution of sodium(I) chloride is cooled at ordinary pressure the first solid to appear is the salt. This occurs when the solution becomes saturated as the temperature is decreased. The cooling of the solution is represented by RS in Fig. 12–4. With further cooling more sodium(I) chloride separates until the concentration falls to 23·6 per cent of salt. This again occurs at −21°C, when the salt and ice crystallize together. The curve DCB represents the variation in concentration of the solution with temperature; that is, DCB is actually the solubility curve of sodium(I) chloride shown in Fig. 12–1, except that axes have been interchanged. There is a discontinuity in the curve at C, representing 0·15°C. Above this temperature the anhydrous salt is deposited, while below it hydrated crystals (NaCl . 2H₂O) are formed.

The mixture of (hydrated) salt and ice which crystallizes at −21°C from either a dilute solution or a concentrated solution is called a *eutectic mixture*. The temperature at which the mixture is deposited is the *eutectic temperature*. At any point along AB equilibrium exists between the salt solution, water vapour, and ice. At points along DC and CB saturated solution and water vapour are in equilibrium with anhydrous salt and hydrated salt respectively. At B saturated solution, ice, water vapour, and hydrated salt are in equilibrium.

Eutectic mixtures can also be formed by two metals or two organic solids. Providing the two substances are completely miscible when melted together and the pure substances separate on cooling, the phase equilibrium diagrams have the form in Fig. 12–5 for tin and lead. Tin melts at 232°C, and lead at 327°C. The eutectic mixture, containing 64 per cent of tin and 36 per cent of lead, melts at 183°C. Thus, if we cool a molten mixture of tin and lead containing less than 36 per cent of lead,

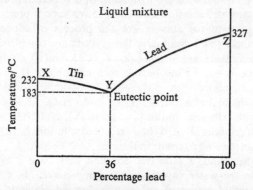

*Fig.* 12–5. Phase equilibrium diagram for tin–lead mixtures

pure tin is first deposited (curve XY). If the mixture contains more than 36 per cent of lead, the latter separates first (curve YZ). XY and YZ can be regarded as solubility curves, XY representing the solubility of tin in molten lead between 183°C and 232°C, and YZ the solubility of lead in molten tin between 183°C and 327°C. At Y, the eutectic point, the liquid is saturated with respect to both metals.

**Efflorescence and Deliquescence.** A hydrated salt is a loose chemical compound of the salt and water. It shows a definite vapour pressure at a given temperature. According to the value of the pressure of the water vapour with which it is in contact, it is capable of transformation into the anhydrous salt, a different hydrate, or a solution. If two barometer tubes filled with mercury are inverted over mercury and some sodium(I) sulphate(VI)-10-water inserted in one, the mercury falls slowly in this tube and finally becomes stationary. This is due to the hydrated salt giving off water molecules into the vacuum, leaving the anhydrous salt. The difference between the heights of the two mercury columns represents the vapour pressure of the salt $Na_2SO_4.10H_2O$. A similar process normally occurs if the hydrated salt is exposed to air. Water molecules both leave the crystals and enter the crystals from the surrounding air. The vapour pressure of the crystals, however, is about 2100 Pa as compared with about 1600 Pa for the water vapour in air (this value of course varies with the humidity of the atmosphere). Therefore, more water molecules leave than enter the crystals, which are thus converted into the anhydrous salt. This phenomenon is called *efflorescence*. The salt $Na_2CO_3 . 10H_2O$ is also efflorescent, and becomes coated on exposure to air with the white monohydrate, $Na_2CO_3 . H_2O$.

When a hydrate has a lower vapour pressure than the water vapour of the air more water molecules impinge on the surface of the crystal than leave it. This may lead to a condensing of water vapour on the surface of the crystal, with the formation of a superficial saturated solution. If the saturated solution also has a lower vapour pressure than that of water vapour in the atmosphere the process of condensation and solution continues, and eventually the crystals are completely dissolved. This is described as *deliquescence*. Calcium(II) chloride-6-water, $CaCl_2 . 6H_2O$, is one of the best known examples. If anhydrous calcium(II) chloride is exposed to the air it first combines with moisture to give in turn the hydrates $2H_2O$, $4H_2O$, and $6H_2O$. Deliquescence may also occur with salts like ammonium nitrate(V), $NH_4NO_3$, which do not have water of crystallization, if their saturated solutions have lower vapour pressures than the water vapour in the air.

On the other hand, a saturated solution may have a higher vapour pressure than the water vapour in the atmosphere. In this case, if a minute amount of saturated solution is formed water is immediately lost to the air and the solid hydrate re-formed. Thus, although a hydrate may itself have a lower vapour pressure than the moisture in the air, it is possible to expose it to the air without deliquescence taking place if the vapour pressure of its saturated solution is greater than that of the water vapour in the air. This is the case with the salt, $CuSO_4 . 5H_2O$, which is unchanged on exposure to air. At ordinary temperatures the vapour pressure of the crystals is about 700 Pa, while that of the saturated solution is 2100 Pa. In dry climates copper(II) sulphate(VI)-5-water effloresces in contact with the air.

Substances which absorb moisture without giving rise to a saturated solution are *hygroscopic*. Examples of hygroscopic substances are concentrated sulphuric(VI) acid, calcium(II) oxide, and absolute ethanol.

**Hydrates of Copper(II) Sulphate(VI).** At p. 251 we specified two variable factors (temperature and pressure) for a one-component system such as a liquid in equilibrium with its vapour. In a system made up of two components (*e.g.*, a salt and water) another variable factor is introduced, namely, the composition of the system, or the proportion of the two components present. If we fix any one of the three factors we can show graphically how the other two vary. This is actually what we do when we draw a solubility curve. We assume the pressure to be standard (although this is not usually stated), and we plot the variation of solubility with temperature.

A hydrated salt in equilibrium with water vapour is a two-component system. As we saw in the last section, the composition of this system may

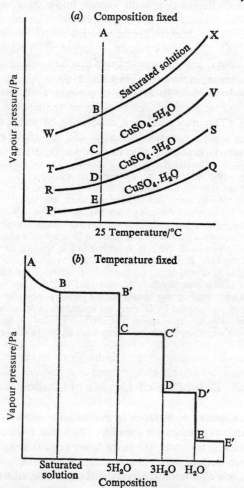

*Fig.* 12–6. Phase equilibrium diagrams for copper(II) sulphate(VI) and water (not drawn to correct scale)

change by formation of the anhydrous salt, a different hydrate, or a solution. Copper(II) sulphate(VI) forms three hydrates with $5H_2O$, $3H_2O$, and $H_2O$ respectively. If we fix the composition of the system as $CuSO_4$. $5H_2O$ in equilibrium with water vapour, we can plot the variation with temperature of the vapour pressure of this particular hydrate as shown by curve TV in Fig. 12–6a. PQ, RS, and WX are the corresponding vapour pressure curves for the monohydrate, the trihydrate, and the saturated solution.

If, instead of fixing the composition, we fix the temperature and plot the variation of composition with vapour pressure, we obtain the step-wise graph in Fig. 12–6b. This is particularly useful for explaining what happens when the pressure of water vapour in contact with the copper(II) sulphate(VI) is varied at a fixed temperature.

Suppose we start with an unsaturated solution of the copper salt in equilibrium with water vapour at 25°C. The solution and vapour are in a closed vessel connected to a suction pump (to remove water vapour) and to a pressure gauge (to measure the pressure of the water vapour). The conditions at the start are represented by point A in both diagrams. If water vapour is now drawn off in a succession of small amounts, and the system is left to come to equilibrium each time, the equilibrium vapour pressure first decreases as more water evaporates from the solution and the concentration of the latter increases. This continues until the solution becomes saturated (point B in both diagrams). Removal of more water vapour results in depositing of crystals of $CuSO_4.5H_2O$. The vapour pressure now remains constant (BB') while the saturated solution is chang-ing into the pentahydrate. This change is completed at B'.

With further removal of water vapour no change occurs in the composi-tion of the system until the vapour pressure has fallen to the value represented by C, when the trihydrate begins to form. While pentahy-drate is being converted to trihydrate the equilibrium vapour pressure again remains constant (CC'). From C' to D trihydrate alone is present as the solid phase, and there is a further fall in vapour pressure to D. DD' represents the constant vapour pressure as $CuSO_4.3H_2O$ changes to $CuSO_4.H_2O$. From D' to E the solid consists of the monohydrate. From E to E' the monohydrate is converted to the anhydrous salt, the vapour pressure again being constant. At values of the vapour pressure less than that represented by E' only the anhydrous salt exists.

If we start with the anhydrous copper(II) sulphate(VI) and add water vapour to the container the same changes occur, but in the reverse order.

## Solubility of Liquids in Liquids

**Immiscible Liquids.** If benzene is shaken with water and the liquids allowed to settle two layers are formed. The upper layer is benzene and contains practically no water, while the lower layer is water and contains practically no benzene. Benzene and water are *immiscible* liquids. Water is also immiscible with tetrachloromethane and carbon disulphide.

When immiscible liquids are together each contributes its vapour pressure to the total vapour pressure, which thus equals the sum of the separate vapour pressures at the given temperature. As a liquid boils when the vapour pressure equals the pressure of the atmosphere above it, the boiling point of a mixture of immiscible liquids is lower than the boiling point of either liquid separately. If a mixture of water and benzene (b.p. 80°C) is distilled, the mixture boils at a constant temperature below 80°C, yielding a distillate of constant composition. The relative masses,

$m_1$ and $m_2$, of the two liquids in the distillate are given by

$$\frac{m_1}{m_2} = \frac{p_1 M_r}{p_2 M_r'}$$

where $p_1$, $p_2$, $M_r$ and $M_r'$ are the respective vapour pressures and relative molecular masses of the liquids.

This principle is often used in steam distillation for the extraction of organic substances of high boiling point, as many of these tend to decompose near their own boiling point. When steam is passed into the impure substance the latter distils over with water below 100°C at atmospheric pressure. Although the partial pressure of the vapour of the organic substance will be relatively small at the distillation temperature, the distillate will contain a fairly large proportion of this substance owing to its comparatively high relative molecular mass.

**Partially Miscible Liquids.** When we shake a few drops of ethoxyethane with a lot of water we obtain a homogeneous liquid consisting of a solution of ethoxyethane in water. Also if we shake a few drops of water with a lot of ethoxyethane, we get a solution of water in ethoxyethane. If, however, we shake about equal volumes of the two liquids together, two layers are formed. The upper layer consists of a saturated solution of water in ethoxyethane, the lower layer of a saturated solution of ethoxyethane in water. Liquids like water and ethoxyethane which have a limited solubility in each other are said to be *partially miscible*. The two solutions are described as *conjugate solutions*.

Conjugate solutions boil at a definite temperature and yield a distillate of constant composition. We consider such solutions to be immiscible liquids, each of which contributes its vapour pressure to the total vapour pressure. The total vapour pressure may be greater than the vapour pressure of either pure solvent or it may lie between the vapour pressures of the two. In the first case the boiling point is less than the boiling point of either solvent and in the second case it is between the two. In either case steam distillation can be used with advantage to extract a liquid of high boiling point with which water is partially miscible.

Liquids which are partially miscible at one temperature may become completely miscible at another. Thus equal amounts of phenol and water yield two layers at ordinary temperatures. If, however, the liquids are heated separation into two layers ceases at 68°C. This is because the mutual solubility of phenol and water increases with rise of temperature.

**Completely Miscible Liquids.** Two liquids are *completely miscible* if they form one solution when mixed in any proportions. Ethanol and water are completely miscible and so are ethoxyethane and benzene. The vapour pressure of a mixture of two completely miscible liquids is not constant at a given temperature, but varies with the proportion of the two liquids. There are three types of variation. The vapour pressure may be somewhere between the vapour pressures of the separate liquids, it may reach a minimum, or it may reach a maximum. Correspondingly, the boiling point varies with the composition so that it is somewhere between the separate boiling points, reaches a maximum, or reaches a minimum. We shall consider these cases in turn.

*Case (i). The Boiling points of all Mixtures are between the Boiling points of the Constituents.* This is so with methanol–water mixtures. When a mixture of these liquids is distilled at atmospheric pressure changes occur progressively in the boiling point of the liquid, the composition

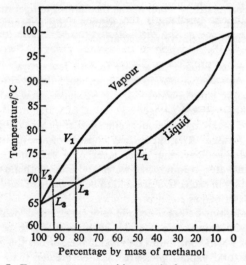

*Fig.* 12–7. Temperature–composition graphs for methanol and water

of the liquid and the composition of the vapour given off. We can follow these changes with the help of the temperature–composition graphs in Fig. 12–7.

Suppose that the mixture being distilled contains 50 per cent by mass of both methanol and water. This mixture is represented by $L_1$ in Fig. 12–7. Boiling commences when the vapour pressure of the liquid equals the atmospheric pressure (assumed to be standard). This occurs at a temperature $(t_1)$ somewhere between the boiling points of the pure liquids, and, as seen from Fig. 12–7, this temperature is about 77°C. The vapour evolved has *not* the same composition as the liquid, but *always contains a higher proportion of the more volatile constituent than the liquid in the flask at the same temperature.* The composition of the vapour given off at the temperature $t_1$ is represented by $V_1$. If this vapour is condensed it gives a liquid $L_2$, containing about 82 per cent of methanol. If this liquid is put into another distilling flask and distilled it begins to boil at a temperature $t_2$ of about 60°C, and the vapour given off has the composition $V_2$, corresponding to a still higher proportion of methanol. When condensed the vapour gives a liquid $L_3$, containing about 93 per cent of methanol. It is clear that by repeating the process a sufficient number of times we can ultimately obtain practically pure methanol. The liquids represented by $L_1$, $L_2$, $L_3$, etc., have progressively lower boiling points, and finally, when the liquid is pure methanol, the boiling point is 65°C. The upper curve in Fig. 12–7 shows the composition of the vapour which is in equilibrium with the boiling mixture at the various temperatures.

Since the liquid remaining in the flask becomes weaker in the more volatile constituent (methanol) the boiling point of the liquid rises and continues to rise as the distillation proceeds. The lower curve in Fig. 12–7 shows the variation in the boiling point of the liquid in the flask as distillation progresses. *The composition of the liquid in the flask alters in the ascending direction of the temperature–composition curve.* The final drops of liquid consist of practically pure water and the boiling point is then approximately 100°C.

The number of distillations required to obtain two pure liquids from a mixture of equal proportions of the two depends on the difference in the boiling points of the liquids. If the liquids are methanol and water it is possible to obtain the separate constituents in three distillations by means of the following procedure. During the first distillation the. receiver is changed for each rise of about 5°C in the reading of the thermometer. The first and last fractions (which represent the bulk of the distillate) are separately redistilled, and this time the receiver is changed for each 2°C rise in temperature. Again the first and last fractions respectively are redistilled. The first few drops of distillate collected from the *first* fraction consist of practically pure methanol, while the final few drops from the *last* fraction consist of practically pure water.

*Fig.* 12–8.
A Hempel
column

The separation of two completely miscible liquids by utilizing the difference in their boiling points is *fractional distillation*. The method just described for carrying out the process is slow and yields only small amounts of the pure constituents. In practice fractional distillation is performed more efficiently with the help of a fractionating column, or still-head. This is a device whereby the redistillations are combined in one operation. Fractionating columns have various forms, one of the best known being the Hempel column (Fig. 12–8). This is a plain tube, 30–40 cm long, packed with small coils of aluminium or stainless steel. It is inserted into the neck of the distillation flask. During distillation the temperature decreases progressively from the bottom to the top of the column and the ascending vapour is partially condensed. Condensation of the vapour is assisted by the large surface area of the filling and by the liquid already condensed. *The vapour of the less volatile constituent of the mixture condenses more readily than that of the more volatile constituent*, and therefore the higher the vapour ascends into the column the richer it becomes in the more volatile constituent. For the column to work at maximum efficiency an equilibrium must be established between ascending vapour and descending liquid at the different levels. The filling of the column is designed to promote the setting-up of these equilibria.

In the fractional distillation of petroleum and other large-scale uses of the process the whole of the liquid mixture is vaporized at once and the

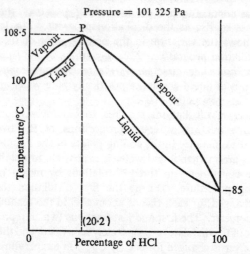

*Fig.* 12–9. Boiling point–composition diagram for hydrochloric acid and water (temperature axis not to scale)

vapours are condensed at different levels in a fractionating tower. The same general principles apply to this as to the Hempel column.

*Case (ii). The Boiling point is a Maximum at a Certain Composition.* A mixture of minimum vapour pressure has a maximum boiling point because the liquid has to be heated to a higher temperature to make the vapour pressure equal to that of the atmosphere. A typical equilibrium diagram for vapour and liquid in this case is shown in Fig. 12–9 for aqueous solutions of hydrochloric acid. The diagram has two pairs of curves, each similar to the curves in Fig. 12–7. A mixture of water with hydrochloric acid has a maximum boiling point at 108·5°C, when the mixture contains 20·2 per cent of the acid. This is represented by point P in Fig. 12–9.

On distilling a solution of hydrochloric acid containing less than 20·2 per cent HCl (composition to the left of P) boiling commences below 108·5°C, and the vapour given off contains a higher percentage of water than does the solution. (Notice that in this case the vapour does *not* contain a higher percentage of the more volatile constituent than the liquid.) The liquid becomes richer in acid (ascending direction of the boiling point curve) until the solution contains 20·2 per cent HCl. It then distils unchanged at 108·5°C.

Similarly, on distilling a solution which contains more than 20·2 per cent HCl (composition to the right of P) the vapour contains more hydrogen chloride than the original liquid and the boiling point rises. The concentration of acid in the solution decreases until it is once more 20·2 per cent, when the mixture distils unchanged in composition at 108·5°C.

The mixture containing 20·2 per cent HCl (about 6M) is called a *constant boiling mixture*, or *azeotropic mixture*. It is the final product of distillation of hydrochloric acid solution at standard pressure whatever the initial concentration of the acid. A solution containing *less* than 20·2 per cent HCl can be separated by fractional distillation into pure water and the azeotropic mixture. A solution containing *more* than 20·2 per cent HCl can be made to yield pure hydrogen chloride and the azeotropic mixture by fractional distillation.

A number of the common acids form azeotropic mixtures of maximum boiling point with water. Some examples are given in Table 12–1.

**Table 12–1.** MAXIMUM BOILING POINT AZEOTROPIC MIXTURES OF ACIDS AND WATER AT STANDARD PRESSURE

|  | Percentage of acid | Boiling point /°C |
| --- | --- | --- |
| Nitric(V) acid-water | 68 | 120 |
| Sulphuric(VI) acid-water | 98·3 | 330 |
| Hydrobromic acid-water | 47·5 | 126 |
| Hydriodic acid-water | 57 | 127 |
| Methanoic acid-water | 77 | 107 |

Azeotropic mixtures resemble eutectic mixtures in having a fixed composition at a definite pressure. Their composition varies, however, with the pressure at which distillation is carried out.

*Case (iii). The Boiling point is a Minimum at a Certain Composition.* This case is conveniently illustrated by propan-1-ol and water. The boiling point–composition diagram for mixtures of these liquids at standard pressure is shown in Fig. 12–10. A mixture of the alcohol and water has a maximum vapour pressure when the liquid contains 72 per cent by mass of the alcohol. This mixture is therefore one of minimum boiling point (88°C), and is a constant boiling mixture. This mixture is represented by Q in Fig. 12–10.

On distilling a mixture of propan-1-ol and water which contains less than 72 per cent of the alcohol (composition to the left of Q) boiling begins above 88°C. The vapour given off contains a bigger percentage of alcohol than the solution, which becomes weaker in alcohol (ascending direction of the boiling point curve). We can thus separate the mixture into a distillate of lower boiling point and a residual liquid of higher boiling point. By repeated distillation of the liquid we obtain collected distillates of increasing strength in alcohol and lower boiling point. The limits of concentration and boiling point obtainable in this way are 72 per cent of the alcohol and 88°C. Beyond this we cannot go.

Suppose a mixture contains more than 72 per cent of propan-1-ol (composition to the right of Q). Such a mixture boils above 88°C,

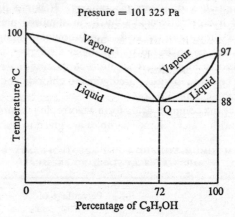

*Fig.* 12–10. Boiling point–composition diagram for propan-1-ol and water

yielding a vapour which contains less alcohol than the original solution. Again, by condensing the vapour and repeating the distillation we can obtain ultimately the azeotropic mixture. Thus mixtures of propan-1-ol and water can be separated by fractional distillation into the azeotropic mixture and that constituent which is present in excess. Here we obtain the azeotropic mixture from the vapour given off, whereas in the case of hydrochloric acid and water the azeotropic mixture is left in the flask.

Ethanol and water also form an azeotropic mixture of minimum boiling-point. At standard pressure the mixture contains 95·6 per cent by mass of alcohol and has a boiling point of 78·15°C compared with 78·4°C, the boiling point of pure alcohol. Again this mixture (known as rectified spirit) represents the limit of concentration which can be reached by fractional distillation of a dilute aqueous solution. The remaining water can be removed by chemical treatment with calcium(II) oxide and metallic calcium.

**Distribution of a Solute between Two Solvents.** When we shake a solute with two immiscible solvents A and B we find that at a given temperature if the substance has the same relative molecular mass in each solvent,

$$\frac{\text{Concentration in A}}{\text{Concentration in B}} = \text{a constant}$$

This is the *distribution law*, or *partition law*, and the constant is the *distribution coefficient*, or *partition coefficient*, at the given temperature. Thus when iodine is shaken with tetrachloromethane and water together, the iodine dissolves in both liquids. If we allow the liquids to stand, a dynamic equilibrium is established between them, so that the rate at which iodine molecules pass from one solution to the other is the same in both directions. Titration with standard sodium(I) thiosulphate(VI) solution then shows that the ratio of the concentrations of iodine in the two layers at a given temperature is constant, no matter what mass of iodine and what proportions of the two solvents are used. At 20°C the ratio of the concentration in tetrachloromethane to that in water is 85:1, which is approximately the same as the ratio of the solubilities of iodine in the two liquids separately at this temperature. We can determine the distribution coefficient practically as now described.

**Experiment.** The solutions required are M/20 and M/200 sodium(I) thiosulphate(VI)-5-water ($Na_2S_2O_3.5H_2O$) and 25 cm³ of a solution of iodine in tetrachloromethane of concentration 15–20 g dm⁻³.

Put the 25 cm³ of iodine solution into a 250-cm³ glass bottle and add about 200 cm³ of distilled water. Insert a glass stopper into the bottle and shake the liquids every half minute or so for 10–15 minutes. Then stand the bottle in a trough of water at room temperature and allow the layers to settle (10 minutes). Remove 5 cm³ of the lower tetrachloromethane layer by pipette to a conical flask and add some potassium(I) iodide solution to facilitate transfer of iodine from the organic solvent. Titrate with M/20 sodium(I) thiosulphate(VI) solution, swirling the flask round and adding starch solution as indicator near the end point. Titrate 50 cm³ of the upper aqueous layer with M/200 sodium(I) thiosulphate(VI) solution.

Add 5 cm³ of fresh tetrachloromethane and 50 cm³ of distilled water to the shaking bottle (this is best done before the first titrations are started). Shake the bottle and allow the layers to come to equilibrium as before. Again titrate 5 cm³ of the lower layer and 50 cm³ of the upper layer. This gives a second set of readings. Obtain a third and fourth set by repeating the procedure.

It is not necessary to calculate the actual concentrations of iodine in the two layers. The concentrations are proportional to the volumes of M/200 thiosulphate(VI) solution required to decolorize equal volumes of the solutions. In this experiment 5 cm³ of tetrachloromethane solution are titrated with M/20 thiosulphate(VI), while 50 cm³ of aqueous solution are titrated with M/200 thiosulphate(VI). Hence for comparison the first titre must be multiplied by 100. Table 12–2 shows some specimen results at 17°C.

Table 12–2

| (CCl$_4$ layer) Vol. of M/20 thio. $v_1$/cm$^3$ | 100$v_1$/cm$^3$ | (H$_2$O layer) Vol. of M/200 thio. $v_2$/cm$^3$ | $\dfrac{c_1}{c_2}$ |
|---|---|---|---|
| 12·8 | 1280 | 15·0 | 85·3 |
| 10·1 | 1010 | 11·7 | 86·3 |
| 7·9 | 790 | 9·4 | 84·0 |
| 6·4 | 640 | 7·6 | 84·2 |

The distribution law only holds when the solute is in the same molecular condition in both solvents. Thus if we shake an aqueous solution of hydrogen chloride with benzene the ratio of the concentrations of gas in the two solvents is not constant. This is because of the large dissociation of the hydrogen chloride into ions in aqueous solution.

Sometimes a solute undergoes association in one solvent while remaining in the simple molecular condition in the other. In this case the relative molecular mass is higher in the first solvent. If association is complete— that is, all the molecules are in the associated condition—the relative molecular mass in the first solvent is a simple multiple of that in the second solvent. If the relative molecular mass in the first solvent is $n$ times that in the second solvent, and $c_1$ and $c_2$ are the concentrations respectively in these solvents,

$$\frac{\sqrt[n]{c_1}}{c_2} = \text{a constant}$$

Association does not usually occur in aqueous solution, but is common in organic solvents like tetrachloromethane and benzene. Thus the relative molecular mass of ethanoic acid or benzenecarboxylic acid dissolved in benzene is double that corresponding to the simple molecule CH$_3$COOH or C$_6$H$_5$COOH. We can show this for benzenecarboxylic acid by the following experiment.

Experiment. Dissolve about 5 g of the acid in about 100 cm$^3$ of benzene and put the solution into a 250-cm$^3$ shaking bottle. Add about 100 cm$^3$ of distilled water. Shake the liquids well and allow them to come to equilibrium in a trough of water at room temperature as in the previous experiment. Also, make two solutions of sodium(I) hydroxide (NaOH) of concentrations approximately M/2 and M/50. The first solution can be made by diluting the bench alkali (2M) to roughly four times its volume. The second solution is made by diluting 10 cm$^3$ of the first solution *accurately* to 250 cm$^3$. Put the two solutions into separate burettes.

When the liquids in the shaking bottle have come to equilibrium transfer 20 cm$^3$ of the upper benzene layer and 20 cm$^3$ of the lower aqueous layer (using separate pipettes) to two conical flasks, replacing the liquids removed with 20 cm$^3$ of distilled water and 20 cm$^3$ of fresh benzene. Add some water (about 50 cm$^3$) to the 20 cm$^3$ of benzene solution in the conical flask and shake well to remove the acid from the organic solvent. Titrate the benzene solution, shaking frequently, with the M/2 alkali and the aqueous solution with the M/50 alkali, using phenolphthalein as indicator.

Repeat the procedure with the decreasing quantities of the acid in the shaking bottle, to obtain three or four sets of titrations. Set out the results as in the specimen table below. In each case the concentrations of acid are proportional to the volumes of M/50 alkali required to neutralize 20 cm³ of the solution. Since M/2 alkali is used for titration of the benzene layer the titre in this case must be multiplied by 25.

### Table 12–3

Temperature = 19°C

| (C₆H₆ layer) Vol. of M/2 NaOH $v_1$/cm³ | $25v_1$/cm³ | (H₂O layer) Vol. of M/50 NaOH $v_2$/cm³ | $\dfrac{c_1}{c_2}$ | $\dfrac{\sqrt{c_1}}{c_2}$ |
|---|---|---|---|---|
| 16·2 | 405 | 18·6 | 21·8 | 1·08 |
| 12·8 | 320 | 15·8 | 20·3 | 1·14 |
| 10·4 | 260 | 13·9 | 18·7 | 1·16 |
| 8·1 | 202·5 | 12·6 | 16·1 | 1·13 |

We can deduce from the results that benzenecarboxylic acid exists in solution in benzene as $(C_6H_5COOH)_2$ molecules. This can be confirmed by determining the relative molecular mass of the acid in benzene by the cryscopic method.

**Graphical Determination of Degree of Association ($n$).** If in the general case $\sqrt[n]{c_1}/c_2$ is constant, $c_1/c_2^n$ is also constant. Denoting the latter constant by $K$ and taking logarithms, we have,

$$\log c_1 = \log K + n \log c_2$$

Since the last equation has the form $y = a + bx$ we should obtain a straight-line graph when $\log c_1$ is plotted against $\log c_2$, and the value of $n$ should then be given by the slope of this line. Fig. 12–11 shows the result of plotting the logarithms of $c_1$ and $c_2$, using the four sets of values obtained in the last experiment. The slope of the graph line is very nearly 2, which agrees with the value of $n$ deduced numerically.

**Applications of the Distribution Law.**

*Distinction between a Bromide and an Iodide.* In confirming the presence of a bromide or an iodide in qualitative analysis we treat a solution of the salt with chlorine water or acidified sodium(I) chlorate(I) solution. If the solution is dilute a reddish-brown solution of bromine or iodine is formed. If we shake this with tetrachloromethane the latter dissolves nearly all the bromine or iodine, forming a lower layer which is red for bromine and violet for iodine.

*Desilverization of Lead (Parkes's Process).* Silver is more soluble in molten zinc than in molten lead. Hence, when zinc is melted with the argentiferous lead most of the silver is dissolved in the zinc. The zinc solution, mixed with a little lead, separates on cooling on the surface of the liquid and is skimmed off. The silver is then obtained by distilling off the zinc and 'cupelling' away the lead. The latter process consists of heating the metal to red heat in a shallow bone-ash crucible (called a 'cupel') and directing a blast of air on to the surface. The lead is oxidized to litharge, which spills over the side of the crucible.

*Extraction of Substances by Ethoxyethane or Benzene.* In preparing many organic compounds we obtain the desired substance either in aqueous

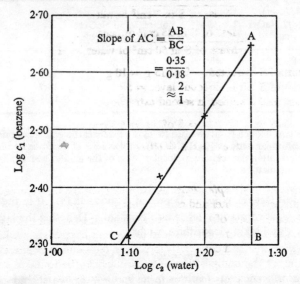

*Fig.* 12–11. A graphical method of deducing degree of association of benzenecarboxylic acid in benzene

solution or disseminated in water in the form of oily droplets. This is so when phenylamine has been steam-distilled. Phenylamine is more soluble in ethoxyethane than in water. If we shake the mixture of phenylamine and water with ethoxyethane most of the phenylamine dissolves in the latter and this solution can be separated from the aqueous solution by tap funnel. The ethoxyethane is then removed from the phenylamine by distillation.

In practice we want to extract as much as possible of the substance which is being prepared. If the amount of extracting liquid is limited it is better for this purpose to divide it into several portions and use each separately rather than use the whole of the liquid in one extraction. An illustration will make this clear.

Let the partition coefficient of a solute S between benzene and water be 4, so that

$$\frac{\text{Concentration in benzene}}{\text{Concentration in water}} = \frac{4}{1}$$

Suppose we have a solution containing 15 g of S in 50 cm³ of water, and that 50 cm³ of benzene are available for extraction.

If we use the 50 cm³ benzene in one extraction the amount of S extracted = ⅘ of 15 g = 12 g. If we use the benzene in two equal portions, we have for the first extraction

$$\frac{\text{Mass of S in 1 cm}^3 \text{ of benzene}}{\text{Mass of S in 1 cm}^3 \text{ of water}} = \frac{4}{1}$$

and

$$\frac{\text{Mass of S in 25 cm}^3 \text{ of benzene}}{\text{Mass of S in 50 cm}^3 \text{ of water}} = \frac{2}{1}$$

Amount of S extracted = ⅔ of 15 g = 10 g
Amount of S left in aqueous layer = 5 g
Amount of S obtained in second extraction

$$= \frac{2}{3} \text{ of 5 g} = 3 \cdot 3 \text{ g}$$

Total amount of S extracted = (10 + 3·3) g

$$= 13 \cdot 3 \text{ g}$$

*Investigation of Complex Ions.* A knowledge of the partition coefficient of a solute between water and another solvent is frequently used to investigate the composition of a complex ion in aqueous solution. Thus when excess of aqueous ammonia is added to aqueous copper(II) sulphate(VI)-5-water ammonia and copper(II) ions form tetraamminecopper(II) ions, $Cu(NH_3)_4^{2+}$. The composition of these ions can be found if we know the partition coefficient of ammonia between water and trichloromethane.

When the deep-blue solution, containing both 'free' and 'fixed' ammonia, is shaken with trichloromethane and left to settle two layers are formed. The upper, deep-blue, aqueous layer contains the complex ion and free ammonia in equilibrium with free ammonia in the lower trichloromethane layer. If $N$ molecules of ammonia combine with one copper(II) ion we can represent the system as follows:

| Aqueous layer | $NH_3 \rightleftharpoons Cu(NH_3)_N^{2+}$ |
|---|---|
| | ⥮ |
| Trichloromethane layer | $NH_3$ |

The partition coefficient of free ammonia between water and trichloromethane at 20°C is 25. Hence, if the concentration of ammonia in the trichloromethane is found by withdrawing a portion of this layer and titrating with standard hydrochloric acid, we can obtain the concentration of free ammonia in the aqueous layer by multiplying by 25. We can also determine the *total* concentration of ammonia (free and fixed) in the aqueous layer by titrating a known volume of the upper layer with standard acid (the equilibrium is displaced to the left during the titration). By subtracting the concentration of free ammonia in the aqueous layer from the total ammonia concentration in this layer we obtain the concentration of fixed ammonia. If we have used copper(II) sulphate(VI) solution of known concentration in the experiment we know the concentration of copper in the complex ion since practically all the copper is converted into the complex ion. (The virtual absence of copper(II) ions is shown by sodium(I) carbonate giving no precipitate of copper(II) carbonate with the deep-blue liquid.) We can thus deduce the number $(N)$ of molecules of ammonia which have combined with one copper(II) ion.

**Example.** *25 $cm^3$ (excess) of ammonia solution were added to 25 $cm^3$ of $M/10$ $CuSO_4$ . $5H_2O$ solution, and the resulting deep-blue solution shaken with trichloromethane. After the layers had settled, 50 $cm^3$ of the trichloromethane layer needed 25·5 $cm^3$ of $M/20$ hydrochloric acid ($HCl$) for neutralization; 20 $cm^3$ of the blue aqueous layer were neutralized by 33·3 $cm^3$ of $M/2$ hydrochloric acid. If the partition coefficient of ammonia between water and trichloromethane at the existing temperature was 25·0, calculate the formula of the complex ion.*

(*Note.* The calculation is most easily performed by expressing the concentrations in mole per $dm^3$.)

Concentration of free ammonia ($NH_3$) in the chloroform layer

$$= \frac{25 \cdot 5}{50} \times \frac{1}{20} = 0 \cdot 0255 \text{ mol dm}^{-3}$$

Concentration of free ammonia in the aqueous layer

$$= 0 \cdot 0255 \times 25 = 0 \cdot 6375 \text{ mol dm}^{-3}$$

Concentration of total ammonia in the aqueous layer

$$= \frac{33 \cdot 3}{20} \times \frac{1}{2} = 0 \cdot 8325 \text{ mol dm}^{-3}$$

Concentration of fixed ammonia in the aqueous layer

$$= 0 \cdot 8325 - 0 \cdot 6375 = 0 \cdot 195 \text{ mol dm}^{-3}$$

By addition of an equal volume of ammonia to the copper(II) sulphate(VI) solution the concentration of the latter was halved—that is, concentration of copper(II) salt $= M/20 = 0 \cdot 05$ mol of copper ($Cu$) per $dm^3$.

Assuming that all the copper was converted into the complex ion, we have

$$\frac{\text{Concentration of fixed ammonia}}{\text{Concentration of copper}} = \frac{0 \cdot 195}{0 \cdot 05} = 3 \cdot 9$$

$$= 4 \text{ approximately}$$

That is, the formula of the complex ion is $Cu(NH_3)_4^{2+}$.

## Solubility of Gases in Liquids

The amount of a gas which dissolves in a liquid depends on

* The nature of the gas and the liquid
* The temperature
* The pressure

The solubility is expressed in different ways, the two most common being by means of Ostwald's *coefficient of solubility* and Bunsen's *absorption coefficient*.

*The* **coefficient of solubility** *of a gas in a liquid is the number of cm³ of the gas which dissolve in one cm³ of the liquid at the given temperature.*

*The* **absorption coefficient** *is the volume, reduced to s.t.p., of the gas which dissolves in one cm³ of the liquid at the given temperature when the pressure of the gas is 101 325 Pa.*

If $s$ is the coefficient of solubility of a gas and $a$ is the absorption coefficient at the kelvin temperature $T$, the relation between the two coefficients is given by

$$s = a \times \frac{T}{273}$$

Clearly at 0°C both coefficients have the same value.

**Nature of the Gas and Liquid.** Gases vary greatly in their solubility. The most soluble in water are those which react with water to give ions or which form hydrogen bonds with water molecules. Hydrogen chloride is largely converted into ions,

$$HCl + H_2O \rightleftharpoons H_3O^+ + Cl^-$$

Ammonia also forms ions in aqueous solution, but to a much smaller extent than hydrogen chloride (ammonia solution is a very weak alkali). The great solubility of the gas is chiefly caused by hydrogen bonding.

$$\begin{array}{c} H \\ | \\ O-H\cdots NH_3 \rightleftharpoons NH_4^+ + OH^- \end{array}$$

Note that 'ammonium hydroxide,' $NH_4OH$, does not exist. This formula would require the nitrogen atom to have a covalency of five and hence ten electrons in its outer quantum shell. This is impossible for a 'first row' element like nitrogen.

The coefficient of solubility, or absorption coefficient, includes the gas which forms the true solution and the gas which undergoes chemical reaction with the water. The coefficients of solubility in water of some common gases are given in Table 12–4.

**Table 12–4.** COEFFICIENTS OF SOLUBILITY IN WATER

| Gas | Solubility coefficient at 0°C | 15°C |
|---|---|---|
| Ammonia | 1050 | 727 |
| Hydrogen chloride | 507 | 458 |
| Sulphur dioxide | 80 | 47 |
| Chlorine | 4·6 | 2·84 |
| Carbon dioxide | 1·7 | 1·0 |
| Oxygen | 0·05 | 0·034 |
| Nitrogen | 0·023 | 0·017 |

The solubility of a given gas varies in different solvents. Thus at 15°C oxygen is seven times more soluble in ethanol than in water. Ammonia, however, is less soluble in ethanol than in water. (Why?) A gas is usually less soluble in an aqueous solution of another substance than in pure water unless it reacts chemically with the substance in solution (*e.g.*, carbon dioxide and sodium(I) hydroxide solution). The smaller solubility is explained by combination between water molecules and the molecules or ions of the dissolved substance, so that less water is available to act as a solvent for the gas.

**Temperature.** The solubility of nearly all gases decreases with rise of temperature. (The only exceptions are helium and neon. The absorption coefficient of helium in water increases above 40°C, while that of neon increases above 49°C.) This is in accordance with Le Chatelier's principle, since the dissolving of a gas is nearly always exothermic. Sometimes we use decreased solubility at higher temperatures in collecting gases (*e.g.* nitrous oxide) over warm water.

**Pressure.**

Henry's law states that *at a given temperature the mass of gas which dissolves in a given volume of solvent is proportional to the pressure.*
It follows that the *volume* of gas dissolved is independent of the pressure.

(Hence there is no reference to pressure in the definition of Ostwald's coefficient of solubility.)

At pressure $p$ let mass of gas dissolved by volume $v$ of solvent be $m$.

Then at pressure $2p$ mass of gas dissolved by volume $v$ of solvent will be $2m$.

But a mass $2m$ of gas at a pressure of $2p$ has the same volume as a mass $m$ of gas at a pressure of $p$.

Therefore, the volume of gas dissolved is the same at both pressures.

We can deduce Henry's law by applying the kinetic theory to a gas and its saturated solution contained in a closed vessel. Since the solution is saturated the number of molecules of gas leaving the solution per second is equal to the number entering the solution per second. If the pressure of the gas above the solution is doubled we shall have twice the original number of molecules entering the solution per second and, when equilibrium is once more attained, twice the number of molecules leaving the solution per second. The solution will therefore contain eventually twice the number of molecules under the doubled pressure—that is, the mass of gas dissolved is doubled.

Soda water (an aqueous solution of carbon dioxide under pressure) illustrates the greater solubility of a gas at higher pressures, the excess of gas being released when the liquid is exposed to the atmosphere. Similarly tap water often has for some seconds a milky appearance owing to the evolution of 'air' dissolved under extra pressure in the water pipes. This effect is of great importance to divers who breathe compressed air from cylinders. Even at the moderate depth of 60 m the pressure is 6 atm, and the amounts of oxygen and nitrogen dissolved by the blood are correspondingly increased. Excess of oxygen in the blood stream may cause convulsions, while excess of nitrogen acts like ethanol and produces 'drunkeness'. Furthermore, if the diver ascends too quickly the excess of nitrogen forms bubbles which may result in paralysis through partial blocking of the blood stream (a condition known as 'the bends'). Deep-sea divers therefore breathe a mixture of 2 per cent oxygen and 98 per cent helium, which is much less soluble than nitrogen.

Henry's law does not hold for very soluble gases nor at high pressures. As we have seen, the solubility of very soluble gases is increased by chemical reaction with water. The law is therefore applicable only to that part of the gas which forms a true solution. If this condition is made the law holds fairly well. For example, sulphuric(IV) acid is unstable above 40°C, and sulphur dioxide forms a true solution above this temperature. Henry's law then holds for sulphur dioxide. At high pressures the mass of gas dissolved is less than would be expected from Henry's law. Thus, at a pressure of 10 atm the amount of carbon dioxide dissolved at ordinary temperatures is about 11 per cent less than the amount calculated from the solubility at 1 atm. This is due to intermolecular forces coming into play at higher pressures, as with exceptions to Boyle's law (Ch. 10).

**Solubility of Gaseous Mixtures.** We saw earlier that each gas in a mixtures of gases behaves independently of the others and the total pressure is the sum of the partial pressures of the separate gases (Dalton's law of partial pressures). This independent behaviour also applies to the dissolving of gases from a mixture contained above a common solvent. Each gas establishes its own equilibrium, so that an equal number of its molecules enter and leave the liquid surface per second. The number depends only on the particular solubility of the gas at the given temperature and on its partial pressure. Thus the *mass* of any gas dissolved from a mixture by 1 cm$^3$ of solvent is proportional to its coefficient of solubility and its partial pressure. The *volume* dissolved is constant *providing it is measured at the partial pressure of the gas.*

Consider a sparingly soluble gas X dissolving in water at a fixed temperature, firstly, when it is by itself and, secondly, when it is mixed with another sparingly soluble gas Y. We can suppose that X (or X and Y) is contained in a barometer tube over mercury at a known pressure and that a few drops of water are introduced into the tube. Some of X will dissolve, but since it is sparingly soluble we can assume that its pressure (or partial pressure) remains constant. Let the volume of X dissolved by 1 cm$^3$ of water at the given temperature be $v$ . Then the amount of X dissolved by 1 cm$^3$ of water under different conditions of pressure can be deduced as now shown,

(i) *X alone at a pressure of* 100 000 Pa

Volume of X dissolved by 1 cm$^3$ of water $= v$

Mass of X dissolved by 1 cm$^3$ of water $= m$

(ii) *X alone at a pressure of* 50 000 Pa

Volume of X dissolved by 1 cm$^3$ of water

$$= v \text{ at } 50\,000 \text{ Pa}$$
$$= v/2 \text{ at } 100\,000 \text{ Pa}$$

Mass of X dissolved by 1 cm$^3$ of water $= m/2$

(iii) *Equal volumes of X and Y at a total pressure of* 100 000 Pa

Volume of X dissolved by 1 cm$^3$ of water

$$= v \text{ at a partial pressure of } 50\,000 \text{ Pa}$$
$$= v/2 \text{ at a pressure of } 100\,000 \text{ Pa}$$

Mass of X dissolved by 1 cm$^3$ of water $= m/2$

(iv) *20 cm$^3$ of X and 30 cm$^3$ of Y at a total pressure of* 80 000 Pa

Partial pressure of $X = \dfrac{20}{50} \times 80\,000 \text{ Pa} = 32\,000 \text{ Pa}$

Volume of X dissolved by 1 cm$^3$ of water

$= v$ at a partial pressure of 32 000 Pa

$= \dfrac{v \times 32\,000}{100\,000}$ at a pressure of 100 000 Pa

Mass of X dissolved by 1 cm$^3$ of water $= \dfrac{m \times 32\,000}{100\,000}$

**Example.** (i) *Assuming that dry purified air contains 21 per cent by volume of oxygen and 79 per cent of nitrogen, calculate the percentage volumetric composition of air at 101 300 Pa and 15°C when three-fourths saturated with moisture.* (ii) *Calculate the composition of the gas which is dissolved when the air is shaken in a closed vessel with an equal volume of water. Tension of aqueous vapour at 15°C = 1708 Pa. Solubility coefficient of oxygen = 0·04, and that of nitrogen = 0·02.* (J.M.B.)

(i) Partial pressure of water vapour

$$= \tfrac{3}{4} \times 1708 \text{ Pa} = 1281 \text{ Pa}$$

Partial pressure of oxygen

$$= \frac{21}{100} \times (101\,300 - 1281) \text{ Pa} = 21\,000 \text{ Pa}$$

Partial pressure of nitrogen

$$= \frac{79}{100} \times (101\,300 - 1281) \text{ Pa} = 79\,000 \text{ Pa}$$

Percentage volume of water vapour

$$= \frac{1281}{101\,300} \times 100 = 1\cdot3$$

Percentage volume of oxygen

$$= \frac{21\,000}{101\,300} \times 100 = 20\cdot7$$

Percentage volume of nitrogen

$$= \frac{79\,000}{101\,300} \times 100 = 78\cdot0$$

(ii) Volume of oxygen dissolved by 1 cm$^3$ of water

$= 0\cdot04$ cm$^3$ at a partial pressure of 21 000 Pa

$= 0\cdot04 \times \dfrac{21\,000}{101\,300}$ cm$^3$ at a pressure of 101 300 Pa

$= 0\cdot0083$ cm$^3$

Volume of nitrogen dissolved by 1 cm³ of water

$$= 0.02 \text{ cm}^3 \text{ at a partial pressure of } 79\ 000 \text{ Pa}$$

$$= 0.02 \times \frac{79\ 000}{101\ 300} \text{ cm}^3 \text{ at a pressure of } 101\ 300 \text{ Pa}$$

$$= 0.0156 \text{ cm}^3$$

Percentage of oxygen dissolved

$$= \frac{0.0083 \times 100}{0.0083 + 0.0156} = 34.7$$

Similarly, percentage of nitrogen dissolved = 65·3.

We see that, whereas the proportion of nitrogen to oxygen in ordinary air is approximately 4 to 1, the proportion in 'air' dissolved in water is approximately 2 to 1.

When a solution of a gas is made the dissolved molecules are in equilibrium with the molecules of gas above the solution. If the latter molecules are removed the partial pressure of the gas is reduced to zero, and the dissolved molecules can no longer stay in solution. Thus we can remove even a very soluble gas like ammonia from solution by passing a stream of air or other gas of small solubility through or over the solution. For the same reason gases are expelled from solution by boiling, the place of air in this case being taken by steam. An exception, however, is a solution of hydrogen chloride, which, as we have seen, yields on boiling a mixture of water and hydrogen chloride of constant composition.

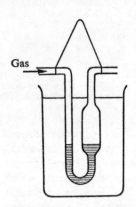

Gas

Fig. 12–12. Apparatus used for finding solubility of a gas

**Determination of the Solubility of a Gas in Water.**

(1) *By Chemical Analysis of a Saturated Solution.* This method applies to gases (such as ammonia, sulphur dioxide, and hydrogen chloride) which are very soluble in water. All very soluble gases can be estimated volumetrically. We can also use the method for some gases (*e.g.*, hydrogen sulphide, chlorine, and carbon dioxide) which have a relatively small solubility, but are more than sparingly soluble.

The method consists of preparing a saturated solution of the gas in the apparatus shown in Fig. 12–12. The apparatus is made by bending a 10-cm³ pipette to the appropriate shape. About 3 cm³ of distilled water are introduced, and the apparatus is suspended by a wire loop in a beaker of water at room temperature (or maintained at the temperature required).

A slow stream of gas is passed through the apparatus *in the direction shown* for about 20 minutes. The open ends are then sealed with short lengths of rubber tubing and glass rod, after which the apparatus is removed, dried, and weighed. The saturated solution is run out into a known volume (excess) of a standard solution of a suitable substance (sulphuric(VI) acid for ammonia, sodium(I) hydroxide for hydrogen chloride, etc.), and the apparatus is again weighed. The difference in mass gives the mass of saturated solution.

We find the excess of standard solution remaining by titration with a standard solution of acid, alkali, or other reagent as appropriate. From the difference caused by the saturated solution we calculate the mass of gas dissolved. Subtracting this from the mass of saturated solution, we arrive at the mass (and hence the volume) of water. After finding the mass of gas dissolved by 1 cm³ of water, we multiply this by $101\,300/p$, where $p$ is the barometric pressure. This gives the mass of gas dissolved by 1 cm³ of water at a pressure of 101 300 Pa. We then calculate the volume at s.t.p. from the fact that the molar volume at s.t.p. is 22·4 dm³. The result is the *absorption coefficient* of the gas. The coefficient of solubility at the temperature of the experiment is found from the relation given at p. 293.

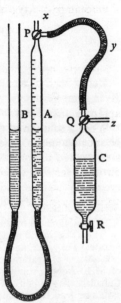

Fig. 12–13. Apparatus used for measuring coefficient of solubility

(2) *By Direct Measurement*. In this method we measure directly the volume of gas absorbed by a known volume of water at a given temperature. The method can be used for gases of moderate or fairly small solubility, such as carbon dioxide, dinitrogen oxide ($N_2O$), ethene, oxygen and nitrogen. The apparatus (Fig. 12–13) consists of a gas burette A connected to a levelling tube B and an absorption pipette C. P and Q are three-way taps. The burette and levelling tube contain mercury, which is first made to fill A completely by raising B. The absorption pipette is filled completely with distilled water which has been boiled to remove any dissolved gases. The tube $x$ is connected to the gas supply and air is swept from tubes $x$, $y$, and $z$. By lowering B about 100 cm³ of gas are drawn into A and the volume ($V_1$) is noted after levelling. A and C are now put into communication. By opening the tap R and raising B gas takes the place of water in the absorption pipette until about 25 cm³ of water have been run out of C. We note this volume ($v$), so that when we subtract it from the volume ($V$) of water filling the pipette, we know the volume of water remaining.

The pipette is shaken to accelerate the dissolving of the gas, and as the mercury in A rises B is raised to maintain atmospheric pressure. The pipette is left for an hour or two in a thermostat at a known temperature to allow equilibrium to be reached between gas and liquid. When there is no further rise of mercury the volume ($V_2$) of gas left in A is measured. Leaving out of consideration the volume of gas always present in the connecting tube $y$, we have the following results:

Original volume of gas    $= V_1$

Volume of remaining gas $= (V_2 + v)$

Volume of gas dissolved $= \{V_1 - (V_2 + v)\}$

Volume of solvent       $= (V - v)$

Hence we can calculate the coefficient of solubility of the gas at the given temperature.

(3) *By Gas Chromatography*. The solubilities in water of sparingly soluble gaseous hydrocarbons have been investigated by means of gas chromatography. This method is described in Chapter 13. The absorption coefficients found for methane, ethane, propane, and butane at 25°C were 0·034, 0·045, 0·031, and 0·024 respectively.

## EXERCISE 12

**Solubility of Solids.**

1. What is meant by the 'solubility' of a solid in a liquid? Outline the methods by which you would measure the solubility at room temperature of (i) silver(I) nitrate(V) in water, and (ii) iodine in ethanol.

2. 20 cm³ of a saturated solution of lead(II) chloride at 15°C needed 13·2 cm³ of M/10 AgNO₃ solution to precipitate all the chlorine as silver(I) chloride. Calculate the solubility of lead(II) chloride at 15°C. (Assume that the density of saturated lead(II) chloride solution is equal to that of water.)

3. Define the following terms: (*a*) saturated solution, (*b*) supersaturated solution, (*c*) transition temperature.

The following data give points on the solubility curve of a substance X, 53 g of which neutralize 36·5 g HCl.

| Temperature/°C | 10 | 20 | 30 | 40 | 50 | 60 |
|---|---|---|---|---|---|---|
| Solubility/(g per 100 g water) | 12·5 | 21·5 | ? | ? | 47·0 | 46·4 |

At 30°C and 40°C respectively 100 g of saturated solution are equivalent to 106·3 cm³ and 123·6 cm³ 5M HCl. Calculate the solubility of X in water at these temperatures and on the graph paper provided plot the results with the above data so as to obtain the solubility curve from 10 to 60°C. Comment on the shape of the curve and indicate whether the process of dissolving X at 30°C and at 40°C is exothermic or endothermic.    (J.M.B.)

4. Describe an experiment to find the solubility of ammonium chloride in water at room temperature.

Construct solubility curves from the following solubility data:

| Temperature/°C | KCl/(g per 100 g) | KClO₃/(g per 100 g) |
|---|---|---|
| 0 | 28 | 3 |
| 20 | 34 | 7 |
| 50 | 42 | 19 |
| 100 | 56 | 59 |

Use the curves to show how you would obtain at least 7 g of pure chloride and 20 g of pure chlorate(V) from 60 g of a mixture containing equal masses of both. (S.U.)

**Vapour pressure of Salt Hydrates.**

5. (Part question.) Three barometer tubes are filled with mercury and inverted over mercury. Water is added to the first tube until the space above the mercury is saturated. Crystals of sodium(I) sulphate(VI)-10-water, $Na_2SO_4.10H_2O$, and zinc(II) chloride-1-water, $ZnCl_2.H_2O$, are separately placed in the other two tubes. What will be the final *relative* levels of the mercury in each of the three tubes? Discuss how this explains either the efflorescent or deliquescent property of these salts. (S.U.)

6. (Part question.) The vapour pressures at 50°C of mixtures of copper(II) sulphate(VI) and water are as follows:

| Molecules of $H_2O$ per molecule of $CuSO_4$ | 0·6 | 1·3 | 2·5 | 3·4 | 4·8 | 5·5 |
|---|---|---|---|---|---|---|
| Vapour pressure/Pa | 600 | 4130 | 4130 | 6000 | 6000 | 12 000 |

On the graph paper provided, plot the vapour pressure against the number of molecules of $H_2O$ per molecule of $CuSO_4$, and from the graph deduce the formulae of the hydrates of copper(II) sulphate(VI).

State what substances are present in the mixtures represented by the various points plotted on the graph. (J.M.B.)

**Steam Distillation.**

7. (Part question.) When an organic compound was distilled in steam at 101 300 Pa pressure, the temperature of distillation was 99°C. The vapour pressure of water at 99°C = 97 700 Pa. 80·0 per cent by mass of the distillate was found to be water. Calculate the relative molecular mass of the organic compound. H = 1; O = 16. (J.M.B.)

8. (Part question.) Explain briefly the physico-chemical basis of steam distillation. The vapour pressure of water at 95°C is 84 700 Pa. A water-insoluble organic liquid X, of relative molecular mass 160, steam distils at 95°C under an atmospheric pressure of 101 300 Pa. Calculate the mass of water collected in the steam distillate during the collection of 40 g of X. State the law which you use in your calculation. H = 1·0; O = 16·0. (W.J.E.C.)

9. (Part question.) Chlorobenzene, which is insoluble in water, steam-distils at 91°C under an atmospheric pressure of 100 300 Pa. A sample of the steam-distillate contains 23·7 g of chlorobenzene for every 10 g of water. Calculate the vapour pressures of water and chlorobenzene at 91°C. C = 12; H = 1; O = 16; Cl = 35·5. (W.J.E.C.)

**Distribution Law**

10. (Part question.) In the following calculation, define any terms or laws that you use:

An aqueous solution contains 4 g of A per $dm^3$. When 500 $dm^3$ of this solution were shaken with 100 $cm^3$ of pentan-1-ol, 1·5 g of A were extracted. What will be the mass of A obtained from the aqueous solution by shaking with a further 100 $cm^3$ of pentan-1-ol? (Assume that the molecular state remains the same in both solvents.) (S.U.)

11. State the law defining the equilibrium distribution of a solute between two immiscible liquids. Outline *very briefly* how you would attempt to determine experimentally the partition coefficient for the distribution of butanedioic acid, $C_2H_4(COOH)_2$, between ethoxyethane and water at 0°C.

25 $cm^3$ portions of a 0·05 M solution of $I_2$ in tetrachloromethane were

separately shaken with (*a*) 50 cm$^3$ and (*b*) 200 cm$^3$ of water at 25°C. After equilibrium had been attained, 25 cm$^3$ portions of the aqueous layers required (i) 2·74 cm$^3$ and (ii) 2·53 cm$^3$, respectively, of 0·01 M Na$_2$S$_2$O$_3$ solution to react completely with the iodine. From these two results, calculate the mean value of the partition coefficient of iodine between the two solvents.

State *very briefly* how and why the distribution of iodine would be affected by the addition of potassium(I) iodide to the aqueous layer.          (W.J.E.C.)

**12.** (Part question.) The following results of five different experiments, all at the same temperature, show the distribution of ethanoic acid between water and trichloromethane, expressed in mole per dm$^3$ of each solvent:

| Water layer | 4·90 | 5·70 | 8·52 | 9·82 | 10·63 |
|---|---|---|---|---|---|
| Trichloromethane layer | 0·30 | 0·40 | 0·90 | 1·20 | 1·40 |

The relative molecular mass of ethanoic acid in water is known to be 60. Calculate the apparent relative molecular mass of ethanoic acid in trichloromethane.          (S.U.)

**Solubility of Gases.**

**13.** At 15°C and 96 000 Pa pressure 100 g of water dissolve 12·80 g of sulphur dioxide. Calculate (i) the absorption coefficient, (ii) the coefficient of solubility of sulphur dioxide at this temperature.

**14.** (Part question.) Describe briefly how you would measure the solubility of carbon dioxide in water at room temperature and pressure.

10 dm$^3$ of oxygen-free water were shaken in a closed vessel at 0°C with 1 dm$^3$ of oxygen, initially at 101 300 Pa pressure. Calculate the volume of oxygen (converted to s.t.p.) which dissolved and also the final pressure of the undissolved oxygen. Assume that at 0°C 1 cm$^3$ of water dissolves 0·05 cm$^3$ of oxygen at a pressure of 101 300 Pa.          (J.M.B.)

**15.** State Henry's law and Dalton's law relating to the solubility of gases in liquids. In what cases are these laws not obeyed? Give reasons. What is meant by the expression 'absorption coefficient' of a gas in a liquid?

If the absorption coefficients of two inert gases, X and Y, in water at 40°C are 0·028 and 0·014 respectively, calculate the amounts of X and Y (expressed in cm$^3$ at s.t.p.) dissolved in 1 dm$^3$ of water at 40°C from the following gas mixture: X at 20 260 Pa, Y at 40 520 Pa, oxygen at 19 860 Pa, water vapour at 7 300 Pa.          (J.M.B.)

**16.** 100 cm$^3$ of ethanol are shaken with a mixture of 60 per cent of oxygen and 40 per cent of methane by volume of 15°C and 101 300 Pa pressure. Assuming that the pressure and composition of the mixture remain constant, calculate the amount of each gas dissolved (expressed in cm$^3$ at standard pressure). The coefficients of solubility of oxygen and methane in ethanol at 15°C at 0·24 and 0·51 respectively. The vapour pressure of ethanol at 15°C is 4000 Pa.

## MORE DIFFICULT QUESTIONS

**17.** Outline the methods you would use to determine the solubility in water of the following at the temperature of the laboratory: (i) crystals of iron(II) sulphate(VI)-7-water, (ii) sodium(I) hydrogencarbonate, (iii) phenylamine.

**18.** Define *saturated vapour pressure* and *partial pressure*.

Explain, with an example, the operation of steam distillation. The saturated vapour pressures for phenylamine and water at various temperatures are as follows:

| $t/°C$ | 70 | 80 | 90 | 100 | 110 |
|---|---|---|---|---|---|
| $p/$Pa (phenylamine) | 1 400 | 2 400 | 3 900 | 6 100 | 9 200 |
| $p/$Pa (water) | 31 200 | 47 300 | 70 100 | 101 300 | 143 300 |

(*a*) At what temperature will steam distillation of phenylamine be possible under atmospheric pressure?

(*b*) What proportion by mass of phenylamine and water would you expect to be present in the distillate? H = 1; C = 12; N = 14; O = 16. (O. and C.)

**19.** An acid, Q, and water are completely miscible. Using the data below, plot on one sheet of graph paper boiling-point/composition curves for mixtures of Q and water. Compositions of Q are expressed as mole per cent.

| b.p./°C at 1 atm | | 118 | 114 | 108 | 104 | 102 | 100 |
|---|---|---|---|---|---|---|---|
| Q (mole per cent) | liquid | 100 | 90·0 | 70·0 | 50·0 | 30·0 | 0 |
| | vapour | 100 | 84·0 | 58·0 | 38·0 | 18·0 | 0 |

Use this diagram to illustrate the principles of fractional distillation and the distillation column.

On another sheet of graph paper, using the same scale, sketch roughly the curve which would be obtained if Q and water formed a constant-boiling mixture containing 70 mole per cent of Q, having a maximum boiling point at 128°C.

State the products which could be obtained from the fractional distillation of aqueous solutions containing 90, 70, and 50 mole per cent of Q in the second case. (S.U.)

**20.** 50 cm$^3$ of 1·5 M ammonia (NH$_3$) solution were shaken with 50 cm$^3$ of trichloromethane in a separating funnel. After the layers had settled 20 cm$^3$ of the trichlormethane layer were withdrawn and titrated with M/20 hydrochloric acid (HCl). 22·9 cm$^3$ of the acid were required for neutralization. Find the partition coefficient of ammonia between water and trichloromethane at the temperature of the experiment.

# CHAPTER 13

# The colloidal state and adsorption

## The Colloidal State

**Dialysis.** If a very small amount of starch is dissolved in boiling water and the liquid is cooled the resulting solution looks similar to a solution of sodium(I) chloride. The starch solution, however, differs in some ways from the salt solution. One difference is that if we place the former on one side of a cellophane membrane with water on the other side, the starch will not diffuse through the membrane. Under similar circumstances the salt passes through the membrane quite readily. This can be shown as follows.

> **Experiment.** Leave some cellophane soaking in water for 24 hours (this enables results to be obtained more quickly). Boil a little starch with water and, after cooling, put 2–3 cm$^3$ of the liquid into a test tube. Add about the same amount of a sodium(I) chloride solution. Put a piece of the cellophane over the end of the tube and fasten it to the sides with a rubber band. Turn the tube upside down and leave it standing (or clamp it) in a small beaker containing water, so that the latter is below the level of the rubber band (Fig. 13–1). After $\frac{1}{2}$ hour test some of the water in the beaker for the presence of starch by means of dilute iodine solution. No blue colour is formed, indicating that no starch has diffused through the membrane. Also, test some of the water with silver(I) nitrate(V) solution. A white precipitate of silver(I) chloride shows that some sodium(I) chloride has passed through the cellophane.

Experiments similar to this were first conducted about the middle of the last century by Graham, using a membrane of parchment paper. Graham noted that salts like sodium(I) chloride in aqueous solution would diffuse easily through the membrane, while substances like starch, gelatin, and white of egg would diffuse only very slowly or not at all. Graham found that the relative times required for the diffusion of equal amounts of some common substances were as follows: hydrochloric acid, 1; sodium(I) chloride, 2·3; sucrose, 7; egg albumin, 49. As the substances which diffused readily were usually crystalline in the solid state Graham called them *crystalloids*, while he referred to substances in the second group as *colloids* (Greek *kolla*, glue). This classification, however, is no longer recognized. Substances which behave as crystalloids in one solvent may behave as colloids in another—*e.g.*, sodium(I) chloride forms a colloidal solution in benzene. Also, some substances (*e.g.*, silver and gold) which can act as colloids in water have a crystalline structure.

Separating substances by the different rates at which they diffuse through a membrane is called *dialysis*, and the apparatus is a *dialyser*.

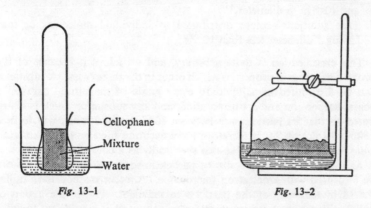

Fig. 13–1                                   Fig. 13–2

For separations on a larger scale a dialyser of the kind shown in Fig. 13–2 is used. This has a glass vessel shaped like a bell jar but open at the top and bottom. A sheet of cellophane or parchment is securely fastened over the bottom by string or rubber bands (frequently a cellophane bag alone is used as a dialyser). The mixture of substances (*e.g.*, sodium(I) chloride and starch in water) is introduced into the vessel, which is then stood in a trough of water. To effect a complete separation we must change the water in the trough from time to time.

**Explanation of Dialysis: Size of Particles.** Graham attributed the difference in the rates of diffusion of salts and substances like starch and egg albumin to the smaller size of the particles of the former enabling them to pass more freely through the pores of parchment paper. Graham's explanation is now known to be correct. The particles of a colloidal solution are intermediate in size between those of a coarse suspension (such as muddy water) and those of an ordinary solution. We can easily see the particles of a coarse suspension under a microscope, and if the suspension is allowed to stand the particles settle, or sediment, under the action of gravity. The ions or molecules of an ordinary solution are not visible under a microscope, and show no tendency to settle. The particles of a colloidal solution are also invisible under a microscope, but we can detect those of some colloidal solutions with an electron microscope. With an optical microscope particles as small as $10^{-4}$ mm diameter can be seen, whereas an electron microscope will reveal particles with a diameter of $10^{-6}$ mm. Accordingly we distinguish the following solid–liquid or liquid–liquid systems:

* Coarse suspensions or emulsions—particles or droplets with a diameter larger than about $10^{-7}$ m.
* Colloidal solutions—particles or droplets roughly between $10^{-7}$ m and $10^{-9}$ m in diameter.[1]
* True solutions—solute distributed as individual molecules or ions having a diameter less than $10^{-9}$ m.

This classification is quite arbitrary, and we adopt it because of the broad differences in properties which occur in the three cases. A substance can be distributed in a liquid in every grade of magnitude between a coarse suspension and a true solution, and any substance which is distributed so that its particles are between $10^{-7}$ m and $10^{-9}$ m will form a colloidal solution. It is therefore more accurate to refer to the *colloidal state* than to describe a substance specifically as a colloid.

As indicated above, the size of particle in some colloidal solutions can be measured with an electron microscope. Another method is to utilize the slight tendency of the particles to sediment. Under the action of gravity this tendency is very small indeed, although it is sufficient to make the system heterogeneous, rather than homogeneous. The rate of sedimentation is greatly increased if a colloidal solution is whirled round in an ultracentrifuge, and by measuring the rate under known conditions we can find the approximate size of the particles. This confirms that these have the dimensions which have been specified.

The particles of a colloidal solution may be very large single molecules (macromolecules) or they may be aggregates of small molecules, atoms, or ions. Proteins and plastics invariably consist of macromolecules, and so do many carbohydrates (*e.g.*, starch). Strictly speaking, many macromolecules cannot be said to have a diameter since they have a long chain-like form. The particles in colloidal sulphur are clusters of a few hundred sulphur molecules ($S_8$) held together by van der Waals attraction. Colloidal silver(I) chloride and colloidal gold consist of tiny crystals. Soap forms a true solution when dilute, but a colloidal solution when concentrated. In the latter the particles are clusters of octadec-

[1] The sizes of particles were formerly expressed in *microns* or *millimicrons*. A micron is $10^{-6}$ m, or 1 micrometre (1$\mu$m), and a millimicron is $10^{-9}$ m, or 1 nanometre (1 nm). Micron and millimicron are not SI units.

anoate, and similar ions resulting from van der Waals attraction between the long alkanoic chains. Aggregates of molecules, atoms, or ions in a colloidal solution are *micelles*.

**Disperse Systems.** A colloidal solution is only one example of a large class of systems known as *disperse systems*. These consist of two phases, one of which is distributed throughout the other (rather like currants in a cake). The distributed substance is the *disperse phase* and the continuous substance the *dispersion medium*. Either of these can be solid, liquid, or gas and hence there should be nine types of disperse system (compare solutions at p. 271). In practice only eight are known. The one which does not exist is the gas–gas system, which would necessitate bubbles of one gas being distributed in another gas. The eight types of disperse system are given in Table 13–1.

Table 13–1. TYPES OF DISPERSE SYSTEM

| Disperse phase | Dispersion medium | Examples |
|---|---|---|
| Solid | Solid | Gold-bearing rock |
| Solid | Liquid | Milk, custard |
| Solid | Gas | Smoke |
| | | |
| Liquid | Solid | Jelly, butter |
| Liquid | Liquid | Hair cream |
| Liquid | Gas | Mist, cloud |
| | | |
| Gas | Solid | Charcoal, pumice stone |
| Gas | Liquid | Foam, whipped cream |

In some of the examples in Table 13–1 only part of the disperse phase is colloidal. The remainder consists of larger aggregates which settle rapidly when the system is left standing. Thus milk can be separated into cream, a coarse suspension of fat particles, and skimmed milk, a colloidal solution.

True solutions are not included in disperse systems, a true solution constituting a single phase. However, disperse systems in which the dispersion medium is a liquid are analogous to solutions, and are called *sols*. Thus we have

Solution = solvent + solute
Sol = dispersion medium + disperse phase

The most common sols are those in which water is the dispersion medium. These are known as *hydrosols*. *Aerosols*, which are widely used in the form of sprays, are disperse systems in which air is the dispersion medium.

**Classification of Sols.** Some colloid systems are reversible, while others are not. Thus dried milk (obtained by careful evaporation of milk after removal of cream) can be 'reconstituted' merely by mixing it with water. Similar examples of reversible hydrosols are seen in dried eggs and dried blood plasma. Rubber dispersed in benzene is a reversible sol in which the dispersion medium is not water. Hydrosols of inorganic substances such as sulphur and gold are irreversible. When they are coagulated they cannot be reconstituted by mixing the precipitated substance with water.

Reversibility or irreversibility of a sol depends on whether there is an attraction or not between the disperse phase and the dispersion medium. If the disperse phase is a protein (as in eggs and blood plasma) hydrogen bonding takes place between water molecules and the amino groups (—NH₂ or —NH—) of the protein. Hydrogen bonding likewise occurs between water molecules and the hydroxyl groups of carbohydrates like starch. With rubber dispersed in benzene attraction between the molecules is provided by the van der Waals force. There are no similar forces of attraction when sulphur or gold is dispersed in water, so that these systems are irreversible. Accordingly we divide sols into two classes as follows:

*Lyophilic sols* (literally, 'solvent-loving' sols) are reversible sols such as those of egg albumin and starch.

*Lyophobic sols* ('solvent-hating' sols) are irreversible sols such as those of sulphur and gold.

If water is the dispersion medium these terms become respectively *hydrophilic* and *hydrophobic*. The name *suspensoid* is sometimes used for a sol when the disperse phase is a solid, and *emulsoid* when it is a liquid.

One direct result of the attraction between the disperse phase and dispersion medium in lyophilic sols is the high viscosity of the latter. This is perhaps best seen with gums and glues dispersed in water. The viscous character of these liquids is partly due to the increase in size of the disperse particles by hydration and partly to the decrease in the amount of 'free' water. The viscosity of a lyophobic sol is not appreciably different from that of the pure liquid medium.

Many substances which form lyophilic sols can be obtained in a jelly-like, or *gel*, condition. Thus when gelatin is mixed with warm water it forms a colloidal solution, but when this is cooled it sets to a semi-solid mass. This change is known as *gelation*. The gel condition is an intermediate stage in the coagulation of the sol. In this condition the gelatin probably forms a net-like structure containing water in the open spaces. It is often very difficult to remove the water completely from such a gel. Possibly separate particles are only formed by the gelatin at higher temperatures or when the amount of water is large.

The formation of the gel condition, while typical of lyophilic sols, is not confined to them. Iron(III) hydroxide, aluminium(III) hydroxide, and silicic(IV) acid are all precipitated in the gel form from a hydrosol, but we cannot reconvert them to the sol by mixing with water.

## Preparation of Sols

*Lyophilic sols* are readily made by stirring the solid (usually in the form of a gel or resin) with the liquid. The dispersion of the solid is often assisted by heating. Examples of this method are the dispersion of gelatin in water, shellac in ethanol (to obtain French polish), and nitrocellulose in 3-methylbutyl ethanoate ('isoamyl acetate') (to make nail varnish). The method is essentially similar to the one used for the preparation of true solutions. The 'dissolving' in organic solvents is a much slower process, however, and the sols obtained are more viscous. They are usually quite stable.

*Lyophobic sols* have to be prepared by special methods. These fall into two categories—*dispersion* methods and *condensation* methods. In the first the solid (or liquid) in bulk form is broken down to the right size of particle (or droplet) in the presence of the dispersion medium. In the second, molecules, atoms, or ions are built up to the size of particle required. As the disperse phase is insoluble in the dispersion medium careful control has to be maintained over conditions (*e.g.*, concentration and temperature). Otherwise the building-up process goes too far and produces a precipitate.

Lyophilic sols tend to be unstable. In particular they are coagulated by electrolytes (the reason for this is explained presently). To 'protect' the sols small amounts of a stabilizer are often added to the dispersion medium. The most common stabilizer for hydrosols is gelatin, which forms a thin layer on the surface of the particles and inhibits the action of an electrolyte.

### Dispersion Methods.

*Mechanical Dispersion.* Some suspensoids are made by feeding the powdered solid and the liquid into a 'colloid mill'. This consists of metal plates (or cones) placed very close together and rotating at high speed in opposite directions. The shearing action of the plates reduces the size of the solid particles still further and after a time a colloidal solution is formed. This method is used to make 'colloidal graphite' (a lubricant) and certain kinds of printing ink.

*Peptization.* Some substances which cannot be made into colloidal solutions by means of water alone are so obtained if very small amounts of another substance are added. This, if gelatinous silicic(IV) acid is mixed with water containing a trace of sodium(I) hydroxide a colloidal solution is formed. Similarly, a small quantity of hydrochloric acid is effective in changing aluminium(III) hydroxide to a sol condition. This action is analogous to that of the digestive enzyme pepsin on food and is called *peptization*, the substance added being termed the *peptizing agent*. Peptization is brought about partly by lowering the surface tension of water and partly by adsorption of ions.

*Bredig's Arc Dispersion Method.* This is often used to prepare colloidal solutions of the noble metals, silver, gold, and platinum. An arc is

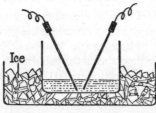

*Fig.* 13–3.  Bredig's arc dispersion method

struck between two gold wires held with the ends about 2 mm apart in distilled water containing a little alkali, the water being kept cool by surrounding it with ice. The heat of the arc vaporizes some of the metal, and the vapour then condenses into particles of colloidal size. Thus both dispersion and condensation are involved in this method. If the resulting liquid is filtered from the coarser particles of gold, a clear bright-red filtrate is obtained which is a colloidal solution of gold. Radioactive gold sols have been used in the treatment of cancer.

**Condensation Methods.**

*Change of Solvent.* A few drops of an ethanolic solution of octadec-anoic (stearic) acid or hexadecanoic (palmitic) acid are stirred into about 100 cm$^3$ of distilled water. The acids are insoluble in water and a milky-looking suspensoid results. If octadecenoic (oleic) acid, which is a liquid, is used the product is an emulsoid.

*Double Decomposition.* Hydrogen sulphide is passed through a saturated solution of arsenic(III) oxide until the liquid smells of gas. A clear yellow colloidal solution of arsenic(III) sulphide results. This can be kept by adding a further small amount of arsenic(III) oxide to remove excess of hydrogen sulphide.

$$As_2O_3 + 3H_2S \rightarrow As_2S_3 + 3H_2O$$

If hydrochloric acid is added to a concentrated solution of sodium(I) silicate(IV) a gelatinous precipitate of silicic(IV) acid is formed:

$$2Na^+ + SiO_3^{2-} + 2H^+ + 2Cl^- \rightarrow H_2SiO_3 \downarrow + 2Na^+ + 2Cl^-$$

When the sodium(I) silicate(IV) solution is dilute the same reaction occurs, but no precipitate appears. When the clear liquid is placed in a dialyser sodium(I) ions and chloride ions pass through the membrane and a colourless sol of silicic(IV) acid remains in the dialyser.

The most important sol made by double decomposition is the one used in the preparation of photographic film, paper, and plates. This is obtained by adding together measured volumes of standard silver(I) nitrate(V) solution and standard potassium(I) bromide solution containing a known mass of gelatin. Silver(I) bromide is 'precipitated' in the form of colloidal particles.

$$Ag^+ + NO_3^- + K^+ + Br^- \rightarrow AgBr + K^+ + NO_3^-$$

The unwanted potassium(I) and nitrate(V) ions are removed by dialysis. More gelatin is added together with small amounts of special substances to improve light sensitivity. The viscous liquid is then coated on to nitrocellulose film, paper, or glass and dried. Photographic 'emulsion' is thus a colloidal suspension of silver(I) bromide grains in the gelatin gel.

*Hydrolysis.* By pouring a few cm³ of concentrated iron(III) chloride solution into an excess of water at 80°C we obtain a sol, reddish-brown in colour, of iron(III) 'hydroxide'. This is produced as a result of hydrolysis.

*Oxidation—Reduction Methods.* Sulphur hydrosols are readily formed by reaction between hydrogen sulphide in aqueous solution and various oxidizing agents. If the gas is bubbled through hot water a brown sol is produced as a result of oxidation by the oxygen dissolved in the water. We can obtain a yellow sol by passing hydrogen sulphide through a dilute solution of sulphur dioxide.

$$2H_2S + SO_2 \rightarrow 2H_2O + 3S$$

Colloidal sulphur is frequently encountered in qualitative analysis when hydrogen sulphide is used in the presence of an oxidizing agent (*e.g.*, chromate(VI) or iron(III) ions). It can be removed by boiling (to coagulate the sulphur) and filtering through two filter papers folded together.

Silver sols and gold sols can be prepared by treating dilute solutions of silver(I) nitrate(V) or gold(III) chloride with organic reducing agents like tannic acid, 2,3-dihydroxybutanedioic (tartaric) acid, or ethanal. A colloidal solution of silver is brown.

When a sol has been made it often contains ions which would cause it to coagulate on standing. It must therefore be purified. Methods of purification consist essentially of filtration. Ordinary filter paper cannot be used because it will only retain particles with a diameter larger than 0·001 mm. Dialysis is a form of filtration in which the pore size of the parchment or cellophane membrane is small enough for colloidal particles to be held back while ions and small molecules of a true solution pass through. Dialysis, however, is a slow process. In *electrodialysis* it is accelerated by putting the sol in a tube with a membrane on each side. The two outer compartments contain pure water in which electrodes are placed. When a potential difference is applied to these the unwanted ions are attracted through the membranes to one or other of the electrodes.

Another method of purification is to employ *ultrafiltration*. Unglazed porcelain or filter paper impregnated with collodion is used as the filtering material, and increased pressure or suction is applied to speed up the separation.

## Properties of Sols

**Optical Properties.** We can distinguish a colloidal solution from a true solution by the *Tyndall effect*. This was actually discovered by Faraday (1857), but is named after John Tyndall who investigated it more fully (1869). When a beam of light passes through a true solution (which should be made in distilled water and filtered) we cannot distinguish the path of the beam and the liquid appears clear. If we repeat the experiment with a sol we see the track of the beam as a cloudy pencil or cone of light. This can be shown by covering the glass of an electric

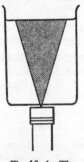

*Fig.* 13–4. The Tyndall effect

torch with a piece of black paper or aluminium foil, in which a small hole is pierced, and shining the torch upwards or sideways through the liquid (Fig. 13–4). The liquid is viewed from a direction at right angles to the beam.

The Tyndall effect is caused by the scattering of light by colloidal particles or droplets. In the same way minute dust particles in a room scatter sunlight and render its path visible. Molecules or ions in a true solution are too small to scatter light. The light waves scattered most are the shorter ones, that is, those towards the blue end of the spectrum. Hence, when we view a colloidal solution from a direction at right angles to the incident light, it appears blue. Seen by transmitted light it has the complementary colour, reddish-brown. This can be tested by adding a few drops of milk to a beaker full of water.

Scattering of light takes place similarly with other disperse systems. The blue colour of the sky is caused by sunlight scattered by dust and fine ice crystals in the upper atmosphere. On the other hand, the setting sun looks red because it is observed by direct sunlight, from which the blue constituents have been removed by scattering. Coarse disperse systems appear white or grey if both disperse phase and dispersion medium are colourless. This is because the larger particles, droplets, or bubbles reflect sunlight as a whole. Examples are clouds, hair cream, foam, and milk in bulk. Smoke rising from the end of a cigarette is blue, the particles being colloidal in size. Exhaled smoke is grey because moisture has condensed on the particles, thus increasing their size.

The particles of a sol show Brownian movement, although not in the same way as described in Chapter 2, where we saw how the phenomenon can be demonstrated with a coarse suspension. In the experiment described the particles themselves are visible in motion. The particles in a sol are too small to be visible, but we can detect their presence by concentrating a powerful beam of light into the sol and observing the latter through a high-power microscope from a direction at right angles to the beam. This arrangement, which is essentially similar to that in Fig. 2–7, constitutes an *ultramicroscope*. The colloidal particles reveal themselves as brilliant disks of light dancing here and there almost as if they were alive. The more vigorous movement is due to the smaller size of the particles.

**Colligative Properties.** As explained in Chapter 4, these include the related properties of lowering of vapour pressure, depression of freezing point, elevation of boiling point, and osmotic pressure, all of which depend on the number of particles in solution. They are well marked with true solutions, but not with sols, where vapour pressure, freezing point, and boiling point differ only slightly from those of the dispersion medium itself. This is readily understood since the extent to which the values are changed by a given mass of 'solute' in a given mass of 'solvent' is inversely proportional to the relative molecular mass of the former. Osmotic pressure is in a somewhat different category because of its much greater magnitude. Although the osmotic pressures of sols are thousands of times smaller than those of true solutions of similar concentration, they are still measurable. We can thus find the relative molecular masses of macromolecules from the osmotic pressures of the sols which they form.

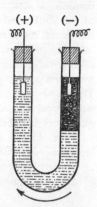

*Fig.* 13–5. Electrophoresis of a gold sol

**Electrical Properties.** The distributed particles or droplets in a lyophobic sol are electrically charged. We can show this by putting a red gold sol in a U-tube (Fig. 13–5) and inserting platinum electrodes connected to the d.c. mains (220 volt) through a 600-ohm resistance. The liquid round the cathode soon becomes colourless, and after ½ hour the colourless zone extends to a depth of 7·5 or 10 cm in the cathode limb. At the same time the red colour becomes deeper on the anode side and some gold is deposited on the anode. These changes clearly indicate that the particles of the gold sol are negatively charged, unlike gold ions which are positively charged and attracted to the cathode.

Other sols in which the distributed particles are negatively charged (as revealed by their movement to the anode) are those of sulphur, arsenic(III) sulphide, silicic(IV) acid, silver, and platinum. In others *e.g.*, iron(III) hydroxide and aluminium(III) hydroxide sols, the particles are positively charged and travel to the cathode. The movement of the disperse phase as a whole towards one or other of the electrodes is called *electrophoresis.* The phenomenon is also exemplified by the particles of a coarse suspension. Thus clay particles suspended in water are negatively charged and move towards the anode.

The disperse phase in a lyophilic sol may, or may not, be electrically charged. If it is charged the sol undergoes electrophoresis in the same way as a lyophobic sol. If it is neutral, as with starch or gelatin in pure water, the sol is unaffected by an electric field.

A colloidal solution as a whole is electrically neutral. We should therefore expect that if the disperse phase is positively charged the dispersion medium would be negatively charged, and vice versa. If this is so, the dispersion medium should tend to move in the opposite direction to the disperse phase under the influence of an electric field. Under the conditions used for electrophoresis there is no appreciable movement of the liquid medium. If, however, the disperse phase is kept stationary the liquid medium moves towards the electrode of sign opposite to its own charge. We can show this by inserting a plug of wet clay in the bend of a U-tube, adding distilled water to the same level in the two limbs, and then applying a potential difference by means of two electrodes. The clay is negatively charged, while the water is positively charged. The water level rises on the cathode side and sinks on the anode side. This phenomenon is called *electro-osmosis.*

There is little doubt that the charge on colloid particles is due to 'adsorption' of ions from the liquid, adsorption being brought about by unsatisfied valencies on the surface of the particles (see later). Ions of different sign are usually present when the sol is made. If they are

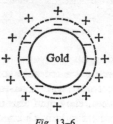

*Fig.* 13–6

absent, or if they are completely removed by prolonged dialysis, the sol is unstable and precipitation soon occurs. Adsorption of positive or negative ions takes place preferentially, so that the particles are charged positively or negatively. In making a gold sol by Bredig's method a little sodium(I) hydroxide or other alkali is necessary. OH⁻ ions are adsorbed in preference to Na⁺ ions and form a strongly adhering layer round the gold particles, giving them a negative charge. This layer is responsible for the stability of the sol. The distributed particles are prevented from coming together and forming larger aggregates by repulsion between the similar charges. The Na⁺ ions, while freer to move, will generally remain in the vicinity of the negatively charged particles. The presence of sodium(I) ions is the reason for the liquid medium having a positive charge. The distribution of ions round a particle of colloidal gold can thus be represented (in two dimensions) as in Fig. 13–6. The layer of adsorbed hydroxyl ions (negative signs) is enclosed within the broken line and surrounding sodium(I) ions (positive signs) are outside this line.

There are no definite rules as to whether positive or negative ions are adsorbed by different kinds of disperse phase. When the disperse phase is crystalline in character, it frequently shows preferential adsorption for one of its own ions. Thus arsenic(III) sulphide ($As_2S_3$) sols, prepared from arsenic(III) oxide and hydrogen sulphide, adsorb $S^{2-}$ ions and iron (III) hydroxide sols, obtained by partially hydrolysing iron(III) chloride, adsorb $Fe^{3+}$ ions. Such adsorptions are consistent with a tendency for the crystal to increase its size by adding on its own ions. Alternatively there may be vacant sites in the surface of the crystal which can be filled by the added ions.

The ions adsorbed by the disperse phase may be positive ions under one set of conditions and negative ions under another. When lyophilic sols are made from proteins the nature of the charge on the disperse phase depends on the pH (degree of acidity or alkalinity) of the liquid medium. At one value of pH the colloidal particles adsorb $H^+$ ions from the medium and are positively charged, while at another value they adsorb $OH^-$ ions and are negatively charged. At a certain intermediate value they do not undergo electrophoresis. At this point (termed the *isoelectric point*) they are electrically neutral. The pH value at the isoelectric point varies for different proteins. This allows us to separate the proteins in a mixture such as blood plasma. By suitable adjustment of the pH one protein will move towards the anode on electrophoresis, while another will move towards the cathode. Even when two proteins travel towards the same electrode they do so at different speeds, so that they can still be separated. An ingenious form of electrophoresis cell designed for the separation of proteins by the Swedish chemist Tiselius is now used in medical research laboratories all over the world.

### Coagulation.

*Lyophobic sols.* When coagulation of a sol occurs the particles or droplets of the disperse phase come together and form larger masses, which settle under the action of gravity. As we have already seen, this is normally prevented by mutual repulsion between the similarly charged particles or droplets. If the charge is removed, precipitation follows. Thus in electrophoresis when the disperse phase comes into contact with an oppositely charged electrode it is deposited on the surface of the electrode. One application of this principle is to get rid of smokes and mists, which are a nuisance in some industries. As the suspended particles or droplets are colloidal in character they are electrically charged with respect to the dispersion medium, and precipitation can be effected by passing the smoke or mist between metal plates charged to a high potential difference (75 000 volt). The particles or droplets are attracted to, and settle on, the plate of opposite charge. This is the principle of the Lodge–Cottrell electrostatic precipitator, which is widely used in industry.

It removes dust from blast furnace gases and particles of arsenic(III) oxide from sulphur dioxide in the contact process for manufacturing sulphuric(VI) acid.

Another illustration of the same principle is the attraction of particles of dust or smoke to a wall above hot water pipes or a radiator. Here the wall acquires a static charge owing to warm air passing over its surface.

A second method of bringing about coagulation of a lyophobic sol is to mix it with another sol of opposite charge. For example, when colloidal iron(III) hydroxide is added to colloidal gold the positive charge of the former neutralizes the negative charge of the latter and both are precipitated. Precipitation is complete, however, only when the charges just neutralize each other, that is, at the isoelectric point. Coagulation does not occur when two positive colloids or two negative colloids are added to each other.

While small amounts of electrolytes are essential for the stabilization of lyophobic sols, the addition of larger amounts results in precipitation. Thus, although the distributed particles in an iron(III) hydroxide sol and an arsenic(III) sulphide sol carry charges of opposite sign, both sols can be coagulated by adding sodium(I) chloride solution. This is because the solution furnishes both anions and cations and the charge on the disperse phase is neutralized by adsorption of the oppositely charged ions. The precipitating effect of different anions or cations is not the same, however. The charge on a magnesium ion ($Mg^{2+}$) is twice as large, and that on an aluminium ion ($Al^{3+}$) is three times as large, as the one on a sodium(I) ion. Hence precipitation of negatively charged particles should occur more easily with aluminium(III) ions than with magnesium(II) ions and more easily with magnesium(II) ions than with sodium(I) ions. This is expressed by the **Hardy–Schulze rule**:

> *The precipitating effect of an ion on a disperse phase of opposite charge increases with the valency of the ion.*

The relative concentrations (in mole per $dm^3$) of different cations required to coagulate the same amount of a negative arsenic(III) sulphide sol are given below:

| | | | | | |
|---|---|---|---|---|---|
| $Na^+$ | 1·00 | $Mg^{2+}$ | 0·014 | $Al^{3+}$ | 0·0018 |
| $K^+$ | 0·98 | $Ba^{2+}$ | 0·014 | $Ce^{3+}$ | 0·0016 |
| $H^+$ | 0·98 | $Ca^{2+}$ | 0·013 | | |

The precipitating effects of the monovalent, divalent, and trivalent cations are approximately in the ratio 1:70:600. In the precipitation of a positive colloid the precipitating effect depends on the valency of the anion added. Thus phosphates(V) bring about precipitation more easily than sulphates(VI) and sulphates(VI) more easily than chlorides.

An everyday example of the precipitating effect of electrolytes on colloidal solutions is seen at the mouth of a river. The latter carries along in suspension negatively charged clay particles, which are deposited as mud when they meet the relatively high concentrations of sodium(I), magnesium(II), and calcium(II) ions in sea water. In time this may produce an extensive delta. A further example is the use of aluminium(III) sulphate(VI) in the clarification of water for industrial purposes

Some lyophilic sols (*e.g.*, those of sulphur and iron(III) hydroxide in water) can be coagulated by boiling. This increases the thermal energy of both the adsorbed ions and the molecules of the liquid medium. Collisions between the two may then result in some of the adsorbed ions becoming 'free'. If this happens to a sufficient number the charge left on the particles is too small to prevent precipitation.

*Lyophilic sols.* As mentioned earlier, the disperse phase in lyophilic sols usually consists of macromolecules of proteins, carbohydrates, or plastics. These sols are more stable and show greater resistance to coagulation than lyophobic sols. Although the disperse phase may be electrically charged, relatively large amounts of electrolytes are needed to bring about coagulation. The reason for this lies in the attraction between the disperse phase and dispersion medium. Thus, if the disperse phase is a protein and the dispersion medium is water, the attraction between the water molecules and the numerous hydrophilic groups of the protein molecules results in the latter becoming covered with a layer of water molecules. Ions may then be adsorbed on to the hydrated molecules. The ions may be $H^+$ ions or $OH^-$ ions, depending on the pH of the aqueous medium. The particles of the disperse phase may thus have *two* protective

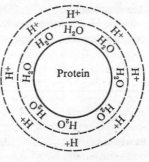

*Fig.* 13-7

layers (Fig. 13-7). Even if the adsorbed ions are removed by adding an electrolyte aggregation is still hindered by the layer of water molecules. This explains why hydrosols of starch and gelatin are quite stable at their isoelectric points.

Large amounts of electrolytes cause precipitation of lyophilic sols because they reduce the amount of free water, some of the latter being used for hydration of the added ions. This is known as the *salting-out effect*. The clotting of blood in small cuts is accelerated by putting common salt on the injury. Even more effective (owing to the trivalent $Al^{3+}$ ions) is 'alum' (p. 35). Another method of reducing the amount of free water is to add ethanol or propanone. These are liquids which have a strong attraction for water and compete with the disperse phase for the water available. If either ethanol or propanone is present, much smaller amounts of electrolytes bring about precipitation. When this occurs the disperse phase is obtained as a gel, and in all cases the sol can be reformed by mixing the gel with water.

The coagulation of some lyophilic sols (*e.g.*, white of egg) by boiling is probably the result of chemical change in the disperse phase. The substance precipitated cannot be reconverted to the sol condition.

**Surface Area of Colloids.** Owing to their small size the particles in a colloidal solution have a very large surface area. This explains their high powers of adsorption. Examples have already been given of the adsorption of ions. In addition colloidal particles readily adsorb dyestuffs. In dyeing cloth various metal hydroxides (*e.g.*, those of aluminium(III) and chromium(III)) are used as mordants to make the dye fast. A less welcome application is the carrying-down of zinc(II) ions with aluminium(III) hydroxide and iron(III) hydroxide in qualitative analysis (hence the necessity of the thorough washing of precipitates). As a result of the large surface area sols frequently act as catalysts. The decomposition of hydrogen peroxide is accelerated by many colloidal solutions. Enzymes are colloidal in character and their operation probably depends on adsorption processes.

Table 13–2 summarizes the chief differences between an ordinary solution, a lyophobic sol, and a lyophilic sol.

**Table 13–2**

| Solution | Lyophobic sol | Lyophilic sol |
|---|---|---|
| Reversible | Irreversible | Reversible |
| Low viscosity | Low viscosity | High viscosity |
| Solute passes through parchment readily | Very slow diffusion of disperse phase | Very slow diffusion of disperse phase |
| Not sedimented by ultracentrifuge | Sedimented by ultracentrifuge | Sedimented by ultracentrifuge |
| Tyndall effect and Brownian movement not shown | Shows Tyndall effect and Brownian movement | Shows Tyndall effect and Brownian movement |
| No electrophoresis | Electrophoresis | Electrophoresis (except at isoelectric point) |
| High osmotic pressure | Low osmotic pressure | Low osmotic pressure |
| Not coagulated by electrolytes | Coagulated by electrolytes at low concentration | Coagulated by electrolytes at high concentration |

## Adsorption

*Adsorption* is a process in which a substance becomes attached to the surface of a solid or liquid. This process must be distinguished from *absorption*, in which the substance is distributed throughout the solid or liquid. Thus calcium(II) oxide absorbs moisture, forming calcium(II) hydroxide, and freshly distilled water absorbs air, giving a solution.

The substances adsorbed at a solid surface may be gases, liquids, or other solids (from solution). The amount of material adsorbed by a given mass of adsorbent depends on the physical condition of the latter, most material being adsorbed when the solid is porous or finely divided. This confirms the theory that adsorption is a surface phenomenon, since the surface area exposed is much greater when the particles are smaller. (A cube of side 1 cm long has a surface area of 6 $cm^2$. If it is subdivided into cubes of side 0·000 001 cm, which may be the size of particles in very finely divided metals, the surface area is 6 000 000 $cm^2$.

Common adsorbents are carbon (charcoal), which has a large internal surface area, and finely divided metals like platinum and nickel. If freshly warmed carbon (wood charcoal) is inserted into a tube containing ammonia gas over mercury, the mercury rises and fills the tube almost completely. This form of carbon also adsorbs other gases, the action being most marked with gases which liquefy easily (*e.g.*, $CO_2$, $H_2S$, $SO_2$). It is used in gas masks, for removing benzene vapour from coal gas, and in recovering the vapours of solvents used in the rubber and celluloid industries. For these purposes the adsorbent is 'activated' by blowing steam through it under pressure so that tarry matter is removed from the internal surfaces. This increases the amount of gas which the material can take up. Carbon (animal charcoal) is used to decolorize brown sugar solutions in the manufacture of white sugar.

Adsorption may be either a physical process or a chemical process, although in some cases both occur simultaneously. In physical adsorption the fundamental cause is the relatively weak attraction due to the van der Waals force (or, sometimes, hydrogen bonding). This is illustrated by the adsorption of inert gases (*e.g.*, $N_2$) by carbon (charcoal). In a closed vessel dynamic equilibrium exists between the adsorbed gas and the gas in contact with the solid. The amount of gas adsorbed is greater the lower the temperature, this effect being in accordance with Le Chatelier's principle since heat is evolved in the adsorption.

Chemical adsorption (*chemisorption*) is due to the existence on the adsorbent of unsatisfied valencies, whereby molecules or ions of another substance become attached to the surface atoms or ions of the solid. The attachment takes place by covalent or ionic bonds as with ordinary chemical compounds. In chemisorption of gases diatomic molecules may split into single atoms, which become separately attached. Carbon (charcoal) consists of hexagonal rings of carbon atoms joined together to form very small crystals. It is prepared by the destructive distillation of complex organic molecules. Many of the carbon hexagons are inevitably left broken and incomplete, particularly at the surface, so that some carbon atoms are in a different state of combination than others. The chemisorption of oxygen by carbon (charcoal) can be represented as follows:

$$
\begin{array}{c}
\text{O} \qquad \text{O} \qquad \overset{\text{O}}{\underset{\|}{\text{C}}} \quad \overset{\text{O--O}}{} \quad \overset{\text{O}}{\underset{\|}{\text{C}}} \\
\overset{\|}{\text{C}} \quad \overset{\|}{\text{C}} \\
\text{C} \quad \text{C} \quad \text{C} \quad \text{C}
\end{array}
$$

The resulting 'oxide' of carbon cannot be given a formula, but its formation is, nevertheless, true chemical combination. If oxygen is adsorbed on carbon (charcoal) at $0\,^{\circ}\text{C}$ and the unadsorbed gas is pumped off, both carbon monoxide and carbon dioxide (but no oxygen) are given off on heating. Other examples of chemisorption are oxygen on platinum and hydrogen on nickel, both of which are important in connection with catalysis. As we might expect the heat evolved in chemisorption is much larger than in physical adsorption.

**Adsorption Indicators.** In volumetric analysis use is sometimes made of indicators which depend for their action on the preferential adsorption of ions. Thus in the estimation of chlorides, bromides, and iodides by standard silver(I) nitrate(V) solution fluorescein (or eosin) solution is added to the solution which is titrated. Fluorescein is the sodium(I) salt of an organic acid, and we can represent its chemical formula by NaFl. In aqueous solution it is dissociated into ions ($\text{NaFl} \rightleftharpoons \text{Na}^+ + \text{Fl}^-$). When silver(I) nitrate(V) solution is added to a chloride solution containing fluorescein, part of the precipitate of silver(I) chloride formed is in the colloidal state and has great adsorbing power. Adsorption, however, is selective, and when a slight excess of silver(I) nitrate(V) has been added positive silver(I) ions are adsorbed in preference to negative nitrate(V) ions. The adsorbed silver(I) ions at once attract the negative fluorescein ions and form a rose-pink precipitate (adsorbed) of silver(I) fluorescein ($\text{Ag}^+ + \text{Fl}^- \rightarrow \text{AgFl}$). The end point of the titration is therefore marked by the white precipitate of silver(I) chloride turning rose-pink. No precipitate is formed if silver(I) nitrate(V) solution is added to fluorescein solution alone, because the silver(I) fluorescein salt is soluble. This shows that the rose-pink precipitate formed in the titration is adsorbed.

**Chromatography.** This is a method of analysing mixtures of substances, and its particular virtue is that it can bring about separations which are difficult or impossible by other methods. 'Chromatography' means 'colour writing'. The name arose from some of the early experiments with coloured mixtures. Thus in 1903 the Russian botanist Tswett separated chlorophyll into its constituents by passing a solution of chlorophyll in petroleum ether down a tube packed with calcium(II) carbonate (chalk). He got a series of coloured bands, each being a separate constituent of the chlorophyll. Nowadays the method is applied to a wide variety of mixtures, coloured and uncoloured.

Chromatography is divided into various types according to the way in which it is carried out. In some cases adsorption is involved directly, in others only indirectly. All types, however, have the same basic requirements and depend on similar principles. The system always contains a stationary phase (solid or liquid) to adsorb or absorb the mixture being separated, and a moving phase (liquid or gas), which passes over the stationary phase and competes with it for the constituents of the mixture. The general principles of separation are most easily explained by gas chromatography and we shall therefore consider this type first.

*Gas Chromatography.* Suppose we have a tube (Fig. 13–8) containing

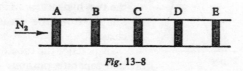

*Fig.* 13–8

portions of activated carbon (charcoal) at A, B, C, D, and E. If we add a drop of benzene to the material at A the liquid is adsorbed and an equilibrium is established between molecules on the internal surface and molecules in the pores of the solid. If we now pass a slow stream of nitrogen through the tube as shown, the benzene vapour in the pores is removed and some of the molecules vaporize from the surface to maintain the equilibrium. These in turn are swept away by the nitrogen and the action continues until all the adsorbed benzene has been removed from the adsorbent at A. The process is similar to that involved in removing a very soluble gas from solution by passing an inert gas over the solution.

The nitrogen and benzene vapour pass on to the carbon (charcoal) at B. Since this contains no benzene it will adsorb some of the hydrocarbon from the mixture, and, if the latter were moving sufficiently slowly, an equilibrium would be established at B between benzene molecules on the internal surface and molecules in the pores. However, the concentration of benzene vapour in the nitrogen decreases to zero as benzene is removed from the adsorbent at A. Therefore the concentration of benzene in the adsorbent at B increases from zero to a maximum and then falls to zero again. This is indicated by the benzene curve in the concentration–time graph in Fig. 13–9. The building-up and falling-away of the benzene concentration is repeated in turn at C, D, and E.

Now suppose that instead of putting benzene on to the carbon (charcoal) at A we add a mixture of benzene and methylbenzene. Both hydrocarbons are adsorbed. When the nitrogen stream is passed the increase and decrease in concentration occurs with both substances at B, C, D and E. Methylbenzene, however, has a higher boiling point than benzene and is less volatile, so that adsorbed methylbenzene is less easily removed. Thus the methylbenzene maximum in the concentration-time graph occurs later than for benzene. For adsorbent layers near to A the two curves will overlap, but if there are many layers of adsorbent the curves will become increasingly separated in time with greater distance from A. Hence benzene and methylbenzene can be completely separated, the two vapours arriving at the end of the tube in separate portions and in concentrations varying with time as shown by the curves. Furthermore, if we first calibrate the apparatus with known masses of substances, we can estimate the amounts of the two hydrocarbons in the original sample from the areas under the curves.

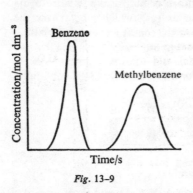

*Fig.* 13–9

In practice the tube does not contain separate portions of the fixed phase, but is packed uniformly. Other adsorbing agents used besides carbon (charcoal) are silicon(IV) oxide ('silica gel') and aluminium(III) oxide. Sometimes the fixed phase is a film of oil adsorbed on 'silica gel' or similar solid support. Adsorption then enters into the process only indirectly. In a more recent development the packing is dispensed with and the liquid is coated directly on to the glass sides of a capillary tube.

The gas passed through the column is called the *carrier gas*. Nitrogen, hydrogen, and argon are common as carrier gases. There are various methods of detecting the arrival of the separated fractions at the end of the tube. One of the simplest is to pass the issuing gas over an electrically heated wire. A change in composition of the gas causes a change in the resistance of the wire. The change is recorded automatically by a pen moving over a revolving drum, and produces curves of the type in Fig. 13–9. The different fractions can also be collected and identified.[1]

Gas chromatography is used to separate and identify the constituents of complex mixtures of liquids and gases such as petroleum fractions and gases produced in biological processes. It has the advantage that only

[1] For a simple method of demonstrating gas chromatography, using a detergent powder as the fixed phase and fuel gas as the carrier gas, see *Gas Chromatography as a Class Experiment* by I. W. Williams, *The School Science Review*, vol. XLV, No. 156, p. 402.

minute amounts of material are required. Thus it has been used to measure the solubilities of slightly soluble hydrocarbon gases in water.

*Elution Chromatography.* This depends on the different extents to which substances are adsorbed from solution by a finely divided inert solid such as aluminium(III) oxide or kieselguhr. If a mixture of two substances in solution is added to a tube packed with one of these solids adsorption of both substances occurs, but if more solvent is added the substances pass down the column at different rates and are separated. By continuing this washing, or *elution*, process with the same solvent or with different solvents in turn, both substances can be run out of the tube as separate solutions. The substances can then be obtained by evaporation.

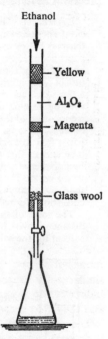

Elution chromatography can be illustrated by means of a mixed solution of methyl orange and magenta dye (rosaniline hydrochloride) and a column of aluminium(III) oxide.[1] (The latter is specially made for chromatography so that it has the right grain size.) The tube (Fig. 13–10) is first filled with a slurry, or paste, of the oxide and ethanol. Two or three drops of the mixed solution are then added to the tube. When further portions of ethanol are added separation occurs into a yellow band and a magenta, or purple, band. With more ethanol the latter can be washed down and eventually out of the column. The methyl orange layer can be eluted subsequently by means of water.

*Fig.* 13–10

We can also apply elution chromatography to mixtures of inorganic chemicals, but then the adsorbed layers are often colourless. They are identified either before or after elution by means of chemical tests. The process is used to separate and identify the constituents of vegetable extracts in pharmacy, for the separation of vitamin A from fish-liver oil, and in the detection of adulterants in foods and wines.

*Paper Partition Chromatography.* This takes advantage of the fact that filter paper, even when 'dry', contains up to 12 per cent of water, which is strongly adsorbed on the cellulose fibres. The adsorbed water constitutes the fixed phase in this type of chromatography. As the paper merely acts as a support for the water adsorption is not involved directly. The adsorbed water behaves as an immiscible liquid towards another liquid (and even towards water itself) which passes over the paper. If

---

[1] Class experiments on elution chromatography can be carried out with test tubes with a hole in the bottom and plugged with glass wool.

the second liquid contains a solute which is soluble in water the solute distributes itself between the moving solvent and the adsorbed water according to the partition coefficient between the two. This has given rise to the term 'paper partition chromatography'. If the moving solvent contains two or more substances which are soluble in water the difference in their partition coefficients permits us to separate them.

Paper partition chromatography is readily demonstrated with mixtures of coloured inks, indicators, etc. A simple method of illustrating the process is as follows:

> **Experiment.** Choose a cork to fit a boiling tube (15 cm × 2·5 cm) and cut the cork down the middle so that a strip of filter paper gripped between the two halves can be suspended in the tube. Cut out a strip of filter paper about 15 cm long and 1·5 cm wide. Make a pencil mark on the paper about 2·5 cm from one end and place on the mark a small drop of a mixture of phenolphthalein and methyl orange, using a glass tube drawn out to a fine jet. This should give a spot of liquid about 5 mm across. Dry the spot by holding the paper some distance above a small flame. Add a second drop of the mixture at the same place and dry again. Put 3–4 cm³ of '880' ammonia solution into the boiling tube and suspend the strip of paper from the split cork so that it is not in contact with the sides of the tube and so that the dried spot is about 1 cm above the surface of the liquid. Leave the tube standing until the solvent front has risen nearly to the cork. The phenolphthalein will travel up the paper more quickly than the methyl orange, the two indicators forming a *chromatogram* consisting of a red patch followed by a yellow patch (Fig. 13–11).
>
> The ratio of the distance moved by a solute in a given time to the distance moved by the solvent front is the *retention factor* ($R_f$) of the solvent for the solute.

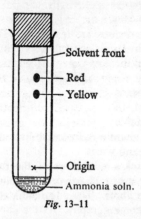

*Fig.* 13–11

- Solvent front
- Red
- Yellow
- Origin
- Ammonia soln.

This method is called *ascending* paper chromatography because the solvent travels upwards. Sometimes the paper is suspended with its upper end in a trough of solvent, so that the solvent front moves down the paper. This is known as *descending* paper chromatography. It is immaterial which method is used.

To separate more complex mixtures *two-way* chromatography is used. The mixture is placed at one corner of a square sheet of filter paper and a solvent is run up (or down) the paper. This separates some of the constituents, but not others. The paper is dried and after it has been turned at right angles a second solvent is passed up (or down) the paper. Components not separated by the first solvent are separated by the second one.

**Ion Exchange.** This is widely used both for softening water and for 'demineralizing' water (that is, removing ions completely from water).

In the process certain ionic materials of special composition exchange ions for similarly charged ions from an aqueous solution. Some of these materials (*e.g.*, clay and zeolites) are naturally occurring silicates(IV) Others are synthetic resins, like phenol-methanal and poly(phenylethene), into which appropriate acidic or basic groups have been introduced.

All ion-exchange materials have one feature in common. They have a three-dimensional network structure made up of cross-linked atoms or groups. The network forms a single giant multiply charged ion, in which the charges may be positive or negative. At various sites throughout the structure are attached small ions of opposite charge, such as $Na^+$ or $Cl^-$. The latter are soluble in water, while the large composite ion is insoluble. We classify the materials as *cationic* or *anionic* according to whether the small soluble ions (which undergo exchange) are cations (like $Na^+$) or anions (like $Cl^-$). Those of the first type contain acidic groups such as $-COOH$, $-SO_3H$, or $-OH$, in which the hydrogen may have been replaced by sodium. The small ion is then $H^+$ or $Na^+$, while we can represent the polymerized insoluble ion as $R^-$.

We are not sure of the exact manner in which the small ions are attached to the oppositely charged network, but it resembles adsorption. Thus, when hard water containing $Ca^{2+}$ ions is in contact with a cationic material containing $Na^+$ ions, an equilibrium is established between the material and the solution as follows:

$$2(R^-\text{---}Na^+) + Ca^{2+} \rightleftharpoons (2R^-\text{---})Ca^{2+} + 2Na^+$$

Here the dotted lines signify that the ions are bound to the solid. The position of equilibrium depends on the relative attractions of the composite anion for the two cations. The attraction is larger the bigger the charge on the cation and the smaller its size (the cations are hydrated). The equilibrium point also depends on the concentrations of the cations. With the concentration of $Ca^{2+}$ ions found in hard water the position of the equilibrium is far to the right. This means that practically all the $Ca^{2+}$ ions are removed from solution. The reaction can be reversed, however, if the concentration of the $Na^+$ ions is increased sufficiently. Thus, if a concentrated solution of sodium(I) chloride is now passed through the exchanger, the bound $Ca^{2+}$ ions are replaced by $Na^+$ ions. In this way a 'spent' material can be regenerated.

In an anion exchanger the polymerized ion is a multiply charged cation (due to the presence of basic amino groups or quaternary ammonium groups). The small soluble anions are usually $OH^-$ ions. The latter can be exchanged for anions such as $Cl^-$ or $SO_4^{2-}$; *e.g.*,

$$R^+\text{---}OH^- + Cl^- \rightleftharpoons R^+\text{---}Cl^- + OH^-$$

There are two ways of demineralizing water. One is to pass the water first through a cation exchanger, which replaces ions like $Ca^{2+}$ and $Na^+$ by $H^+$. The water is then passed through an anion exchanger, which substitutes OH⁻ ions for Cl⁻, etc. The H⁺ ions and OH⁻ ions combine to form water. Material in the first exchanger is regenerated with dilute sulphuric(VI) acid, that in the second by sodium(I) hydroxide solution. The second, and more common way, is to use a single exchanger. This contains a mixed bed of both types of resin, so that cations and anions are removed from the water simultaneously. The ion-exchange process produces the purest water obtainable.

## EXERCISE 13

**1.** Describe the preparation of *either* an iron(III) hydroxide sol *or* an arsenic (III) sulphide sol. In what ways does such a colloidal solution differ from an ordinary solution and how can these differences be demonstrated experimentally?
(W.J.E.C.)

**2.** Explain the terms (*i*) dialysis, (*ii*) reversible colloid, (*iii*) peptization, (*iv*) disperse phase. Give one example of each term.

What would you expect to see happen when a colloidal solution of silver is treated as follows: (*v*) a beam of light is passed through it, (*vi*) an electrolyte is added to it? (Lond.)

**3.** What methods are available for the preparation of colloidal solutions? Give one example of the use of each method.

Explain carefully why colloidal arsenic(III) sulphide particles are negatively charged, and describe an experiment whereby you could demonstrate the nature of this charge. (Lond.)

**4.** Explain, giving examples where appropriate, what is meant by the following terms: lyophobic colloid, lyophilic colloid, protected colloid, gel.

Describe how you would prepare an aqueous silver sol and how you would determine the sign of the charge on the colloidal silver. (S.U.)

**5.** What do you understand by the terms sol, gel, Brownian motion, Tyndall cone?

Give three examples of the differences you would expect to find between lyophilic and lyophobic colloids. (O. and C.)

**6.** (*a*) What do you understand by *chromatography*? Give two examples of its application in modern analysis. (*b*) Explain the principles of *ion exchange*, giving two examples of its application. (O. and C.)

# CHAPTER 14

# Thermochemistry—basic ideas of thermodynamics

## Thermochemistry

**Chemical Energy.** Energy is defined as the capacity to do work. There are numerous forms of energy, such as *kinetic energy, heat energy,* and *electrical energy.* Another form is *chemical energy,* for chemical substances have the capacity to do work, although this does not become apparent until the energy is converted into one of the other forms. The charging and discharging of an accumulator illustrate the interchange of electrical and chemical energy. Usually, however, when chemical energy is converted into another form of energy it appears as heat, and we can use the heat evolved as a measure of the chemical energy liberated. We cannot measure the absolute amount of energy associated with, say, a certain mass of phosphorus, just as it is impossible to measure he absolute potential energy of a stone. We can find the change of energy when the stone is raised or lowered through a certain height, and similarly we can measure the change in chemical energy when the phosphorus is treated in a certain way, for example, burned in oxygen or chlorine. In all chemical reactions, therefore, we are concerned only with the *difference* in energy between the substances in their initial and final states.

**Law of Conservation of Energy.** If we pass an electric current through acidified water a certain quantity of electricity liberates 2 g of hydrogen and 16 g of oxygen. When the same quantity of electricity is transformed into heat approximately 286 kJ are produced. If the 2 g of hydrogen and 16 g of oxygen are now reconverted to water by burning, approximately 286 kJ are again obtained. We conclude that during the electrolysis this amount of energy has been stored as chemical energy in the hydrogen and oxygen, and during the burning the same amount of energy is liberated again.

This illustrates the *law of conservation of energy*, which states that *in any isolated system the total amount of energy remains constant.* By an 'isolated' system we mean one in which energy is neither given to, nor received from, the surroundings. In practice it is impossible to realize such a system, and therefore the law of conservation of energy cannot be verified rigidly. In investigating energy changes in some chemical systems we can obtain an approximation to an isolated system by enclosing the substances in a vacuum flask, but even then some heat transfers between the system and the flask and also the surrounding air.

Although the law of conservation of energy cannot be verified we think it is true for several reasons. One is that the more closely a system approximates to an isolated system the more exactly the law holds. Also, when another form of energy is converted into heat a fixed amount of the first form is always required to produce a certain amount of the second, although part of the heat obtained is wasted in the ways described above. Again, if energy were not conserved we could devise a process to create energy for nothing. For example, suppose that more energy is given out in combining 2 g of hydrogen with 16 g of oxygen than is required to decompose the resulting water. If the operations were carried out in this order we should arrive back at our starting point with a balance of energy, and by repeating the operations we could obtain unlimited energy. The fact that nobody has ever succeeded in inventing a process or a machine which creates energy is strong evidence of the truth of the law.

The interconvertibility of matter and energy can be reconciled with the law of conservation of energy by regarding matter as another form of energy. The relation between the two is given by Einstein's equation, $E = mc^2$.

**Exothermic and Endothermic Reactions.** Nearly all chemical reactions are accompanied by the evolution or absorption of heat. In an *exothermic* reaction heat is given *out* to the environment; in an *endothermic* reaction heat is taken *in* from the environment. To express a reaction more completely we indicate whether it is exothermic or endothermic by showing the *heat of reaction.* This is *the amount of heat evolved or absorbed when the reaction occurs between molar quantities of the substances as represented by the equation.*

In some cases it is more important to look at the energy change in a reaction from the standpoint of the environment, while in others the system is of chief concern. For example, in the burning of fuels we are interested in the capacity of different materials to heat the air in a room or water in a boiler. In thermodynamics and in chemistry generally the *system* is of greater significance. If heat is evolved in a reaction the system loses energy, and therefore the heat change is given a negative sign. On the other hand, if heat is absorbed the system gains energy, and this is shown by means of a positive sign.

The heat of a reaction varies with the temperature at which it is carried out. The temperature is usually chosen to be 25°C, or 298 K, so that this is both the initial temperature of the reactants and the final temperature of the products. Sometimes this condition cannot be fulfilled and it is necessary to calculate the heat of reaction at the standard temperature from the value found at a different temperature. This requires a knowledge of the molar heat capacities of the substances concerned.

The heat of a reaction may also depend on whether the reaction takes place at constant pressure or constant volume. Most reactions occur at constant pressure (usually one atmosphere), in which case the energy change is represented by $\Delta H$ (delta *H*); for a reaction at constant volume the symbol used is $\Delta U$ (see p. 339). Thus when 12 g of carbon (graphite) are burned completely in oxygen, the initial and final temperatures being 298 K, we have

$$C \text{ (graphite)} + O_2 \rightarrow CO_2 \quad \Delta H(298 \text{ K}) = -393\ 000 \text{ J mol}^{-1}$$

An example of an endothermic reaction is the following:

$$C \text{ (graphite)} + 2S \rightarrow CS_2 \quad \Delta H(298 \text{ K}) = +106\ 000 \text{ J mol}^{-1}$$

As the joule is inconveniently small as a unit for denoting the heat changes in chemical reactions, the kilojoule (written kJ), which is equal to 1000 joule, is often used instead. Thus $393\ 000 \text{ J mol}^{-1}$ equals 393 kJ $\text{mol}^{-1}$.

The apparatus used to measure the heat of a reaction depends on the type of reaction. The determination is usually carried out in some form of calorimeter, and the heat is measured by the temperature changes produced in a known amount of water. Some determinations are described in the following pages.

**Effect of Physical State on Heat of Reaction.** The heat of a reaction depends on the physical state of the substances taking part. When there may be ambiguity as to the physical state of the substances we indicate the physical state in the equation. Sometimes abbreviations are used— 's' for solid, 'l' for liquid, and 'g' for gas. At a temperature just below 100°C we have

$$H_2(\text{gas}) + \tfrac{1}{2}O_2(\text{gas}) \rightarrow H_2O \text{ (liquid)}$$
$$\Delta H = -286 \text{ kJ mol}^{-1}$$

The specific latent heat of vaporization of water at 100°C is 42 kJ mol$^{-1}$, this energy being absorbed by the system from outside.

$$H_2O(\text{liquid}) \rightarrow H_2O(\text{gas})$$
$$\Delta H = +42 \text{ kJ mol}^{-1}$$

Accordingly when 2 g of hydrogen are burned just above 100°C less energy is given out by the system.

$$H_2(g) + \tfrac{1}{2}O_2(g) \rightarrow H_2O(g)$$

$$\Delta H = -244 \text{ kJ mol}^{-1}$$

If there is no indication in the equation of the physical states of the substances we assume that they are in their 'standard' states, that is, when they are pure, under a pressure of 1 atm and at a temperature of 298 K. The latter temperature is chosen because it is normally used for thermostats, in which many experiments are carried out. When $\Delta H$ refers to standard states it usually carries a superscript and is written $\Delta H^{\ominus}$.

The heat evolved in neutralizing a given amount of sodium(I) hydroxide solution by hydrochloric acid is greater when hydrogen chloride gas is passed into the solution than when hydrochloric acid solution is used. This is because hydrogen chloride dissolves in water with evolution of heat.

$$NaOH(aq) + HCl(aq) \rightarrow NaCl(aq) + H_2O$$

$$\Delta H = -57{\cdot}3 \text{ kJ mol}^{-1}$$

$$NaOH(aq) + HCl(gas) \rightarrow NaCl(aq) + H_2O$$

$$\Delta H = -130 \text{ kJ mol}^{-1}$$

The abbreviation 'aq' after a formula indicates that the substance is dissolved in a large amount of water, so that increasing the dilution would not produce any further heat change.

A reaction can be endothermic at one temperature and exothermic at another, owing to a difference in physical state of one or more of the substances at the two temperatures. Thus, the combination of hydrogen and iodine is endothermic at lower temperatures and exothermic at higher temperatures. This is due to heat absorbed as latent heat of sublimation of iodine, the vapour thus being richer in energy than the solid.

$$H_2(g) + I_2(s) \rightarrow 2HI(g) \qquad \Delta H = +53{\cdot}6 \text{ kJ mol}^{-1}$$

$$H_2(g) + I_2(g) \rightarrow 2HI(g) \qquad \Delta H = -11 \text{ kJ mol}^{-1}$$

### Hess's Law of Constant Heat Summation.

*If a system A is converted into a system B the heat absorbed or evolved is independent of the method of passing from A to B.*

This important generalization was discovered by Hess in 1840. Starting from ammonia gas and hydrogen chloride gas, we can make a solution of ammonium chloride in two ways:

- The gases can be allowed to react and the resulting ammonium chloride dissolved in water.
- The gases can first be dissolved in water and the solutions added to each other.

According to Hess's law the heat change is the same in both cases if molar proportions of the gases are taken. This can be expressed:

$$\Delta H^{\ominus}$$

$NH_3(g) + HCl(g) \rightarrow NH_4Cl(s)$      $Q_1$

$NH_4Cl(s) + aq \rightarrow NH_4Cl(aq)$      $Q_2$

Total heat change $= Q_1 + Q_2$

$NH_3(g) + aq \rightarrow NH_3(aq)$      $Q_3$

$HCl(g) + aq \rightarrow HCl(aq)$      $Q_4$

$NH_3(aq) + HCl(aq) \rightarrow NH_4Cl$      $Q_5$

Total heat change $= Q_3 + Q_4 + Q_5$

Then, by Hess's law,

$$Q_1 + Q_2 = Q_3 + Q_4 + Q_5$$

This law is of great use in calculating heats of reaction when direct determination is not possible.

## Heats of Formation and Combustion

**Endothermic and Exothermic Compounds.** An *endothermic* compound is one which is formed from its elements with absorption of heat. An *exothermic* compound is one which is formed from its elements with evolution of heat. As endothermic compounds represent the storing of chemical energy they are usually reactive bodies. When exothermic compounds are made energy is dissipated, and these are therefore more stable and less reactive. As we have seen with hydrogen iodide, a compound which is endothermic at ordinary temperatures may be exothermic at higher temperatures.

*The heat of formation of a compound is the heat change when one mole of the compound is formed from its elements.*

Heat of formation, which normally refers to formation of a compound under standard conditions, is thus a heat of reaction and is given the symbol $\Delta H_f^{\ominus}$. When 44 g of carbon dioxide are formed from 12 g of carbon (graphite) and 32 g of oxygen under standard conditions, 393 kJ are evolved. The carbon dioxide is poorer by this amount of energy than the original carbon and oxygen. The heat of formation, $\Delta H_f^{\ominus}$, of carbon dioxide is therefore $-393$ kJ mol$^{-1}$. The heats of formation of some common substances are shown in the table given below.

|  | $\Delta H_f^{\ominus}$/kJ mol$^{-1}$ |
|---|---|
| $Na + \frac{1}{2}Cl_2 \rightarrow NaCl$ | $-418$ |
| $H_2 + \frac{1}{2}O_2 \rightarrow H_2O$ | $-286$ |
| $\frac{1}{2}H_2 + \frac{1}{2}Cl_2 \rightarrow HCl$ | $-92$ |
| $C + 2H_2 \rightarrow CH_4$ | $-74.9$ |
| $\frac{1}{2}N_2 + \frac{3}{2}H_2 \rightarrow NH_3$ | $-50.2$ |
| $\frac{1}{2}N_2 + \frac{1}{2}O_2 \rightarrow NO$ | $+90$ |
| $2C + H_2 \rightarrow C_2H_2$ | $+211$ |
| $C(graphite) \rightarrow C(diamond)$ | $+1.8$ |

The heats of formation of compounds can sometimes be obtained directly by measuring the heat change in the combination of known masses of the elements. At other times we deduce the values indirectly as explained shortly.

We cannot find the energy associated with a certain mass of an element. Hence for simplicity we consider all elements have zero energy. When an element has two allotropes, *e.g.*, carbon (graphite) and carbon (diamond), at ordinary temperatures we regard the more stable form—in this case carbon (graphite)—as the one having zero energy. So a less stable form has a heat of formation relative to a more stable form.

### Heat of Combustion.

*The heat of combustion of a substance is the heat evolved when one mole of the substance is completely burned in oxygen under specified conditions (reactants and products usually being in their 'standard' states).*

Heats of combustion (like heats of formation) are represented by $\Delta H$ since they are also particular heats of reaction. As heat is evolved in all combustions the values of $\Delta H$ are always negative. Thus we have

|  | $\Delta H^{\ominus}$/kJ mol$^{-1}$ |
|---|---|
| $C(graphite) + O_2(g) \rightarrow CO_2(g)$ | $-393$ |
| $CH_4(g) + 2O_2(g) \rightarrow CO_2(g) + 2H_2O(l)$ | $-891$ |
| $C_2H_6(g) + 3\frac{1}{2}O_2(g) \rightarrow 2CO_2(g) + 3H_2O(l)$ | $-1560$ |
| $C_2H_2(g) + 2\frac{1}{2}O_2(g) \rightarrow 2CO_2(g) + H_2O(l)$ | $-1310$ |
| $CH_3OH(l) + 1\frac{1}{2}O_2(g) \rightarrow CO_2(g) + 2H_2O(l)$ | $-715$ |

To find the heat of combustion of carbon (graphite)—that is, heat of formation of carbon dioxide—a known mass of 'sugar charcoal' (one of the purest forms of carbon) is burned in compressed oxygen, and the temperature rise in a known mass of water noted. The burning is done in a calorimetric bomb (Fig. 14-1) which is designed to withstand high pressures. It consists of a steel vessel and lid fitted with two valves, $V_1$

and $V_2$. A weighed piece of carbon is contained in a small platinum crucible, C. The bomb is filled with oxygen at a pressure of about 20 atm and immersed in a known amount of water, which is kept stirred. The carbon is ignited by passing an electric current through a small coil of

iron wire, W, touching the carbon. The rise in temperature is read to $0.01\,°C$ by an accurate thermometer. The heat given out in burning 12 g of carbon is found after allowing for the heat capacity of the bomb and heat evolved in burning of the iron wire. Since the reaction occurs at constant volume the heat of combustion found is the $\Delta U$ value. This is corrected to the $\Delta H$ value as explained at p. 339.

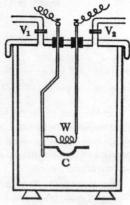

Fig. 14-1. A calorimetric bomb

**Indirect Calculation of Heats of Formation.** When the heat of formation of a substance cannot be found directly it is often possible. to find its value indirectly by application of Hess's law. The following examples illustrate the method.

**Example 1.** *Find the standard heat of formation of carbon monoxide if the standard heats of combustion, $\Delta H^{\ominus}$, of carbon and carbon monoxide are $-393$ kJ mol$^{-1}$ and $-285$ kJ mol$^{-1}$ respectively.*

Let $\Delta H_f^{\ominus}$ be the standard heat of formation of carbon monoxide. We can pass from 12 g of carbon and 32 g of oxygen to 44 g of carbon dioxide in two ways:

(i)     $C(s) + \frac{1}{2}O_2(g) \rightarrow CO(g)$     $\Delta H_f^{\ominus} = x$ kJ mol$^{-1}$

$CO(g) + \frac{1}{2}O_2(g) \rightarrow CO_2(g)$     $\Delta H^{\ominus} = -285$ kJ mol$^{-1}$

Total heat change $= (x - 285)$ kJ mol$^{-1}$

(ii)     $C(s) + O_2(g) \rightarrow CO_2(g)$     $\Delta H^{\ominus} = -393$ kJ mol$^{-1}$

Total heat change $= -393$ kJ mol$^{-1}$

$\therefore (x - 285)$ kJ mol$^{-1} = -393$ kJ mol$^{-1}$

$x = -108$ kJ mol$^{-1}$

Thus the standard heat of formation, $\Delta H_f^{\ominus}$, for carbon monoxide is $-108$ kJ mol$^{-1}$, which means that 108 kJ are evolved in the formation of 1 mole of carbon monoxide from its elements under standard conditions.

A shorter method of making the same calculation is to insert the values of standard heats of formation in the equation for the burning of carbon monoxide, the heat change for the reaction being included on the *left-hand* side of the equation. Thus

$$CO(g) + \tfrac{1}{2}O_2(g) \rightarrow CO_2(g)$$
$$x + 0 + (-285) \text{ kJ mol}^{-1} = -393 \text{ kJ mol}^{-1}$$
$$x = -108 \text{ kJ mol}^{-1}$$

**Example 2.** *Calculate the heat of formation of methane, given that its heat of combustion is* $-891 \text{ kJ mol}^{-1}$. *The heats of formation of carbon dioxide and water are* $-393 \text{ kJ mol}^{-1}$ *and* $-286 \text{ kJ mol}^{-1}$ *respectively. (The values quoted are for substances in their standard states.)*

Let the heat of formation of methane in its standard state be $x$ kJ mol$^{-1}$. One mole of methane, $CH_4$, consists of 12 g of carbon and 4 g of hydrogen. We can pass from 12 g of carbon and 4 g of hydrogen to 44 g of carbon dioxide and 36 g of water in two ways:

(i)     $C(s) + 2H_2(g) \rightarrow CH_4(g)$     $\Delta H_1 = x \text{ kJ mol}^{-1}$

$CH_4(g) + 2O_2(g) \rightarrow CO_2(g) + 2H_2O(l)$     $\Delta H_2 = -891 \text{ kJ mol}^{-1}$

Total heat change $= (x - 891) \text{ kJ mol}^{-1}$

(ii)     $C(s) + O_2(g) \rightarrow CO_2(g)$     $\Delta H_3 = -393 \text{ kJ mol}^{-1}$

$2H_2(g) + O_2(g) \rightarrow 2H_2O(l)$     $\Delta H_4 = 2(-286) \text{ kJ mol}^{-1}$

Total heat change $= -393 + 2(-286) \text{ kJ mol}^{-1}$

$\therefore$  $(x - 891) \text{ kJ mol}^{-1} = -393 + 2(-286) \text{ kJ mol}^{-1}$

$$x = -74 \text{ kJ mol}^{-1}$$

Hence $\Delta H_f^{\ominus}$ for methane is $-74$ kJ mol$^{-1}$; that is, 74 kJ are evolved in forming one mole of $CH_4$ under standard conditions.

Using the shorter method of setting out the calculation, we have for the combustion of 1 mole of methane:

$$CH_4(g) + 2O_2(g) \rightarrow CO_2(g) + 2H_2O(l)$$
$$x + 0 + -891 \text{ kJ mol}^{-1} = -393 + 2(-286) \text{ kJ mol}^{-1}$$
$$x = -74 \text{ kJ mol}^{-1}$$

### Calculation of Bond Energies.

*C—H Bond.* Having found the heat of formation of methane as shown in the last section, we can use it to calculate the energy of formation of the C—H bond in methane. Again we use Hess's law.

In calculating bond energies we remember that the energy required is for the formation of the bond from *free gaseous atoms*. In this case we have to know the heat of sublimation (or atomization) per mole of carbon, C, and the heat of dissociation per mole of hydrogen, $H_2$, into free atoms. The latter is equal numerically to the bond energy of formation of the hydrogen molecule given at p. 165. Since energy is absorbed in the sublimation of carbon and in the dissociation of hydrogen molecules $\Delta H$ is positive in both cases.

We can pass from 12 g of solid carbon (graphite) and 4 g of gaseous hydrogen to 16 g of methane gas in two ways:

(*a*) The carbon and hydrogen are changed from their normal states into free gaseous atoms, which are then combined to give methane.

$$C(s) \rightarrow C(g) \qquad \Delta H = 720 \text{ kJ mol}^{-1}$$
$$2H_2 (g) \rightarrow 4H \qquad \Delta H = 2 \times 431 \text{ kJ mol}^{-1}$$
$$C(g) + 4H(g) \rightarrow CH_4(g) \; \Delta H = x \text{ kJ mol}^{-1}$$
$$\text{Total heat change} = (x + 720 + 862) \text{ kJ mol}^{-1}$$

(*b*) The 12 g of carbon and 4 g of hydrogen are directly combined (theoretically) to give methane. The heat of formation, $\Delta H$, of methane from carbon and hydrogen *in their standard states* has been found previously (p. 334) to be $-74 \text{ kJ mol}^{-1}$.

$$C(s) + 2H_2(g) \rightarrow CH_4(g) \qquad \Delta H = -74 \text{ kJ mol}^{-1}$$
$$\text{Then} \quad (x + 720 + 862) \text{ kJ mol}^{-1} = -74 \text{ kJ mol}^{-1}$$
$$x = -1\,656 \text{ kJ mol}^{-1}$$

The heat change involved in forming 1 mole of methane from free carbon atoms and free hydrogen atoms is thus $-1\,656$ kJ. Since a molecule of methane contains four C—H bonds, the energy of formation, $\Delta H$, for one C—H bond is $-1\,656/4$, or $-414 \text{ kJ mol}^{-1}$.

*C—C Bond.* Here we form ethane from solid carbon and gaseous hydrogen in two ways, and equate the heat changes.

(*a*)
$$2C(s) \rightarrow 2C(g) \qquad \Delta H = 2 \times 720 \text{ kJ mol}^{-1}$$
$$3H_2(g) \rightarrow 6H(g) \qquad \Delta H = 3 \times 431 \text{ kJ mol}^{-1}$$
$$2C(g) + 6H(g) \rightarrow C_2H_6(g) \; \Delta H = x \text{ kJ mol}^{-1}$$
$$\text{Total heat change} = (x + 1\,440 + 1\,293) \text{ kJ mol}^{-1}$$

(*b*) The heat of formation, $\Delta H$, of ethane from carbon and hydrogen in their standard states can be determined in the same way as that of methane.

$$2C(s) + 3H_2(g) \rightarrow C_2H_6(g) \qquad \Delta H = -88 \text{ kJ mol}^{-1}$$
$$\text{Then} \quad (x + 1\,440 + 1\,293) \text{ kJ mol}^{-1} = -88 \text{ kJ mol}^{-1}$$
$$x = -2\,821 \text{ kJ mol}^{-1}$$

The ethane molecule contains six C—H bonds and one C—C bond. Using the energy value already found for the C—H bond, we have

Heat of formation, $\Delta H$, of C—C bond

$$= -2\,821 - (-414 \times 6) \text{ kJ mol}^{-1}$$
$$= -337 \text{ kJ mol}^{-1}$$

Similarly, we can find the heat of formation of the C=C bond in ethene and that of the C ≡ C bond in ethyne.

## Heats of Solution and Neutralization

**Heat of Solution.** The dissolving of a solid in water may be accompanied either by the evolution or absorption of heat. Which occurs depends on the separate heat changes attending each part of the dissolving process. We have seen previously that the dissolving of ionic compounds like sodium(I) chloride in water involves the overcoming of lattice energy, the oppositely charged ions being held together in the crystal with considerable strength. The necessary energy is largely derived from hydration, or solvation, of the ions, which become surrounded by an envelope of water molecules. For most ionic compounds the energy of hydration of the ions is somewhat smaller than the lattice energy. The balance is taken from the thermal energy of the solvent, which therefore becomes cooled.

The dissolving in water of many molecular solids (*e.g.*, carbamide and sucrose involves overcoming intermolecular forces like hydrogen bonding and van der Waals attraction. Again energy may be supplied by hydration of the molecules (usually by means of hydrogen bonding). In some cases there may also be an energy change due to ionization of the solute. Whether the overall change is exothermic or endothermic is again determined by the energies of the different steps.

The extent to which ions or molecules become separated and hydrated in aqueous solution varies with the relative amounts of solute and solvent. The two steps become complete only in dilute solution. Hence the quantity of heat evolved or absorbed depends on the amount of water in which a given amount of solute is dissolved.

> The **heat of solution** *of a substance is the heat change when 1 mole of the substance is dissolved in a specified number of mole of water.*

Thus when 1 mole (58·5 g) of sodium(I) chloride (NaCl) is dissolved in 10 mole (180 g) of water ($H_2O$) 2008 J are absorbed.

$$NaCl + 10H_2O \rightarrow NaCl(10H_2O) \qquad \Delta H = +2008 \text{ J mol}^{-1}$$

If a concentrated solution is diluted there is a further heat change (the *heat of dilution*) depending on the amount of water added. The heat of dilution gradually decreases, so that eventually increasing the dilution produces no further heat change. In practice this occurs when there are 800–1000 mole of water to 1 mole of solute. We then say that the substance is at 'infinite dilution', and the heat of solution is expressed thus:

$$NaCl + aq \rightarrow NaCl(aq) \qquad \Delta H = +4980 \text{ J mol}^{-1}$$

With salts we frequently need to specify whether the salt is in a hydrated or anhydrous state and, if hydrated, the precise hydrate concerned when more than one occur. As the process of hydration is generally exothermic, some salts have a negative value for $\Delta H$ when the anhydrous salt is used and a positive value when the hydrated form is used.

**Determination of Heat of Solution.** To find the heat of solution of 1 mole of a solute in a certain number of mole of water, a suitable fraction of these amounts is taken. The water is placed in a copper calorimeter, A in Fig. 14–2, fitted with a stirrer and thermometer reading to $\frac{1}{10}$ degC. The calorimeter is surrounded by an empty copper cylinder, B, which is surrounded in turn by a double-walled copper vessel, C, containing water. The vessels are prevented from touching by corks. The purpose of the outer vessels is to prevent gain or loss of heat by radiation. As rapid dissolving is essential the solid is finely powdered, and the weighed amount contained in a thin sealed bulb is added to the water in A. The bulb is broken in the water by means of the thermometer and the solution is stirred until there is no further change in temperature. The specific heat capacity of the solution is determined (if the solution is dilute this may be assumed to be equal to that of water) and the heat of solution is calculated after allowing for the heat evolved or absorbed by the calorimeter and the glass bulb.

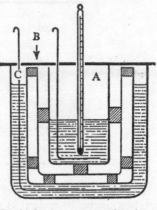

### Heat of Neutralization.

*This term refers to the heat evolved in the reaction between that mass of an acid which furnishes one mole of $H^+$ (or $H_3O^+$) ions and that mass of an alkali which provides one mole of $OH^-$ ions.*

Fig. 14–2. Apparatus used to determine heat of solution

The heats of neutralization of strong acids (*e.g.*, hydrochloric acid) by strong alkalis (*e.g.*, sodium(I) hydroxide) are approximately constant, being equal to about 57 300 J.

$$NaOH + HCl \rightarrow NaCl + H_2O \qquad \Delta H = -57 \cdot 3 \text{ kJ mol}^{-1}$$

All the substances, except water, represented in the above equation are highly dissociated into ions in dilute solution. Therefore the reaction is shown more accurately by the equation

$$Na^+ + OH^- + H_3O^+ + Cl^- \rightarrow Na^+ + Cl^- + 2H_2O$$

or $\qquad OH^- + H_3O^+ \rightarrow 2H_2O \qquad \Delta H = -57 \cdot 3 \text{ kJ mol}^{-1}$

The only change which occurs is the combination of 17 g of hydroxyl ions with 19 g of oxonium ions. This combination is attended by the evolution of 57·3 kJ. Obviously we can represent any other neutralization of a strong acid by a strong alkali in a similar manner. One might expect that the heat of neutralization would be constant whatever the acid and alkali employed, but this is true only when the acid and alkali are completely dissociated into ions. Only dilute solutions of strong acids and alkalis fulfil this conditions (salts are always highly dissociated in dilute solution). An aqueous solution of a weak acid like ethanoic acid contains very few oxonium ions. When the solution is mixed with sodium(I) hydroxide solution the main reaction which occurs is:

$$CH_3COOH + OH^- \rightarrow CH_3COO^- + H_2O \qquad \Delta H = -552 \text{ kJ mol}^{-1}$$

The heat evolved in this reaction, which requires removal of hydrogen ions from ethanoic acid molecules, is less than in the case of a strong acid. Some further heats of neutralization are shown in Table 14–1.

Table 14–1. HEATS OF NEUTRALIZATION OF ACIDS AND ALKALIS

| Acid | Alkali | Heat of neutraliza-tion/kJ mol$^{-1}$ |
|---|---|---|
| $HNO_3$ | NaOH | 57·1 |
| $\frac{1}{2}H_2SO_4$ | NaOH | 66·5 |
| HCl | KOH | 57·3 |
| $HNO_3$ | KOH | 57·3 |
| $HNO_3$ | $\frac{1}{2}Ca(OH)_2$ | 58·4 |
| $\frac{1}{2}H_2S$ | NaOH | 16·0 |
| HCN | NaOH | 12·0 |
| HCl | $NH_3(aq)$ | 51·5 |

The heat of neutralization of hydrochloric acid by sodium(I) hydroxide solution can be determined as now described.

**Experiment.** A copper calorimeter can be used, but the determination is more easily made with a vacuum flask. First find the heat capacity of the flask to the nearest JK$^{-1}$ as follows. Put 100 cm$^3$ of water into the flask and take its temperature with an ordinary thermometer. Heat 200 cm$^3$ of water in a beaker to 60°C, keep the temperature steady for 2 or 3 minutes, and then pour the hot water into the vacuum flask. Stir the mixture carefully (the glass of the inside container is thin and fragile), and note the temperature. Calculate the heat capacity of the flask from the equation:

Heat gained by flask and cold water = heat lost by hot water

Prepare about 200 cm$^3$ of semi-molar NaOH solution and a similar amount of semi-molar HCl solution. Rinse out the vacuum flask with a little of the alkali and put into it 125 cm$^3$ of this solution. Rinse a conical flask with the hydrochloric acid and put into the flask 125 cm$^3$ of the acid. Insert thermometers reading to $\frac{1}{10}$°C into each liquid, start a stop-clock, and note the temperatures of the liquids at the end of each minute for 6 minutes. As there is often a slight gain or loss of heat owing to radiation, we can deduce the exact temperature of each liquid at the moment of mixing as shown in Fig. 14–3.

At the seventh minute pour the acid into the alkali, stir well, and note the temperature at the eighth minute and for several minutes afterwards. Again plot the temperature readings and times, and draw a line through the readings. Produce the line backward to meet the ordinate at the time of mixing, and take the temperature at the point of intersection as the highest temperature produced. In Fig. 14–3 $t_1$ is the temperature of the sodium(I) hydroxide if the solutions are mixed at the seventh minute, $t_2$ that of the hydrochloric acid, and $t_3$ the maximum temperature after mixing.

Before mixing the average temperature is $(t_1 + t_2)/2$. After mixing the flask contains 250 cm³ of one-fourth molar NaCl solution at a temperature $t_3$. If $\rho$ and $c$ are the density and specific heat capacity of the salt solution, and $C$ the heat capacity of the flask, the heat evolved is obtained by substituting in the following expression. (Note that the density of one-fourth molar NaCl solution is 1·01 g cm⁻³. The specific heat capacity may be taken as 4·18 J g⁻¹ K⁻¹.)

$$\left(t_3 - \frac{t_1 + t_2}{2}\right)(250\rho c + C)$$

But 125 cm³ of semi-molar acid and semi-molar alkali contain 1/16th of the molar masses. The heat of neutralization is therefore obtained by multiplying the experimental value of the heat evolved by sixteen.

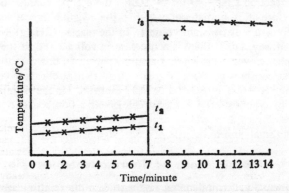

*Fig.* 14–3

**Effect of Volume Changes on Heat of Reaction.** Chemical changes involving gases may have two heats of reaction, according to whether the reaction is carried out at constant volume or constant pressure. When the reaction is accompanied by an increase in volume work has to be done in increasing the volume against the external pressure. If the reaction is exothermic the heat evolved at constant pressure is therefore less than the heat evolved at constant volume. The heat change at constant pressure is denoted by $\Delta H$, and that at constant volume by $\Delta U$.

In the reaction $Zn + H_2SO_4 \rightarrow ZnSO_4 + H_2$ there is an increase in volume equal to the volume, $v$, of hydrogen produced. This increase takes place against the atmospheric pressure, $p$, and absorbs an amount of energy $pv$, which equals $RT$ for each mole of hydrogen ($H_2$) formed at temperature $T$. In general, if $n$ is the increase in the number of mole of gas,

$$\Delta H = \Delta U + nRT$$

For one mole of gas $R$ has the value 8·3 J mol⁻¹ K⁻¹ (p. 45).

**Example.** *What is the heat of formation,* $\Delta H$, *of carbon monoxide at constant pressure at* $500°C$ *if its heat of formation,* $\Delta U$, *at constant volume at this temperature is* $109$ *kJ mol*$^{-1}$?

$$C(s) + \tfrac{1}{2}O_2(g) \rightarrow CO(g) \qquad \Delta U = -109 \text{ kJ mol}^{-1}$$

Energy absorbed by increase in volume $= 8\cdot3n$ J mol$^{-1}$ K$^{-1}$

$$= 8\cdot3 \times \tfrac{1}{2} \times 773$$
$$= 3\cdot2 \text{ kJ mol}^{-1} \text{ K}^{-1}$$

Therefore heat of formation, $\Delta H$, at constant pressure

$$= -109 + 3\cdot2 \text{ kJ mol}^{-1}$$
$$= -105\cdot8 \text{ kJ mol}^{-1}$$

In the reaction $C(s) + O_2(g) \rightarrow CO_2(g)$ there is no change in volume, and therefore it is immaterial whether the reaction is carried out at constant pressure or constant volume. In the reaction $2H_2(g) + O_2(g) \rightarrow 2H_2O(g)$ at, say, $110°C$ there is a decrease in volume and, if the pressure is atmospheric, work is done *by* the atmosphere. In this case the heat of reaction at constant pressure is greater than that at constant volume by $8\cdot3n$ J mol$^{-1}$ K$^{-1}$ (*i.e.*, by $8\cdot3 \times 1 \times 383$ J mol$^{-1}$ K$^{-1}$ in the example given).

Heats of combustion found by calorimetric bomb are $\Delta U$ values. These are corrected to $\Delta H$ values as described above. The corrections are usually small and have little practical significance.

## Basic Ideas of Thermodynamics

Fundamentally thermodynamics is the study of the relation between heat energy and the mechanical work which can be obtained from it. This branch of science developed during the last century in connection with the design of steam engines. Today it has applications to physical and chemical changes of all kinds.

Thermodynamics provides an entirely different method of looking at chemical changes from the usual one, which is from the point of view of the behaviour of individual atoms and molecules. Thus, when we write a chemical equation we represent the behaviour of billions of atoms or molecules by the behaviour of just a few of these particles. Thermodynamics is concerned with the energy changes in reactions, and it explains these changes in terms of the statistical behaviour to be expected from vast numbers of atoms or molecules.

**First Law of Thermodynamics.** As stated earlier, the equivalence of the different forms of energy indicates that energy can be neither created nor destroyed, the latter statement being essentially the law of conservation of energy. When it is expressed in more precise form it is also the *first law of thermodynamics: The total amount of energy in an isolated system remains constant.*

By 'system' we mean any material or collection of materials on which we wish to focus our attention. A system may consist simply of a test-tube containing chemicals, or it may be a blast-furnace, the sun, or a spiral nebula. An 'isolated' system is one to which, or from which, transfer of energy is impossible. A completely isolated system may exist only in the imagination, although the universe is usually assumed to be one. In practice, transfer of energy (usually in the form of heat) takes place between a system and its surroundings. If the amount of energy transferred is small compared with the total energy change, the system approaches an isolated one. This is the case when a chemical reaction is carried out in a vacuum flask or a well-lagged vessel. A laboratory can be regarded as an approximately isolated system if no external heating or cooling source is used, the energy lost by a reaction vessel and its contents then being gained by the rest of the laboratory.

**Second Law of Thermodynamics.** Since the total amount of energy in an isolated system is constant, any change which occurs can only consist of a redistribution of energy. The second law is concerned with this redistribution. To explain the law it is necessary to introduce the concept of *entropy*.

If we consider a molecular substance we can define its entropy as a measure of the randomness, disorder, or 'mixed-upness' of its molecules. The randomness may be thought of as applying to the distribution of the molecules. Only in the case of a pure crystalline substance at absolute zero would the entropy be zero. This is because at absolute zero the molecules as a whole would be at rest, and their positions relative to each other would be fixed. Above absolute zero the molecules vibrate in a number of ways depending on the molecular complexity. This causes some disorder in their relative positions. Other modes of motion are also possible. Molecules may rotate, and in the case of a liquid or gas they can move in straight lines. All these modes of motion produce molecular disorder, which increases as the temperature rises and the motion becomes more violent.

Alternatively we can interpret entropy in terms of the energy states of the molecules. The energies of vibration, rotation, and translation are quantized, but owing to the random character of molecular collisions not all molecules possess the same energy at the same temperature. At low temperatures collisions are mainly between low-energy molecules, while at high temperatures the collisions are chiefly high-energy ones. Hence the molecules have a much greater variety of energy levels at the high temperature. We can thus identify an increase in entropy with an increase in the number of ways in which the energy of a substance is distributed amongst its molecules.

The entropy of a substance can be altered in other ways besides changing its temperature. Entropy becomes larger when the volume increases, when a solid melts or dissolves, and when a liquid vaporizes. In chemical systems an increase in entropy is associated with a decrease in molecular complexity or an increase in the number of gaseous molecules. In general, entropy increases when atoms or molecules are removed to greater distances from each other.

The *second law of thermodynamics* states that *any process which occurs in Nature is accompanied by an increase in total entropy.* There are many common illustrations of the working of this law. Two gases placed in contact mix spontaneously. Water diffuses through a semipermeable membrane into an aqueous solution of sucrose. When living creatures die their bodies are converted by decay into simple gaseous products such as carbon dioxide and ammonia.

We must emphasize that it is the *total* entropy change which must be examined. More orderly arrangements of matter are often produced from less orderly ones; that is, the entropy decreases. Examples include the oxidation of metals in air, the combination of ammonia and hydrogen chloride to form solid ammonium chloride, and the synthesis of sugars in plants. In each of these cases, however, we are considering only part of an isolated system and ignoring the increase in entropy which takes place in the remainder. Thus the formation of both the metal oxide and the ammonium chloride are attended by evolution of heat, which increases the entropy of the surrounding air. The radiant energy of sunlight falling on a green leaf is only partially used in the synthesis of sugars. Much of the energy is frittered away in heating other materials in the leaf, and again this heat is communicated to the air. The second law of thermodynamics tells us that in every case the decrease in entropy in one part of an isolated system is more than offset by an increase in entropy in another part.

As stated earlier, transformation of any other kind of energy into heat takes place quantitatively. Heat, however, is never converted quantitatively into other forms of energy, nor are other forms into each other. When coal is used to generate electricity for lighting a series of energy transformations occurs. By burning the coal under a boiler we obtain compressed steam. This drives a turbine, which rotates an armature in a magnetic field and produces the current for lighting. At each stage some of the useful energy is wasted (*e.g.*, in the hot gases which escape, in overcoming friction, and in heating the filament of the bulb). The energy transformations can be summarized as now shown.

Chemical → useful heat → mechanical → electrical → light
+ + + +
waste heat    waste heat    waste heat    waste heat

In any change involving energy a tax has to be paid to Nature in the form of waste heat, which merely increases the entropy of the universe. Thus our stock of useful, available, or 'organized' energy is constantly being dissipated as useless, unavailable, or 'disorganized' energy. This trend is described as the *degradation of energy*. The first and second laws of thermodynamics were neatly summed up by Clausius as follows: "The energy of the universe is constant; the entropy of the universe tends to a maximum." If the universe really is a closed system (which cannot be proved) all change must ultimately cease when the universe reaches its maximum entropy.

The absolute entropy, $S$, of a substance at a given temperature and pressure can be calculated. Its value is expressed in 'entropy units' (joule per degree per mole). Usually changes of entropy (represented by $\Delta S$) are more important than absolute values. It can be shown that if $q$ is heat absorbed by one mole of a substance at the kelvin temperature $T$, and if the heat is used only to increase the entropy, this increase is given by

$$\Delta S = \frac{q}{T}$$

or

$$q = T\Delta S$$

Thus, when 18 gram of ice are melted at 0°C and 101 325 Pa pressure 6025 joule are absorbed. The increase in entropy is therefore 6025/273 = 22·1 J K$^{-1}$.

**Conversion of Heat into Useful Work.** Any contrivance which turns heat energy into mechanical energy is a heat engine, and in principle any substance which can supply heat can be used to operate such an engine. As we have seen, however, it is never possible to obtain 100 per cent conversion of heat into useful work. The total heat energy of a substance is called its *enthalpy*, and is represented by $H$. Of this only part, represented by $G$, could be converted into useful work. $G$ is called the *free energy*. It is not possible to find the absolute values of $H$ and $G$, but again we are interested, not in absolute values, but in the changes which $H$ and $G$ undergo. If $H_1$ and $G_1$ are the enthalpy and free energy before a change, and $H_2$ and $G_2$ are these quantities after the change,

$$\Delta H = H_1 - H_2 \quad \text{and} \quad \Delta G = G_1 - G_2$$

For changes carried out at constant pressure and constant temperature (*e.g.*, in an open vessel in a thermostat) the difference between $\Delta H$ and $\Delta G$ is the amount of energy involved in the entropy change. Thus if $T$ is the kelvin temperature

$$\Delta H - \Delta G = T\Delta S$$

or

$$\Delta G = \Delta H - T\Delta S$$

This equation also applies to systems consisting of more than one substance (as is usually the case for chemical systems). The enthalpies, free energies, and entropies of such systems can be obtained by adding together those of the constituents.

The equation $\Delta G = \Delta H - T\Delta S$ is extremely important. It enables us to systematize physical and chemical changes and to predict which are probable and which improbable. It will be noted that we say 'probable' and 'improbable', and not 'possible' and 'impossible'. Theoretically any change is possible when molecules with a wide variety of energies collide in random fashion. Thus, if we have a sugar solution, it is conceivable that enough sugar molecules might come together with just the right energies and in just the right manner to re-form a visible sugar crystal. The event is extremely unlikely, however. A 'probable' change is one which we should expect to occur on thermodynamical grounds.

The second law of thermodynamics establishes one condition for a probable change—an increase in the total entropy of the universe. As a rule, however, we are not interested in the universe, but in what is likely to happen in our test-tube or flask. This is a non-isolated system and, as we have seen, the entropy of such a system may increase or decrease. We now establish a second condition for a probable change. *A change is probable only if the value of $\Delta G$ is negative.* Furthermore, the higher the negative value of $\Delta G$, the greater is the tendency for the change to occur. The driving force behind a process is thus the tendency for the free energy to diminish.

In earlier chapters we drew attention to two opposing tendencies in regard to matter on the atomic scale. One is the tendency for particles to come together owing to mutual attraction and form more orderly arrangements. This results in lowering of potential energy. The second tendency is for the particles to become more widely separated, or more disordered, owing to their thermal energy. We can now identify these two tendencies with the terms $\Delta H$ and $T\Delta S$ respectively. For a given temperature $\Delta G$ has a higher negative value the higher the negative value of $\Delta H$ (heat evolved) and the higher the positive value of $\Delta S$ (entropy increased). If we re-write our equation in the form $-\Delta G = -\Delta H + T\Delta S$, we see that it stands for the following:

Tendency to change = tendency to minimum potential energy
+ tendency to maximum entropy

At ordinary temperatures $T\Delta S$ is relatively unimportant compared with $\Delta H$. Hence the majority of chemical reactions which take place at ordinary temperatures are exothermic ones. $T\Delta S$ increases, however, with rise of temperature, and at high temperatures the entropy term is the chief factor in determining whether a reaction is probable or not.

**Applications of the Equation** $\Delta G = \Delta H - T\Delta S$. In practice a negative value for $\Delta G$ can arise in different ways, as now explained.

(i) *$\Delta H$ is zero and $\Delta S$ is positive.* In some changes heat is neither evolved nor absorbed. In this case the driving force of the change comes solely from the increase in entropy. Some examples are the following:

A perfect gas expanding into a vacuum.

Mixing together of two gases (or two liquids) at constant total volume.

Transfer of solvent to a solution through a semipermeable membrane.

(ii) $\Delta H$ *is negative and* $\Delta S$ *is positive.* When an exothermic reaction is accompanied by an increase in entropy, the two factors combine to give a high negative value for $\Delta G$. Examples are the decomposition of fused ammonium nitrate(V), the liberation of ethyne from calcium(II) dicarbide and water, and the burning of fuels such as octane.

$$NH_4NO_3(l) \rightarrow N_2O(g) + 2H_2O(g)$$

$$CaC_2(s) + 2H_2O(l) \rightarrow Ca(OH)_2(s) + C_2H_2(g)$$

$$C_8H_{18}(l) + 12\tfrac{1}{2}O_2(g) \rightarrow 8CO_2(g) + 9H_2O(g)$$

(iii) $\Delta H$ *is negative and* $\Delta S$ *is negative.* Here $-T\Delta S$ is positive. Hence for a change to be probable it must be exothermic and $\Delta H$ must be more negative than $-T\Delta S$. This is the case for the burning of magnesium in air and for the combination of hydrogen and oxygen to form water-vapour. In both cases there is a decrease in the number of gaseous molecules.

$$2Mg(s) + O_2(g) \rightarrow 2MgO(s)$$

$$2H_2(g) + O_2(g) \rightarrow 2H_2O(g)$$

The transition of sulphur (monoclinic) to sulphur (rhombic) below the transition temperature also comes in this category. The transition is exothermic, and there is a decrease in entropy because the atoms are brought closer together (the rhombic form has a higher density than the monoclinic form).

It should be noted that thermodynamics tells us nothing about the *rate* of a probable change. The combination of hydrogen and oxygen is thermodynamically favoured at ordinary temperatures, but the gases can be left together indefinitely at room temperature without appreciable change. We must think of the combination as actually taking place, but at an extremely low rate. To make the reaction proceed more rapidly we can employ a catalyst (spongy platinum). This does not affect the free energy change (which is independent of the path followed).

(iv) $\Delta H$ *is positive and* $\Delta S$ *is positive.* Here the change is endothermic and $\Delta G$ can only be negative if $T\Delta S$ is more positive than $\Delta H$. This may not be the case at ordinary temperatures, but since $T\Delta S$ increases with rise of temperature the condition may be fulfilled at higher temperatures. Thus calcium(II) carbonate and ammonium chloride split up when strongly heated.

$$CaCO_3(s) \rightarrow CaO(s) + CO_2(g)$$

$$NH_4Cl(g) \rightarrow NH_3(g) + HCl(g)$$

For any reaction which has $\Delta H$ positive and $\Delta S$ negative $\Delta G$ must be positive. The reaction is therefore unlikely to occur. Examples of improbable reactions are the converse of those shown in section (ii).

**Equilibrium.** Interesting possibilities arise in the two cases when both $\Delta H$ and $\Delta S$ are negative or positive. We have seen that in the combination of hydrogen and oxygen $\Delta H$ and $\Delta S$ are both negative and the combination is explained by $\Delta H$ being larger than $T\Delta S$. The latter increases, however, with rise of temperature, and at high temperatures $\Delta H < T\Delta S$. The tendency now is for the reaction to go in the opposite direction, $\Delta G$ now being negative for the reverse change. At some intermediate temperature $\Delta H = T\Delta S$, and $\Delta G$ is zero. At this temperature (which varies with the pressure) there is no tendency for the composition of the system to change, which means that equilibrium exists.

We see that the equilibrium state is one of minimum free energy. It must also be one of maximum entropy for the system and its surroundings under the given conditions. If this were not so the system would still be capable of changing so as to increase the total entropy.

Theoretically it should always be possible (at a suitable temperature and pressure) to establish an equilibrium for those systems in which $\Delta H$ and $\Delta S$ have the same sign. In practice this may require the system to be in a closed vessel so that gaseous reactants do not escape. Other systems in which $\Delta H$ and $\Delta S$ are negative in the forward direction are the following:

$$\text{Water} \rightleftharpoons \text{ice}; \quad S_{monoclinic} \rightleftharpoons S_{rhombic}$$
$$N_2 + 3H_2 \rightleftharpoons 2NH_3; \quad 2SO_2 + O_2 \rightleftharpoons 2SO_3$$

### EXERCISE 14

*(Relative atomic masses are given at the end of the book)*

*Note.* Except where other values are specified use the heats of combustion of carbon and hydrogen now given.

$$C(s) + O_2(g) \rightarrow CO_2(g) \quad \Delta H = -393 \text{ kJ mol}^{-1}$$
$$H_2(g) + \tfrac{1}{2}O_2(g) \rightarrow H_2O(l) \quad \Delta H = -286 \text{ kJ mol}^{-1}$$

Corrections for volume changes are required only in Questions 15 and 16.

**1.** Calculate the standard heat of formation of propane ($C_3H_8$) if its standard heat of combustion, $\Delta H^{\ominus}$, is $-2\ 213$ kJ mol$^{-1}$.

**2.** Calculate the standard heat of combustion of ethene ($C_2H_4$) if its standard heat of formation, $\Delta H_f^{\ominus}$, is $+52\cdot3$ kJ mol$^{-1}$.

**3.** The heat of formation of carbon dioxide is given by

$$C + O_2 \rightarrow CO_2 \quad \Delta H = -393\cdot3 \text{ kJ mol}^{-1}$$

Explain fully what this statement means.

The heats of formation of carbon dioxide, water, and benzene are $-393\cdot3$, $-285\cdot8$, and $+48\cdot1$ kJ mol$^{-1}$ respectively. Calculate the heat of combustion of benzene.

Outline briefly an experiment to verify your result.    (S.U.)

4. Calculate the standard heat of formation of methanol if its standard heat of combustion is $-714$ kJ $mol^{-1}$.

5. The standard heat of formation of ethanol is $-274$ kJ $mol^{-1}$. What is its standard heat of combustion?

6. What is the standard heat of formation of ammonia if its standard heat of combustion is $-379 \cdot 5$ kJ $mol^{-1}$?

7. The standard heat of combustion of sulphur in oxygen is $-297 \cdot 1$ kJ $mol^{-1}$, and in dinitrogen oxide ($N_2O$) is $-443 \cdot 5$ kJ $mol^{-1}$. Calculate the standard heat of formation of dinitrogen oxide.

8. Define *heat of combustion* and *heat of formation*. State Hess's law.

(*a*) For the reaction $3C_2H_2$ (gas) $\rightarrow C_6H_6$ (gas), calculate the heat of reaction, stating whether the heat is absorbed or evolved.

(*b*) Calculate the heat of combustion of ethyne gas. You are given the following heats of formation: $C_6H_6$ (gas) $82 \cdot 9$ kJ $mol^{-1}$ (absorbed); $CO_2$ (gas) $393 \cdot 3$ kJ $mol^{-1}$ (evolved); $C_2H_2$ (gas) $226 \cdot 8$ kJ $mol^{-1}$ (absorbed); $H_2O$ (gas) $241 \cdot 8$ kJ $mol^{-1}$ (evolved). (O. and C.)

9. (Part question.) The heats of combustion of benzene and of ethyne are $3\,260$ kJ $mol^{-1}$ and $1\,300$ kJ $mol^{-1}$ respectively (evolved). The heats of formation of carbon dioxide and of water are $393$ kJ $mol^{-1}$ and $285$ kJ $mol^{-1}$ respectively (evolved). Calculate the heat of formation of ethyne.

Calculate also the heat change (per mole of benzene) when ethyne polymerizes to benzene. What deductions are possible regarding the relative stability of these two compounds?

(For all heat changes which you calculate, state precisely whether the heat is evolved or absorbed.) (J.M.B.)

10. (Part question.) The heats of formation of iron(III) oxide and of aluminium(III) oxide are $-816$ kJ $mol^{-1}$ and $-1\,590$ kJ $mol^{-1}$ respectively. (In each case the heat is evolved.) Calculate the heat change in the reaction

$$Fe_2O_3 + 2Al \rightarrow Al_2O_3 + 2Fe$$

and state whether this heat is evolved or absorbed. (J.M.B.)

## MORE DIFFICULT QUESTIONS

11. The heats of formation, $\Delta H$, of phosphorus(V) oxide ($P_2O_5$) from phosphorus (white) and phosphorus (violet) are $-765 \cdot 7$ kJ $mol^{-1}$ and $-729 \cdot 7$ kJ $mol^{-1}$ respectively. Calculate the heat of formation per mole of the violet form (P) from the white form. Comment on the significance of your answer.

12. Experiment showed that a certain mixture of methane and ethane required for complete combustion exactly three times its own volume of oxygen. Calculate the heat given out when $134 \cdot 4$ $cm^3$ of the original mixture has been completely oxidized by oxygen. (The heats of formation of methane and ethane are $-90 \cdot 8$ kJ $mol^{-1}$ and $-129 \cdot 7$ kJ $mol^{-1}$ respectively. The heats of combustion of carbon and hydrogen are $-408 \cdot 4$ kJ $mol^{-1}$ and $-286 \cdot 2$ kJ $mol^{-1}$ respectively. The minus sign denotes heat given out in all cases.) (S.U.)

13. Calculate the heat of formation of sodium(I) oxide from the data given:

$$Na_2O + H_2O \rightarrow 2NaOH \qquad \Delta H = -205 \text{ kJ } mol^{-1}$$
$$NaOH + aq \rightarrow NaOH \text{ (aq)} \qquad \Delta H = -56 \cdot 5 \text{ kJ } mol^{-1}$$
$$Na + H_2O + aq \rightarrow NaOH \text{ (aq)} + \tfrac{1}{2}H_2 \qquad \Delta H = -410 \text{ kJ } mol^{-1}$$
$$H_2 + \tfrac{1}{2}O_2 \rightarrow H_2O \qquad \Delta H = -285 \cdot 8 \text{ kJ } mol^{-1}$$

14. Discuss the use of Hess's law in calculating heats of reaction that are difficult to determine experimentally.

Calculate the heats evolved or absorbed in each of the gas reactions:

$$I + H_2 \rightarrow HI + H$$
$$H + I_2 \rightarrow HI + I$$

(The heat absorbed in forming 1 mole of HI(g) from hydrogen (g) and iodine (s) is 25 kJ mol$^{-1}$. The heat absorbed in dissociating 1 mole of $H_2$(g) into atoms is 435 kJ mol$^{-1}$, and in dissociating one mole of $I_2$ (g) is 151 kJ mol$^{-1}$; the heat absorbed in subliming one mole of $I_2$(s) is 63 kJ mol$^{-1}$.)    (C.L.)

**15.** The heat of formation, $\Delta H$, of ammonia gas at constant pressure at 350°C is $-50 \cdot 2$ kJ mol$^{-1}$. Calculate the heat of formation of ammonia at constant volume at this temperature.

**16.** The heat of combustion, $\Delta H$, of methanol at constant pressure at 25°C is $-716$ kJ mol$^{-1}$. The heats of combustion of carbon and hydrogen given before Question 1 are at constant pressure. Calculate the heat of formation of methanol at constant volume at 25°C.

# CHAPTER 15

# Chemical equilibrium and chemical kinetics

**Reversible Reactions.** When steam is passed over heated iron hydrogen and iron(II) diiron(III) oxide are formed. If, however, hydrogen is passed over heated iron(II) diiron(III) oxide, steam and iron are produced. This is an example of a reversible reaction, which is expressed by the following:

$$3Fe + 4H_2O \rightleftharpoons Fe_3O_4 + 4H_2$$

The direction of the reaction depends on the conditions. If these remove the hydrogen as fast as it is formed, the reaction proceeds from left to right. If the steam is constantly removed no metal oxide remains. By having steam and iron in a closed vessel a state is reached in which all four substances exist together in equilibrium. The reaction is then a balanced one. There are many examples of reversible or balanced reactions in chemistry, *e.g.*

$$N_2 + 3H_2 \rightleftharpoons 2NH_3$$
$$C_2H_5OH + HCl \rightleftharpoons C_2H_5Cl + H_2O$$
$$N_2 + O_2 \rightleftharpoons 2NO$$

Other reactions are described as irreversible, since they appear to proceed in one direction only, *e.g.*

$$2KClO_3 \rightarrow 2KCl + 3O_2$$
$$PCl_5 + 4H_2O \rightarrow 5HCl + H_3PO_4$$
$$CH_3COOC_2H_5 + OH^- \rightarrow CH_3COO^- + C_2H_5OH$$

The equilibrium in a balanced reaction is a *dynamic* one, and not a *static* one—that is, the substances are still reacting together, but the velocities of the forward and backward reactions have become equal.

The dynamic character of the equilibrium can be shown by means of the reversible reaction between chlorine and water:

$$Cl_2 + H_2O \rightleftharpoons H^+ + Cl^- + HOCl$$

To the system is added sodium(I) chloride containing chloride ions which have been made radioactive by bombardment of the salt with neutrons as described in Chapter 5. When the solution is evaporated it is found that the chlorine gas given off is radioactive. This can only be because some of the radioactive chloride ions have taken part in the backward reaction.

**Activation Energy and Reversibility of Reactions.** To understand why some reactions are reversible and others apparently irreversible we must look more closely into the manner in which chemical reactions take place. When we examine the smallest scale on which a reaction can occur we find that there are two types of elementary reactions. ('Elementary' reactions are those which take place in a single step as distinct from reactions (described later) which occur in a series of steps.) In the first type only a single molecule is required, and this either decomposes or undergoes rearrangement of its atoms into a new molecule:

$$A \rightarrow B + C \quad \text{or} \quad A \rightarrow B$$

In either case the breaking of chemical bonds or their rearrangement requires absorption of energy, which is usually supplied in the form of heat (or sometimes light). The energy which the molecule must acquire before it can react is called the *activation energy* of the reaction.

The activation energy, $E_a$, for a reaction can be calculated from the rate constants (see p. 357) for the reaction at two different temperatures. Thus is can be shown that, if $k_1$ and $k_2$ are the rate constants at kelvin temperatures $T_1$ and $T_2$ respectively,

$$\log_{10} \frac{k_2}{k_1} = \frac{E_a}{2 \cdot 303 R} \left( \frac{T_2 - T_1}{T_1 T_2} \right)$$

where $R$ is the molar gas constant (approximately $8 \cdot 3$ J mol$^{-1}$ K$^{-1}$).

The second type of elementary reaction consists of two molecules colliding together and either combining or producing two new molecules. A typical example is the combination of hydrogen and iodine vapour at, say, 400°C to give hydrogen iodide. This is a reversible reaction, which is exothermic in the forward direction.

$$H_2(g) + I_2(g) \rightleftharpoons 2HI(g) \quad (\Delta H = -11 \cdot 3 \text{ kJ mol}^{-1})$$

One might suppose that the forward reaction occurs whenever a molecule of hydrogen collides with a molecule of iodine, but this is not so. The number of collisions per second which occur between hydrogen molecules and iodine molecules of given concentrations at a given temperature can be calculated from the kinetic theory and it is far too large to account for the observed rate of disappearance of the hydrogen and iodine. Thus only a small fraction of the collisions actually result in a chemical reaction. The reason for this is that the reaction involves the breaking of bonds between the hydrogen atoms on one hand and the iodine atoms on the other. Only if two molecules collide with sufficient energy are the bonds broken and new ones formed. This is often expressed by saying that there is an *energy barrier* which has to be surmounted. The combined energy which the molecules must have in order to react is again the activation energy of the reaction. If the energy of the colliding molecules is less than this activation energy they merely rebound from each other like two billiard balls.

Another factor concerned in whether reaction takes place or not is the orientation of the molecules towards each other when they collide. The

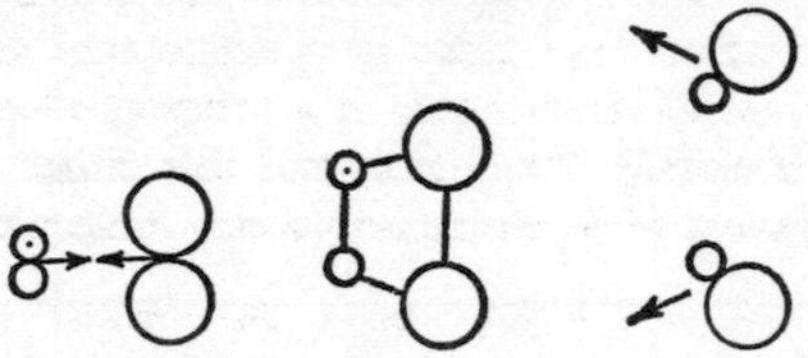

$$
\begin{array}{ccc}
\text{H} \quad \text{I} & \text{H}\text{------}\text{I} & \text{H}\text{---}\text{I} \\
| + | \;\rightleftharpoons & | \qquad | & \rightleftharpoons \; + \\
\text{H} \quad \text{I} & \text{H}\text{------}\text{I} & \text{H}\text{---}\text{I}
\end{array}
$$

(Activated complex)

*Fig.* 15–1

most favourable orientation is that in which the axes of the molecules are parallel (Fig. 15–1), but since collisions occur in random fashion every orientation is possible. Less favourable orientations will require a higher activation energy for reaction, while some orientations may be so unfavourable that no reaction takes place at all.

The breaking of the H—H bonds and I—I bonds and the formation of the new H—I bonds are not consecutive processes, but take place simultaneously. If a hydrogen molecule and an iodine molecule approach each other with their axes parallel the two processes start as soon as the electrical field of one molecule affects that of the other molecule. Bonds begin to form between the hydrogen and iodine atoms, while the H—H and I—I bonds stretch and become weaker. With a still closer approach an *activated complex* is formed, in which all four atoms are joined together by weak bonds of abnormal length. The bonds between the hydrogen atoms and iodine atoms then decrease to the normal H—I bond length, and the bonds between the two hydrogen atoms and the two iodine atoms are completely broken. The activated complex is to be regarded as a temporary phase of the reactants, and not as a definite compound capable of isolation. Its duration is extremely brief (of the order of $10^{-10}$ second).

Stretching of bonds requires expenditure of energy, and therefore the formation of the activated complex is accompanied by an increase in the potential energy of the system. This is shown in Fig. 15-2, where the potential energy of the system is plotted against the reaction path. The increase in potential energy, $E_1$, corresponding to the formation of the activated complex is derived from the kinetic energies of the reacting particles. $E_1$ represents the energy barrier, or the activation energy, of the

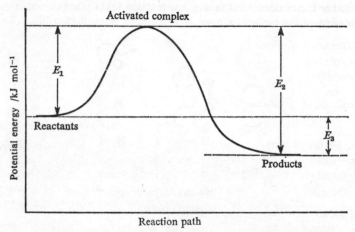

*Fig.* 15-2. Potential energy diagram for an exothermic reaction

forward reaction. If the reacting particles possess just the right amount of energy to form the activated complex there is an equal chance of the latter decomposing into either reactants or products. If the energy is greater the system passes through the intermediate stage and then loses energy, $E_2$, as the new H—I bonds shorten. The difference, $E_3$, between $E_1$ and $E_2$ is the heat evolved in the forward reaction ($\Delta H = -11 \cdot 3$ kJ mol$^{-1}$).

The activated complex can also be reached from the opposite direction. This occurs when two hydrogen iodide molecules collide with an energy equal to at least $E_2$, the activation energy of the reverse reaction. Since $E_2$ is larger than $E_1$ a smaller fraction of hydrogen iodide molecules will surmount the energy barrier in the reverse direction in 1 second. However, when the system has come to equilibrium the total number of molecules changing per second is the same in both directions. This means that in the equilibrium mixture there must be a larger proportion of hydrogen iodide molecules than hydrogen and iodine molecules. It is found experimentally that at 400°C the equilibrium mixture contains about 79 per cent of hydrogen iodide molecules.

We see that the position of equilibrium in a balanced reaction depends on the relative activation energies of the forward and backward reactions. The higher the (average) activation energy of the backward reaction compared with that of the forward reaction, the smaller will be the proportion of products which succeed in changing back to original reactants. If the activation energy of the backward reaction is relatively *very* high this reaction will not occur appreciably. The reaction can then be described as irreversible. Such a reaction is theoretically reversible, but it is very unlikely to occur or it occurs only under exceptional conditions.

**Conditions Affecting Chemical Change.** The chief factors affecting chemical reactions are the following:

* Physical state of the reactants
* Temperature
* Concentration
* Pressure
* Catalysts
* Light

A chemical reaction can be influenced by the prevailing conditions in two ways. There may be an effect (i) on the rate at which the reaction proceeds and (ii), when the reaction is a balanced one, on the proportions of reactants and resultants in the final equilibrium mixture. It is often of the utmost importance to distinguish the effect of a given set of conditions on (i) and (ii). The manufacture of many substances involves the use of balanced reactions and obviously the manufacturer must adjust his conditions so that, other costs being equal, the maximum yield may be obtained in the minimum time.

## Physical State of the Reactants

The rate of a chemical action is frequently affected by whether the substances are in solution or in solid form; if in solution, by the solvent employed, and, if solid, by the state of aggregation of the particles. If solid silver(I) nitrate(V) is mixed with solid sodium(I) chloride no action occurs, while in aqueous solution reaction is immediate. Again chloroethane (or bromoethane) and aqueous silver(I) nitrate(V) yield a precipitate of silver(I) halide only very slowly. In aqueous-ethanolic solution precipitation of the halide occurs very rapidly. The more finely divided a metal the more quickly does it burn in oxygen or chlorine. Even lead burns when sprinkled into oxygen in the form of minute particles.

The influence of physical state on the velocity of a reaction usually has its explanation in the kinetic theory, since the bringing of the reacting substances into more intimate contact will facilitate their interaction.

## Temperature

Temperature influences the rate of a chemical reaction and, in the case of a balanced reaction, the final position of equilibrium.

**Effect on Reaction Rate.** Except in nuclear changes the rate at which a reaction goes to completion or reaches equilibrium is always greater at a higher temperature. It is a common practice in the laboratory to heat substances together to make them react more quickly. Thus, in titrating

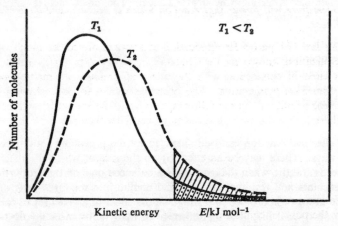

*Fig.* 15–3. Distribution of molecular kinetic energies at different temperatures

aqueous potassium(I) manganate(VII), $KMnO_4$, with acidified ethanedioic acid solution, $(COOH)_2$, the latter is warmed to about 60°C before adding the potassium(I) manganate(VII). Many systems if left at ordinary temperatures, would take years or even centuries to come to equilibrium. One example is the hydrogen-oxygen system.

The increase in rate of reaction with rise of temperature is not simply due to increase in the average molecular velocity resulting in a greater number of collisions per second. For many reactions the reaction rate is roughly doubled by a rise in temperature of 10°C. However, for a gas or mixture of gases of concentration 1 mole per $dm^3$ the increase in collision frequency for a rise of 10°C is only about 1 per cent. The explanation for the observed increase in reaction rate is to be found in the increased number of molecules which acquire the activation energy. The curves in Fig. 15–3 illustrate Maxwell's distribution law of molecular velocities (or kinetic energies) of a gas at temperatures $T_1$ and $T_2$, $T_2$ being higher than $T_1$. The number of molecules which possess more than a given energy $E_a$ at the lower temperature is represented by the dotted area under the first curve. At the higher temperature the number is represented by the lined area under the second curve. The number of molecules with an energy equal to, or greater than, $E_a$ increases rapidly with rise of temperature. This can also be shown by calculation (see next section).

**Effect on Position of Equilibrium.** In nearly all balanced reactions the equilibrium point varies with the temperature. The direction of the change is given by **van't Hoff's law of mobile equilibrium:**

*If a system is in equilibrium raising the temperature will favour that reaction which is accompanied by absorption of heat, and lowering the temperature will favour that reaction which is accompanied by evolution of heat.*

This law is a particular application to temperature of the much wider generalization known as Le Chatelier's principle (p. 252), which has been cited in connection with the effect of pressure on melting point and transition temperature. The principle applies to both physical and chemical equilibria. In the following examples the forward reactions are exothermic and the backward reactions endothermic.

$$2H_2 + O_2 \rightleftharpoons 2H_2O \qquad \text{Heat evolved}$$
$$N_2 + 3H_2 \rightleftharpoons 2NH_3 \qquad \text{in forward}$$
$$2SO_2 + O_2 \rightleftharpoons 2SO_3 \qquad \text{reaction}$$

We can predict from these laws that an increase in temperature will move the point of equilibrium in these reactions to the left and a decrease in temperature will move it to the right. The effect will be the opposite in the following cases where the forward reaction is endothermic and the backward reaction exothermic—that is, increase of temperature will cause the point of equilibrium to be moved to the right and vice versa:

$$N_2O_4 \rightleftharpoons 2NO_2 \qquad \text{Heat absorbed}$$
$$PCl_5 \rightleftharpoons PCl_3 + Cl_2 \qquad \text{in forward}$$
$$N_2 + O_2 \rightleftharpoons 2NO \qquad \text{reaction}$$

A practical illustration of the effect of temperature change on equilibrium point is provided by the $N_2O_4$ equilibrium, in which the dinitrogen tetraoxide ($N_2O_4$) molecules are colourless and the nitrogen dioxide ($NO_2$) molecules deep brown. If a small sealed glass globe containing the two species of molecule in equilibrium at room temperature is gently warmed, the colour darkens as more $N_2O_4$ molecules dissociate into $NO_2$ molecules. On the other hand if the bulb is immersed in freezing water the colour becomes paler as more $NO_2$ molecules form $N_2O_4$ molecules.

Esterification of ethanol by ethanoic acid is one of the few chemical reactions in which there is scarcely any evolution or absorption of heat (the esterification is very slightly endothermic). The position of equilibrium should therefore be practically independent of the temperature, and this is found to be true. It must be remembered, however, that a higher temperature increases the speed at which equilibrium is reached in this case as in others.

The explanation of the effects of change of temperature on position of equilibrium again lies in the change in the number of molecules possessing activation energy. In a balanced reaction there are different activation energies for the forward and backward reactions. Raising the temperature increases the number of molecules with these activation energies, but the number is not increased in the same proportion in the two cases. The larger relative increase occurs for the larger activation energy (this can be deduced from Fig. 15-3 or it can be calculated as shown shortly). Thus in the $H_2 + I_2 \rightleftharpoons 2HI$ equilibrium the rates of both the forward and backward reactions are increased by rise of temperature, but that of the backward reaction is increased to the larger extent. This means that at a higher temperature fewer hydrogen iodide molecules are present at equilibrium. Thus in practice the percentage of hydrogen iodide molecules in the equilibrium mixture falls from 79 per cent at 400°C to 76 per cent at 500°C.

The fraction of the total number of molecules which have an energy at least equal to $E_a$ at kelvin temperature $T$ is given by $e^{-E_a/RT}$, where e is the base of natural logarithms (2·72), and $R$ is the molar gas constant (approximately 8·3 J mol$^{-1}$ K$^{-1}$). This expression is called the *Boltzmann factor*. The activation energy for the combination of hydrogen and iodine at 400°C is 174 kJ mol$^{-1}$ and therefore the fraction of molecules with this energy is given by

$$2\cdot72^{-174\,000/8\cdot3\times673} = \frac{1}{2\cdot72^{174\,000/8\cdot3\times673}}$$

$$= 3 \times 10^{-14}$$

In the same way we can calculate the factor for the activation energy at 500°C. The activation energy for the dissociation of hydrogen iodide is 185 kJ mol$^{-1}$. The table now given shows the values of the Boltzmann factor for the two values of the activation energy at 400°C and 500°C.

| $E_a$/kJ mol$^{-1}$ | *At* 400°C | *At* 500°C |
|---|---|---|
| 174 | $3 \times 10^{-14}$ | $170 \times 10^{-14}$ |
| 185 | $4 \times 10^{-15}$ | $300 \times 10^{-15}$ |

Thus the bigger the activation energy the larger the *relative* increase with rise of temperature in the number of molecules possessing this energy.

**Combined Effect on Reaction Rate and Position of Equilibrium.** Unless the two effects of temperature on a balanced reaction are carefully distinguished, the facts sometimes appear at variance with van't Hoff's law and Le Chatelier's principle. Although the combination of hydrogen and oxygen is exothermic, a mixture of the two gases can be left indefinitely at ordinary temperatures without any measurable change. If, however, the temperature is raised sufficiently the gases explode. This is explained by the fact that the rate of reaction, which is very slow at ordinary temperatures, is tremendously increased at higher temperatures. If the mixture could be left sufficiently long at the lower temperature the yield of water vapour from the same volumes of hydrogen and oxygen would theoretically be greater than at the higher temperature. This is borne out by the fact that steam dissociates into hydrogen and oxygen at high temperatures, and the amount of dissociation is greater the higher the temperature.

## Concentration

The concentrations of the substances taking part in a reaction again affect both the rate of reaction and the position of equilibrium (if the reaction is reversible). For convenience concentrations are usually expressed in mol $dm^{-3}$ instead of in mol $m^{-3}$ (the SI unit). So far we have used the term 'rate of reaction' without defining it. Actually three 'rate' terms are distinguished nowadays, as explained below.

*Rate of Formation of Products.* In the reaction $A \rightarrow B + C$ the rate at which B is being produced at any instant is expressed in calculus notation by $dn_B/dt$, where $n_B$ is the amount of B. Rate of formation is measured in mol $sec^{-1}$, although other time units (minutes, hours, etc.) may be used when more convenient.

*Rate Reaction.* The modern meaning of this term is most easily understood by considering a reaction like the reduction of aqueous iron(III) chloride by aqueous tin(II) chloride. The reaction is written ionically:

$$2Fe^{3+} + Sn^{2+} \rightarrow 2Fe^{2+} + Sn^{4+}$$

Since two $Fe^{2+}$ ions are formed for each $Sn^{4+}$ ion the rate of formation of $Fe^{2+}$ ions is clearly double that of $Sn^{4+}$ ions. Similarly the rate of disappearance of $Fe^{3+}$ ions is twice that of $Sn^{2+}$ ions. But since the rate of formation of $Sn^{4+}$ ions is equal to the rate of disappearance of $Sn^{2+}$ ions we have

$$-\tfrac{1}{2} . \, dn_1/dt = -dn_2/dt = \tfrac{1}{2}dn_3/dt = dn_4/dt$$

where $n_1$, $n_2$, etc. stand for the amounts (in mole) of $Fe^{3+}$ ions, $Sn^{2+}$ions, etc., and the minus signs indicate a decrease in the number of mole. Any one of the four quantities shown represents the rate of the reaction. In the case of the general reaction

$$aA + bB \rightarrow pP + qQ$$

the rate of the reaction is given by any one of the following:

$$-\frac{1}{a} \cdot \frac{dn_A}{dt} = -\frac{1}{b} \cdot \frac{dn_B}{dt} = \frac{1}{p} \cdot \frac{dn_P}{dt} = \frac{1}{q} \cdot \frac{dn_Q}{dt}$$

Rate of reaction is again expressed by number of mole per unit of time, that is, by mol sec$^{-1}$, mol min$^{-1}$, etc. (as appropriate).

*Rate of Increase in Concentration of Products.* In the course of a reaction which takes place at a fixed volume the concentrations of the products increase. The rate ($r_P$) of increase in concentration of a product P in the general reaction given above is obtained by dividing the rate of formation of P by the volume $V$. Thus we have

$$r_P = \frac{dn_P/dt}{V} = dc_P/dt$$

where $c_P$ is the concentration of P. This rate quantity was formerly used for rate of reaction. The usual unit for $r_P$ is mol dm$^{-3}$ sec$^{-1}$ or mol dm$^{-3}$ min$^{-1}$.

**Effect on Rate of Increase in Concentration of Products.** An increase in concentration of the reactants usually produces a faster increase in the concentration of the products. There is no general rule, however, by which the quantitative effects can be deduced. These effects have to be determined experimentally. If the reaction is of the elementary type A → B + C, it is found that the rate at which the concentration of B (or C) increases is directly proportional to the concentration of A; that is,

$$r_B = d[B]/dt = k[A]$$

where $k$ is a constant (called the *rate constant* of the reaction) and the square bracket denotes 'molar concentration of'. A reaction of this kind is the decomposition of ethyl methanoate vapour to ethene and methanoic acid when heated at constant temperature and pressure.

$$HCOOC_2H_5 \rightarrow C_2H_4 + HCOOH$$

In an elementary reaction of the type A + B → C + D the rate of increase in the concentration of C (or D) is found to be proportional to the product of the concentrations of A and B; that is,

$$r_C = d[C]/dt = k[A][B]$$

where $k$ again is the rate constant of the reaction. An example of this type of reaction is the combination of hydrogen and iodine vapour to form hydrogen iodide ($H_2 + I_2 \rightarrow 2HI$).

The decomposition of hydrogen iodide by heat ($2HI \rightarrow H_2 + I_2$) is an elementary reaction of the type 2A → B + C, which depends on collision between two similar A molecules. If we double the concentration of A we have not only double the number of activated molecules to collide with a given activated molecule, but also double the number of activated molecules to be collided with. Hence the rate of increase in the concentration of B (or C) is proportional to the *square* of the concentration of A. Thus

$$r_B = d[B]/dt = k[A]^2$$

The great majority of reactions, however, are not elementary. They proceed in a series of steps, each of which consists of one or other of the elementary steps illustrated above. The manner in which the overall reaction depends on the various concentrations cannot be deduced from the equation, but has to be determined by experiment, as explained shortly.

### Effect on Position of Equilibrium. Law of Mass Action.

*The law of mass action* states that *in a reversible reaction there is a fixed relationship at a given temperature between the molar concentrations of the products and those of the reactants in the equilibrium mixture.*

'Molar concentration' was formerly called 'active mass', which gave rise to the name 'law of mass action'. The latter is an empirical law deduced from the results of thousands of experiments on different kinds of reversible reactions. Thus, if in general a reversible reaction is represented by the equation

$$A + B \rightleftharpoons C + D$$

it is found experimentally that at equilibrium

$$K_c = \frac{[C]\,[D]}{[A]\,[B]}$$

where [A], [B], etc., refer to the molar concentrations of the substances in the equilibrium mixture. $K_c$ is the *concentration equilibrium constant*.

When ethanol is added to ethanoic acid an equilibrium mixture containing alcohol, acid, ester, and water results.

$$C_2H_5OH + CH_3COOH \rightleftharpoons CH_3COOC_2H_5 + H_2O$$

The value of $K_c$ is given by the expression

$$K_c = \frac{[ester]\,[water]}{[alcohol]\,[acid]}$$

When two molecules of one substance take part in the forward or backward reaction, the position of equilibrium depends on the square of the molar concentration of that substance. Thus for the equilibrium between hydrogen, iodine vapour, and hydrogen iodide ($H_2 + I_2 \rightleftharpoons 2HI$) we have

$$K_c = \frac{[HI]^2}{[H_2]\,[I_2]}$$

We can express the law of mass action in its most general form as follows:

$$pA + qB + rC \rightleftharpoons xD + yE + zF$$

$$\text{and} \quad K_c = \frac{[D]^x [E]^y [F]^z}{[A]^p [B]^q [C]^r}$$

**Displacement of Position of Equilibrium.** We have seen that for the system $A + B \rightleftharpoons C + D$ to be in equilibrium the concentrations of the four substances must be such that

$$K_c = \frac{[C] [D]}{[A] [B]}$$

If the concentration of any one of the substances is altered the concentrations of the other substances must change so as to keep $K_c$ constant. Thus, if [D] is increased by adding D from outside, [C] will decrease by combination of C with D to give more A and B. Thus $K_c$ will be kept constant by decrease in [C] and increase in [A] and [B]. Conversely a decrease in [D] will be followed by an increase in [C] and a decrease in [A] and [B]. In both cases the effects are in accordance with Le Chatelier's principle since the changes are such as to minimize the change in [D].

An illustration of the effect of change in concentration on equilibrium point is provided by the system formed when a solution of bismuth(III) chloride is diluted with water.

$$BiCl_3 + H_2O \rightleftharpoons 2HCl + BiOCl \downarrow$$

If a drop or two of hydrochloric acid is added to the system, the equilibrium point is displaced to the left, and the precipitate of bismuth(III) chloride oxide redissolves. Addition of more water causes the precipitate to appear once more.

In the reversible reaction of iron with steam

$$3Fe + 4H_2O \rightleftharpoons Fe_3O_4 + 4H_2$$

the hydrogen is removed when steam is passed over heated iron, until eventually the latter is completely converted into iron(II) diiron(III) oxide. The reverse process occurs of hydrogen is passed over the heated oxide so that the water vapour is swept away.

The preparation of many important chemicals depends on the principle of removing one of the products in a balanced reaction. The iron and steam reaction is itself used in the manufacture of hydrogen. In the preparation of calcium(II) oxide from calcium(II) carbonate (limestone) carbon dioxide is removed in a current of air:

$$CaCO_3 \rightleftharpoons CaO + CO_2$$

In the preparation of esters of organic acids concentrated sulphuric (VI) acid is often added to the reaction mixture. The mineral acid acts partly as a catalyst and partly as an absorbent for the water formed, so that the position of equilibrium is displaced in the direction of ester formation. For the reaction between ethanol and ethanoic acid to form ethyl ethanoate we have

$$C_2H_5OH + CH_3COOH \rightleftharpoons CH_3COOC_2H_5 + H_2O$$

Alkalis reverse the reaction by combining with the ethanoic acid.

The preparation of many gases also depends on displacement of an equilibrium. When an ammonium salt is warmed with an alkali ammonia gas is given off. This is brought about by disturbance of the following equilibrium by vaporization of ammonia from the solution:

$$NH_4^+ + OH^- \rightleftharpoons NH_3 \uparrow + H_2O$$

The backward reaction occurs when ammonia gas is dissolved in water.

## Pressure

Since pressure has a negligible effect on the volumes of solids and liquids, it affects only those reactions in which gases are involved.

**Effect on Rate of Increase in Concentration of Products.** Increased pressure in a gaseous system means an increased concentration of the reactants. Hence more collisions per second occur between the reacting particles and a more rapid increase in concentration of the products.

**Effect on Position of Equilibrium.** The proportions of resultants and reactants in a balanced reaction may or may not be altered by a change in pressure. This depends on whether the reaction is accompanied by a change in volume or not. When the reaction occurs with change of volume the effect on the equilibrium point can be predicted by means of Le Chatelier's principle. When increased pressure is applied to a system in equilibrium the system reacts so as to oppose the increase in pressure— that is, the equilibrium point moves in the direction which is accompanied by decrease of volume. Conversely, a decrease of pressure favours that reaction which is accompanied by increase of volume. If the volume of the products is the same as the volume of the initial substances pressure has no effect on the position of equilibrium, although of course the rate at which equilibrium is attained is greater at higher pressures. The following examples will serve to illustrate the application of the general rule.

Forward reaction favoured by increased pressure:

$$N_2 + 3H_2 \rightleftharpoons 2NH_3$$
$$2SO_2 + O_2 \rightleftharpoons 2SO_3$$
$$2CO + O_2 \rightleftharpoons 2CO_2$$

Forward reaction favoured by decreased pressure:

$$PCl_5 \rightleftharpoons PCl_3 + Cl_2$$
$$N_2O_4 \rightleftharpoons 2NO_2$$
$$2NO_2 \rightleftharpoons 2NO + O_2$$

Equilibrium point not affected by pressure:

$$H_2 + I_2 \rightleftharpoons 2HI$$
$$N_2 + O_2 \rightleftharpoons 2NO$$

**Heterogeneous Systems.** A heterogeneous system is one in which the different substances are not all in the same physical state. Thus, in the dissociation of calcium(II) carbonate we have two solids and one gas in equilibrium:

$$CaCO_3 \rightleftharpoons CaO + CO_2$$

The law of mass action can be applied to heterogeneous systems of this type in the same way as to homogeneous systems. In practice a solid behaves in a reversible reaction as if it had a constant concentration. If we apply the law of mass action to dissociation of calcium(II) carbonate we obtain

$$K_c = \frac{[CaO]\,[CO_2]}{[CaCO_3]}$$

As the concentrations of calcium(II) oxide and calcium(II) carbonate during the reaction are constant, we can write $K_c = c_1/c_2 \times [CO_2]$. It follows that the concentration, or pressure, of carbon dioxide is constant for any particular temperature. The value of this pressure at a given

temperature is called the *dissociation pressure* at that temperature. That the pressure of carbon dioxide remains constant for a given temperature, provided that all three reacting substances are present, can be seen by heating calcium(II) carbonate in a closed vessel in a thermostat to a temperature of 600° to 800°C. The vessel (Fig. 15–4) is joined to a manometer and connected either to a suction pump or to a source of carbon dioxide under pressure. If carbon dioxide is removed from the system when equilibrium has been reached, the pressure is soon restored by further dissociation of calcium(II) carbonate. If carbon dioxide is pumped into the vessel calcium(II) oxide combines with the gas until the original pressure is regained. This apparatus

*Fig. 15–4.*
Dissociation of calcium(II) carbonate

can be used to measure the dissociation pressure of carbon dioxide at different temperatures. The dissociation of barium(II) peroxide is analogous to that of calcium(II) carbonate:

$$2BaO_2 \rightleftharpoons 2BaO + O_2$$

With steam and iron together in a closed vessel we have the equilibrium

$$3Fe + 4H_2O \rightleftharpoons Fe_3O_4 + 4H_2$$

Applying the law of mass action again, we have

$$K_c = \frac{[Fe_3O_4]\,[H_2]^4}{[Fe]^3\,[H_2O]^4}$$

As before, the concentration of a solid is constant. Therefore

$$\frac{[H_2]}{[H_2O]} = \text{a constant}$$

That is, for a given temperature the ratio of the partial pressure of hydrogen to that of steam remains constant, being independent of the amounts of iron and iron oxide present in the system.

## Catalysts

A **catalyst** is defined as *a substance which alters the rate of a reaction, but remains unchanged in quantity at the end of the reaction or at equilibrium.*

Catalytic actions are *homogeneous* when the catalyst and reacting substances are in the same physical state, and *heterogenous* when they are in different physical states. Thus esterification of ethanol by ethanoic acid in the presence of sulphuric(VI) acid illustrates homogeneous catalysis, while the combination of hydrogen and oxygen under the influence of platinum is an example of heterogeneous catalysis.

**General Properties of Catalysts.** (1) A catalyst is unchanged in quantity at the end of a reaction. However, it may be altered physically. For example, if granular manganese(IV) oxide catalyses the decomposition of potassium(I) chlorate(V) it remains as a fine powder afterwards.

(2) A catalyst alters the *reaction rate* but never the *final position of equilibrium* in a balanced reaction. A catalyst may be *positive* or *negative*— that is, it may increase or decrease the rate of a reaction. The majority of catalysts are positive, but a few reactions are retarded by a suitable substance. Thus, the rate at which aqueous hydrogen peroxide decomposes is increased by alkalis but decreased by acids. Phosphoric(V) acid is usually added to commercial hydrogen peroxide to retard decomposition under the action of light.

(3) A small amount of catalyst usually influences a large amount of reacting substances. The effect of a catalyst may be proportional to its amount (within limits). Thus, when an acid catalyses the esterification

of ethanol, the rate of esterification is proportional to the amount of acid, provided this is small.

(4) Catalytic power is more or less specific in character. It does not follow that because a certain substance will catalyse one reaction it will also act as a catalyst for another. It frequently does happen that the same substance will catalyse several reactions—for example, finely divided platinum and nickel are catalysts in many gaseous reactions. Also, the same chemical action is often catalysed by different substances. Besides manganese(IV) oxide, oxides of copper, iron, and lead are catalysts (although less effective) in the decomposition of potassium(I) chlorate(V). It appears that particular types of chemical substance are required as catalysts for particular reactions.

(5) The efficiency of a catalyst is often increased or decreased by traces of other substances. The iron catalyst used in the Haber process for ammonia is activated by addition of a small amount of aluminium(III) oxide. A substance which improves the efficiency of a catalyst is called a *promoter*. On the other hand, small quantities of foreign matter may render the catalyst useless, and in many industrial processes great care has to be taken to prevent such matter from reaching the catalyst. The most harmful substances are often those which have a poisonous effect on the human body, and by analogy they are said to poison the catalyst. Catalyst poisons, or *inhibitors*, include hydrogen sulphide, arsenic(III) oxide, hydrogen cyanide, and mercury salts.

**Autocatalysis.** In some reactions one of the products acts as a catalyst. Thus the rate of oxidation of ethanedioic acid by potassium(I) manganate(VII) is accelerated by the manganese(II) sulphate(VI) formed in the reaction. If manganese(II) sulphate(VI) is added the reaction can be carried out in the cold. Similarly, the decomposium of arsine ($AsH_3$) by heat is catalysed by the metallic arsenic formed. This type of catalytic action is known as *autocatalysis*.

**Explanations of Catalytic Action.** Basically there is only one explanation of catalysis, namely, that it depends on lowering the activation energy necessary for a reaction to occur. This may be achieved, however, in two ways. The first is characteristic of homogeneous catalysis, the second of heterogeneous catalysis.

*Homogeneous Catalysis.* This is usually explained by the *intermediate compound theory*, according to which the catalyst takes a definite part in the reaction by being converted into an intermediate compound, which is subsequently reconverted to the original substance. It is significant that in a large number of examples of catalytic action the catalyst contains atoms or ions which can exist in different stages of oxidation. Oxidation of sulphur dioxide to sulphur(VI) oxide is catalysed by nitrogen oxide(NO).

Although the details are probably more complex (they are still uncertain) the change can be explained by the series of reactions below:

$$2NO + O_2 \rightarrow 2NO_2$$
$$NO_2 + SO_2 \rightarrow SO_3 + NO$$
$$SO_3 + H_2O \rightarrow H_2SO_4$$

As nitrogen oxide is constantly re-formed, a small amount only is needed to convert a large amount of sulphur dioxide to sulphur(VI) oxide. Nitrogen oxide also catalyses oxidation of carbon monoxide by oxygen to carbon dioxide at 500°C.

Not surprisingly, we find that ions of transition metals show marked catalytic power (see Chapter 8). These elements, more than any others, have a highly variable valency and readily form ions in different oxidation states.

*Heterogeneous Catalysis.* Although some examples of heterogeneous catalysis can be explained by the intermediate compound theory this type of catalysis usually depends on adsorption. The adsorption theory of heterogeneous catalysis was prompted by the great increase in catalytic power evident in such metals as platinum and nickel when they are in the finely divided state. In this condition a metal exposes a tremendously increased surface area to the reacting substances. Surface catalysis, however, is not confined to finely divided substances. It has been proved that in some reactions the glass walls of the containing vessel act catalytically; for example, the rate of combination of ethene and bromine becomes considerably slower if the glass walls of the reaction vessel are covered with paraffin wax. Thus some reactions which at first sight appear homogeneous are actually heterogeneous.

Platinum crystallizes in the cubic system. Atoms inside the crystal are surrounded by other platinum atoms and have all their valencies satisfied. Atoms at the surface are situated differently, however. Only some of their valencies are satisfied in connecting them to other atoms in the surface and interior. Thus on the surface, edges, and corners of the crystal a number of valencies are left over. The number of such valencies is greater the bigger the surface area, and it is increased by any unevenness in the surface. (It has been demonstrated that even a well-polished metal surface contains many pits and elevations.)

Free valencies at a metal surface are able to attach atoms of other substances, as with carbon (graphite) (p. 320). Thus when platinum catalyses the combination of hydrogen and iodine vapour it is supposed that the bonds between the atoms are broken and the elements are chemisorbed in the atomic form. If atoms of hydrogen and iodine occupy adjacent positions and possess sufficient energy of vibration they may come close enough to form an absorbed activated complex. The latter consists of a hydrogen atom and an iodine atom joined to each other and to platinum atoms by abnormally long weak bonds as illustrated below.

$$
\begin{array}{cccc}
& \text{H---I} & & \\
\text{H} & | & | & \text{I} \\
| & | & | & | \\
\text{---Pt---Pt---Pt---Pt---} \\
| & | & | & | \\
\text{---Pt---Pt---Pt---Pt---} \\
| & | & | & |
\end{array}
$$

The activated complex has only a fleeting existence. Providing the hydrogen and iodine atoms have sufficient energy the bond between them shortens, the bonds with the metal are broken, and the newly formed molecule of hydrogen iodide vaporizes. This leaves vacant sites for the adsorption of further atoms. The metal surface assists the combination also by absorbing the heat liberated in the combination. If the energy were not dissipated in this way it would cause the new molecule to split up again. The poisoning of a catalyst is due to the metal surface having a greater attraction for the molecules of the poisoning agent than for those of the reacting molecules.

*Activation Energy and Catalysis.* For the uncatalysed combination of hydrogen and iodine the activated complex has the form given at p. 351. For the platinum-catalysed combination it has the form shown above. If the activation energies required to produce the two complexes are calculated, the value in the catalysed reaction is only about $92 \text{ kJ mol}^{-1}$ as compared with about $177 \text{ kJ mol}^{-1}$ for the uncatalysed reaction. Here is the fundamental explanation of catalytic action, whether heterogeneous or homogeneous. As noted earlier, the lower the activation energy the bigger is the fraction of molecules which possess this energy, and the more rapid is the reaction.

The function of catalysts is thus to provide alternative routes of reaction with lower energy barriers. In the homogeneous catalysis by nitrogen oxide of the combination of sulphur dioxide and oxygen two energy barriers are involved. These occur in the combination of the nitrogen oxide with oxygen and in the reaction between nitrogen dioxide and sulphur dioxide. Both of these, however, are lower than the energy barrier encountered in the direct combination of sulphur dioxide and oxygen.

Note that if a reaction is reversible a catalyst decreases the activation energy of both the forward and backward reactions. Since the position of equilibrium is not altered by the catalyst the forward and backward reactions must be accelerated to the same extent.

## Light

All radiations, from X-rays to infrared rays, can exert a chemical effect. Photography, which is based on this fact, makes use of the susceptibility of silver halides to the action of these waves. Some reactions take place very slowly, or not at all, in the absence of light. Thus the process of photosynthesis in the leaves of plants is arrested at night. The following are examples of reactions accelerated by light:

> The decomposition of hydrogen peroxide.
> The combination of hydrogen and chlorine.
> The substitution of hydrogen in methane by chlorine.

The combination of hydrogen with chlorine (or bromine) takes place differently from that of hydrogen with iodine. The rate of reaction between hydrogen and chlorine is not directly proportional to the concentrations of hydrogen and chlorine, but depends on the intensity of light. To explain the effect of light in this and similar reactions involving chlorine, Nernst in 1916 put forward his theory of a *chain reaction*. According to this theory absorption of a suitable quantum of light energy ruptures the bond between the two atoms in a chlorine molecule, and free atoms are produced.[1]

$$Cl - Cl + h\nu \rightarrow 2Cl \cdot \tag{1}$$

A free chlorine atom is extremely reactive. When it comes sufficiently close to a hydrogen molecule it captures a hydrogen atom, including one of the two bond electrons, and leaves a free hydrogen atom. This also is very reactive, and in the same way it can remove a chlorine atom from a chlorine molecule to give once more a free chlorine atom. Thus we have

$$Cl \cdot + H_2 \rightarrow HCl + H \cdot \tag{2}$$

$$H \cdot + Cl_2 \rightarrow HCl + Cl \cdot \tag{3}$$

Step (2) and (3) can now be repeated over and over again, so that absorption of a single quantum of light energy may result in the combination of many thousands of hydrogen and chlorine molecules. The sequence is eventually stopped by two of the free atoms colliding and forming a stable molecule (*e.g.*, $H \cdot + H \cdot \rightarrow H_2$).

Experimental evidence in support of the above mechanism comes from the exposure of chlorine to ultraviolet light or visible light of short wavelength. After exposure the gas shows the absorption spectrum of atomic chlorine. Also, if the chlorine is then passed into a blackened bulb containing hydrogen, the gases combine in complete darkness.

A reaction which takes place through the agency of light is called a *photochemical* reaction. The photochemical substitution of hydrogen in

---

[1] A quantum of energy, $h\nu$, is the product of Planck's constant, $h$, and the frequency, $\nu$ (nu), of the radiation.

methane by chlorine similarly takes place by a chain reaction. The steps are as follows:

$$Cl_2 + h\nu \rightarrow 2Cl\cdot$$
$$Cl\cdot + CH_4 \rightarrow HCl + CH_3\cdot$$
$$CH_3\cdot + Cl_2 \rightarrow CH_3Cl + Cl\cdot$$

## Some Important Balanced Reactions

The following paragraphs deal with some reversible reactions of commercial application. Details of the processes are not given, but it is intended to show how the conditions of chemical change discussed above are applied to the best advantage.

### Contact Process for Sulphuric(VI) Acid

$$2SO_2 + O_2 \rightleftharpoons 2SO_3 \qquad (\Delta H = -188 \text{ kJ mol}^{-1})$$

Applying the law of mass action to this equilibrium we have

$$K_c = \frac{[SO_3]^2}{[SO_2]^2\,[O_2]}$$

The yield of sulphur(VI) oxide would therefore be increased by increasing the concentration of either sulphur dioxide or oxygen. The cheaper reactant is oxygen (the air costs nothing), and hence excess of oxygen is used. In practice, however, the amount of oxygen which can be mixed with the sulphur dioxide is limited by the diluting effect of the nitrogen present in air. By a process of trial and error the mixture which gives the best yield for a given amount of sulphur dioxide is found. This is the *optimum* mixture.

The forward reaction is exothermic, and by van't Hoff's law of mobile equilibrium is favoured by a low temperature. At ordinary temperatures the reaction occurs too slowly to be profitable, and so the temperature has to be increased. That temperature is found which, with the catalyst, gives the best yield in a given time. This optimum temperature is about 500°C. The catalyst used is vanadium(V) oxide ($V_2O_5$). If the sulphur dioxide is made by burning iron(II) disulphide (pyrites) it is purified from arsenic(III) oxide, which poisons the catalyst.

Increased pressure, which would theoretically increase the yield of sulphur(VI) oxide according to Le Chatelier's principle, is not used as it is uneconomical. Under working conditions 97–98 per cent conversion is obtained.

### Haber Process for the Fixation of Atmospheric Nitrogen

$$N_2 + 3H_2 \rightleftharpoons 2NH_3 \qquad (\Delta H = -100 \text{ kJ mol}^{-1})$$

At ordinary temperatures and pressures a mixture of nitrogen and hydrogen shows no tendency to combine. Gaseous ammonia is almost completely decomposed by electric sparks. The equilibrium point is normally, therefore, well to the left.

The combination of the gases is exothermic, and by van't Hoff's law is favoured by low temperatures. Again, owing to the velocity of combination being slow at low temperatures, an optimum temperature of about 550°C and a catalyst are used. The catalyst is finely divided iron which is made by reduction of the oxide $Fe_3O_4$. The catalyst is activated by mixing a small percentage of aluminium(III) oxide with the iron oxide before reduction. The action of aluminium(III) oxide as a promoter is explained as follows. The two oxides are isomorphous, so that when the iron oxide is reduced the second oxide is evenly dispersed throughout the catalyst and prevents the crystals of iron from joining together. The catalytic power of the iron depends on the number and smallness of the iron crystals.

In accordance with Le Chatelier's principle the yield of ammonia is favoured by increased pressure, and in practice a pressure of about 350 atm is used. Since a larger pressure means a larger concentration of gases the effect of increased pressure can also be deduced from application of the law of mass action:

$$K_c = \frac{[NH_3]^2}{[N_2][H_2]^3}$$

No economic advantage is gained by increasing or decreasing the proportion of nitrogen to hydrogen, and the gases are used in the volume proportion of 3:1, as represented by the equation.

**Bosch Process for Hydrogen.** In this process hydrogen is manufactured from steam and water gas. The latter is first obtained by blowing steam through white-hot coke (at the high temperature the reaction goes almost to completion):

$$C + H_2O \rightleftharpoons CO + H_2 \quad (\Delta H = +133 \text{ kJ mol}^{-1})$$

The water gas is now mixed with more steam and passed over a heated catalyst of iron(III) oxide ($Fe_2O_3$), with chromium(III) oxide ($Cr_2O_3$) as a promoter. The following reaction, which is reversible, takes place:

$$CO + H_2O \rightleftharpoons CO_2 + H_2 \quad (\Delta H = -41 \text{ kJ mol}^{-1})$$

The carbon dioxide is removed by dissolving it in water under pressure and any unchanged carbon monoxide by treatment with ammoniacal copper(I) methanoate.

Since the forward reaction is exothermic, the yield of hydrogen is favoured by a low temperature (van't Hoff's law), but as the rate of reaction is slow at low temperatures an optimum temperature of about 450°C is used. There is no change in the number of molecules, and therefore pressure has no influence on the yield (Le Chatelier's principle). By increasing the proportion of steam (the cheaper reactant) to water gas the equilibrium point will be displaced from left to right in accordance with the law of mass action. Usually the water gas is mixed with two to three times its own volume of steam.

## Chemical Kinetics

**Kinetic Order of a Reaction.** *Chemical kinetics* is the study of rates of reaction (and allied 'rate' quantities) and how these depend on the factors affecting them. This study is important practically because the knowledge gained can be used to adjust the reaction conditions to the greatest advantage. It is also important theoretically because it often throws light on the mechanism by which a reaction occurs (chemical reactions are seldom as simple as they appear from the usual equations).

The *order of a reaction* expresses the manner in which the rate of increase in concentration of the products depends on the concentrations of the reactants. A *first-order* reaction is one in which the rate of increase in concentration of any one of the products is directly proportional to the first power of the concentration of a single reactant. Thus when ethyl methanoate vapour is heated

$$HCOOC_2H_5 \rightarrow C_2H_4 + HCOOH$$

the rate of increase in concentration of ethene (as measured by increase of pressure at constant volume) is proportional to the concentration of the ester. This can be expressed as follows:

$$\frac{d[C_2H_4]}{dt} = k_1[HCOOC_2H_5]$$

where $k_1$ is the rate constant.

In a *second-order* reaction the rate of change in concentration of a product is proportional to the product of the concentrations of two reactants or to the square of the concentration of a single reactant. Thus in the combination of hydrogen and iodine ($H_2 + I_2 \rightleftharpoons 2HI$) the rate of increase in concentration of hydrogen iodide at any instant is given by

$$\frac{d[HI]}{dt} = k_2[H_2][I_2]$$

The forward reaction is said to be first-order with respect to hydrogen and first-order with respect to iodine, the overall forward reaction being of second-order. Similarly the reverse reaction, the dissociation of hydrogen iodide, is a second-order reaction:

$$\frac{d[H_2]}{dt} = \frac{d[I_2]}{dt} = k_3[HI]^2$$

Combination of nitrogen oxide with oxygen ($2NO + O_2 \rightarrow 2NO_2$) is found to be a reaction of the *third-order*. The reaction is second-order with respect to nitrogen oxide and first-order with respect to oxygen.

$$\frac{d[NO_2]}{dt} = k_4[NO]^2\,[O_2]$$

These illustrations show that the order of reaction as a whole is obtained by adding together the orders with respect to the separate reactants. We can express the general case as follows. Let the reactants be numbered 1, 2, 3, etc., and let their concentrations at any instant be $c_1$, $c_2$, $c_3$, etc. If the rate of increase in concentration of a product P is separately proportional to $c_1^a$, $c_2^b$, $c_3^c$, etc., we have

$$r_P = d[P]/dt = kc_1^a c_2^b c_3^c \ldots$$

and the kinetic order of the reaction is equal to $a + b + c + \ldots$

If the reaction yields two products, P and Q, in the proportion of one mole of P to 2 mole of Q, $r_Q = 2r_P$. In order to keep a constant value for $k$ and so that P can refer to any one of the products we must amend the rate law given above to the following:

$$r_P = d[P]/dt = pkc_1^a c_2^b c_3^c \ldots$$

where p is the stoichiometric coefficient of P.

*The order of a reaction cannot be deduced from the equation for the reaction.* It can only be found experimentally. In some cases (elementary one-step reactions) the kinetic order agrees with the number of molecules of reactants in the equation, but usually this is not so.[1] Thus, while combination of hydrogen and iodine is a second-order reaction, that of hydrogen and chlorine (or bromine) has no simple reaction order. This is due to the chain mechanism initiated by light, as explained earlier.

The decomposition of hydrogen peroxide is usually represented by the equation

$$2H_2O_2 \rightarrow 2H_2O + O_2$$

Under suitable experimental conditions (see p. 374) this reaction is found to be of the first, not second, order, showing that it does not take place simply by two molecules of hydrogen peroxide colliding together.

$$\frac{-d[H_2O_2]}{dt} = k[H_2O_2]$$

---

[1] Formerly a first-order reaction was described as *unimolecular*, a second-order reaction as *bimolecular*, etc. These terms are no longer used in connection with chemical kinetics. Nowadays 'molecularity' is applied only to elementary reactions or the elementary stages of more complex reactions. Thus an elementary reaction is unimolecular if the activated complex is formed from a single molecule, bimolecular if it is formed from two molecules, etc. Molecularity is nearly always one or two, but may be (rarely) three.

When a reaction occurs in stages the rate of change in concentration of a reactant or product is governed by the rate for the slowest stage. Thus a *possible* explanation of the first-order character of the hydrogen peroxide reaction is that the rate of decrease in concentration of the peroxide depends on the separate rates of the following changes:

$$\text{(i) } H_2O_2 \rightarrow H_2O + O \qquad \text{(ii) } O + O \rightarrow O_2$$

Probably the second step is almost instantaneous, so that the rate of change in concentration is controlled by the first and slower stage.

Again, in dilute solution the hydrolysis of ethyl ethanoate is a first-order reaction:

$$CH_3COOC_2H_5 + H_2O \rightarrow CH_3COOH + C_2H_5OH$$

The reason is that the amount of water taking part in the reaction is small compared with the total amount of water present, so that there is no sensible change in the amount of water. There are many reactions of the second-order which becomes first-order when a large excess of one of the reactants is used. Such reactions are sometimes described as 'pseudo first-order.'

**First-order Reactions.** In general, a first-order elementary reaction may be written:

$$A \rightarrow B + C$$

Suppose we start with a concentration of A of $a$ mole per $m^3$ and that, after time $t$, $x$ mole of A have reacted, giving $x$ mole of B and $x$ mole of C. If the reaction is elementary, the rate at which the concentration of B (or C) is increasing at any instant is proportional to the concentration of A. Using calculus notation, we may write

$$\frac{dx}{dt} = k(a - x)$$

where $k$ is the rate constant. The expression may be rearranged and integrated as follows:

$$\frac{dx}{a - x} = k dt$$

$$-\log_e (a - x) = kt + c \text{ (a constant)}$$

To find $c$ we put $t = 0$; in this case $x = 0$ and

$$c = -\log_e a$$

$$\therefore \quad -\log_e (a - x) = kt - \log_e a$$

$$kt = -\log_e (a - x) + \log_e a$$

$$k = \frac{1}{t}\log_e \frac{a}{a - x}$$

$$= \frac{2 \cdot 303}{t} \log_{10} \frac{a}{a - x}$$

We can therefore determine the rate constant by finding the change in concentration after a known interval of time.

The method of finding the change in concentration depends upon the particular reaction. In some reactions we can use a physical quantity (such as pressure) which changes uniformly during the reaction. In other cases (*e.g.*, hydrolysis of an ester) the reaction proceeds so slowly at ordinary temperatures that we can estimate one of the substances present by titration. Since in the expression for the rate constant of a first-order reaction the concentrations are involved as a ratio, we need not determine the actual concentrations. The ratio of the concentrations is given by the relative values of the pressures, titrations, or whatever quantities are used in following the reaction. This applies only to reactions of the first order.

The units in which $k$ must be expressed can be deduced from the last equation. We have

$$\text{Units of } k = \frac{\text{concentration}}{\text{time} \times \text{concentration}} = \frac{1}{\text{time}} = \text{time}^{-1}$$

Hence the rate constant, $k$, for a first-order reaction is a number per unit of time. It is in fact the fraction of the molecules which change in unit time. Thus suppose that $k = 0.01$, and that the original concentrations, $a$, is 1 mol m$^{-3}$. If the time is measured in seconds, the concentration after 1 second will be 0.99 mol m$^{-3}$, after 2 seconds (0.99 − 0.0099) mol m$^{-3}$, and so on.

Another feature of first-order reactions is that *the time required for any specified fraction of the given substance to change is independent of the initial concentration*. Thus in the reaction $A \rightarrow B + C$ the time, $t'$, required for one half of A to disappear is evidently the same whether we start with 1 mole of A per m$^3$ or 2 mole of A per m$^3$. In both cases

$$t' = \frac{2.303}{k} \log \frac{a}{a - a/2}$$

$$= \frac{2.303}{k} \log 2$$

$$= \frac{0.693}{k}$$

Conversely, if the time taken for a specified fraction of a substance to disappear is the same whatever the initial concentration the substance is behaving according to the first-order rate law. An interesting illustration is seen in the case of radioactive elements. These elements have a definite half-life which is independent of the initial amount of the element. In the case of naturally occurring radioactive elements, the half-life has been utilized to calculate the age of minerals, as the rate at which the radio-active change occurs is independent of physical factors such as temperature and pressure. Thus in the transformation

$$\text{Uranium} \rightarrow \text{radium} \rightarrow \text{lead}$$

uranium has a half-life of $4 \cdot 5 \times 10^9$ years, while that of radium is 1600 years. The proportions of uranium, radium, and lead occurring in some of the oldest rocks indicate that these rocks have an age of nearly 2000 million years. The earth must therefore have been in existence for at least this period.

Radiocarbon dating is used to determine the age of more recent organic remains found in geological strata and archaeological excavations. This method depends on the fact that a radioactive isotope of carbon of mass number 14 is created continually in the atmosphere by bombardment of carbon dioxide by cosmic rays. The resulting carbon-14 becomes evenly distributed throughout the atmosphere, the oceans, and all living organisms, a constant ratio being maintained between carbon-14 and its disintegration product, carbon-12. When plants and animals die renewal of their radiocarbon ceases, and the carbon-14 present decays at a known rate (the half-life is 5568 years). We can find the age of the specimen by measuring the radioactivity of the remaining carbon-14. This method is reliable for ages of up to about 30,000 years.

*To Show that the Decomposition of Hydrogen Peroxide is a First-order Reaction.*[1]

$$2H_2O_2 \rightarrow 2H_2O + O_2$$

The decomposition is slow at ordinary temperatures, but is catalysed by colloidal platinum, alkalis, manganese(IV) oxide, etc. A suitable catalyst for the investigation is colloidal manganese(IV) oxide made by adding a little potassium(I) manganate(VII) solution to the hydrogen peroxide in slightly alkaline solution (as described below). The reaction flask is placed in a thermostat at 25°C and the course of the reaction is followed by withdrawing 10-cm³ samples of the solution at known time intervals and titrating with standard potassium(I) manganate(VII) solution in the presence of dilute sulphuric(VI) acid. This method of working is permissible because of the slowness of the reaction even in the presence of the catalyst. In the absence of a thermostat the investigation can be carried out at room temperature as now described.

**Experiment.** Put in a conical flask 10 cm³ of 6-volume hydrogen peroxide, 150 cm³ of distilled water, and 50 cm³ of a borate buffer solution. The latter is made by dissolving $3 \cdot 1$ g of boric(III) acid in 25 cm³ of molar NaOH solution and adding distilled water to bring the volume up to

---

[1] The experimental procedure is taken by permission from *Experimental Physical Chemistry*, by W. G. Palmer, 2nd edn. (Cambridge University Press).

250 cm$^3$. Clamp the flask in a trough containing a large volume of water at room temperature and add 10 cm$^3$ of M/250 KMnO$_4$ solution (see p. 440). Brown colloidal manganese(IV) oxide is formed in the flask. Insert a plug of cotton wool into the neck of the flask to exclude dust and allow the reaction to settle down. Withdraw the first 10-cm$^3$ sample of the reaction mixture after about 5 minutes (this is $t_0$), add dilute sulphuric(VI) acid, and titrate with M/250 KMnO$_4$ solution. Withdraw further samples at 10-minute intervals for 80 or 100 minutes.

If $v_0$ is the volume of the permanganate solution required at the beginning of the reaction ($t_0$), and $v_t$ the volume required after time $t$, the concentrations of hydrogen peroxide at these times are proportional to $v_0$ and $v_t$ respectively. These titrations therefore represent $a$ and $a - x$, while $v_0 - v_t$ represents $x$. There are three ways of using the results to show that the reaction is first-order. The first two yield the value of the rate constant, $k$. To illustrate the three methods a table is constructed as now shown (the values of $x$ are used only in the third method).

**Table 15-1**

| $t$/min | $a$ ($v_0$/cm$^3$) | $a - x$ ($v_t$/cm$^3$) | $x$ (($v_0 - v_t$)/cm$^3$) | $\log \dfrac{a}{a-x}$ | $\dfrac{k/\text{min}^{-1}}{\dfrac{2 \cdot 303}{t} \log \dfrac{a}{a-x}}$ |
|---|---|---|---|---|---|
| 0 | 27·3 | — | — | — | — |
| 20 | 27·3 | 17·0 | 10·3 | 0·2041 | $2\cdot34 \times 10^{-2}$ |
| 40 | 27·3 | 10·3 | 17·0 | 0·4232 | $2\cdot43 \times 10^{-2}$ |
| 60 | 27·3 | 6·3 | 21·0 | 0·6368 | $2\cdot44 \times 10^{-2}$ |
| 80 | 27·3 | 3·8 | 23·5 | 0·8564 | $2\cdot43 \times 10^{-2}$ |
| 100 | 27·3 | 2·3 | 25·0 | 1·0745 | $2\cdot47 \times 10^{-2}$ |

(i) *Calculation of the Rate Constant.* The various values obtained for $k$ as shown in the last column are nearly constant, indicating that the reaction is first-order. The average value for $k$ (0·0242 min$^{-1}$) shows that about one fortieth of the hydrogen peroxide molecules decompose per minute.

(ii) *Graphical Evaluation of the Rate Constant.* The integrated rate expression for $k$ can be rearranged as follows:

$$\frac{\log\{a/(a - x)\}}{t} = \frac{k}{2\cdot303} = \text{a constant}$$

Hence the reaction can be shown to be first-order by plotting $\log\{a/(a - x)\}$ against $t$, when a straight-line graph should be obtained. The slope of the line is then equal to $k/2\cdot303$, and we find $k$ by multiplying the slope by $2\cdot303$.

Fig. 15–5 shows the graph obtained from the data given in the table. A good straight-line plot (AC) results.

$$\text{Slope of AC} = \frac{AB}{BC} = \frac{1 \cdot 05}{100} = 0 \cdot 0105$$

$$k = 0 \cdot 0105 \times 2 \cdot 303$$

$$= 2 \cdot 42 \times 10^{-2} \, min^{-1}$$

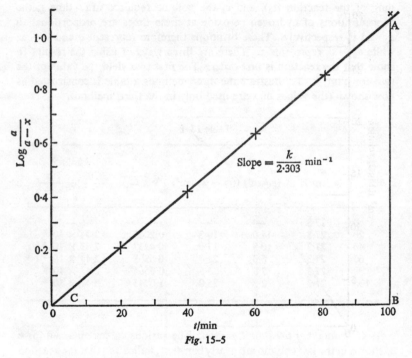

*Fig.* 15–5

We can compare the efficiencies of different catalysts by the different rate constants obtained with the respective catalysts.

(iii) *Half-period Method.* The times, $t'$, required for the initial concentration of hydrogen peroxide to be reduced to one half, from one half to one quarter, and from one quarter to one eighth are found. For a first-order reaction these times should be equal. We find the times graphically by plotting $x$ (the number of mole of hydrogen peroxide decomposed) against $t$. The data in the table yield the curve in Fig. 15–6. The times specified are approximately equal.

*Hydrolysis of an Ester.*

$$HCOOCH_3 + H_2O \rightarrow HCOOH + CH_3OH$$

Hydrolysis of methyl methanoate is really a balanced reaction, but goes practically to completion in the presence of a large excess of water. The reaction is catalysed by mineral acids. Since the rate of increase in concentration of each of the products is approximately proportional to the hydrogen ion concentration furnished by an acid, we can use the reaction to compare the 'strengths' of acids at equal concentrations. Weak organic acids provide too small a hydrogen ion concentration to make any appreciable difference to the rate constant.

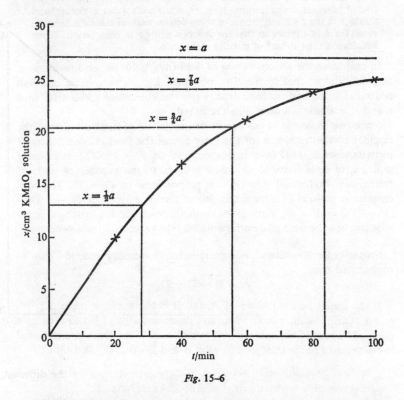

*Fig.* 15–6

In general the acid-catalysed hydrolysis of an ester is kinetically a third-order reaction, the rate of increase in concentration of the products being proportional to [ester][$H_2O$][$H^+$(aq)]. With a large excess of water and acid, however, the last two concentrations remain practically constant, so that the rate is proportional only to [ester]. The reaction is then first-order.

Methyl methanoate is more convenient to use than ethyl ethanoate because it hydrolyses more rapidly. With the former the reaction can be carried out at room temperature and a thermostat is not required.

**Experiment.** Put 10 cm$^3$ of methyl methanoate (which has been newly redistilled) into a conical flask. Add 200 cm$^3$ of approximately 0·5 molar HCl solution. Stopper the flask and swirl the liquids round to mix them. At once withdraw 10 cm$^3$ of the mixture and run it into about 100 cm$^3$ water. This slows down the reaction so much that it practically ceases. Before titrating the mixture clamp the reaction flask in a large trough of water at room temperature and start a stop-clock. Now titrate the mixture with approximately 0·5 molar NaOH solution, using phenolphthalein as indicator.

At intervals of 10 minutes after starting the reaction withdraw further 10-cm$^3$ portions of the reaction mixture, dilute with water as before, and titrate with the alkali. After six or seven determinations leave the reaction vessel for a few hours so that the reaction can go to completion. Then withdraw a final 10 cm$^3$ of mixture and titrate.

In each case the concentration of acid (hydrochloric, and methanoic, acid) is proportional to the titre obtained. Let $V_0$ = volume of alkali neutralizing the hydrochloric acid, $V_1$ = the volume of alkali after time $t$, and $V$ = final volume of alkali required.

Since one molecule of ester gives one molecule of methanoic acid, the original concentration, $a$, of the ester equals the final concentration of methanoic acid—that is, $a$ is proportional to $V - V_0$. The number of mole, $x$, of ester changed at time $t$ is equal to the number of mole of methanoic acid formed—that is, $x$ is proportional to $V_1 - V_0$. The concentration $(a - x)$ of ester remaining at time $t$ is given by $(V - V_0) - (V_1 - V_0) = V - V_1$. Any of the methods described for hydrogen peroxide can now be used to demonstrate that the reaction is first-order.

**Second-order Reactions.** A second-order elementary reaction can be represented thus:

$$A + B \rightarrow C + D$$

If the initial concentration of A and B is the same, $a$ mole per m$^3$, and if at any instant $x$ mole of A have disappeared, $x$ of B will also have gone. The rate of increase in concentration of C (or D) is proportional to the product of the concentrations of A and B—that is, after time $t$

$$\frac{dx}{dt} = k(a - x)^2$$

By integration of this expression we obtain

$$k = \frac{1}{t} \cdot \frac{x}{a(a - x)}$$

The units in which $k$ should be expressed can be deduced from the last equation. We have

$$\text{Units of } k = \frac{\text{concentration}}{(\text{concentration})^2 \times \text{time}} = \frac{1}{\text{concentration} \times \text{time}}$$

$$= \text{concentration}^{-1} \text{ time}^{-1}$$

It will be seen that if the concentration is in mol m$^{-3}$ and the time is in seconds, the units of $k$ are $(mol\ m^{-3})^{-1}\ s^{-1}$ or $m^3\ mol^{-1}\ s^{-1}$.

In the case of a second-order reaction the time, $t'$, taken for the reaction to be half completed depends upon the initial concentration, for if $x = a/2$

$$t' = \frac{1}{k} \cdot \frac{a/2}{a \times (a/2)}$$

$$= \frac{1}{ka}$$

The time taken for the reaction to be half completed is thus inversely proportional to the initial concentration, or

$$a \times t' = \frac{1}{k} = \text{a constant}$$

The three general methods described for first-order reactions can also be used to show that a reaction is second-order.

(1) When $x$ has been found for various values of $t$ a series of values for $k$ can be calculated. These values should agree within the limits of experimental error.

(2) In the case of a second-order reaction we have

$$\frac{x/a(a - x)}{t} = k$$

Therefore a straight-line graph should be obtained by plotting $x/a(a - x)$ against $t$. The value of $k$ is given directly by the slope of the straight line.

(3) The half-period method is applied by plotting $x$ against $t$ as described for first-order reactions. In this case $a \times t'$ should be constant.

When the initial concentrations of A and B are not the same, but $a$ and $b$ mol m$^{-3}$ respectively, the rate of change is expressed by

$$\frac{dx}{dt} = k(a - x)\,(b - x)$$

This expression can also be integrated to yield an equation which gives the value of $k$ in terms of $a$, $b$, $x$, and $t$.

*Saponification of an Ester.*

$$CH_3COOC_2H_5 + OH^- \rightarrow CH_3COO^- + C_2H_5OH$$

Although hydrolysis of ethyl ethanoate by aqueous sodium(I) hydroxide is not an elementary reaction, it is second-order unless the alkali is present in large excess, when it becomes first-order. An ester is decomposed more rapidly in alkaline than in acid solution, and therefore the solutions used are more dilute than those described at p. 378.

Equal volumes of 0·025 molar NaOH solution and 0·025 molar $CH_3COOC_2H_5$ solution are mixed and kept in a thermostat at 25°C. The change in alkali concentration is followed by withdrawing portions of the mixture at known intervals of time and adding them to an excess of standard hydrochloric acid. The concentration of alkali in the mixture is found by back-titrating the excess of acid with standard alkali. Any of the methods previously described can be used to show that the reaction is second-order. This experiment is used to compare the 'strengths' of alkalis (p. 466).

**Third-order Reactions.** Third-order reactions are comparatively rare. In the gas phase examples are limited to a few reactions of nitrogen oxide; *e.g.*,

$$2NO + H_2 \rightarrow N_2O + H_2O$$

The rate of increase in concentration of dinitrogen oxide is given by

$$d[N_2O]/dt = k(a - x)^3$$

when the initial concentrations, $a$, are the same. It can be shown that the time, $t'$, required for the reaction to be half completed in this case is given by

$$t' = \frac{3}{2ka^2}$$

That is, $t'$ is inversely proportional to the square of the initial concentration, or $a^2 \times t'$ is a constant.

## Determination of Equilibrium Constants

### Liquid Systems. Esterification of Ethanol

$$\underset{(a-x)/v}{CH_3COOH} + \underset{(b-x)/v}{C_2H_5OH} \rightleftharpoons \underset{x/v}{CH_3COOC_2H_5} + \underset{x/v}{H_2O}$$

By the law of mass action the equilibrium constant $K_c$ is given by

$$K_c = \frac{[CH_3COOC_2H_5] \, [H_2O]}{[CH_3COOH] \, [C_2H_5OH]}$$

Suppose that $a$ mole of ethanoic acid are mixed with $b$ mole of ethanol, and the total volume of the mixture is $v$ (in $m^3$). One mole of acid combines with one mole of alcohol, and if at equilibrium $x$ mole of each have reacted the concentrations in $mol \, m^{-3}$ of acid, alcohol, ethyl ethanoate, and water will be

$$(a - x)/v, \quad (b - x)/v, \quad x/v, \quad \text{and} \quad x/v$$

respectively. Then

$$K_c = \frac{(x/v) \, (x/v)}{\{(a - x)/v\}\{(b - x)/v\}} = \frac{x^2}{(a - x) \, (b - x)}$$

Hence we can determine the equilibrium constant from the composition of the final equilibrium mixture. In general the values of equilibrium constants, like those of rate constants, vary with the temperature. Rate constants, however, always increase with rise of temperature, whereas equilibrium constants may increase or decrease with rise of temperature according to whether the forward reaction is endothermic or exothermic. Again, rate constants may be increased at a fixed temperature by employing a catalyst, but equilibrium constants are not affected by catalysts. A large value for $K_c$ signifies that most of the reactants have been converted to products, while a small fractional value means that relatively small amounts of products are present in the equilibrium mixture. The units of $K_c$ vary with the nature of the reaction. From the expression given for $K_c$ in the esterification of ethanol we see that here the equilibrium constant has no units.

To determine the equilibrium constant for esterification of ethanol by ethanoic acid, known amounts of the alcohol and acid are put into a tube, which is then sealed. At ordinary temperatures the reaction is very slow, but at 60° to 70°C the system comes to equilibrium in a few hours. As the reaction is almost thermally neutral, raising the temperature to 60° or 70°C has no appreciable effect on the position of equilibrium. In fact, the French chemist Berthelot found that with the same number of mole of both acid and alcohol the percentage conversion to ester increased only from 65·2 per cent at 10°C to 66·5 per cent at 220°C. The sealed tube is therefore left in a water bath at this temperature for 7 or 8 hours. The tube is broken in cold water and remaining ethanoic acid estimated by titration with standard sodium(I) hydroxide solution with phenolphthalein as indicator. By using the value of $x$ found, $K_c$ can be calculated as indicated above. Further determinations are then made with different proportions of acid and alcohol to show that $K_c$ is approximately constant.

A classical series of experiments on the esterification of ethanol by ethanoic acid at 100°C was performed by Berthelot in 1862. He found that if 1 mole of ethanoic acid was mixed with 1 mole of ethanol approximately two thirds of acid and alcohol were converted into ester and water at equilibrium. Using these values in the above expression, we see that

$$K_c = \frac{(\frac{2}{3})^2}{(1 - \frac{2}{3})(1 - \frac{2}{3})} = 4$$

When the equilibrium constant for the esterification has been determined it can be used to calculate the amount of ester which should be formed from given masses of alcohol and acid. The method is illustrated in the example now given.

**Example.** (i) *When 8·28 g of ethanol were heated with 60 g of ethanoic acid 49·74 g of the acid remained at equilibrium (Berthelot). Calculate the value of $K_c$.*

(ii) *What mass of ethyl ethanoate should be present in the equilibrium mixture formed from* 13·8 g *of ethanol and* 12 g *of ethanoic acid?* ($H = 1$, $C = 12$, $O = 16$.)

(i) Relative molecular mass of ethanol = 46
Initial mass of ethanol = 8·28 g

$$= \frac{8 \cdot 28}{46} \text{ mol} = 0 \cdot 18 \text{ mol}$$

Relative molecular mass of ethanoic acid = 60
Initial mass of ethanoic acid = 60 g = 1 mol
Final mass of ethanoic acid = 49·74 g

$$= \frac{49 \cdot 74}{60} \text{ mol} = 0 \cdot 829 \text{ mol}$$

Number of mole of ethanoic acid which have reacted = $x$
$= 1 - 0 \cdot 829 = 0 \cdot 171$

$$K_c = \frac{0 \cdot 171^2}{0 \cdot 829 \, (0 \cdot 18 - 0 \cdot 171)} = 3 \cdot 92$$

(ii) 13·8 g of ethanol $= \dfrac{13 \cdot 8}{46} \text{ mol} = 0 \cdot 3 \text{ mol}$

12 g of ethanoic acid $= \dfrac{12}{60} \text{ mol} = 0 \cdot 2 \text{ mol}$

Number of mole of ethyl ethanoate formed = the number of mole ($x$) of acid (or alcohol) which disappear.

Using the value found for the equilibrium constant in (i) we have

$$3 \cdot 92 = \frac{x^2}{(0 \cdot 2 - x) \, (0 \cdot 3 - x)}$$

or,   $2 \cdot 92 x^2 - 1 \cdot 96 x + 0 \cdot 235 = 0$

The values of $x$ in the general quadratic equation

$$a x^2 + b x + c = 0$$

are given by

$$x = \frac{-b \pm \sqrt{(b^2 - 4ac)}}{2a}$$

By substitution we obtain $x = 0 \cdot 514$ or $0 \cdot 158$. The first of these values is inadmissible, since the number of mole of alcohol or acid which react cannot exceed the initial amounts. Hence $x = 0 \cdot 158$ mol.

Amount of ethyl ethanoate formed $= 0.158$ mol

Mass of ethyl ethanoate formed $= 0.158 \times 88$ g $= 13.9$ g

**Gaseous Systems.** The equilibrium constant in a gaseous system can be expressed in two ways. If it is obtained from the concentrations of the substances in mole per m$^3$ it is called the *concentration equilibrium constant* ($K_c$). It is often more convenient, however, to derive the equilibrium constant from the partial pressures of the gases present, and in this case it is called the *pressure equilibrium constant* ($K_p$). The two equilibrium constants may, or may not, be equal. This depends on whether the reaction takes place without, or with, a change in the total number of molecules.

**1. Number of Molecules Remains Constant. The Hydrogen Iodide Equilibrium**

(i) $K_c$.
$$2HI \rightleftharpoons H_2 + I_2$$
$$\underset{(a-x)/v \quad x/2v \quad x/2v}{}$$

Suppose we start with $a$ mole of hydrogen iodide and at equilibrium $x$ mole have dissociated. Since 1 mole on dissociation furnishes $\frac{1}{2}$ mole of hydrogen and $\frac{1}{2}$ mole of iodine vapour, we shall have at equilibrium $x/2$ mole of both hydrogen and iodine. If $v$ is the volume, the concentrations of hydrogen iodide, hydrogen, and iodine will be $(a - x)/v$, $x/2v$, and $x/2v$ respectively. Then

$$K_c = \frac{[H_2]\,[I_2]}{[HI]^2} = \frac{(x/2v)\,(x/2v)}{\{(a - x)/v\}^2}$$

$$= \frac{x^2}{4(a - x)^2}$$

The equilibrium point is thus independent of the volume and therefore of the pressure (as can be predicted from Le Chatelier's principle). Again $K_c$ has no units.

The degree of dissociation of the hydrogen iodide is the fraction of the original molecules which have undergone dissociation when equilibrium is reached. In the above case the degree of dissociation is $x/a$. If $a = 1$ (that is, if we start with 1 mole of hydrogen iodide), the degree of dissociation is represented by $\alpha$. Thus if we wish to express $K_c$ in terms of the degree of dissociation we can write for the equilibrium

$$2HI \rightleftharpoons H_2 + I_2$$
$$\underset{(1-\alpha)/v \quad \alpha/2v \quad \alpha/2v}{}$$

$$\therefore \quad K_c = \frac{\alpha^2}{4(1 - \alpha)^2}$$

The hydrogen iodide equilibrium can also be approached from the opposite direction, that is, by heating hydrogen and iodine together.

$$H_2 + I_2 \rightleftharpoons 2HI$$
$$\frac{a-x}{v} \quad \frac{b-x}{v} \quad \frac{2x}{v}$$

If we start with $a$ mole of hydrogen and $b$ mole of iodine vapour in a total volume of $v$ (in m³), and if at equilibrium $x$ mole of hydrogen and iodine have disappeared, the concentrations of the substances present at equilibrium are as shown above since the volume remains constant. Then by the law of mass action

$$K_c = \frac{[HI]^2}{[H_2][I_2]} = \frac{4x^2}{(a-x)(b-x)}$$

The value for the equilibrium constant given by this equation is of course the reciprocal of that for the dissociation of hydrogen iodide at the same temperature.

(ii) $K_p$. Calculation of $K_p$ for the combination of hydrogen and iodine is carried out as follows. At equilibrium there are present $a - x$ mole of hydrogen, $b - x$ mole of iodine, and $2x$ mole of hydrogen iodide. The total number of mole present is $(a - x) + (b - x) + 2x$, which equals $a + b$. Then, if $p$ is the total pressure at equilibrium,

$$\text{Partial pressure of hydrogen} = p_1 = \frac{a-x}{a+b}P$$

$$\text{Partial pressure of iodine} = p_2 = \frac{b-x}{a+b}P$$

$$\text{Partial pressure of hydrogen iodide} = p_3 = \frac{2x}{a+b}P$$

Since the partial pressures of the gases are proportional to their molar concentrations we can write

$$K_p = \frac{p_3^2}{p_1 p_2} = \frac{\left(\dfrac{2x}{a+b}P\right)^2}{\left(\dfrac{a-x}{a+b}P\right)\left(\dfrac{b-x}{a+b}P\right)}$$

$$= \frac{4x^2}{(a-x)(b-x)}$$

Thus for this reaction (and all others in which the number of molecules remains the same) $K_p$ is equal to $K_c$.

The dissociation of hydrogen iodide and the combination of hydrogen and iodine were investigated by Bodenstein in 1897. Known amounts of hydrogen and iodine were sealed in a glass bulb and kept in a thermostat until equilibrium was reached. The bulb was then removed and was rapidly cooled to 'freeze' the equilibrium (the rate of reaction is very slow at ordinary temperatures). The iodine present in the equilibrium mixture was estimated by breaking the bulb below excess of standard sodium(I) thiosulphate(VI) solution and back-titrating the excess with standard iodine solution.

**Example.** *When 6·22 $cm^3$ of hydrogen were heated with 5·71 $cm^3$ of iodine in a sealed tube at 356°C it was found that 9·60 $cm^3$ of hydrogen iodide were present at equilibrium (Bodenstein). Calculate* (i) *the equilibrium constant,* (ii) *the volume of hydrogen iodide in the equilibrium mixture formed by heating together 6·41 $cm^3$ of hydrogen and 10·40 $cm^3$ of iodine at 356°C.*

(i) Since the number of mole of each gas present is proportional to the volume, we have at equilibrium

$$\text{Number of mole of HI} = 2x = 9\cdot60$$
$$\text{Number of mole of } H_2 = a - x = 6\cdot22 - 4\cdot80$$
$$= 1\cdot42$$
$$\text{Number of mole of } I_2 = b - x = 5\cdot71 - 4\cdot80$$
$$= 0\cdot91$$

$$K_c = \frac{4x^2}{(a - x)(b - x)} = \frac{4 \times (4\cdot80)^2}{1\cdot42 \times 0\cdot91}$$
$$= 71\cdot32$$

(ii) As in part (i) the number of mole of HI $= 2x$. Using the value of $K_c$ found above we have

$$71\cdot32 = \frac{4x^2}{(6\cdot41 - x)(10\cdot40 - x)}$$
$$\text{or} \quad x^2 - 17\cdot81x + 70\cdot63 = 0$$

By substitution in the general quadratic equation given at p. 382 we obtain $x = 11\cdot85$ or $5\cdot95$, so that $2x = 23\cdot7$ or $11\cdot9$.

The first of these two values for $2x$ is inadmissible, since the volume of hydrogen iodide cannot be greater than the original volume of hydrogen and iodine together.

Therefore the volume of hydrogen iodide present at equilibrium $= 11\cdot9$ $cm^3$.

**Carbon Monoxide–Steam Equilibrium.** This equilibrium is concerned in the manufacture of hydrogen from water gas and steam by the Bosch process. We can calculate the percentage composition of the system at equilibrium if we know the proportions of water gas and steam and the equilibrium constant of the following action at 450°C:

$$\underset{a-x}{CO} + \underset{b-x}{H_2O} \rightleftharpoons \underset{x}{CO_2} + \underset{x}{H_2}$$

$$\text{At } 450°C \quad K_c = \frac{[CO_2][H_2]}{[CO][H_2O]} = 7\cdot3$$

The number of mole of each gas is proportional to its volume. Suppose we have 2 volumes of water gas (containing 1 volume each of carbon monoxide and hydrogen) mixed with 5 volumes of steam. Let $x$ be the volume of carbon monoxide converted into carbon dioxide. Now $x$ volumes of carbon monoxide combine with $x$ volumes of steam to give $x$ volumes of carbon dioxide and $x$ volumes of hydrogen. Substituting in the expression for the equilibrium constant, we have at equilibrium

$$\frac{x(1+x)}{(1-x)(5-x)} = 7\cdot3$$

This gives us a quadratic equation, from which the value of $x$ can be calculated in the usual way. The value of $x$ is 0·94. The volume ratios of the gases present in the equilibrium mixture are therefore:

$$CO_2 \quad 0\cdot94 \qquad CO \quad 0\cdot06$$
$$H_2 \quad 1\cdot94 \qquad H_2O \quad 4\cdot06$$

If the steam is condensed the approximate percentages of carbon dioxide, hydrogen, and carbon monoxide in the final mixture of gases are obtained as follows:

$$\text{Total number of volumes} = 0\cdot94 + 0\cdot06 + 1\cdot94 = 2\cdot94$$

$$\text{Percentage } CO_2 = \frac{0\cdot94}{2\cdot94} \times 100 = 32 \text{ per cent}$$

$$\text{Percentage } H_2 = \frac{1\cdot94}{2\cdot94} \times 100 = 66 \text{ per cent}$$

$$\text{Percentage } CO = \frac{0\cdot06}{2\cdot94} \times 100 = 2 \text{ per cent}$$

**2. Number of Molecules Increases. The Dissociation of Phosphorus Pentachloride.** First we shall derive $K_c$ in terms of the degree of dissociation and the volume at equilibrium.

$$\underset{(1-\alpha)/v}{PCl_5} \rightleftharpoons \underset{\alpha/v}{PCl_3} + \underset{\alpha/v}{Cl_2}$$

According to the law of mass action we have

$$K_c = \frac{[PCl_3][Cl_2]}{[PCl_5]}$$

If we start with 1 mole of phosphorus pentachloride and the degree of dissociation is $\alpha$, we shall have formed $\alpha$ mole of both phosphorus trichloride and chlorine. If the volume at equilibrium is $v$ (in m³), the concentrations of the three substances will be $(1 - \alpha)/v$, $\alpha/v$, and $\alpha/v$ respectively. Then

$$K_c = \frac{(\alpha/v)\,(\alpha/v)}{(1 - \alpha)/v} = \frac{\alpha^2}{(1 - \alpha)v}$$

Thus the degree of dissociation at a given temperature varies with the volume and therefore with the pressure. The equation shows that $K_c$ is expressed in mol m⁻³.

Starting again with 1 mole of phosphorus pentachloride, we can derive $K_p$ in terms of the degree of dissociation, $\alpha$, and the total pressure, $p$, in N m⁻² at equilibrium as now shown.

At equilibrium total number of mole $= (1 - \alpha) + \alpha + \alpha = 1 + \alpha$.

Partial pressure $(p_1)$ of $PCl_5 = \dfrac{\text{number of mole of } PCl_5}{\text{total number of mole}} \times p$

$$= \frac{1 - \alpha}{1 + \alpha} p$$

Partial pressure $(p_2)$ of $PCl_3 = $ partial pressure of $Cl_2$

$$= \frac{\alpha}{1 + \alpha} p$$

$$K_p = \frac{p_2^2}{p_1} = \frac{\left(\dfrac{\alpha}{1 + \alpha} p\right)^2}{\dfrac{1 - \alpha}{1 + \alpha} p}$$

$$= \frac{\alpha^2 p}{(1 - \alpha)(1 + \alpha)} = \frac{\alpha^2 p}{1 - \alpha^2}$$

The relation between $K_c$ and $K_p$ can be obtained by dividing $K_p$ by $K_c$. We have

$$\frac{K_p}{K_c} = \frac{\alpha^2 p}{(1 - \alpha)(1 + \alpha)} \div \frac{\alpha^2}{(1 - \alpha)v} = \frac{pv}{(1 + \alpha)}$$

Now $1 + \alpha$ is the total number of mole at equilibrium. If $p$ is in Pa and $v$ is taken to be the volume in m³ occupied by one mole at the kelvin temperature $T$, $1 + \alpha$ corresponds to $n$ in the general gas equation $pv = nRT$, or $pv/n = RT$. Hence

$$\frac{K_p}{K_c} = RT \quad \text{or} \quad K_p = K_c RT,$$

where $R$, the gas constant, has the value $8\cdot31$ J mol$^{-1}$ K$^{-1}$. The above relation between $K_p$ and $K_c$ holds for all gaseous equilibria in which the number of molecules on the right of the equation is one more than on the left, *e.g.*,

$$COCl_2 \rightleftharpoons CO + Cl_2$$
$$N_2O_4 \rightleftharpoons 2NO_2$$

In practice the degree of dissociation, $\alpha$, in equilibria of this type can be found by relative density measurement (see Chapter 3).

**Dissociation of Dinitrogen Tetraoxide**

$$\underset{(1-\alpha)/v}{N_2O_4} \rightleftharpoons \underset{2\alpha/v}{2NO_2}$$

The expressions for $K_c$ and $K_p$ differ slightly from those obtained for the dissociation of phosphorus pentachloride. Verify that the two constants are given by

$$K_c = \frac{4\alpha^2}{(1-\alpha)v} \quad \text{and} \quad K_p = \frac{4\alpha^2 p}{1-\alpha^2}$$

**3. Number of Molecules Decreases. Combination of Nitrogen and Hydrogen.** One mole of nitrogen combines with 3 mole of hydrogen, and in the Haber process the gases are used in this proportion. If $x$ mole of nitrogen disappear and $v$ is the volume at equilibrium, the concentrations of the gases at equilibrium are as follows:

$$\underset{\frac{1-x}{v}}{N_2} + \underset{\frac{3(1-x)}{v}}{3H_2} \rightleftharpoons \underset{\frac{2x}{v}}{2NH_3}$$

By applying the law of mass action to the equilibrium we find that

$$K_c = \frac{4x^2 v^2}{9(1-x)^4}$$

Similarly, if $p$ is the total pressure at equilibrium, we can deduce from the partial pressures of nitrogen, hydrogen, and ammonia that

$$K_p = \frac{4x^2(4-2x)^2}{27(1-x)^4 p^2}$$

If $x$ is much less than unity we can simplify this to

$$K_p = \frac{64x^2}{27p^2}$$

Here $x$ is proportional to $p$. This is approximately true over a limited range of temperature and pressure. Thus at 750°C the equilibrium percentage of ammonia at 100 atm pressure is $1\cdot54$, while at 200 atm pressure it is $2\cdot99$.

## EXERCISE 15
(*Relative atomic masses are given at the end of the book*)

**1.** Describe how (*a*) the position, (*b*) the rate of attainment of a chemical equilibrium are influenced by the experimental conditions.

Illustrate your answer by reference to reactions between the following pairs of substances: (i) oxygen and sulphur dioxide, (ii) hydrogen and iodine, (iii) steam and iron. (C.L.)

**2.** Discuss *concisely* the effect of alterations in (i) concentration, (ii) pressure, (iii) temperature, on the following exothermic reaction.

$$N_2 + 3H_2 \rightleftharpoons 2NH_3$$

What other considerations have to be studied to make the reaction an economic proposition? (Lond.)

**3.** What do you understand by catalysis? Name and discuss *four* features which are characteristic of catalytic reactions. Illustrate your answer by reference to reactions of industrial importance, *other than the manufacture of sulphuric (VI) acid.* (Lond.)

**4.** (Part question.) Define the following terms, quoting *one* example of each: (*a*) homogeneous catalysis, (*b*) heterogeneous catalysis, (*c*) catalyst poison, (*d*) autocatalysis.

What influence has a catalyst on the composition of the final equilibrium mixture of a reaction?

Mentioning essential experimental details, describe concisely how you would find out whether a given mineral acid acted as a catalyst for the hydrolysis of ethyl ethanoate by water. (J.M.B.)

### Chemical Kinetics

**5.** A certain volume of hydrogen peroxide solution was decomposed in the presence of platinum. The amount of hydrogen peroxide remaining after time $t$ was found by withdrawing aliquot portions of the solution, adding dilute sulphuric(VI) acid, and titrating with potassium(I) manganate(VII) solution. The volumes, $v$, of the latter required were as follows:

| $t$/min | 0 | 5 | 10 | 15 | 20 |
|---|---|---|---|---|---|
| $v$/cm$^3$ | 12·3 | 9·2 | 6·9 | 5·2 | 3·9 |

Show graphically that the decomposition of hydrogen peroxide is a first-order reaction.

**6.** A certain amount of methyl ethanoate was hydrolysed in the presence of excess of 0·05 molar HCl solution at 25°C. When 25-cm$^3$ portions of the reaction mixture were removed and titrated with sodium(I) hydroxide solution after time $t$ the volumes, $v$, of alkali required for neutralization were as follows:

| $t$/min | 0 | 20 | 75 | 119 | ∞ |
|---|---|---|---|---|---|
| $v$/cm$^3$ | 24·4 | 25·8 | 29·3 | 31·7 | 47·2 |

Show that the reaction is a first-order reaction.

**7.** Aqueous solutions of ethyl ethanoate ($CH_3COOC_2H_5$) and potassium(I) hydroxide(KOH) in equimolar proportions were mixed and kept in a thermostat at 25°C. At time $t_1$ after the start of the reaction the concentration of alkali left was 0·0128 molar and the rate of increase in concentration of ethanol was 0·0025 mol dm$^{-3}$ min$^{-1}$. At time $t_2$ the concentration of alkali was 0·0083 molar. Find the rate of increase in concentration of ethanol at time $t_2$.

**8.** The disintegration (rate) constant for the conversion of radioactive phosphorus into silicon is $3·85 \times 10^{-3}$, the time being measured in seconds. What fraction of the original material will be left after 180 seconds?

9. Describe how the rate of decrease in concentration of methyl ethanoate brought about by sodium(I) hydroxide solution can be determined experimentally. How is the *rate constant* obtained from these results?

Distinguish between *order* and *molecularity* in a chemical reaction.

Comment on the *order* and *molecularity* of the hydrolysis of methyl ethanoate by sodium(I) hydroxide solution (*a*) with equimolar concentrations of sodium(I) hydroxide and methyl ethanoate; (*b*) with a fifty-fold excess of sodium(I) hydroxide. (O. and C.)

## Chemical Equilibrium

10. (Part question.) 0·196 g nitrogen and 0·146 g hydrogen are heated together in a closed vessel, provided with a manometer and containing a suitable catalyst, until equilibrium is established, the temperature being kept constant throughout. At the end of the experiment the pressure in the vessel is found to be 90 per cent of its value at the beginning. Calculate the percentage composition *by volume* of the resulting gas mixture. H = 1; N = 14. (J.M.B.)

11. Explain what is meant by the terms *active mass*, *rate constant*, and *equilibrium constant* in their application to chemical reactions and show how they are related to each other. Illustrate your answer by reference to the reaction of ethanol with acetic acid.

Ethanol (18·4 g) and ethanoic acid (12·0 g) were mixed and heated at 60°C until equilibrium was attained. At this stage 1·86 g of ethanoic acid remained unchanged. Calculate the equilibrium constant for the reaction at 60°C.

Give *one* example (excluding the above-mentioned substances) of a reversible reaction which occurs in aqueous solution and indicate briefly how its reversibility may be demonstrated qualitatively. (W.J.E.C.)

12. When equimolar proportions of hydrogen and iodine were heated together at a certain temperature the system at equilibrium was found to contain 0·0017 mol dm$^{-3}$ of hydrogen ($H_2$), 0·0017 mol dm$^{-3}$ of iodine ($I_2$) and 0·0118 mol dm$^{-3}$ of hydrogen iodide (HI). Calculate the equilibrium constant for the reaction $H_2 + I_2 \rightleftharpoons 2HI$ at this temperature.

13. Discuss briefly the effect of changes of temperature, pressure, and concentration of reactants on the following reversible reactions (*a*) and (*b*). You may assume that both reactions are exothermic in the forward direction.

(*a*) $2HI \rightleftharpoons H_2 + I_2$.

(*b*) $N_2 + 3H_2 \rightleftharpoons 2NH_3$.

1 mole of hydrogen and $\frac{1}{4}$ mole of iodine are heated together at 450°C. Given that the equilibrium constant for (*a*) is 0·02, calculate the number of mole of hydrogen iodide present in the equilibrium mixture at that temperature. (C.L.)

14. At 700°C the equilibrium constant in the reaction

$$CO + H_2O \rightleftharpoons CO_2 + H_2$$

is 1·4. Calculate the percentage by volume of gases in the equilibrium mixture at this temperature when carbon monoxide and steam are mixed in the following proportions: (i) equal volumes, (ii) one volume of carbon monoxide to two volumes of steam.

15. Explain what is meant by the terms 'rate constant' and 'equilibrium constant.'

Nitrogen and oxygen combine at high temperatures, with absorption of heat, according to the equation:

$$N_2 + O_2 \rightleftharpoons 2NO$$

The equilibrium constant for this reaction, at 2680 K and 1 atm total pressure, is 3·6 × 10$^{-3}$.

Equal volumes of nitrogen and oxygen are mixed, at 2680 K and 1 atm total pressure, and allowed to react until equilibrium is reached. Calculate the fraction of the original nitrogen which is used in the reaction, and the fraction (by volume) of nitrogen oxide (NO) in the equilibrium mixture.

Is this yield increased, decreased, or unchanged when (a) the pressure is increased to 10 atm, (b) the temperature is raised to 2780 K, (c) a catalyst is added? Give reasons.                                                                           (J.M.B.)

## MORE DIFFICULT QUESTIONS

**16.** What do you understand by the *activation energy* of a chemical reaction and how is it determined?

Explain how a knowledge of the activation energies of the forward and reverse reactions enables one to predict the effect of *temperature* on (a) the rate constants of and (b) the *position of equilibrium* of

$$C_2H_4 + H_2 \rightleftharpoons C_2H_6$$

(activation energy of forward reaction $= 172$ kJ mol$^{-1}$
activation energy of reverse reaction $= 305$ kJ mol$^{-1}$).                    (O. and C.)

**17.** In acid solution, bromate(V) ions slowly oxidize bromide ions to bromine.

$$BrO_3^- + 5Br^- + 6H^+ \rightarrow 3Br_2 + 3H_2O$$

Use the following experimental data to determine how the initial rate of this reaction depends on the concentrations of bromate(V), bromide, and hydrogen ions. Comment on your results (a) in relation to the equation, (b) in relation to the law of mass action.

| Mixture | Molar $BrO_3^-$ (cm$^3$) | Molar $Br^-$ (cm$^3$) | Molar $H^+$(aq) (cm$^3$) | Water (cm$^3$) | Relative rate of increase in [$Br_2$] |
|---|---|---|---|---|---|
| A | 100 | 500 | 600 | 800 | 1 |
| B | 50 | 250 | 600 | 100 | 4 |
| C | 100 | 250 | 600 | 50 | 8 |
| D | 50 | 125 | 600 | 225 | 2 |

(C.L.)

**18.** (Part question.) A small sample of *gold* was irradiated in a nuclear reactor and the activity of the *radioactive gold* produced was measured with a Geiger counter at various intervals of time. The same experiment was carried out with a small sample of *bromine* and the results obtained are given below.

*Time in hours*
0   0·1  0·2  0·5   1    2    5   10   24   48   72   96

*Disintegrations per minute*

| | | | | | | | | | | | | |
|---|---|---|---|---|---|---|---|---|---|---|---|---|
| Radioactive gold | 299 | — | — | — | — | — | 285 | 271 | 232 | 181 | 139 | 108 |
| Radioactive bromine | 901 | 795 | 719 | 577 | 482 | 435 | 405 | 367 | 279 | 178 | 112 | 71 |

(a) From a suitable graph deduce the half-life of *radioactive gold*.
(b) From a similar graph comment on the decay of *radioactive bromine*.
                                                                           (O. and C.)

**19.** At 274°C and a total pressure of 1 atm phosphorus pentachloride vapour is 87·4 per cent dissociated into phosphorus trichloride and chlorine. What is the value of $K_p$? Calculate the degree of dissociation at the same temperature at a total pressure of 2 atm and find the partial pressures of the three vapours.

**20.** (Part question.) At 1500°C and 101 300 Pa pressure carbon dioxide is 40 per cent dissociated into carbon monoxide and oxygen. Calculate the equilibrium constant for this dissociation in terms of partial pressure. (J.M.B.)

**21.** (Part question.) At 40°C and under a pressure of 1 atm dinitrogen tetraoxide contains 60 per cent by volume of $NO_2$ molecules. Calculate the percentage dissociation of $N_2O_4$ and the equilibrium constant in terms of the partial pressures of the two gases. Deduce (i) the percentage dissociation at the same temperature when the gases are under a pressure of 6 atm, (ii) the pressure at which dinitrogen tetraoxide would be dissociated to the extent of 80 per cent, at this temperature.                                           (J.M.B.)

# CHAPTER 16

# The ionic theory

We have said previously that science is primarily concerned with investigating *how* matter behaves, and, secondly, with trying to explain *why* it behaves as it does. The *ionic theory* is an attempt to explain the facts which have been discovered in connection with the electrical properties of chemical compounds either in the fused state or in solution. The chief facts with which we are concerned in this chapter are the following:

* Some compounds will conduct a current when they are fused or dissolved in water, while others will not.
* The passing of the current is accompanied by chemical changes at the electrodes.
* The conductivity of a solution varies with the concentration of the dissolved substance.
* Solutions which are good conductors show 'abnormalities' in regard to their colligative properties (osmotic pressure, freezing-point depression, etc.).
* The conductivity of a solution varies with the nature of the dissolved substance.

We shall examine each of these aspects of electrical behaviour in turn and see what explanations are given for them.

## Conductivities of Compounds when Fused or Dissolved in Water

**Electrolytes and Non-electrolytes.** Some compounds conduct a current both when they are fused and when dissolved in water. Examples are sodium(I) chloride and zinc(II) chloride, the electrolysis of which in the fused state has been described earlier. Other compounds give rise to conducting solutions in water, but are non-conductors in the fused state.

In this category we have most acids and some bases. Thus an aqueous solution of hydrogen chloride is an excellent conductor, but pure liquid hydrogen chloride does not conduct electricity. Similarly ammonia solution is a conductor (a poor one in this case), while liquid ammonia is practically a non-conductor. Finally we have compounds like sucrose and ethanamide, which are non-conductors both when fused or dissolved in water. Compounds which conduct electricity either in the fused state or in aqueous solution are called *electrolytes*. In general these are acids, alkalis, or salts. Compounds like sucrose which are non-conducting in both states are called *non-electrolytes*.

Electrolytes are sometimes sub-divided into *true electrolytes* and *potential electrolytes*. The first are compounds such as sodium(I) chloride which are conductors both when fused and when dissolved in water, while the second comprise compounds like hydrogen chloride which are conductors only in aqueous solution.

Electrolytes are described as 'strong' or 'weak' according to whether their aqueous solutions are good or poor conductors. There is no sharp dividing line, however, between the two groups. Strong acids such as hydrochloric acid and strong alkalis like sodium(I) hydroxide are also strong electrolytes. Weak acids like ethanoic acid and weak alkalis like ammonia solution are weak electrolytes. Nearly all salts (mercury(II) chloride and cadmium(II) chloride are exceptions) are strong electrolytes.

**Explanation of Conductivity.** The passage of a current through a fused electrolyte or its aqueous solution must take place differently from the passage of a current through a metallic conductor such as a copper wire. In the first case the passing of the current is invariably attended by a chemical change, while in the second the conductor remains unaltered chemically. According to modern theory a current in a copper wire consists of a one-way flow of *electrons* through the wire, in the opposite direction to what our sign convention indicates as the direction of the current. A current in a fused electrolyte or its aqueous solution is explained by the ionic theory as a two-way flow of *ions* towards the electrodes of opposite polarity. In the fused electrolyte the ions are bare, while in the solution they are hydrated—that is, they are surrounded by a sheath of attached water molecules.

Practical evidence of the two-way movement of ions in solution is provided by experiments on the *migration* of ions. This term refers to the movement of similarly charged ions as a body towards the oppositely charged electrode. If the ions are coloured their movement can be observed under appropriate conditions. An experiment which illustrates migration of ions is described at p. 417.

*All* the ions in an aqueous solution contribute towards the carrying of the current, the share of each kind depending on its concentration and its speed in the electrical field. Even the water makes a slight contribution to the conductivity, because water molecules are ionized to a small extent into hydrogen and hydroxyl ions:

$$H_2O \rightleftharpoons H^+ + OH^-$$

As, however, only 1 in about 550 million water molecules is dissociated in this way, the conductivity due to water can usually be neglected in comparison with that due to the ions of the electrolyte.

**History of the Ionic Theory.** Like the theory of atoms, the ionic theory has been modified in the course of time as new facts have been discovered. Faraday himself was one of the first to suggest a theory of conductivity, and some of the terms which he coined (such as 'ion,' 'anode,' and 'cathode') are in use to this day.

At the time of Faraday all electrolytes were supposed to consist of individual molecules, which were composed of two parts. Thus,

*Acids:* hydrogen + an acid radical
*Salts:* metal + an acid radical
*Alkalis:* metal + hydroxyl (OH)

Faraday suggested that the effect of an electric current on an electrolyte in solution was to break up some of its molecules into the two constituent parts, one part being positively charged and the other negatively charged. He called the electrified parts *ions* (Greek, 'to go'). The ions were attracted to the electrode of opposite sign, gave up their charges, and became ordinary atoms. The current was thus carried through the solution by the ions themselves.

Faraday's theory implied that the breaking up of the molecules of the electrolyte was brought about only when a potential difference was applied to the electrodes and that part of the electrical energy was consumed in breaking up the molecules. It was proved, however, by Clausius (1857) that none of the electrical energy was expended in this way. Clausius put forward the theory that the molecules of an electrolyte were at once 'ionized' to a small extent on dissolving in water owing to the molecules colliding together in the liquid. As the ions were discharged at the electrodes during electrolysis more molecules ionized. In this manner an equilibrium was maintained between the number of ions and the number of unionized molecules; *e.g.*,

$$HCl \rightleftharpoons H^+ + Cl^-$$

The next contribution was made by a Swedish chemist, Arrhenius (1887), as a result of investigations into the conductivity of solutions of electrolytes. Arrhenius was able to show that the extent of ionization, or 'dissociation,' was much greater than Clausius had supposed. He maintained that the proportion of electrolyte in the form of ions was increased by diluting the solution, and by adding enough water the electrolyte could be made to dissociate completely into ions. The fraction of the electrolyte which had dissociated into ions at any particular dilution was called by Arrhenius the 'degree of ionization or dissociation.'

At first Arrhenius's theory met with considerable opposition. It was argued that, in view of the large amount of energy which is evolved when sodium combines with chlorine, it was most unlikely they could be separated again merely by dissolving sodium(I) chloride in water. Arrhenius replied that ions and atoms must be thought of as quite different particles. The presence of an electrical charge on an atom profoundly altered its chemical nature, so that separation of the salt into ions was entirely different from separation into atoms of sodium and chlorine.

Arrhenius's theory gradually came to be accepted because it convincingly explained the facts connected with conductivity. As we shall see later a solution of a strong electrolyte shows abnormalities with regard to freezing point depression and other colligative properties. Arrhenius was able to show that these abnormalities were related quantitatively to the degree of dissociation calculated from the conductivity of the solution.

In the present century Arrhenius's version of the ionic theory has been modified because of discoveries concerning the electronic structures of compounds. X-ray diffraction shows that salts like sodium(I) chloride and strong alkalis like sodium(I) hydroxide consist of oppositely charged ions even when solid. Arrhenius was therefore mistaken in believing that these compounds were composed of molecules, which became partly dissociated into ions on dissolving in water. As explained earlier, modern theory emphasizes the importance of the solvent in promoting dissociation of ionic compounds. The action of water is twofold: to 'hydrate' the ions (which releases energy), and to separate and insulate the ions from one another. The second is a gradual process depending on the amount of water present. The difference between the Arrhenius theory and the modern theory of dissociation of sodium(I) chloride in dilute aqueous solution is thus represented by the two equations,

$$\text{(Arrhenius's theory)} \quad NaCl \rightleftharpoons Na^+ + Cl^-$$
$$\text{(Modern theory)} \quad Na^+Cl^- + aq \rightleftharpoons Na^+(aq) + Cl^-(aq)$$

(Although the second equation is a more accurate representation of what takes place, for simplicity we still use equations of the first type to show the dissociation of ionic compounds in solution.)

The strong acids ($HCl$, $H_2SO_4$, and $HNO_3$) resemble salts and strong alkalis in being highly dissociated into ions in aqueous solution, but differ in being covalent, and not ionic, compounds in the pure state. Unlike ionic compounds they are not solids of high melting point. Acids form ions in aqueous solution owing to chemical reaction with the water, water molecules being able to sever the covalent bond which exists between the hydrogen atom and the rest of the molecule. With hydrochloric acid the hydrogen atom leaves its electron, which it has been sharing with a chlorine atom, and forms a co-ordinate covalent bond with an oxygen atom.

$$H\!-\!\overset{\cdot\cdot}{O}\!: + H\!-\!Cl \rightleftharpoons \left[ H\!-\!\overset{\cdot\cdot}{O}\!-\!H \right]^{+} + Cl^{-}$$
$$\quad\; | \qquad\qquad\qquad |$$
$$\quad\; H \qquad\qquad\qquad H$$

$$\text{or}\quad H_2O + HCl \rightleftharpoons H_3O^{+} + Cl^{-}$$

As we have seen earlier this chemical change is brought about by the attraction of a lone pair of electrons in the oxygen atom for the hydrogen atom. Although, strictly speaking, hydrogen ions exist in aqueous solution as 'oxonium' ions, $H_3O^{+}$ or $H_2O . H^{+}$, we usually represent the reaction as a simple dissociation:

$$HCl \rightleftharpoons H^{+} + Cl^{-}$$

The ionic theory attributes the general acid properties of acids in aqueous solution to the hydrogen ions produced. Thus a solution of dry hydrogen chloride in dry benzene evolves neither hydrogen with zinc, nor carbon dioxide with calcium(II) carbonate. The extent to which an acid forms hydrogen, or oxonium ions determines the 'strength' of the acid in aqueous solution, 'Strong' acids have a high degree of dissociation (about 78 per cent for a molar solution of HCl), whereas 'weak' acids are only slightly dissociated (about 0·4 per cent for molar $CH_3COOH$ solution).

The typical properties of alkalis in solution—*e.g.*, soapy feel, caustic action, etc.—are due to hydroxyl ($OH^{-}$) ions. Strong alkalis are ionic compounds, and their dissociation in water takes place in a similar manner to that of sodium(I) chloride. Thus, ignoring hydration of ions, we can write

$$Na^{+}(OH)^{-} \rightleftharpoons Na^{+} + OH^{-}$$

Weakly alkaline solutions are formed when compounds like ammonia and methylamine are dissolved in water. These are covalent compounds, which produce a small concentration of hydroxyl ions by abstracting protons from water molecules, *e.g.*,

$$NH_3 + H_2O \rightleftharpoons NH_4^{+} + OH^{-}$$

Here the reaction is caused by the attraction of the lone pair of electrons in the nitrogen atom for the hydrogen atom.

## Chemical Effects of a Current

**Electrolysis.** This term refers to the overall chemical change which takes place at the electrodes when a current passes through a fused electrolyte or its aqueous solution. The change may, or may not, be the same in the two cases. Electrolyte, solvent, and electrodes may all be concerned in the electrolytic reaction. If the electrodes take no part they are described as *inert* electrodes. Platinum and carbon (graphite) usually behave as inert electrodes, although platinum is attacked by liberated chlorine and carbon by liberated oxygen. If inert electrodes are used the overall reaction usually (but not always) consists of decomposition. The substance decomposed may be the electrolyte (*e.g.*, copper(II) chloride solution) or the solvent (*e.g.*, sodium(I) sulphate(VI) solution). Sometimes the reaction products are derived partly from the electrolyte and partly from the solvent (*e.g.*, sodium(I) chloride solution).

Table 16–1 shows the products obtained in some typical examples of

**Table 16–1.** PRODUCTS OF ELECTROLYSIS OF SOME COMMON ELECTROLYTES

| Electrolyte | Electrodes | At cathode | At anode |
|---|---|---|---|
| Acidified water | Pt | $H_2$ | $O_2$ |
| Fused NaCl | C | Na | $Cl_2$ |
| Aqueous NaCl (conc.) | C | $H_2$ | $Cl_2$ |
| Fused NaOH | Pt | Na | $O_2$ |
| Aqueous NaOH | Pt | $H_2$ | $O_2$ |
| Aqueous $CuSO_4$ | Cu | Cu | — |
| Aqueous $CuSO_4$ | Pt | Cu | $O_2$ |
| Aqueous $AgNO_3$ | Ag | Ag | — |
| Aqueous $AgNO_3$ | Pt | Ag | $O_2$ |
| Concentrated HCl | C | $H_2$ | $Cl_2$ |
| Aqueous $Na_2SO_4$ | Pt | $H_2$ | $O_2$ |

electrolysis. The reasons for the appearance of particular products at the electrodes are explained in the next chapter. For the present we may note that, in general, *metals and hydrogen are liberated at the cathode, non-metals (except hydrogen) at the anode.* An exception to the general rule occurs in the electrolysis of fused hydrides of alkali metals and alkaline-earth metals. In these hydrides (*e.g.*, $Na^+ H^-$) the hydrogen is present as a negative ion and is liberated at the anode.

Electrolysis is of great importance in industry. Among the chief applications are the following:

*Extraction of Elements.* Both metals (*e.g.*, Na, K, Mg, Ca, Al, Zn) and non-metals (*e.g.*, $H_2$, $F_2$, $Cl_2$) are obtained by electrolysis of fused compounds or their aqueous solutions.

*Purification of Metals.* Copper and gold are refined electrolytically.

*Electroplating*—*e.g.*, plating with silver, gold, chromium, and nickel.

*Anodic Oxidation of Aluminium.* Electrolysis of dilute sulphuric(VI) acid is used to grow a tough oxide film on aluminium articles for some purposes (see p. 442).

*Preparation of Important Compounds*—*e.g.*, sodium(I) hydroxide and sodium(I) chlorate(V).

**Faraday's Laws of Electrolysis.** The quantitative aspects of electrolysis were first investigated by Faraday, and the laws which are obeyed were discovered about 1833.

**1.** *The mass of any substance liberated by a current is proportional to the quantity of electricity which has passed.*

'Quantity' of electricity is expressed in coulomb and is measured by multiplying the strength of the current in ampere by the time in second. Hence, according to Faraday's first law

$$m \propto I \times t$$
$$\text{or} \quad m = EIt$$

where $m$ is the mass of substance liberated and $E$ is a constant.

When $m$ is expressed in gram the constant $E$ is the mass in gram of the the substance liberated by the passage of 1 coulomb (C) of electricity— *i.e.*, 1 ampere for 1 second—and is called the *electrochemical equivalent* of the substance. For silver the electrochemical equivalent is $0.001\ 118$ g $C^{-1}$. Hence

$$m = 0.001\ 118\ \text{g}\ C^{-1} \times It$$

**2.** *If the same quantity of electricity is passed through solutions of different electrolytes the masses of the different substances liberated are proportional to their chemical equivalent masses.*

The quantity of electricity which liberates the equivalent mass in gram of a substance is called the *Faraday constant*, $F$ ($96\ 487$ C $mol^{-1}$ on the $^{12}C = 12$ standard). The use of this law in fixing equivalent masses has previously been noted. Faraday's laws are not affected by alteration of temperature, concentration, or size of electrodes.

Faraday's second law can be tested by placing a silver coulometer, a water coulometer, and a copper coulometer in series, as illustrated in Fig. 16–1. (The term 'coulometer' has now replaced the older term 'voltameter' because the apparatus is a means of measuring quantity of electricity (coulomb), not volt.) The silver coulometer consists of thin sheets of silver suspended in 10 per cent silver(I) nitrate(V) solution. In the water coulometer we have two large burettes (250 cm³) initially filled with dilute sulphuric(VI) acid and inverted in a trough of this acid. The electrodes are of platinum. The copper coulometer consists of copper plates suspended in a saturated solution of copper(II) sulphate(VI) with a little sulphuric(VI) acid (to prevent formation of basic salt). A direct current of 0·5–0·8 A is passed through the coulometers for about ½ hour, the current strength is read from an ammeter included in the circuit, and the time is recorded by stop-clock. The increases in mass of the silver and copper cathodes are found, and the volumes of hydrogen and oxygen evolved are measured after levelling in a deep vessel of water. The volumes

of the gases are reduced to s.t.p. and their masses calculated. The masses
of the different elements liberated are proportional to their chemical
equivalent masses.

To test Faraday's first law only the water coulometer is needed. Hydrogen is collected for a period of five minutes and the volume is measured
(after levelling). This is repeated for further periods of five minutes, and
a graph is plotted of volume against time. A straight-line graph is obtained.
Further determinations can be made with different values of current.

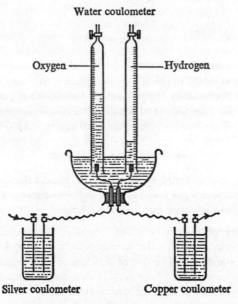

*Fig.* 16–1. Apparatus used to test Faraday's second law of electrolysis

**Example.** *By passing a current of* 0·65 *A for* 35 *minutes through water,
copper, and silver coulometers the following masses of elements were liberated:* 0·0143 *g of hydrogen,* 0·114 *g of oxygen,* 0·449 *g of copper, and*
1·542 *g of silver. (i) Show that these results agree with Faraday's second
law; (ii) calculate the electrochemical equivalent of silver. (H* = 1·008,
*O* = 16·0, *Cu* = 63·5, *Ag* = 107·9.)

(i) Ratios of masses of H:O:Cu:Ag

$$= 0·0143 : 0·114 : \ 0·449 : \ 1·542$$
$$= 1 \qquad :7·97 \ :31·40 \ :107·8$$

The ratios of the equivalent masses obtained from the relative atomic
masses = 1·008:8·0:31·75:107·9.

Hence, within the limits of experimental working, the experimental results agree with Faraday's second law.

(ii) Quantity of electricity passed $= 0.65 \times 2100$ C
$$= 1365 \text{ C}$$

1365 C liberate 1·542 g of silver

$\therefore$ 1 C liberates $\dfrac{1·542}{1365}$ g of silver

$$= 0.001\ 13 \text{ g of silver}$$

**Atomic Nature of Electricity.** It has been mentioned previously that during electrolysis electricity is transferred between the electrodes and the ions in solution in definite units. This is easily shown by calculating the quantities of electricity which are required to liberate 1 mole of different elements. The quantities ($Q$) are calculated simply by dividing the relative atomic masses of the elements by their electrochemical equivalents. When we compare these quantities of electricity with the valencies of the elements we find that the two are directly related, as shown in Table 16–2.

**Table 16–2**

| Element | Electrochemical equivalent ($E/\text{gC}^{-1}$) | Relative atomic mass ($A_r$) | $Q/\text{C}$ | Valency |
|---|---|---|---|---|
| Hydrogen | $1·045 \times 10^{-5}$ | 1·008 | $9·65 \times 10^4$ | 1 |
| Silver | $1·118 \times 10^{-3}$ | 107·9 | $9·65 \times 10^4$ | 1 |
| Oxygen | $8·29 \times 10^{-5}$ | 16·00 | $19·3 \times 10^4$ | 2 |
| Copper | $3·29 \times 10^{-4}$ | 63·54 | $19·3 \times 10^4$ | 2 |
| Gold | $6·81 \times 10^{-4}$ | 197·0 | $28·9 \times 10^4$ | 3 |
| Chromium | $1·80 \times 10^{-4}$ | 52·00 | $28·9 \times 10^4$ | 3 |

This table shows that the quantity of electricity needed to liberate 1 mole of an element is twice as great for a divalent element as for a monovalent element, while for a trivalent element it is three times as great. Faraday's second law of electrolysis can thus be put in the form: *The quantity of electricity required to liberate 1 mole of an element is proportional to its valency.*

From the definition given at p. 61 one mole of different elements contains the same number of atoms. Hence the quantity of electricity required to set free *one atom* of an element at an electrode is twice as great for a divalent element as for a monovalent element, and three times as great for a trivalent element. This means that electricity passes between electrodes and ions in units, the number of units transferred depending on the valency of the ions (or atoms). As Helmholtz wrote in 1881, "We cannot avoid concluding that electricity, positive as well as negative, is divided into definite elementary portions, which behave like atoms of electricity."[1]

> The magnitude of the unit charge ($e$) can be calculated by dividing the number of coulomb required to liberate 1 mole of hydrogen or silver by the number of atoms in this amount of the element. The number of coulomb is 96 487. The number of atoms is given by the Avogadro constant ($N_A$), which is $6 \cdot 02 \times 10^{23}$. Thus we have
>
> $$e = \frac{F}{N_A} = \frac{96\,487}{6 \cdot 02 \times 10^{23}} = 1 \cdot 60 \times 10^{-19}\ C$$
>
> As mentioned earlier, the unit charge calculated as above is the same as the charge on the electron. Since the electronic charge can be found independently, the calculation given above can be reversed, and the Avogadro constant can be determined. Thus, $N_A = F/e$.

**Discharge of Ions at Electrodes.** We have seen that the discharge of a monovalent ion at an electrode is accompanied by the transfer of a quantity of electricity equal to the charge on an electron. We may therefore infer that electrons are directly involved in the discharge. Positively charged cations are discharged at the cathode by receiving electrons from the cathode; negatively charged anions are discharged at the anode by giving up their excess electrons to the anode. In both cases neutral atoms result. In the case of copper(II) chloride solution these changes are represented as follows:

(At the cathode) $\qquad\qquad Cu^{2+} + 2e \to Cu$

(At the anode) $\quad 2Cl^- - 2e \to 2Cl\ (Cl + Cl \to Cl_2)$

The transfer of two electrons to or from an electrode represents the passage of a definite amount of electricity. Obviously the amounts of copper and chlorine liberated will depend on how many times this transfer of two electrons takes place—that is, on the quantity of electricity which passes (Faraday's first law).

Again, the number of electrons lost by the cathode must equal the number of electrons gained by the anode (otherwise there would be a building-up of electrons in some part of the circuit). It follows that one divalent copper ion must be discharged in the same time as two univalent chloride ions—that is, the masses of copper and chlorine liberated will be proportional to their chemical equivalent masses (Faraday's second law).

For the present this is as far as we can go in explaining the liberation

---

[1] The constancy of the charge required to liberate 1 mole of a monovalent element does not *prove* that every ion of hydrogen, silver, etc., carries the same charge. The effect might be a statistical one resulting from the huge numbers of ions involved. Stronger evidence of the 'atomic' nature of electricity comes from Millikan's oil-drop experiment (described in text books of physics).

of ions at electrodes. The reasons for the appearance of particular substances at electrodes when solutions of different electrolytes are electrolysed depend upon an understanding of the electrochemical series, and consideration of this must be postponed until later.

## Electrolytic Conductivity and Concentration

The resistance which a solid conductor, such as a copper wire, offers to the passage of an electric current is measured in ohm ($\Omega$, omega), and is given by the expression

$$R = \rho \times \frac{l}{A}$$

where $\rho$ (rho) is the resistivity, $l$ is the length of the conductor, and $A$ is its area of cross-section. Since length is normally in m and area in $m^2$, the units for resistivity are $\Omega$ m.

In dealing with solutions of electrolytes it is more convenient to use the quantities 'conductance' and 'electrolytic conductivity' than resistance and resistivity.

*Conductance.* If $R$ is the electrical resistance of a conductor, $1/R$ is the conductance (in $\Omega^{-1}$). The SI unit for conductance is the *siemens*, having the symbol S. This is not a plural, the unit being named after Sir William Siemens, a noted electrical engineer of the last century.

*Electrolytic Conductivity.* If $\rho$ is the resistivity of a solution, $1/\rho$ is the electrolytic conductivity of the solution. This is symbolized by $\kappa$ (kappa), and in SI is expressed in $Sm^{-1}$ (that is, in $\Omega^{-1}\,m^{-1}$).

*Dilution.* The concentration of a solution in SI is the number of mole of solute per $m^3$ of solution. The reciprocal of the concentration gives us the 'dilution,' which is thus the number of $m^3$ containing one mole of solute. Since there are 1000 $dm^3$ in 1 $m^3$, a concentration of 0·1 mol $dm^{-3}$ is equivalent to a concentration of 100 mol $m^{-3}$. In this case the dilution is 0·01 $m^3\,mol^{-1}$.

*Molar Conductivity.* In general molar conductivity is represented by $\Lambda$ (lambda) and is obtained either by dividing the electrolytic conductivity by the concentration (in mol $m^{-3}$), or by multiplying the electrolytic conductivity by the dilution (in $m^3\,mol^{-1}$)

$$\Lambda_V = \kappa/c = \kappa \times V$$

We can derive the units for molar conductivity from those used for electrolytic conductivity and dilution.

$$(\kappa/Sm^{-1}) \times (V/m^3\,mol^{-1}) = \Lambda_V/Sm^2\,mol^{-1}$$

The quantity, 'molar conductivity' was designed for a special purpose. Arrhenius attributed the conducting power of an electrolyte to its dissociation into ions, and maintained that the 'degree of dissociation' increased with dilution. If this is true the conducting power of a given amount of electrolyte should also increase with dilution, because there should be a larger number of ions in solution. The electrolytic conductivity, however, depends not only on the number of ions but also on the amount of water present. As the amount of water increases the electrolytic conductivity decreases. There are thus two opposing factors. By using molar conductivity instead of electrolytic conductivity Arrhenius was able to cancel out the effect of the water on the conductivity and investigate the effect of the electrolyte. From his results he calculated the degree of dissociation at different dilutions.

**'Degree of Dissociation' from Molar Conductivity.** When we plot the molar conductivity of a sodium(I) chloride solution against the dilution we get a curve of the type shown in Fig. 16–2. The molar conductivity

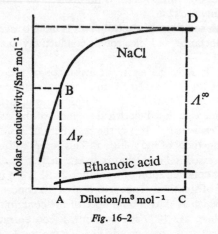

*Fig.* 16–2

increases with dilution to a maximum at which it remains constant no matter how much the dilution is increased. This maximum value is therefore called the molar conductivity at 'infinite dilution.' If, as Arrhenius supposed, the current is carried only by the ions the molar conductivity will be proportional to the number of ions, which, in turn, depends on the degree of dissociation. Hence we assume that when the molar conductivity becomes constant the salt is completely dissociated into ions.

Let $\Lambda_V$ be the molar conductivity at a dilution of $V$. Then, if $\alpha$ is the degree of dissociation, $\Lambda_V$ is proportional to $\alpha$, or

$$\Lambda_V = k\alpha \tag{1}$$

where $k$ is a constant.

At infinite dilution all the salt is dissociated and $\alpha - 1$. If $\Lambda^\infty$ is the molar conductivity at infinite dilution,

$$\Lambda^\infty = k \tag{2}$$

Dividing (1) by (2), we find that

$$\alpha = \frac{\Lambda_V}{\Lambda^\infty}$$

In practice a solution of the electrolyte of known concentration (that is, known dilution) is prepared with 'conductivity water' and placed in a conductivity cell. Conductivity cells have various forms depending on whether the conductivity to be measured is high, medium, or low. One form suitable for solutions of medium conductivity is shown in Fig. 16–3. The vessel is a glass cylinder tapering towards the base and covered

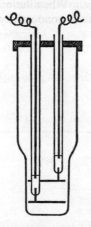

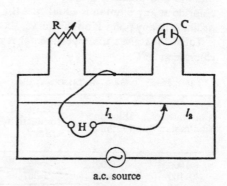

a.c. source

*Fig.* 16–3. Conductivity cell

*Fig.* 16–4. Apparatus used to measure electrolytic conductivity

with an ebonite plate. The electrodes are circular platinum plates attached to platinum wires fused through the ends of two glass tubes. Copper wires are inserted into the tubes and make connection with the platinum by means of a little mercury. A known amount of the standard solution is introduced by a pipette into the conductivity cell, which is then made the unknown resistance in a Wheatstone's Bridge experiment (Fig. 16–4).

C is the conductivity cell containing the solution and R is a resistance box. A direct current would cause electrolysis of the solution, and therefore an alternating current from a valve oscillator or a small induction coil is employed. Since a galvanometer does not detect alternating current we use headphones, H, instead to find the point of balance. This occurs when the buzzing sound in the headphones is a minimum. If at the balance point $l_1$ and $l_2$ are the lengths of bridge wire on each side of the point of contact we have

Resistance of R : resistance of solution $= l_1 : l_2$

The resistance of the solution can now be calculated, and the reciprocal gives the conductance. To find the electrolytic conductivity it would be necessary to measure the area of the electrodes and their distance apart. As these dimensions are kept constant for the cell their joint effect for calculation of the electrolytic conductivity is usually expressed as the 'cell constant' so that we can obtain the electrolytic conductivity simply by multiplying the conductance by this constant. Finally, the molar conductivity is the product of the electrolytic conductivity and the dilution, or the volume of solution containing one mole of the electrolyte.

By withdrawing known amounts of the solution from the cell and adding known volumes of water, we can make further determinations of the molar conductivity at progressively greater dilutions. When there is no further increase in the molar conductivity we plot the latter against dilution and obtain a graph. The degree of dissociation of sodium(I) chloride at any dilution is equal to AB/CD (Fig. 16–2), where AB is the ordinate of any point B on the curve.

Table 16–3 gives the experimental results for solutions of sodium(I) chloride at 25°C.

**Table 16–3.** DEGREE OF DISSOCIATION OF SODIUM(I) CHLORIDE SOLUTIONS

| Concentration/mol m$^{-3}$ | Dilution $V$/m$^3$mol$^{-1}$ | Electrolytic conductivity $\kappa$/Sm$^{-1}$ | Molar conductivity $\Lambda_V$/Sm$^2$mol$^{-1}$ | $\alpha = \dfrac{\Lambda_V}{\Lambda^\infty}$ |
|---|---|---|---|---|
| 1000 | 0·001 | 9·36 | 0·936 × 10$^{-2}$ | 0·74 |
| 100 | 0·01 | 1·067 | 1·067 × 10$^{-2}$ | 0·85 |
| 20 | 0·05 | 0·232 | 1·160 × 10$^{-2}$ | 0·92 |
| 10 | 0·1 | 0·118 | 1·18 × 10$^{-2}$ | 0·94 |
| 1 | 1·0 | 0·0124 | 1·24 × 10$^{-2}$ | 0·98 |
| 0 | ∞ | | 1·264 × 10$^{-2}$ | (1·00) |

**Degree of Dissociation.** Originally degree of dissociation (or ionization) meant the fraction of an electrolyte in solution which had changed into ions. This meaning has disappeared in relation to strong electrolytes because salts and strong bases are completely ionized even in the solid state. It is therefore pointless to talk about their degree of ionization in solution, and this usage has been abandoned. The term 'degree of dissociation' is still used, but it has rather a different interpretation according to the type of electrolyte to which it is applied.

The molar conductivity of a salt like sodium(I) chloride in solution depends upon two factors: the concentration of ions present, and the speed with which the ions move towards the electrodes. Arrhenius assumed that the second factor was constant at any given temperature, and explained the variation in molar conductivity with dilution as due solely to a change in the concentration of ions. In this he was wrong. Actually it is the first factor which remains constant (since sodium(I) chloride is completely ionized) and the second which varies with dilution. Owing to the attraction of opposite charges an ion tends to build up round itself an 'atmosphere' of ions of opposite sign. (This is in addition to the envelope of water molecules with which most ions are solvated.) The 'atmosphere' acts as a drag upon the ion and reduces its velocity and conductivity. At bigger dilutions the ions become increasingly insulated and free from one another's influence, so that their conductivity is increased. The molar conductivity finally becomes constant when sufficient water has been added to make it possible for the ions to behave as completely independent particles.

'Degree of dissociation' as regards sodium(I) chloride and other ionic compounds in solution is thus a measure of the extent to which the ions have become free particles. In this respect it is an important quantity. In order to distinguish this interpretation from the original meaning, however, the vague but non-committal term 'apparent degree of dissociation' is often used. A better term, perhaps, is 'molar conductivity ratio,' which is applied to the ratio $\Lambda_V/\Lambda^\infty$.

Strong acids like hydrochloric acid are rather different from sodium(I) chloride. They are strong electrolytes. Although they have covalent molecules, they are highly ionized in solution and the extent of ionization increases with dilution. In concentrated solution, however, the attraction between oppositely charged ions is appreciable, and the molar conductivity ratio is not a true indication of the proportion of the acid which has been converted into ions. With weaker solutions the mutual interference of the ions is reduced and the molar conductivity ratio gives approximately the fraction of the acid present in solution as ions.

Weak electrolytes like ethanoic acid give comparatively few ions in solution, so that these may be said to be completely free from effect on one another. The degree of dissociation determined by the conductivity method in this case reflects the proportion of the solute which is in the form of ions. It is not possible, however, to determine the degree of dissociation directly in this way. The molar conductivity of ethanoic acid and other weak electrolytes increases very slowly with increasing dilution (Fig. 16-2). The dilution required to obtain the value of $\Lambda^\infty$ is too great to carry out in practice. $\Lambda^\infty$ can be found, however, by an application of Kohlrausch's law (p. 414), so that the molar conductivity ratio can be determined indirectly.

A number of electrochemical phenomena (*e.g.*, the osmotic pressure of a solution of an electrolyte) depend partly on the concentration of ions in solution and partly on the extent to which they have become free from interionic action. It is often convenient to combine these quantities into the single quantity, *activity*. The relation between the mean activity, $a_\pm$, and the molality, $m$, of the ions in a sodium(I) chloride solution is given by

$$a_\pm = \gamma_\pm m$$

where $\gamma_\pm$ is the mean activity coefficient of the ions. The value of $\gamma_\pm$ can be determined from freezing-point or osmotic-pressure data. For very dilute solutions, in which dissociation is practically complete, $\gamma_\pm$ is 1. The activity concept is widely used in more advanced electrochemistry.

**Effect of Temperature on Conductance.** The conductances of solutions of strong electrolytes usually increase by 2–3 per cent for each degree Celsius rise in temperature. This is due to several causes, chief of which is the decreased viscosity of water, which makes it possible for the ions to move more freely. As the increased conductance affects similarly both the molar conductivity at a given dilution and that at infinite dilution, rise of temperature has little effect on the degree of dissociation of strong electrolytes.

With weak electrolytes the influence of temperature on conductance depends on the heat change which accompanies ionization. If the electrolyte dissociates into ions with evolution of heat, the degree of dissociation is less at higher temperatures, in accordance with Le Chatelier's principle. The effect of the lower viscosity of the solvent may then be more than offset by the effect of the smaller number of ions, and the conductance may decrease.

## Colligative Properties of Electrolytes

**Degree of Dissociation from Freezing Point Depression, etc.** We have noted in Chapter 4 that the relative molecular mass of a strong electrolyte like hydrochloric acid cannot be determined from the osmotic pressure, freezing point depression, boiling point elevation, or lowering of vapour pressure of its aqueous solution. Since the magnitudes of these quantities are proportional to the number of particles in solution, it is reasonable to expect larger values in the case of substances which dissociate into ions in solution. Complete dissociation of hydrochloric acid or sodium(I) chloride yields twice the number of particles in solution as compared with one containing only simple HCl or NaCl particles. Osmotic pressure, freezing point depression, etc., should therefore be double the theoretical value calculated from the formulae HCl or NaCl. This is found to be so. For calcium(II) chloride, $CaCl_2$, and iron(III) chloride, $FeCl_3$, the experimental values should be three and four times respectively the theoretical values if the salts are completely dissociated. Again this is found to hold in practice.

If an electrolyte is only partially dissociated into ions its degree of dissociation can be found from the ratio of the observed osmotic pressure, freezing point depression, etc., to the theoretical value. The latter is calculated on the assumption that no dissociation takes place and that the substance exists in solution in the form corresponding to its usual formula. Van't Hoff introduced a factor $g$ (until recently, $i$) to express the ratio between the observed and theoretical values of osmotic pressure. This factor, the *osmotic coefficient*, can be applied equally well to the other colligative properties. Thus

$$g = \frac{\text{observed osmotic pressure}}{\text{theoretical osmotic pressure}}$$

$$= \frac{\text{observed freezing point depression}}{\text{theoretical freezing point depression}} \quad \text{etc.}$$

Consider the dissociation of an electrolyte A which dissociates into ions B, C, D, etc., and let the degree of dissociation be $\alpha$.

$$\underset{1-\alpha}{A} \rightleftharpoons \underset{\alpha}{B} + \underset{\alpha}{C} + \underset{\alpha}{D} + \dots$$

If one molecule dissociates into $N$ ions, the ratio of the number of particles actually present to the number if no dissociation occurred

$$= \frac{1 - \alpha + N\alpha}{1}$$

That is, van't Hoff's factor, $g$, is given by

$$g = \frac{1 + \alpha(N - 1)}{1}$$

$$\therefore \quad \alpha = \frac{g - 1}{N - 1}$$

We can thus determine the degree of dissociation of an electrolyte in solution if we know its relative molecular mass (so that the theoretical freezing point depression can be calculated) and the number of ions furnished on dissociation. The actual freezing point depression is ascertained in the usual way by means of Beckmann's apparatus. Note, however, that the method does not apply to weak electrolytes like ethanoic acid. These substances are dissociated to such a small extent in solution that the actual freezing points are practically the same as the calculated ones.

**Example.** *A solution of calcium(II) nitrate(V) containing* 15 *g of the anhydrous salt in* 1000 *g of water freezes at* $-0.435°C$. *Calculate the degree of dissociation of the salt. (The freezing constant for water per* 1000 *g is* 1·86 *degC mol*$^{-1}$. $Ca = 40, N = 14, O = 16$.)

$$Ca(NO_3)_2 \rightleftharpoons Ca^{2+} + 2NO_3^-$$

We first calculate the theoretical depression $(t)$ of the freezing point of the solution.

Molar mass of calcium nitrate $= 164$ g

$$15 \text{ g} : 164 \text{ g} = t : 1\cdot86°C$$
$$t = 0\cdot169 \text{ degC}$$
$$\text{Then} \quad \alpha = \frac{g-1}{N-1} = \frac{(0\cdot435/0\cdot169)-1}{3-1}$$
$$= 0\cdot79, \text{ or } 79 \text{ per cent}$$

To what extent does the apparent degree of dissociation found by the above method agree with that determined by the conductivity method? There is fairly good agreement if dilute solutions only are considered. This is shown by the comparison given in Table 16–4 of the results obtained by the two methods with hydrochloric acid, sodium(I) chloride solution, and potassium(I) nitrate(V) solution.

**Table 16–4.** COMPARISON OF DEGREES OF DISSOCIATION OBTAINED BY CONDUCTIVITY AND FREEZING POINT METHODS

|  | Concentration/mol dm$^{-3}$ | $\alpha = \dfrac{\Lambda_V}{\Lambda^\infty}$ | $\alpha = \dfrac{g-1}{N-1}$ |
|---|---|---|---|
| HCl | 0·1 | 0·92 | 0·90 |
|  | 0·01 | 0·97 | 0·95 |
| NaCl | 0·1 | 0·85 | 0·87 |
|  | 0·01 | 0·94 | 0·94 |
| KNO$_3$ | 0·1 | 0·82 | 0·79 |
|  | 0·01 | 0·93 | 0·94 |

The general agreement in the results obtained by the two methods was one of the main reasons why the ionic theory of Arrhenius came to be accepted. Nowadays the same faith is not placed in the agreement. The divergencies (which become greater with stronger solutions) are too serious to be explained by experimental error and what agreement exists is more or less incidental. In other words, the two methods do not really measure the same quantity. The molar conductivity ratio is basically a measure of the freedom of movement of the ions in the solution, and is therefore a solute property. Osmotic pressure, freezing point depression, etc., are a measure of the freedom of movement of the solvent molecules, and this is a solvent property.

## Conductivity and Nature of the Electrolytes

**Transport Numbers of Ions.** Earlier in this chapter we saw that the molar conductivity of a solution of an electrolyte at a given temperature depends on the concentration of the ions, and the speed with which the ions move towards the electrodes. If we compare the molar conductivities of solutions of different electrolytes of the same concentration at the same temperature we find considerable variation. Thus the molar conductivities at 25°C of 0·1 molar solutions of sodium(I) hydroxide, potassium(I) chloride, and silver(I) nitrate(V) are 2·21 × 10$^{-2}$, 1·29 × 10$^{-2}$, and 1·09 × 10$^{-2}$ Sm$^2$ mol$^{-1}$ respectively. Since the concentrations of the ions are the same in each case, the variations in molar conductivity must be caused by the different velocities of the ions.

The *relative* speeds at which the anion and cation of a single electrolyte migrate towards the oppositely charged electrodes can be found by measuring the different extents to which the concentration of the electrolyte changes round the electrodes during electrolysis. The measurement is carried out with special cells designed to prevent ordinary diffusion of the ions. The relative rates of migration of anion and cation determine the share which each ion has in carrying the current through the solution. Thus, if the electrolyte is silver(I) nitrate(V), we can write

$$\frac{\text{Fraction of current carried by anion}}{\text{Fraction of current carried by cation}} = \frac{\text{velocity of NO}_3^- \text{ ion}}{\text{velocity of Ag}^+ \text{ ion}}$$

The fractions of the current carried by the ions are called their *transport numbers*. If $t_-$ is the transport number of the anion, the transport number, $t_+$, of the cation will be $1 - t_-$. The transport number of a cation varies with the nature of the anion, and vice versa. It also varies slightly with the concentration of the electrolyte and with the temperature. Some values of the transport numbers of cations for decimolar solutions of different electrolytes at 25°C are:

|       | HCl       | NaCl    | NaOH    | KCl    | AgNO$_3$ |
|-------|-----------|---------|---------|--------|----------|
|       | H$^+$(aq) | Na$^+$  | Na$^+$  | K$^+$  | Ag$^+$   |
| $t_+$ | 0·831     | 0·385   | 0·183   | 0·490  | 0·468    |

Note the high proportions of the current carried by the hydrated H$^+$ ion in hydrochloric acid and by the OH$^-$ ion in sodium(I) hydroxide solution. These ions have relatively high velocities.

**Kohlrausch's Law and Ionic Mobilities.** Kohlrausch investigated the molar conductivity at infinite dilution of a large number of electrolytes and found that in all cases it was the sum of two quantities, one due to the anion and the other to the cation. Each ion contributes its own conductivity to the total independently of the other ions present. We can write

$$\Lambda^\infty = \Lambda_+^\infty + \Lambda_-^\infty$$

where $\Lambda_+^\infty$ and $\Lambda_-^\infty$ are the molar conductivities at infinite dilution of the cation and anion respectively. This is known as *Kohlrausch's law of independent ionic mobilities*.

The fractions which each ion contributes to $\Lambda^\infty$ correspond to the fractions of the current carried by the ions, that is, with their transport numbers. Hence,

$$\Lambda_+^\infty = \Lambda^\infty \times \text{transport number of cation}$$
$$\text{and } \Lambda_-^\infty = \Lambda^\infty \times \text{transport number of anion}$$

Therefore $\Lambda_+^\infty$ and $\Lambda_-^\infty$ can be found if the molar conductivity of the electrolyte at infinite dilution and the transport numbers of the ions are known. Thus, if the transport number of the sodium(I) ion for a solution of sodium(I) chloride is 0·395.

$$\Lambda^\infty(\text{Na}^+, 298\ \text{K}) = (1·264 \times 10^{-2}) \times 0·395 = 0·501 \times 10^{-2}\ \text{Sm}^2\ \text{mol}^{-1}$$

Knowing the molar conductivity at infinite dilution of one ion, we can calculate those of other ions, as now shown.

$$\Lambda^\infty(\text{Cl}^-, 298\ \text{K}) = \Lambda^\infty(\text{NaCl}, 298\ \text{K}) - \Lambda^\infty(\text{Na}^+, 298\ \text{K})$$
$$= (1·264 - 0·501) \times 10^{-2} = 0·763 \times 10^{-2}\ \text{Sm}^2\ \text{mol}^{-1}$$
$$\Lambda^\infty(\text{K}^+, 298\ \text{K}) = \Lambda^\infty(\text{KCl}, 298\ \text{K}) - \Lambda^\infty(\text{Cl}^-, 298\ \text{K})$$
$$= (1·498 - 0·763) \times 10^{-2} = 0·735 \times 10^{-2}\ \text{Sm}^2\ \text{mol}^{-1}$$

These numbers are characteristic of the different ions and depend only on the temperature. Obviously we can reverse the process and predict $\Lambda^\infty$ for any electrolyte by adding the molar conductivities of the ions at infinite dilution. As with the transport number of an ion, $\Lambda_+^\infty$ and $\Lambda_-^\infty$ depend on the velocities of the ions. Although the values do not represent actual speeds of ions, they indicate relative speeds. Some molar ionic conductivities at infinite dilution at 298 K are given below

**Table 16–5.** MOLAR IONIC CONDUCTIVITIES AT INFINITE DILUTION AT 298 K/SM$^2$ MOL$^{-1}$

| Cations | | Anions | |
| --- | --- | --- | --- |
| H$^+$ | $3·498 \times 10^{-2}$ | OH$^-$ | $1·986 \times 10^{-2}$ |
| K$^+$ | $0·735 \times 10^{-2}$ | $\frac{1}{2}$SO$_4^{2-}$ | $0·800 \times 10^{-2}$ |
| NH$_4^+$ | $0·735 \times 10^{-2}$ | Br$^-$ | $0·781 \times 10^{-2}$ |
| Ag$^+$ | $0·619 \times 10^{-2}$ | Cl$^-$ | $0·763 \times 10^{-2}$ |
| $\frac{1}{2}$Cu$^{2+}$ | $0·566 \times 10^{-2}$ | NO$_3^-$ | $0·714 \times 10^{-2}$ |
| Na$^+$ | $0·501 \times 10^{-2}$ | Ac$^-$ | $0·409 \times 10^{-2}$ |

**Electric Mobility of an Ion.** The actual velocity of an ion in solution at a given temperature depends on the potential difference between the electrodes and their distance apart. These factors are combined in the term 'potential gradient'—that is, the fall of potential per unit distance between the electrodes. In SI unit potential gradient occurs when there is a fall of potential of 1 volt per metre between the electrodes. The absolute velocity of an ion in metres per second per unit potential gradient is called the *electric mobility* of the ion at the given temperature.

The value of the electric mobility of an ion increases to some extent with dilution because interionic attraction becomes smaller. The maximum value is reached when the electrolyte is completely dissociated. Theoretically, for ions at infinite dilution,

$$\text{Molar ionic conductivity} = F \times \text{electric mobility}$$

where $F$ is the Faraday constant ($96\,487$ C mol$^{-1}$). This gives a simple method of calculating the speed of an ion under unit potential gradient at a given temperature. Some values of electric mobilities obtained in this way are given in Table 16–6.

**Table 16–6.** VELOCITIES OF IONS IN METRE PER SECOND AT $298$ K UNDER A POTENTIAL GRADIENT OF $1$ V M$^{-1}$

| Cations | | Anions | |
|---|---|---|---|
| H$^+$(aq) | $36\cdot3 \times 10^{-8}$ | OH$^-$ | $20\cdot5 \times 10^{-8}$ |
| K$^+$ | $7\cdot62 \times 10^{-8}$ | Fe(CN)$_6^{4-}$ | $11\cdot4 \times 10^{-8}$ |
| Na$^+$ | $5\cdot19 \times 10^{-8}$ | SO$_4^{2-}$ | $8\cdot27 \times 10^{-8}$ |
| Li$^+$ | $4\cdot01 \times 10^{-8}$ | Br$^-$ | $8\cdot12 \times 10^{-8}$ |
| Ag$^+$ | $6\cdot42 \times 10^{-8}$ | Cl$^-$ | $7\cdot91 \times 10^{-8}$ |
| Ca$^{2+}$ | $6\cdot16 \times 10^{-8}$ | NO$_3^-$ | $7\cdot40 \times 10^{-8}$ |

Notice that ions migrate at quite low speeds. Even with a potential gradient of 1 V cm$^{-1}$ the velocities of most ions are only about $0\cdot0007$ cm s$^{-1}$ at 298 K. Under these conditions an ion takes about ½ hour to travel a distance of 1 cm. The relatively large speeds of the H$^+$(aq) ion and the OH$^-$ ion contribute largely to the high conductivities of aqueous solutions of strong acids and alkalis.

Ions travel at different speeds under similar conditions because of difference in charge and size. Ions with multiple charges tend to move more quickly than singly charged ions because they are attracted more strongly by the oppositely charged electrode. Ions of small size have higher velocities than similarly charged ions of larger size because their passage is less impeded. It must be remembered, however, that ions exist in solution in a solvated form, and not as bare ions. The radius of the hydrated ion may be considerably larger than the crystal radius found by X-ray diffraction. Furthermore, the extent of hydration is usually greater the smaller the size of the bare ion. This may have the effect of reversing the order of sizes in a series of related ions and thus reversing the expected order of ionic velocities. Thus the crystal radii of the alkali metal ions increase in the order Li$^+$, Na$^+$, K$^+$, Rb$^+$, Cs$^+$ and we might expect the

electric mobilities to decrease in this order. Actually they increase in the
order given. This is because the $Li^+$ ion is the most heavily hydrated and
the $Cs^+$ ion the least hydrated.

### Applications of Kohlrausch's Law.

*Indirect Determination of $\Lambda^\infty$ for Weak Electrolytes.* We have said
that the molar conductivity at infinite dilution of a weak electrolyte
cannot be found directly owing to the very large dilution required to
produce the constant or limiting value. An indirect determination is,
however, possible. Thus $\Lambda^\infty$ for ethanoic acid can be determined from the
molar conductivities at infinite dilution of potassium(I) ethanoate,
hydrochloric acid, and potassium(I) chloride. For each electrolyte

$$\Lambda^\infty = \Lambda^\infty_+ + \Lambda^\infty_-$$

If the molar conductivities of the ions are represented by chemical symbols,
we have

$$(K^+ + CH_3COO^-) + (H^+ + Cl^-) - (K^+ + Cl^-)$$
$$= (H^+ + CH_3COO^-)$$

This gives the molar conductivity of ethanoic acid at infinite dilution.

*Solubility of 'Insoluble' Substances.* Many salts (such as AgCl, $CaCO_3$,
and $BaSO_4$) are described as insoluble in water. Actually, no substance
is completely insoluble, but some salts dissolve to such a small extent that
their solubility cannot be determined by the ordinary method. Even
when the solubility of a salt is exceedingly small the conductivity of the
water is measurably increased by the amount dissolved. By measuring
the conductivity we can deduce the solubility of the salt. Since the solution
is very dilute we may assume that the dissolved salt is completely dissociated and Kohlrausch's law can be applied. Consider a saturated solution
of silver(I) chloride. The molar conductivity at infinite dilution can be
calculated from the molar conductivities of the cation and anion at the
temperature of the experiment (usually 298 K):

$$\Lambda^\infty(\text{AgCl, 298 K}) = \Lambda^\infty(Ag^+, \text{298 K}) + \Lambda^\infty(Cl^-, \text{298 K})$$

Let $\kappa =$ the electrolytic conductivity of the solution (after deducting the
electrolytic conductivity due to water). Then, since silver(I) chloride is
completely dissociated in very dilute solution, we have (p. 403).

$$\Lambda^\infty = \Lambda_V = \kappa \times V$$

where $V$ is the volume in $m^3$ containing 1 mole of solute. If $\Lambda^\infty$, $\kappa$, and
the molar mass of AgCl are known we can calculate the volume in $m^3$ of
solution which contains 1 mole of AgCl.

**Example.** *The electrolytic conductivity of a saturated solution of silver(I) chloride at 18°C after deducting the electrolytic conductivity of water is* $1.22 \times 10^{-4}\ Sm^{-1}$. *The molar conductivities of the* $Ag^+$ *and* $Cl^-$ *ions at infinite dilution at 18°C are* $0.540 \times 10^{-2}$ *and* $0.652 \times 10^{-2}\ Sm^2\ mol^{-1}$. *Calculate the solubility of silver(I) chloride at 18°C.* $(Ag = 108, Cl = 35.5.)$

$$\Lambda^\infty = (0.540 \times 0.652) \times 10^{-2} = 1.192 \times 10^{-2}\ Sm^2\ mol^{-1}$$

$$V = \frac{\Lambda^\infty}{\kappa} = \frac{1.192 \times 10^{-2}}{1.22 \times 10^{-4}}\ m^3 = 97.7\ m^3$$

That is, $97.7\ m^3$ of solution contain 1 mole of AgCl.

$$1\ m^3 \text{ of solution contains } \frac{1}{97.7} \text{ mol of AgCl}$$

$$= 0.0102 \text{ mol of AgCl}$$

Since the density of the very dilute solution may be taken as $1000\ kg\ m^{-3}$, the solubility of silver(I) chloride is $1.02 \times 10^{-5}\ mol\ kg^{-1}$.

**Conductimetric Titration.** This is an application of conductance in which the different velocities of ions are used to ascertain the end point in a titration. The apparatus is the usual conductivity apparatus described at p. 405.

Suppose we wish to find the exact concentration of a solution of hydrochloric acid which is known to be about decimolar. $100\ cm^3$ of the acid are placed in a conductivity cell which forms the unknown resistance in a Wheatstone's Bridge. To avoid large variations in resistance of the solution owing to dilution it is advisable to use standard alkali which is much more concentrated (say, 2M) than the acid. Small amounts of the alkali are added from a burette to the acid. After each addition the liquids are well mixed and the point of balance determined. The readings indicate that as the acid is neutralized the resistance increases—that is, the conductance decreases. This is because the fast-moving hydrogen ions (see table at p. 413) are converted into water molecules, their place being taken by slower sodium(I) ions. When the point of neutralization has been passed the conductance increases again, because not only is the concentration of sodium(I) ions still being increased, but fast-moving hydroxyl ions are being added. By plotting the bridge readings (these are proportional to the conductance) against the volume of alkali added, a curve (ABC, Fig. 16–5) can be drawn which clearly indicates the end point of the titration.

If a weak acid like ethanoic acid is titrated with sodium(I) hydroxide by this method the curve (PQR, Fig. 16–5) takes a different form. Owing to the small number of hydrogen ions present in a solution of a weak

# 416   *The ionic theory*

acid, there is a slow increase in conductance as sodium(I) hydroxide is added. At the end point there is a sharp increase in conductance due to the addition of excess of hydroxyl ions.

Conductimetric titration is not confined to acids and alkalis. Thus with solutions of barium(II) hydroxide and magnesium(II) sulphate(VI) a double precipitate is formed, and there is a well-marked point of minimum conductance on the graph corresponding to equivalent proportions of the two electrolytes:

$$Ba(OH)_2 + MgSO_4 \rightarrow BaSO_4 + Mg(OH)_2$$

Other pairs of solutions which can be titrated are copper(II) sulphate(VI) and sodium(I) hydroxide, and sodium(I) ethanoate and hydrochloric acid.

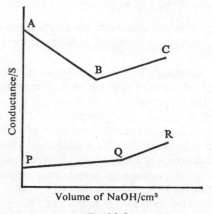

*Fig.* 16–5

**Electrophoresis with Ions on Paper.**[1] 'Electrophoresis' refers to the separation of charged particles in solution by using their tendency to migrate towards the oppositely charged electrodes. The phenomenon has already been described in connection with colloidal particles. Electrophoresis on paper is an adaptation which makes it possible to separate both similarly charged and oppositely charged ions.

In this method a strip of filter paper soaked in a buffer solution is suspended so that its ends dip into separate vessels containing the same buffer solution (Fig. 16–6*a*). The vessels contain platinum electrodes. When a drop of a solution of an electrolyte (or a mixture of electrolytes) is placed on the centre line of the paper and the current is started, the cations move towards the cathode and the anions towards the anode.

---

[1] For a more detailed account of paper electrophoresis see *Teaching Manual for Chromatography and Electrophoresis on Paper* by J. G. Feinberg and Ivor Smith (Shandon Scientific Company, Limited).

Similarly charged ions can be separated if there is sufficient difference in their velocities. If the ions are coloured their movement can be seen. If they are colourless, it may be possible to detect their movement by subsequently drying the paper and 'developing' with a suitable reagent.

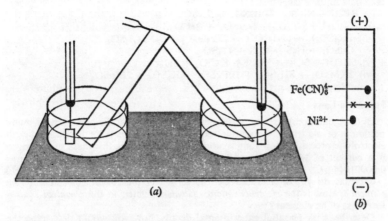

*Fig.* 16–6. (*a*) Apparatus used to demonstrate electrophoresis. (*b*) Electrophoretogram obtained with nickel(II) sulphate(VI) and potassium(I) hexacyanoferrate(II) solutions

**Experiment.** (i) Prepare a buffer solution from equal volumes of about M/10 ethanoic acid and M/25 ammonium ethanoate. Cut a strip of filter paper about 30 cm long and 5 cm wide. Mark in pencil two crosses on the centre line of the paper and about 3 cm apart. Wet the paper with the buffer solution and suspend it from a drawn-out glass tube or rod held in a cork in a clamp (if an ordinary glass tube or rod is used the test drop of liquid tends to spread out across the paper). Allow the ends of the paper to dip into glass vessels containing the buffer solution and standing on an insulating plate of poly(ethene) or glass. Using a tube drawn into a jet, place minute drops of aqueous nickel(II) sulphate(VI) and potassium(I) hexacyanoferrate(II) at the places marked. Insert the platinum electrodes and connect them to a 200-volt supply of d.c. Pass the current for 2 hours. The average distance travelled by the $Ni^{2+}$ ions will be about 2·5 cm and that by the $Fe(CN)_6^{4-}$ ions (in the opposite direction) about 3·5 cm. After switching off the current remove the paper and dry it in the oven. 'Develop' the nickel(II) ions with butanedione dioxime (dimethylglyoxime) (to give a red colour) and the hexacyanoferrate(II) ions with an iron(III) solution (to give a blue colour).

(ii) Repeat the experiment using bromophenol blue, phenol red, and a mixture of the two. Both indicators have coloured anions. These move towards the anode at different rates and a separation can be effected.

Electrophoresis on paper is often used for the separation of mixtures of amino-acids obtained by hydrolysis of proteins. The process is often combined with paper chromatography.

## EXERCISE 16
*(Relative atomic masses are given at the end of the book)*

1. Define *electrolyte, anode, ion, anion, faraday.*
2. Represent the following changes ionically, including only those ions which take part in the change:
   (i) $BaCl_2 + Na_2SO_4 \rightarrow BaSO_4 + 2NaCl$.
   (i) $2NaOH + H_2SO_4 \rightarrow Na_2SO_4 + 2H_2O$.
   (iii) $Fe_2(SO_4)_3 + 6NH_4OH \rightarrow 2Fe(OH)_3 + 3(NH_4)_2SO_4$.
   (iv) $CuSO_4 + H_2S \rightarrow CuS + H_2SO_4$.
   (v) $6KOH + 3Cl_2 \rightarrow 5KCl + KClO_3 + 3H_2O$.
   (vi) $2KMnO_4 + 8H_2SO_4 + 10FeSO_4 \rightarrow K_2SO_4 + 2MnSO_4$
       $+ 5Fe_2(SO_4)_3 + 8H_2O$.

### Faraday's Laws

3. (Part question.) Assuming 100 per cent efficiency for the process how many coulomb of electricity would be required to produce 0·2 g hydrogen by an electrolytic process? How long would it take to produce this mass of hydrogen if a current of 10 A were used? (Electrochemical equivalent of hydrogen = 0·00001044 g $C^{-1}$.)                                             (J.M.B.)
4. What do you understand by the term *electrochemical equivalent of an element*? How is the *electrochemical equivalent* related to the *chemical equivalent* mass of an element?
   Describe, with essential experimental details, how you would determine the electrochemical equivalent of an element such as silver.
   0·406 g of a metal X was deposited from a solution by a current of 1 A flowing for 965 second. The metal formed a volatile chloride of relative vapour density 114. Calculate the relative atomic mass of the metal. (Faraday constant = 96 500 C $mol^{-1}$;  Cl = 35·5.)                                       (O. and C.)
5. (Part question.) A current is passed through a silver and a water voltameter connected in series. A total of 1·950 $dm^3$ of gas, measured dry at 17°C and 98 700 Pa, are evolved from the water voltameter. Calculate the mass of silver deposited on the cathode of the silver voltameter.                          (S.U.)
6. State Faraday's laws of electrolysis and describe *briefly*, with the aid of a labelled diagram, a simple experimental method of illustrating them. Define the meaning of the term *electrochemical equivalent*.
   An electric current was passed for 1 hour through three electrolytic cells connected in series and fitted with platinum electrodes.
   The cells contained aqueous copper(II) sulphate(VI), silver(I) nitrate(V), and dilute sulphuric(VI) acid. During this time 0·106 g of copper was deposited on the cathode of the first cell. Calculate (*a*) the strength, in milliampere, of the current used; (*b*) the mass of silver deposited on the cathode of the second cell; and (*c*) the total volume of gas (measured at 15°C and 100 000 Pa) liberated in the third cell. (Faraday constant = 96 500 C $mol^{-1}$; the molar volume = 22·4 $dm^3$ at s.t.p.; Cu = 63·6; Ag = 108.)                          (W.J.E.C.)
7. (Part question.) The same quantity of electricity is used to liberate iodine (at an anode) and a metal X (at a cathode). The mass of X liberated is 0·617 g, and the iodine is completely reduced by 46·3 $cm^3$ of 0·124 molar $Na_2S_2O_3$ solution. Calculate the chemical equivalent mass of X.                          (C.L.)
8. (Part question.) Calculate the charge on the electron. (Faraday constant = 96 500 C $mol^{-1}$;  Avogadro constant = 6·06 × $10^{23}$ $mol^{-1}$.)          (S.U.)

### Conductivity of Solutions and Degree of Dissociation

9. Describe how you would measure the electrolytic conductivity of an aqueous solution of sodium(I) chloride.

Distinguish between the electrolytic conductivity and the molar conductivity of such a solution. Draw a rough graph to show how the molar conductivity of such a solution changes with dilution, and account qualitatively for the dependence of conductivity on dilution.

At 18°C the molar conductivity of sodium(I) chloride at infinite dilution is $1 \cdot 09 \times 10^{-2}$ Sm$^2$ mol$^{-1}$, whereas the corresponding figure for hydrochloric acid is $3 \cdot 29 \times 10^{-2}$ Sm$^2$ mol$^{-1}$. What interpretation can you place on this difference? (O.L.)

10. Calculate the molar conductivity and the degree of dissociation of the following solutions from the data provided:

| Solution | Electrolytic conductivity/Sm$^{-1}$ | $\Lambda^\infty$/Sm$^2$ mol$^{-1}$ |
|---|---|---|
| (i) 0·1 M KCl | 1·29 | $1 \cdot 499 \times 10^{-2}$ |
| (ii) 0·02M HCl | 0·814 | $4 \cdot 261 \times 10^{-2}$ |
| (iii) 0·005M AgNO$_3$ | 0·064 | $1 \cdot 334 \times 10^{-2}$ |

11. A solution containing 6·66 g of ethanoic acid per dm$^3$ has an electrolytic conductivity of $5 \cdot 21 \times 10^{-2}$ Sm$^{-1}$ at 25°C. The molar conductivity of the acid at infinite dilution at this temperature is $3 \cdot 91 \times 10^{-2}$ Sm$^2$ mol$^{-1}$. Calculate the degree of dissociation.

12. Calculate the degree of dissociation of the following solutions from the freezing point data provided:

| Solute | Concentration (in g per 100 g H$_2$O) | Freezing point/°C |
|---|---|---|
| (i) NaCl | 1·0 | −0·606 |
| (ii) FeCl$_3$ | 1·22 | −0·48 |
| (iii) CuSO$_4$ | 4·8 | −1·04 |
| (iv) CaCl$_2$ | 5·0 | −2·03 |

(The freezing constant of water per 1000 g $= 1 \cdot 86$ degC mol$^{-1}$.)

13. Calculate the freezing points of the following solutions from the data provided:

| Solute | Concentration (in g per 100 g H$_2$O) | Degree of dissociation |
|---|---|---|
| (i) NaCl | 5·85 | 0·83 |
| (ii) HCl | 4·0 | 0·95 |
| (iii) Na$_2$SO$_4$ | 6·5 | 0·62 |

(The freezing constant of water per 1000 g $= 1 \cdot 86$ degC mol$^{-1}$.)

14. (Part question.) What conclusions can be drawn from the following experimental data? A solution containing 54 g of glucose ($C_6H_{12}O_6$) per dm$^3$ has the same osmotic pressure as a solution containing 24 g of magnesium(II) sulphate(VI) (MgSO$_4$) per dm$^3$. (C.L.)

## MORE DIFFICULT QUESTIONS

**15.** What is meant by the Avogadro constant? Describe and explain how this constant can be determined by an electrolytic method.

**16.** The electrolytic conductivity of 0·016 molar ethanoic acid ($CH_3COOH$) at 18°C is 0·0196 $Sm^{-1}$. The molar conductivities at infinite dilution of the $H^+(aq)$ and $CH_3COO^-$ ions at 18°C are 3·15 × $10^{-2}$ and 0·35 × $10^{-2}$ $Sm^2$ $mol^{-1}$ respectively. Calculate the degree of dissociation of the ethanoic acid.

**17.** The osmotic pressure of an aqueous solution which contains 5g of potassium(I) nitrate(V) in 1 $dm^3$ of solution is 220 900 Pa at 15°C. What is the degree of dissociation of the salt?

**18.** The electrolytic conductivity of saturated aqueous silver(I) bromide at 20°C is 5·42 × $10^{-6}$ $Sm^{-1}$. The molar conductivities of the $Ag^+$ and $Br^-$ ions at infinite dilution at 20°C are 0·56 × $10^{-2}$ and 0·70 × $10^{-2}$ $Sm^2$ $mol^{-1}$ respectively. What is the solubility of the salt in mol $kg^{-1}$ at this temperature?

**19.** The electrolytic conductivity of a saturated solution of lead(II) sulphate (VI) in water at 20°C is 3·44 × $10^{-3}$ $Sm^{-1}$. The molar conductivities at infinite dilution of the lead ($\frac{1}{2}Pb^{2+}$) and sulphate ($\frac{1}{2}SO_4^{2-}$) ions at 20°C are 0·63 × $10^{-2}$ and 0·70 × $10^{-2}$ $Sm^2$ $mol^{-1}$ respectively. What is the solubility in g $kg^{-1}$ of lead(II) sulphate(VI) in water at 20°C (i) if the salt is completely dissociated, (ii) if the salt is 92 per cent dissociated?

# CHAPTER 17

# Electrochemical series and redox series

## Electrochemical Series

**Electrolytic Solution Pressure.** When two different metals are introduced into a dilute aqueous solution of an electrolyte they develop a difference of potential, and if they are connected by a wire a current flows between them. If we have zinc and copper in dilute sulphuric(VI) acid current flows externally from the copper to the zinc. At the same time zinc dissolves, forming zinc ions in solution, and hydrogen is evolved at the surface of the copper.

To explain these phenomena, Nernst suggested in 1888 that when a metal is placed in water or a solution of one of its salts the metal exerts an *electrolytic solution pressure*, whereby it tends to throw off positive ions into the liquid. This process, however, is not the same as the ionization of gaseous metal atoms discussed in connection with ionization energy in Chapter 6. In the first place the metal is a crystalline solid, and in the second the particles formed are not bare ions, but hydrated ions. We can divide the process into three stages, each of which is accompanied by an energy change. Firstly, energy has to be expended in converting the solid metal into free 'gaseous' atoms. This energy, which is called *sublimation energy*, carries a positive sign because it increases the energy of the system. Secondly, a further amount of energy, *ionization energy*, has to be supplied to remove electrons from the gaseous atoms and form gaseous ions. This again has a positive sign. Finally, the gaseous ions combine with water molecules, the change being attended by evolution of *hydration energy*. This has a negative sign. All the energy quantities are expressed in kJ mol$^{-1}$. Thus for zinc we have

|  |  | $\Delta H/\text{kJ mol}^{-1}$ |
|---|---|---|
| (1) | Zn (solid) $\rightarrow$ Zn (gas) | $+130$ |
| (2) | Zn (gas) $\rightarrow$ Zn$^{2+}$ (gas) $+$ 2e | $+2653$ |
| (3) | Zn$^{2+}$ (gas) $\rightarrow$ Zn$^{2+}$ (aq) | $-2481$ |

Zn (solid) $\rightarrow$ Zn$^{2+}$ (aq) $+$ 2e     $+302$

We see from this that in the conversion of a mole of metallic zinc into hydrated ions in solution 302 kJ of energy are absorbed.

The throwing-off of positive ions into the liquid means that an excess of electrons is left on the surface of the metal, which therefore acquires a negative charge. Owing to the attraction of opposite charges, the hydrated ions remain close to the metal and an *electric double layer* is formed (Fig. 17–1). After a time an equilibrium is established in which the rate of formation of the ions is equal to the rate at which they are deposited again. Thus for a divalent metal M

$$M(s) \rightleftharpoons M^{2+} (aq) + 2e$$

The formation of the electric double layer is accompanied by a potential difference between the surface of the metal and the liquid. The position of the equilibrium in the reversible reaction given above is of the utmost importance in determining the magnitude of the potential difference.

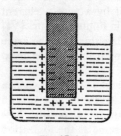

*Fig.* 17–1

The further to the right the position of equilibrium the greater is the electron density on the surface of the metal, and the larger is the potential difference between the metal and the liquid. For a *given* metal the position of equilibrium depends on the concentration of metal ions already present in the liquid. The greater this concentration the further to the left is the position of equilibrium and the smaller is the potential difference. For *different* metals placed in solutions containing the same concentration of their ions at the same temperature the position of equilibrium is governed by the overall energy change in forming hydrated ions from the metal.

It is impossible to measure the absolute potential difference between a metal surface and a liquid because it is necessary to have a second electrode which will also have a potential difference with respect to the liquid. However, *relative* potential differences for different metals under the same conditions can be compared. This is done by using a 'hydrogen electrode' as the second electrode. A description and diagram of this electrode are given at p. 486. A *standard* hydrogen electrode consists essentially of a piece of platinum foil suspended in sulphuric(VI) acid containing 1 g of hydrogen ions per dm$^3$, with hydrogen gas bubbling over the surface of the platinum. Three stages with characteristic energy changes are again concerned in the conversion of hydrogen molecules into hydrated hydrogen ions. In this case, however, the first stage consists of dissociation of molecules into atoms, which involves the dissociation energy of the hydrogen molecule instead of sublimation energy.

$$\Delta H/\text{kJ mol}^{-1}$$

| | | $\Delta H/\text{kJ mol}^{-1}$ |
|---|---|---|
| (1) | $\frac{1}{2}H_2(g) \rightarrow H(g)$ | $+218$ |
| (2) | $H(g) \rightarrow H^+(g) + e$ | $+1305$ |
| (3) | $H^+(g) \rightarrow H^+(aq)$ | $-1070$ |
| | $\frac{1}{2}H_2(g) \rightarrow H^+(aq) + e$ | $+453$ |

Because the overall energy change for a hydrogen electrode differs from that for a metal electrode under the same conditions the potential differences at the electrode surfaces also differ. Table 17–1 shows the energy changes per mole when certain metals and hydrogen ($\frac{1}{2}H_2$) become hydrated ions in solution (values in the last three columns are approximate because heats of hydration of ions cannot be found very accurately).

Table 17–1. ENERGY CHANGES/KJ MOL$^{-1}$ FOR IONIZATION OF METALS AND HYDROGEN IN SOLUTION

| | Sublimation energy | Ionization energy | Hydration energy | Overall energy change ($\Delta H$) | $\Delta H$ relative to hydrogen |
|---|---|---|---|---|---|
| Na | $+109$ | $+494$ | $-398$ | $+205$ | $-248$ |
| Zn | $+130$ | $+2653$ | $-2481$ | $+302$ | $-151$ |
| H | $(+218)$ | $+1305$ | $-1070$ | $+453$ | $0$ |
| Cu | $+339$ | $+2716$ | $-2540$ | $+515$ | $+62$ |
| Ag | $+280$ | $+728$ | $-464$ | $+544$ | $+91$ |

We see from this table that *less* energy has to be supplied to bring about the ionization of zinc than that of hydrogen, but *more* energy is required in the cases of copper and silver. From the overall energy changes we can predict the extents of ionization to be in the order: $Zn > H > Cu > Ag$. This is found to be so in practice. Thus, while a zinc electrode is negative with respect to a hydrogen electrode, the latter is negative with respect to a copper or silver electrode (which means that a copper or silver electrode is positive with respect to a hydrogen electrode).

The potential difference at the platinum surface of a standard hydrogen electrode is arbitrarily taken to be zero. When the electrode is used in conjunction with a metal electrode placed in a solution containing 1 mol dm$^{-3}$ of the metal ion the electromotive force obtained is called the *electrode potential* of the metal.

Electrode potentials are measured by potentiometer. By arranging the elements and their corresponding ions in order of their electrode potentials an *electrochemical series* of the elements is obtained, as shown in Table 17–2.

Table 17–2. THE ELECTROCHEMICAL SERIES

(Electrode potentials ($E^\ominus$) in volt at 298 K)

| Li (Li$^+$) | $-3\cdot04$ | Fe (Fe$^{2+}$) | $-0\cdot44$ | Hg (Hg$^{2+}$) | $+0\cdot85$ |
|---|---|---|---|---|---|
| K (K$^+$) | $-2\cdot92$ | Sn (Sn$^{2+}$) | $-0\cdot14$ | $\frac{1}{2}$Br$_2$ (Br$^-$) | $+1\cdot07$ |
| Ca (Ca$^{2+}$) | $-2\cdot87$ | Pb (Pb$^{2+}$) | $-0\cdot13$ | Pt (Pt$^{2+}$) | $+1\cdot20$ |
| Na (Na$^+$) | $-2\cdot71$ | $\frac{1}{2}$H$_2$ (H$^+$) | $0\cdot00$ | $\frac{1}{2}$Cl$_2$ (Cl$^-$) | $+1\cdot36$ |
| Mg (Mg$^{2+}$) | $-2\cdot37$ | Cu (Cu$^{2+}$) | $+0\cdot34$ | Au (Au$^+$) | $+1\cdot68$ |
| Al (Al$^{3+}$) | $-1\cdot66$ | $\frac{1}{2}$I$_2$ (I$^-$) | $+0\cdot54$ | $\frac{1}{2}$F$_2$ (F$^-$) | $+2\cdot85$ |
| Zn (Zn$^{2+}$) | $-0\cdot76$ | Ag (Ag$^+$) | $+0\cdot80$ | | |

Many electrode potentials cannot be measured directly. For example, it is impossible to place highly reactive metals such as sodium in contact with aqueous solutions of their ions. In cases like these the electrode potentials are determined indirectly or are calculated from thermodynamic data. The values for sodium and other alkali metals can be found by making them into amalgams with mercury. This reduces the reactivity sufficiently for the electrode potential to be measured (allowance being made for the effect of the mercury). Notice that the non-metallic halogen elements can also have electrode potentials. A chlorine electrode is made by bubbling the gas over platinum which is in contact with a solution of chloride ions.

The electrochemical series is an important classification because it expresses the relative tendencies of elements in their normal states to form hydrated ions in solution. As we shall see shortly, a number of chemical properties of elements depend on the strength of this tendency. In general for metals the tendency is greater the higher the position of the metal and its ion in the series. The order in which the metals occur does not correspond exactly with the order given by ionization energies. Thus lithium, which is at the top of the series, is usually regarded as the least electropositive of the alkali metals because it has the highest ionization energy. Lithium owes its high position in the series to the small size of its ion, the Li$^+$ ion being the smallest of the alkali metal ions. For this reason it becomes the most heavily hydrated in solution and a correspondingly large amount of energy is evolved. Similarly calcium occurs above sodium because the double charge on the calcium(II) ion causes greater hydration of the ion.

For the halogens the tendency for the element in its normal state to form negative ions in solution is greater the lower the position of the element in the series. Again this is due (in part) to the decrease in size of the ion from I$^-$ to F$^-$ and the consequent increase in the extent of hydration.

**Electric Potential and Concentration of Ions.** If the concentration of metal ions is not 1 mol dm⁻³, but $c$, the e.m.f., $E$, of the cell formed by the metal and its ions with a standard hydrogen electrode is given by Nernst's equation

$$E = E^{\ominus} \pm \frac{RT}{zF} \log_e c$$

where $E^{\ominus}$ is the electrode potential in volt, $R$ is the gas constant, $T$ is the kelvin temperature, $z$ is the valency of the ion, and $F$ is the Faraday constant. (Strictly speaking, activity should be used instead of concentration, but these can be taken to be the same if the solution is dilute.) The positive sign refers to positive ions and the negative sign to negative ions. At 298K the expression becomes with ordinary logarithms

$$E = E^{\ominus} \pm \frac{0 \cdot 059}{z} \log c$$

The Nernst equation may appear at variance with the statement made at p. 422 that the potential difference between metal and solution is smaller at a higher concentration of ions. However, the charge on a metal electrode in contact with a solution of its ions is always negative, although in comparison with the standard hydrogen electrode the electrode potentials of some metals are conventionally represented as positive. For a copper electrode in a 0·01 molar solution of copper ($Cu^{2+}$) ions $z = 2$ and $\log c = -2$. Thus $E = E^{\ominus} - 0 \cdot 059 = +0 \cdot 34 - 0 \cdot 059 = +0 \cdot 281$ V. For zinc and a solution of $Zn^{2+}$ ions of 0·01 molar concentration we have: $E = -0 \cdot 76 - 0 \cdot 059 = -0 \cdot 819$ V. In both cases the effect of having a solution more dilute than the standard one is to make the metal electrode *more negative* with respect to the solution. Conversely, with a more concentrated solution the electric potential is less.

**Electric Cells.** If a cell is constructed from a metal electrode and a hydrogen electrode, each being in contact with a standard solution of their ions, the metal is always charged negatively if it is above hydrogen, and positively if it is below hydrogen, in the series. As we have seen, this is because a metal above hydrogen has a greater tendency than hydrogen to form hydrated positive ions in solution, while a metal below hydrogen has a smaller tendency. The resultant e.m.f. of any combination of metals from which a cell is constructed can be obtained by subtracting the respective electrode potentials algebraically. Thus a combination of

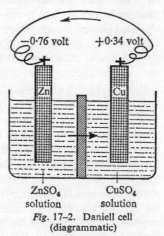

*Fig.* 17–2.  Daniell cell (diagrammatic)

copper and zinc electrodes (immersed in a standard solution of their ions) will produce an e.m.f. of $0 \cdot 34 - (-0 \cdot 76) = 1 \cdot 1$ V, and the copper will be charged positively with respect to the zinc. This arrangement occurs in the Daniell cell (represented in Fig. 17–2). It is true that the solutions employed in this cell do not contain 1 mol dm$^{-3}$ of $Zn^{2+}$ and $Cu^{2+}$ ions. The differences caused by different ionic concentrations, however, are small, so that the e.m.f. approximates to 1·1 V. The cell is represented diagrammatically as follows, the positive pole of the cell being conventionally placed on the right.

$$Zn|Zn^{2+}(aq)|Cu^{2+}(aq)|Cu$$

Similarly, the e.m.f. of the simple cell, consisting of zinc and copper electrodes in dilute sulphuric(VI) acid is approximately 1·1 V.

What is the source of the energy which can be obtained from a Daniell cell? The answer lies in the changes which occur at the electrodes. When the zinc and copper electrodes are not connected the following equilibria exist between the two metal surfaces and the solutions:

$$Zn(s) \rightleftharpoons Zn^{2+}(aq) + 2e \qquad (1)$$

$$Cu(s) \rightleftharpoons Cu^{2+}(aq) + 2e \qquad (2)$$

As noted earlier, the first equilibrium is further to the right than the second, and the electron density on the zinc surface is therefore higher than on the copper surface. If now the electrodes are connected by a wire electrons flow from zinc to copper in an attempt to equalize the electron densities. Removal of electrons in (1) displaces the position of equilibrium to the right and more zinc dissolves. Increase of electrons in (2) displaces the position of equilibrium to the left and copper is deposited. The overall chemical change is represented by the equation

$$Zn(s) + Cu^{2+}(aq) \rightarrow Zn^{2+}(aq) + Cu(s) \qquad (3)$$

From Table 17–1 we see that $\Delta H$ for the forward reaction in (1) is +302 kJ mol$^{-1}$, while $\Delta H$ for the *reverse* reaction in (2) is −515 kJ mol$^{-1}$. It follows that $\Delta H$ for reaction (3) is −213 kJ mol$^{-1}$. This means that there is an evolution of 213 kJ (which appears chiefly as electrical energy) for each mole of zinc dissolved and each mole of copper deposited.

A point of interest is that the sum of the ionization energies and ionic hydration energies of zinc and copper are about the same. Hence the energy difference between the forward reactions in (1) and (2) is essentially the difference between the sublimation energies of the two metals. Fundamentally the driving force for the Daniell cell is this difference in sublimation energies (a difference also reflected in the widely different boiling-points of zinc (907°C) and copper (2580°C)).

In the simple cell the overall reaction is as follows:

$$Zn(s) + 2H^+(aq) \rightarrow Zn^{2+}(aq) + H_2(g)$$

The energy change for the above reaction is the sum of the energy changes for the two simultaneous reactions:

$$Zn(s) \rightarrow Zn^{2+}(aq) + 2e \qquad \Delta H = +302 \text{ kJ mol}^{-1}$$

$$2H^+(aq) + 2e \rightarrow H_2(g) \qquad \Delta H = -906 \text{ kJ mol}^{-1}$$

There is thus an evolution of 604 kJ for the dissolving of a mole of zinc and the giving off of a mole of hydrogen.

**Electrolysis.** If the positive terminal of a battery is connected to the positive copper pole of a Daniell cell and the negative terminal to the negative zinc pole of the cell, a current can be passed through the cell in the opposite direction to the normal one. By including in the circuit a variable resistance the voltage applied to the cell can be varied and, if a voltmeter is placed across the copper and zinc poles, the applied voltage can be measured. With this arrangement continuous electrolysis occurs when the applied voltage is slightly in excess of 1·1 V. Electrolysis results in copper dissolving at the anode and zinc ions being discharged at the cathode. Thus the usual changes in the cell are reversed, and we have for the overall reaction

$$Cu(s) + Zn^{2+}(aq) \rightarrow Cu^{2+}(aq) + Zn(s)$$

It can be shown that the voltage in excess of 1·1 V is used merely to drive the current through the solution, the energy expended in this way producing a corresponding amount of heat energy. Hence we can conclude that 1·1 V are used in overcoming the normal electric potentials of the zinc (0·76 V) and copper (0·34 V). In this case both the zinc and copper are behaving as *reversible electrodes*. The minimum E.M.F. which must be applied between the solution and the electrode to bring about continuous discharge of an ion is called the *discharge potential* of the ion. For most metal ions in molar concentration the discharge potential is approximately equal to the electrode potential. It is, of course, opposite in direction.

The higher the position of a metal in the electrochemical series the bigger is the reversible electrode potential, and therefore the discharge potential, of its ions. Thus $Cu^{2+}$ ions are discharged in preference to $Zn^{2+}$ ions from a solution containing equal concentrations of the two. For the negatively charged ions of halogen elements the lower the position of the element in the series the greater is the e.m.f. required to discharge the ion. We do not know the discharge potentials of the $NO_3^-$ and $SO_4^{2-}$ ions (for simplicity it will now be taken for granted that ions in general are hydrated in solution). The values are very high and these ions are never discharged directly from solution at the anode.

A complication arises where certain gases are concerned. Here the discharge potential of the corresponding ions may differ considerably from the reversible electrode potential. The most important examples are discharge of $H^+(aq)$ ions (which yield hydrogen) and discharge of $OH^-$ ions (which give oxygen). The latter process is irreversible, but the theoretical reversible electrode potential for an oxygen $-OH^-$ ion electrode can be calculated. Its value is $+0.40$ V. This is then the e.m.f. involved in the change

$$\tfrac{1}{2}O_2 + H_2O + 2e \rightleftharpoons 2OH^-$$

assuming that the change can take place in the forward, as well as in the backward, direction.

Discharge potentials found for the $H^+(aq)$ ion and the $OH^-$ ion in practice are larger than the reversible electrode potentials. The difference between the two is called the *overvoltage*, or *overpotential*. Overvoltage is the result of one or more slow steps in the electrode reaction. Thus when dilute sulphuric(VI) acid is electrolysed with platinum electrodes the hydrogen and oxygen liberated adhere to the surface of the electrodes instead of being given off immediately. The films of gas not only set up a back-e.m.f. (which has to be overcome by extra voltage), but make it difficult for $H^+(aq)$ ions and $OH^-$ ions to reach the electrode surface. Overvoltage may also be caused by other factors, *e.g.*, the slowness with which the discharged ions are replaced by fresh ions diffusing or migrating from the rest of the solution. In general the amount of overvoltage depends on the nature of the electrode, the concentration of the solution, the temperature, and the current density (current strength divided by area of the electrode). When an electrode is operating at a higher potential than its reversible potential it is said to be *polarized*.

Although at ordinary temperatures only 1 in about 550 million molecules of water is dissociated into ions ($H_2O \rightleftharpoons H^+ + OH^-$), hydrogen and hydroxyl ions are always present at the electrodes in addition to the ions of the electrolyte. We have seen that if there is a mixture of two positively charged ions at an electrode the ion with the smaller discharge potential loses its charge preferentially. From the positions of lead and hydrogen in the electrochemical series we should expect $H^+(aq)$ ions to be discharged in preference to $Pb^{2+}$ ions. Actually, when an aqueous solution containing $Pb^{2+}$ ions is electrolysed with a lead cathode, lead, and not hydrogen is liberated. This is because the discharge potential of $H^+(aq)$ ions is increased by overvoltage to a higher value than that of $Pb^{2+}$ ions.

The examples given in the following section illustrate how the modern theory of electrolysis explains the products of electrolysis in some typical cases.

### Examples of Electrolysis.

*Fused Sodium(I) Hydroxide.* This is the basis of manufacture of sodium by the Castner process (now almost obsolete). The only ions present at the electrodes are sodium and hydroxyl ions. Metallic sodium is deposited at the iron cathode and oxygen is evolved at the nickel anode.

At the cathode: $Na^+ + e \rightarrow Na$

At the anode: $OH^- - e \rightarrow OH$

$$4OH \rightarrow 2H_2O + O_2$$

*Sodium(I) Hydroxide Solution.* This electrolyte, with nickel or iron electrodes, is used in the manufacture of hydrogen:

$$NaOH \rightleftharpoons Na^+ + \underline{OH^-}$$

$$H_2O \rightleftharpoons \underline{H^+} + OH^-$$

Of the two cations the $H^+(aq)$ ion has the lower discharge potential, even though its discharge potential is increased by overvoltage. Hydrogen gas is therefore evolved at the cathode, the supply of hydrogen ions being maintained by further dissociation of water molecules. At the anode hydroxyl ions are discharged and oxygen is evolved. As equal numbers of $H^+(aq)$ and $OH^-$ ions are discharged, the net result is electrolysis of water, the quantity of sodium(I) hydroxide being unaffected.

*Sodium(I) Chloride Solution.* Sodium(I) hydroxide is manufactured by electrolysis of aqueous sodium(I) chloride in diaphragm cells of the Nelson type, using a carbon anode and iron-cathode:

$$NaCl \rightleftharpoons Na^+ + \underline{Cl^-}$$

$$H_2O \rightleftharpoons \underline{H^+} + OH^-$$

At the cathode $H^+(aq)$ ions are discharged, as they have a lower discharge potential than $Na^+$ ions. At the anode the overvoltage of the $OH^-$ ion raises its discharge potential above that of the $Cl^-$ ion, so that chlorine is liberated. As $H^+(aq)$ and $Cl^-$ ions are constantly removed, further dissociation of water molecules occurs, so that the remaining solution contains the original $Na^+$ ions and accumulated $OH^-$ ions.

In the manufacture of sodium(I) hydroxide by electrolysis of aqueous sodium(I) chloride in the Castner-Kellner cell a mercury cathode is used. Hydrogen ions have a very high overvoltage at a mercury cathode and sodium(I) ions are discharged instead. The sodium dissolves in the mercury to form sodium amalgam.

*Copper(II) Sulphate(VI) Solution.* Electrolysis of this solution with platinum electrodes yields copper at the cathode, while oxygen is liberated at the anode. The ions with the lower discharge potentials are the $Cu^{2+}$ and $OH^-$ ions:

$$CuSO_4 \rightleftharpoons \underline{Cu^{2+}} + SO_4^{2-}$$

$$H_2O \rightleftharpoons H^+ + \underline{OH^-}$$

In electrolysis of copper(II) sulphate(VI) solution with copper electrodes copper is again deposited at the cathode, but no oxygen is liberated at the anode. Instead copper dissolves from the anode. This requires a special word of explanation. We can think of an anode as having the duty to supply electrons to the external circuit. There are three ways in which it might do this:

(a) $SO_4^{2-}$ ions might be discharged, $SO_4^{2-} \rightarrow SO_4 + 2e$

(b) $OH^-$ ions might be discharged, $OH^- \rightarrow OH + e$

(c) Metal atoms on the electrode surface might dissolve as ions and leave their electrons behind, $Cu \rightarrow Cu^{2+} + 2e$

The potential difference between the anode and the solution required for these possibilities to occur is different in each case. With a platinum anode the lowest e.m.f. is required for the discharge of $OH^-$ ions (see above). With a copper anode the conversion of copper atoms to $Cu^{2+}$ ions needs the lowest e.m.f. Hence, with a copper anode no ions at all are discharged at this electrode, and for each atom of copper deposited at the cathode one atom of copper dissolves from the anode.

*Dilute Sulphuric(VI) Acid.* With platinum electrodes hydrogen is formed at the cathode and oxygen at the anode. The ions discharged are the $H^+(aq)$ and $OH^-$ ions:

$$H_2SO_4 \rightleftharpoons \underline{2H^+} + SO_4^{2-}$$

$$H_2O \rightleftharpoons H^+ + \underline{OH^-}$$

With copper electrodes no oxygen is obtained at the anode. Instead, copper dissolves, as explained above for electrolysis of copper(II) sulphate(VI). With carbon electrodes $H^+(aq)$ and $OH^-$ ions are again discharged, but very little oxygen is evolved at the anode. This is because the oxygen is adsorbed internally by the carbon.[1]

*Hydrochloric Acid.* This is an interesting example of electrolysis because it illustrates the effect of change of concentration on the discharge potential of an ion. If the concentrated acid is electrolysed with platinum electrodes the products are hydrogen and chlorine. With dilute acid oxygen is evolved at the anode instead of chlorine. With a solution of intermediate concentration a mixture of oxygen and chlorine is liberated at the anode.

$$HCl \rightleftharpoons H^+ + Cl^-$$

$$H_2O \rightleftharpoons H^+ + OH^-$$

We can explain the variation in anode products as follows. As we saw at p. 428, the electrode potential, $E^{\ominus}$, of an oxygen $-OH^-$ ion electrode

---

[1] The nature of the changes which take place at the anode when solutions of acids are electrolysed with carbon electrodes is complex. Various 'chemisorption' products of carbon and oxygen are formed, according to the acid used. The changes are discussed by H. Thiele in the *Faraday Society Transactions*, vol. xxxiv, pp. 1033–1039. See *The School Science Review*, vol. xxx, No. 110, p. 118.

is theoretically $+0.40$ V, while that of the chlorine $-Cl^-$ ion electrode is $+1.36$ V. However, discharge potentials of ions, like potential differences between a metal electrode and its ions (p. 425) are lower at higher ionic concentrations. Thus with concentrated hydrochloric acid $Cl^-$ ions are discharged by a voltage less than $1.36$ V, while the discharge potential of the $OH^-$ ion is increased by overvoltage at the platinum surface to about $1.40$ V. Hence $Cl^-$ ions are discharged preferentially.

As the concentration of the $Cl^-$ ion decreases, its discharge potential increases and becomes equal to that of the $OH^-$ ions in the solution. Hence over a range of concentrations (depending on the current density) near this point both chlorine and oxygen are liberated. (In the same way, although two metals may have different standard electrode potentials, it may be possible to get their ions discharged at the same time by suitable adjustment of the concentrations.) At still lower concentrations of chloride ion its discharge potential becomes appreciably larger than that of the hydroxyl ion (even with overvoltage). At this stage only $OH^-$ ions are discharged at the anode.

*Solutions of Silver(I) Salts.* If aqueous silver(I) nitrate(V) is electrolysed with platinum electrodes silver is obtained at the cathode and oxygen at the anode. With silver electrodes silver is deposited at the cathode and dissolved from the anode, the changes being analogous to those described for copper(II) sulphate(VI) with copper electrodes.

In silver plating a pure silver anode is employed and the article to be plated is the cathode. As electrolyte potassium(I) dicyanoargentate(I), $KAg(CN)_2$, made by dissolving silver(I) cyanide in aqueous potassium(I) cyanide, is preferred to silver(I) nitrate(V) because the silver deposit adheres better to the article being plated. In the solution equilibrium exists between the complex ion and the simple ions:

$$Ag(CN)_2^- \rightleftharpoons \underline{Ag^+} + 2CN^-$$

The equilibrium point in this reaction is far to the left, but as $Ag^+$ ions are discharged at the cathode a further supply is provided by dissociation of the complex ion. At the anode silver is transferred to the solution as ions $(Ag - e \rightarrow Ag^+)$ in preference to discharge of $CN^-$ or $OH^-$ ions, so that the concentration of silver ions in the solution remains constant.

## Chemical Properties of the Elements

The electrochemical series, like the Periodic Table, enables us to correlate and explain many facts which at first sight appear quite unrelated. Many of the chemical properties of elements are determined by the tendencies of their atoms to form (hydrated) ions in solution. This is particularly the case with metals. Generally speaking, the higher the position of a metal in the electrochemical series the more readily does it give rise to positive ions. For the halogens the lower the position in the series the stronger is the tendency to form negative ions.

**Displacement of an Element from Solution by Another Element.** In general a metal is displaced from aqueous solutions of its salts by a metal which is higher in the electrochemical series. Thus copper(II) ions are displaced by zinc. This reaction is basically the same as the overall reaction in the Daniell cell:

$$Zn(s) + Cu^{2+}(aq) \rightarrow Zn^{2+}(aq) + Cu(s) \qquad \Delta H = -213 \text{ kJ mol}^{-1}$$

As explained previously, the energy evolved in the reaction is the resultant of the energy changes for the two separate reactions

$$Zn(s) \rightarrow Zn^{2+}(aq) + 2e \qquad \Delta H = +302 \text{ kJ mol}^{-1}$$
$$Cu^{2+}(aq) + 2e \rightarrow Cu(s) \qquad \Delta H = -515 \text{ kJ mol}^{-1}$$

Iron also displaces copper from solutions of copper(II) salts, but, as we might expect from the relative positions of the metals in the series, zinc displaces iron from iron(II) salts. Further examples of replacement can be easily demonstrated by introducing strips of various metals (Mg, Zn, Fe, Pb) into solutions of different salts, such as those of lead, mercury, copper, and silver. Aluminium does not normally bring about replacement when it might be expected to do so. This is because the metal is covered with a thin but extremely resistant film of aluminium(III) oxide.

There are some exceptions to the rule that a metal higher in the series displaces a metal lower in the series. It must be borne in mind that the order of the elements in the series is deduced under particular conditions, *e.g.*, a certain concentration of ions. With different concentrations of ions the order of the elements varies slightly. It also depends to some extent on the temperature and the nature of the salt (copper will displace zinc from the complex salt potassium(I) tetracyanozincate(II).

Halogens are displaced from solution by other halogens which are lower in the series. Thus fluorine displaces chlorine, chlorine displaces bromine, and bromine displaces iodine, *e.g.*,

$$Cl_2(g) + 2Br^-(aq) \rightarrow Br_2(l) + 2Cl^-(aq)$$

The order of displacement depends largely on the increase in the heats of hydration of the ions from $I^-$ to $F^-$, the energy evolved being larger the smaller the size of the ion.

**Action of Metals on Water and Acids.** Theoretically any metal above hydrogen in the electrochemical series will displace it from water or aqueous solutions of acids, which contain hydrogen ions. The action of sodium on water can be represented:

$$Na + H^+ + OH^- \rightarrow Na^+ + OH^- + \tfrac{1}{2}H_2$$

In practice the rate at which $H^+(aq)$ ions are discharged and hydrogen gas formed depends upon the concentration of the $H^+(aq)$ ions. In water this is extremely small, but it is high in solutions of the strong mineral acids. Another important factor in the displacement of hydrogen from water is the solubility of the metal hydroxide. The higher metals (K, Ca, Na) readily liberate hydrogen from cold water. All these metals have soluble hydroxides. Other metals (Mg, Zn, etc.) above hydrogen in the series possess hydroxides which are almost insoluble, and the displacement in the cold is quickly arrested by the depositing of a film of hydroxide on the metal. The solubility of magnesium(II) hydroxide increases with rise of temperature, and powdered magnesium will liberate hydrogen from hot water fairly readily. The hydroxides of iron, tin, and lead, however, have such small solubilities that these metals have no appreciable action even with boiling water.

Note that the displacement of hydrogen from steam by metals is outside the province of the electrochemical series. In this case the action occurs not with $H^+(aq)$ ions, but with water molecules. Copper has no action with water, but with steam it is slightly oxidized at very high temperatures.

Metals above hydrogen liberate hydrogen from dilute acids more readily than from water and the ease with which this occurs decreases down the series:

$$M + 2H^+ \rightarrow M^{2+} + H_2$$

The action of sodium and potassium may be so rapid as to be explosive; on the other hand, the dilute acids are attacked only slowly by tin and lead. With nitric(V) acid and concentrated sulphuric(VI) acid, which are oxidizing agents, the reactions are complicated by further reactions between the acids and the hydrogen liberated. Metals below hydrogen in the series are attacked only by oxidizing acids and do not yield hydrogen.

Theoretically, hydrogen should displace from solution metals which are below it in the series. This does occur, but the actions are too slow to be appreciable unless the hydrogen is under pressure or a catalyst is used. Hydrogen will then displace silver and gold from solutions of their salts.

**Combination with Oxygen.** The attraction of metals for oxygen decreases down the electrochemical series. The higher metals burn in air or oxygen, and the resulting oxide cannot be decomposed by heat or reduced by hydrogen or carbon, except at very high temperatures. Intermediate metals, such as lead and tin, form stable oxides when heated in oxygen, but these are readily reduced by hydrogen or carbon on moderate heating. Oxides of metals such as mercury, silver, and gold, which are low in the series, can only be prepared indirectly and are easily decomposed by heat into metal and oxygen.

The position of a metal in the series has a bearing on the method used for its extraction. When the attraction of a metal for oxygen is high (as in the case of the alkali metals) reduction of the oxide by carbon is difficult. Neither, as a rule, can the metal be obtained by electrolysing an aqueous solution of a salt, because hydrogen ions of water are discharged in preference to metal ions. It is therefore necessary to electrolyse the fused chloride or reduce the latter by means of a metal higher in the series. Titanium is manufactured by reducing the chloride ($TiCl_4$) with sodium or magnesium. Although, strictly speaking, the electrochemical series does not apply in this case, it is a matter of experience that compounds like oxides and chlorides of a metal can be reduced by metals occupying higher positions in the series. Another example is the reduction of iron(III) oxide by aluminium.

**Stability of Hydroxides and Salts.** The hydroxides of the alkali and alkaline earth metals are decomposed only at high temperatures. They dissolve in water and form alkaline solutions. The hydroxides of the intermediate metals readily give the oxide on being heated. They are insoluble in water and are mostly amphoteric. The hydroxides of mercury, silver, and gold are too unstable to exist. Carbonates and nitrates(V) show a similar variation in stability, but not to the same degree as the hydroxides.

## Rusting of Iron

The rusting of iron is an illustration of the corrosion of metals. The term 'corrosion' is used to describe the effects of liquids and gases on metals, but it refers particularly to the effects produced by exposing metals to air or water. The atmosphere contains oxygen, moisture, carbon dioxide, sulphur dioxide, hydrogen sulphide, and sodium(I) chloride, all of which can play a part in corrosion of metals. Rust is a hydrated iron(III) oxide of varying composition, which is expressed by the formula $xFe_2O_3 . yH_2O$. It is formed by the drying out of iron(III) hydroxide, $Fe(OH)_3$, and it can be produced in different ways.

**Rusting Due to Differential Oxygen Concentration.** It is well known that iron will not rust if it is kept dry or if oxygen is absent. Rusting is faster in the presence of an electrolyte, like sodium(I) chloride, or a gas, such as carbon dioxide or sulphur dioxide, which can give rise to an electrolyte. It is now known that rusting is electrolytic in character and is caused by small currents in or near the surface of the iron. Usually the currents are due to differences in concentration of dissolved oxygen which prevents the electrolytic solution pressure of the metal from acting equally at different parts of the metal surface.

Suppose we have a drop of water on an iron surface (only half the drop is represented in Fig. 17–3). Oxygen from the air dissolves at the outside of the drop and forms a superficial layer of oxide on the metal. The oxide layer prevents the throwing off of $Fe^{2+}$ ions into the liquid. At the middle of the drop the metal atoms can freely exert their solution pressure, and ions pass into solution leaving an excess of electrons on the metal. There is therefore a higher potential at the outside of the drop than in the middle. A current passes from the outside to the inside of the drop through the metal (actually electrons flow in the opposite direction) and is completed through the liquid from the inside to the outside of the drop. The metal surface at the outside of the drop is a cathodic area, while an anodic area exists in the middle. $H^+$(aq) ions and $OH^-$ ions from the water are attracted to the cathode and anode respectively:

$$2H^+ + 2e \rightarrow H_2$$
$$Fe^{2+} + 2OH^- \rightarrow Fe(OH)_2$$

A precipitate of iron(II) hydroxide is produced by interaction of $Fe^{2+}$ ions and $OH^-$ ions in the neighbourhood of the anode and is converted into iron(III) hydroxide by oxygen in solution diffusing through the liquid:

$$4Fe(OH)_2 + O_2 + 2H_2O \rightarrow 4Fe(OH)_3$$

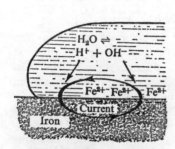

Fig. 17–3. Rusting caused by differential oxygen concentration

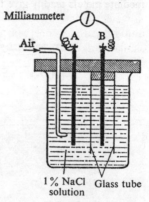

Fig. 17–4. Evans' experiment

The iron(III) hydroxide is not a closely adhering layer and fails to protect the iron below from further action. More iron atoms ionize and, as the oxygen diffusing inwards is constantly removed, the potential difference is maintained. Hence, the action continues. Pitting of iron thus occurs always at the middle of a drop.

The rate at which rusting occurs depends upon how quickly the $H^+$(aq) ions are discharged. This in turn depends upon their concentration and on the conductivity of the liquid. The presence of carbonic acid or sulphuric(VI) acid from the air increases both of these factors and accelerates the action. Dissolved salts also promote corrosion by increasing the conductivity. Thus iron rusts very quickly near the sea owing to the presence of sodium(I) chloride in the atmosphere.

**Evans' Experiment.** A simple method of demonstrating the production of the potential difference due to differential oxygen concentration at iron or steel electrodes was devised by Dr U. R. Evans, of Cambridge University. A modification of the apparatus is shown in Fig. 17–4.

Two similar steel plates, A and B, each about 12·5 cm long and 1 cm broad, are cleaned by rubbing them with emery paper. They are inserted through a wide cork, and one of them, B, passes through a second cork in the top of a wide glass tube which is open at the lower end. A glass tube is also inserted through the wide cork, the lower end of this tube being bent at right angles and drawn out into a jet which lies close to the surface of plate A. The apparatus is placed in a 400-cm³ beaker holding a 1 per cent solution of sodium(I) chloride, so that the wide cork rests on top of the beaker. A milliammeter is connected to the metal plates. When air (from a hand pump) is blown against the plate A the milliammeter shows that a current is passing externally from the aerated plate to the non-aerated plate. When the air stream is stopped the potential difference rapidly falls to zero and the current ceases.

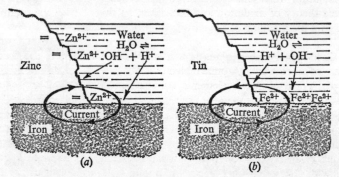

*Fig.* 17–5. (*a*) Corrosion of galvanized iron. (*b*) Corrosion of tinned iron

**Comparison of Zinc and Tin as Protective Coverings.** Zinc and tin are commonly used to protect iron from rusting, but they are not equally efficient. If galvanized iron becomes scratched so as to expose the iron below, rusting still does not occur. If the surface of tinned iron is broken, rusting of the exposed iron is rapid and is facilitated by the presence of the tin. The difference in the actions is connected directly with the relative positions of zinc, iron, and tin in the electrochemical series.

Let us suppose that part of the zinc coating has been removed and that the depression has become filled with water (Fig. 17–5*a*). Zinc is more electropositive than iron and $Zn^{2+}$ ions are thrown off into solution, leaving an excess of electrons on the zinc surface. A current therefore flows from iron to zinc and is completed through the liquid from zinc to iron. The surface of the zinc is an anodic area, while the surface of the iron is a cathodic area. Hydrogen is liberated at the iron. $OH^-$ ions are attracted to the zinc surface, where they meet and interact with the $Zn^{2+}$ ions, giving a precipitate of zinc(II) hydroxide. More $Zn^{2+}$ ions then pass into solution. We see that the rusting is saved by the dissolving of the zinc, which is therefore described as a 'sacrificial' metal.

Iron is more electropositive than tin. In this case, if the tin surface is broken and water is present, $Fe^{2+}$ ions go into solution and the direction of the current is opposite to that obtained with zinc. This will be seen from Fig. 17–5b. This time the exposed iron is an anodic area, towards which $OH^-$ ions are drawn. A precipitate of iron (II) hydroxide is formed from which rust is produced by the series of changes described previously.

## Ionic Theory of Oxidation and Reduction

There is no single theory which accounts satisfactorily for all the chemical reactions which are classed as oxidations and reductions. In the course of time these terms have come to be applied to such a wide variety of reactions that it is often difficult to see any connection between them. The ionic interpretation of oxidation and reduction is applied to reactions in solution when ions are involved in the changes. Before discussing the ionic theory of oxidation and reduction, however, it will be useful if we explain the meaning of the term 'oxidation number'.

**Oxidation Number.** This is a number used to express the oxidation state of an element. Atoms of elements are given an oxidation number of zero. Then when two elements are combined the atoms or ions of the more electropositive element are regarded as being in a positive oxidation state, and those of the more electronegative element in a negative oxidation state. If we arbitrarily assume that the bonds between the atoms are all ionic, the oxidation number of an element is simply the number of electrons given up or received by an atom of that element. In sodium(I) chloride the oxidation number of sodium is I, while that of chlorine is $-I$. Hydrogen in HCl has an oxidation number of I, but in sodium(I) hydride, NaH, its oxidation number is $-I$. Combined oxygen always has an oxidation number of $-II$ (except in peroxides, where it is $-I$). The first number agrees with the formulæ $Na_2O$ and $H_2O$.

Having established the oxidation numbers of a few common elements, we can extend the concept to compounds composed of more than two elements. In any compound the algebraic sum of the oxidation numbers of the atoms present is zero. Thus, knowing that in sulphuric(VI) acid, $H_2SO_4$, each hydrogen atom has an oxidation number of I and each oxygen atom an oxidation number of $-II$, we can deduce that the sulphur atom has an oxidation number of VI, as indicated in the name of the acid. This agrees with the oxidation state of sulphur in $SO_3$ (sulphur-(VI) oxide), from which sulphuric(VI) acid can be obtained merely by combination with water.

Some elements have several oxidation states. Thus the oxidation number of sulphur in $SO_2$ is IV, while in $H_2S$ it is $-II$. Chlorine can have oxidation numbers of $-I$, I, IV, V, and VII.

| I $-$I | I $-$II | IV $-$II | I V $-$II | I VII $-$II |
|--------|---------|----------|-----------|-------------|
| HCl | $Cl_2O$ | $ClO_2$ | $KClO_3$ | $KClO_4$ |

An increase in the oxidation number of an element during a reaction means that the element has been oxidized. Conversely, a decrease in the oxidation number means that the element has been reduced. We shall see examples of these changes in the next section.

**Oxidation and Reduction Involving Ions.** When oxidation and reduction take place in aqueous solution the reacting particles are often ions. Comparing the products of the reaction with the reactants from the ionic point of view, we see that the fundamental change consists of a transference of electrons. Thus the reduction of iron(III) chloride by tin(II) chloride can be represented as follows:

$$2Fe^{3+} + 6Cl^- + Sn^{2+} + 2Cl^- \rightarrow 2Fe^{2+} + 4Cl^- + Sn^{4+} + 4Cl^-$$

$$\text{or} \quad 2Fe^{3+} + Sn^{2+} \rightarrow 2Fe^{2+} + Sn^{4+}$$

The only change is the transfer of two electrons from a tin(II) ion to two iron(III) ions. The oxidation number of the iron is decreased from III to II, while that of the tin is increased from II to IV.

Similarly, the oxidation of an iron(II) salt in solution to an iron(III) salt by chlorine is accompanied by the transfer of an electron from an iron(II) ion to a chlorine atom:

$$2Fe^{2+} + Cl_2 \rightarrow 2Fe^{3+} + 2Cl^-$$

Here, the oxidation number of the iron has been increased from II to III, while that of the chlorine has been decreased from 0 to $-I$.

Again, in the reduction of an iron(III) salt in solution by hydrogen sulphide sulphur is precipitated. A sulphide ion loses two electrons and the oxidation state of the sulphur increases from $-II$ to 0. At the same time the oxidation state of the iron is reduced from III to II:

$$2Fe^{3+} + S^{2-} \rightarrow 2Fe^{2+} + S \downarrow$$

These examples show that from the ionic point of view *oxidation consists of a decrease in the number of electrons belonging to an atom or ion, and reduction is the converse*. If we accept this definition we must include as examples of oxidation and reduction many chemical reactions which are not usually thought of in this way. We must include displacement of hydrogen from dilute sulphuric(VI) acid by zinc ($Zn + 2H^+ \rightarrow Zn^{2+} + H_2$), displacement of copper from copper(II) sulphate(VI) solution by iron ($Fe + Cu^{2+} \rightarrow Fe^{2+} + Cu$), and the displacement of iodine from potassium(I) iodide solution by chlorine ($Cl_2 + 2I^- \rightarrow 2Cl^- + I_2$.) The ionic theory of oxidation and reduction has thus broadened considerably the meaning of these terms.

The agent which gives up electrons to an atom or ion is the reducing agent, while the agent which removes electrons from an atom or ion is the oxidizing agent. In other words, a *reducing agent is an electron donor and an oxidizing agent is an electron acceptor.* (These definitions are easily deduced from an obvious oxidizing agent such as oxygen and an obvious reducing agent such as magnesium. When magnesium and oxygen combine it is the former which gives up electrons to the latter.) In the reactions given in the previous paragraph zinc atoms, iron atoms, and $I^-$ ions are reducing agents and $H^+$ ions, $Cu^{2+}$ ions, and chlorine atoms are oxidizing agents. The relative strengths of reducing agents obviously depend on their readiness to lose electrons, whereas the relative strengths of oxidizing agents depend on their readiness to acquire electrons. Where atoms or simple ions of an element are concerned the readiness to lose or gain electrons varies with the position of the element in the electrochemical series.

Metal elements are reducing agents because their atoms easily lose electrons and form positive ions. The tendency for this to happen is greater the higher the position of the metal in the series. Thus sodium is a stronger reducing agent than iron and iron than copper. Conversely, the strengths of the corresponding ions as oxidizing agents are in the opposite order, because metal ions low in the series accept electrons more readily than those which are higher. Thus silver(I) nitrate(V) oxidizes iron(II) sulphate(VI) in solution, silver being deposited ($Fe^{2+}$ + $Ag^+ \rightarrow Fe^{3+}$ + $Ag$), while copper(II) sulphate(VI) has no action with aqueous iron(II) sulphate(VI). Silver and mercury are easily precipitated from solutions of salts because their ions readily take up electrons.

Non-metal elements are oxidizing agents because their atoms readily take up electrons and become negative ions. The lower the position of the element in the electrochemical series the greater is this tendency. Thus the strength of the halogens as oxidizing agents increases from iodine to fluorine. On the other hand, the order of strength of the corresponding ions as reducing agents is determined by readiness to give up electrons. Of the halogens this is least with the $F^-$ ion and greatest with the $I^-$ ion. We therefore find that hydriodic acid is the strongest reducing agent of the halogen acids, and is followed in order by hydrobromic acid, hydrochloric acid, and hydrofluoric acid.

**Oxidation and Reduction of Complex Anions.** Many of the ions which take part in oxidation and reduction changes are complex ions. Again, however, the reactions can be explained by the ionic theory of electron transfer.

*Oxidation of Thiosulphate(VI) Ions.* In titration of aqueous sodium(I) thiosulphate(VI), $Na_2S_2O_3$, with iodine the thiosulphate(VI) ions are oxidized to tetrathionate ions, and the iodine is reduced to $I^-$ ions. It is convenient to divide the reaction into half-reactions. For the first half-reaction we take the change in which the complex ion is involved. In this case this is the oxidation of the thiosulphate(VI) ion:

$$(a) \quad 2S_2O_3^{2-} \rightarrow S_4O_6^{2-} + 2e$$

In the second half-reaction the oxidation number of the iodine decreases from 0 in $I_2$ to $-I$ (minus one) in $I^-$. Keeping the number of electrons the same as in the first half-reaction, we have

$$(b) \quad I_2 + 2e \rightarrow 2I^-$$

Adding together the equations for the two half-reactions and omitting the electrons which are common to both sides, we obtain

$$2S_2O_3^{2-} + I_2 \rightarrow S_4O_6^{2-} + 2I^-$$

or, putting in the sodium(I) ions (which take no part in the reaction),

$$2Na_2S_2O_3 + I_2 \rightarrow Na_2S_4O_6 + 2NaI$$

The concentrations of solutions of oxidizing and reducing agents are commonly expressed in mol $dm^{-3}$. From the equations given above we see that 1 mole of $Na_2S_2O_3$ (158 g), or 1 mole of $Na_2S_2O_3 . 5H_2O$ (248 g) reacts with 0·5 mole of $I_2$ (0·5 of 254 g). (The reader is reminded that in using the term 'mole' the particles to which the term refers must be specified by means of a formula.) A solution containing 1 mol $dm^{-3}$ of an oxidizing or reducing agent is often described as a 'molar (M)' solution. Strictly speaking, use of the term 'molar' in this context is incorrect (see p. 61) and is now discouraged.

The above method of deducing the reacting proportions of oxidizing agents and reducing agents saves little labour in the case of iodine and sodium(I) thiosulphate(VI). However, if this same method is applied to more complicated examples its greater simplicity as compared with the usual method becomes clear.

*Reduction of Manganate(VII) Ions.* Oxidation of iron(II) sulphate(VI) solution by potassium(I) manganate(VII) solution occurs in the presence of dilute sulphuric(VI) acid. Hydrogen ions play an important part in the half-reaction involving the complex ion, being converted into water by oxygen from the oxidizing agent. At the same time the manganate(VII) ion is reduced to a manganese(II) ion, the oxidation number of the manganese falling from VII to II. Initially we represent the equation for the half-reaction as follows:

$$MnO_4^- + 8H^+ + ne \rightarrow Mn^{2+} + 4H_2O$$

where $n$ is the number of electrons required to balance the equation electrically. By inspection we see that $n$ must be 5. Hence we have

$$(a) \quad MnO_4^- + 8H^+ + 5e \rightarrow Mn^{2+} + 4H_2O$$

The second half-reaction simply consists of deriving the five electrons from five iron(II) ions, which are thereby oxidized to five iron(III) ions, the oxidation number of the iron being increased from II to III.

$$(b) \quad 5Fe^{2+} \rightarrow 5Fe^{3+} + 5e$$

Combining the equations for the two half-reactions, we obtain

$$MnO_4^- + 8H^+ + 5Fe^{2+} \rightarrow Mn^{2+} + 4H_2O + 5Fe^{3+}$$

From this ionic equation we deduce that in acid solution 1 mole of $KMnO_4$ (158 g) oxidizes 5 mole of $FeSO_4$ ($5 \times 152$ g). It is interesting to compare the simple ionic equation with the cumbersome conventional equation:

$$2KMnO_4 + 8H_2SO_4 + 10FeSO_4$$
$$\rightarrow K_2SO_4 + 2MnSO_4 + 5Fe_2(SO_4)_3 + 8H_2O$$

*Reduction of Dichromate(VI) Ions.* Potassium(I) dichromate(VI), $K_2Cr_2O_7$, is a weak oxidizing agent in neutral solution, and for this reason hydrochloric acid or sulphuric(VI) acid is added in titrating it with a solution of an iron(II) salt. Starting with dichromate(VI) ions, in which chromium has an oxidation number of VI, we finish with chromium(III) ions, in which the oxidation number is (III). Initially we write

$$Cr_2O_7^{2-} + 14H^+ + ne \rightarrow 2Cr^{3+} + 7H_2O$$

In this case we find by inspection that $n = 6$. Hence the equation for the half-reaction involving the complex ion is

$$(a) \quad Cr_2O_7^{2-} + 14H^+ + 6e \rightarrow 2Cr^{3+} + 7H_2O$$

As before, we derive our electrons from the reducing agent. For the second half-reaction we have

$$(b) \quad 6Fe^{2+} \rightarrow 6Fe^{3+} + 6e$$

Adding together the equations for the half-reactions we obtain

$$Cr_2O_7^{2-} + 14H^+ + 6Fe^{2+} \rightarrow 2Cr^{3+} + 7H_2O + 6Fe^{3+}$$

In this case we deduce from the ionic equation that 1 mole of $K_2Cr_2O_7$ (294 g) reacts in acid solution with 6 mole of $FeSO_4$ ($6 \times 152$ g). Once again we contrast the relatively simple ionic equation with the conventional equation. which is as follows:

$$K_2Cr_2O_7 + 7H_2SO_4 + 6FeSO_4$$
$$\rightarrow Cr_2(SO_4)_3 + 7H_2O + 3Fe_2(SO_4)_3 + K_2SO_4$$

The reader is recommended to work out the ionic equation, the reacting proportions, and the conventional equation for the reaction which takes place in acid solution between potassium(I) chromate(VI), $K_2CrO_4$, and iron(II) sulphate(VI).

**Anodic Oxidation and Cathodic Reduction.** During electrolysis an anode serves as an agent for removing electrons from a negative ion, while a cathode serves as an agent for adding electrons to a positive ion. Oxidation and reduction are thus processes which normally constitute electrolysis. When a current is passed through fused sodium(I) chloride the chloride ions are oxidized and sodium(I) ions reduced:

$$Cl^- - e \rightarrow \tfrac{1}{2}Cl_2 \qquad Na^+ + e \rightarrow Na$$

Actually this is not usually described as an example of oxidation and reduction, because no other chemical substance is involved in the electron transfer at the electrodes. Nevertheless, the changes are put to advantage in the converting of substances from a lower state of oxidation to a higher, and vice versa. In some cases oxidation at the anode and reduction at the cathode are brought about by products resulting from liberation of ions. Thus in electrolysis of dilute sulphuric(VI) acid with suitable electrodes the oxygen liberated at the anode or the hydrogen liberated at the cathode can be used to oxidize or reduce other substances which are present.

The following are examples of anodic oxidation.

*Preparation of Potassium(I) Hexacyanoferrate(III).* This is manufactured by electrolysis of potassium(I) hexacyanoferrate(II) solution:

$$Fe(CN)_6^{4-} - e \rightarrow Fe(CN)_6^{3-}$$

*Preparation of Peroxodisulphuric(VI) Acid.* This acid is made in solution by electrolysis of ice-cold concentrated sulphuric(VI) acid, which contains hydrogensulphate(VI) ions, using a fine platinum wire as anode:

$$2HSO_4^- - 2e \rightarrow H_2S_2O_8$$

*Manufacture of Chlorates(V) and Chlorates(VII).* By electrolysing a hot (80°C) solution of sodium(I) chloride, using platinum electrodes placed close together, chlorine liberated at the anode is made to combine with $OH^-$ ions left at the cathode:

$$6OH^- + 3Cl_2 \rightarrow ClO_3^- + 5Cl^- + 3H_2O$$

If the electrolysis is prolonged the chlorate(V) ions are further oxidized at the anode to chlorate(VII) ions:

$$ClO_3^- + Cl_2 + H_2O \rightarrow ClO_4^- + 2H^+ + 2Cl^-$$

*Anodizing of Aluminium.* Aluminium is normally covered with a thin coating of its oxide. Although the film has a thickness of only about $2 \cdot 5 \times 10^{-5}$ cm, it is extremely resistant and protects the metal from corrosion in ordinary circumstances. When, however, conditions for corrosion are favourable (as in sea water) the thin oxide film gives insufficient protection. The thickness of the layer is therefore increased by electrolysing dilute sulphuric(VI) acid or chromic(VI) acid with the aluminium object as the anode. In this way the thickness of the oxide film is increased sufficiently to avoid corrosion.

Cathodic reduction is utilized in the electrodeposition of metals from their ions. 'Electrodeposition' is a general term covering the extraction and refining of metals and electroplating. Apart from this, cathodic reduction finds little application.

## Redox Series

Earlier in this chapter we saw that a metal in contact with an aqueous solution of its ions constitutes a reversible electrode and by combining two such electrodes a reversible cell can be constructed. We also saw that conversion of a metal to its ions by loss of electrons is an oxidation, while the reverse change is a reduction, *e.g.*,

$$M \underset{\text{Reduction}}{\overset{\text{Oxidation}}{\rightleftharpoons}} M^{2+} + 2e$$

Other reversible oxidation-reduction systems can be used for making reversible electrodes. Some examples are the following:

$$Fe^{2+} \rightleftharpoons Fe^{3+} + e$$
$$Sn^{2+} \rightleftharpoons Sn^{4+} + 2e$$
$$Fe(CN)_6^{4-} \rightleftharpoons Fe(CN)_6^{3-} + e$$

In all these cases the forward change represents oxidation, and the backward change reduction. The systems are therefore described as *redox* systems, the term 'redox' being an abbreviation of 'reduction-oxidation.'

The tendencies for electrons to be lost or gained in the above systems vary with the nature of the system. This can be utilized for the construction of a cell, in which one particular system plus an inert electrode constitutes one half of the cell. A cell of this type is shown in Fig. 17–6. A mixed solution of iron(II) sulphate(VI) and iron(III) sulphate(VI) is placed in a porous pot, which stands in a large beaker containing a mixture of tin(II) chloride and tin(IV) chloride solutions. Platinum electrodes, X and Y, are introduced into each half-cell, and when the circuit is closed a

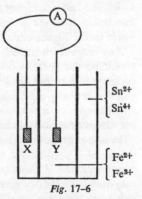

*Fig.* 17–6

current flows through an ammeter included in the circuit. If a voltmeter is substituted for the ammeter an e.m.f. of about 0·5 V is recorded.

With the above all the current flows externally from Y to X; that is, electrons flow from X to Y, X being at a lower potential than Y. The cell is represented diagrammatically as now shown (see p. 426):

$$Pt|Sn^{2+}, Sn^{4+}|Fe^{2+}, Fe^{3+}|Pt$$

Evidently the change which takes place at X is: $Sn^{2+} \rightarrow Sn^{4+} + 2e$, while the corresponding change at Y is: $2Fe^{3+} + 2e \rightarrow 2Fe^{2+}$. This means that the tin(II) ions in one compartment are reducing the iron(III) ions in the other compartment.

As with electric potentials of elements and their ions there is no way of measuring the absolute potential developed by a redox couple such as $Fe^{2+} - Fe^{3+}$ in contact with a platinum electrode. However, if a standard hydrogen electrode is used as the second half-cell, the electric potential relative to the hydrogen electrode can be measured when the solution contains $1 \text{ mol dm}^{-3}$ of both $Fe^{2+}$ and $Fe^{3+}$ ions. This is called the *redox potential* of the redox couple. For any other concentrations of the two ions specified the electric potential, $E$, at 298 K is given by the Nernst equation:

$$E = E^{\ominus} + 0 \cdot 059 \log \frac{[Fe^{3+}]}{[Fe^{2+}]}$$

where $E^{\ominus}$ is the redox potential. Strictly speaking, ionic activities, and not concentrations, should be used.

The electrode potentials of elements and their ions which are given in Table 17–2 are also their redox potentials. Thus the electrochemical series is only part of a much larger oxidation-reduction series, called the *redox series*, obtained by tabulating standard redox potentials.

If a cell is constructed from any two half-cells shown in the redox series, the half-cell higher in the series will be at a lower potential than the one below it in the series. Hence current will flow externally from the lower to the higher one.

Any reagent on the left will theoretically reduce any one on the right, providing the latter is below it in the series. Thus iron(II) ions reduce bromine to $Br^-$, but not iodine to $I^-$. Conversely any reagent on the right will theoretically oxidize any one on the left, providing the latter is above it in the series. For example, in acid solution potassium(I) manganate(VII), but not potassium(I) dichromate(VI), will oxidize $Cl^-$ ions to chlorine.

Two important limitations of the redox series should be noted. Firstly, the series is constructed from redox potentials obtained under standard conditions, and if these conditions do not exist the order of the systems may vary from the one given. Secondly, although the series indicates which redox reactions are possible, it tells us nothing about their *rate*. Theoretically molecular hydrogen should reduce iron(III) ions to iron(II) ions. However, when hydrogen from a Kipp's apparatus is passed into iron(III) chloride solution no apparent reduction occurs. This is due to the extreme slowness of the reaction. Reduction takes place readily if zinc and hydrochloric acid are added to some iron(III) chloride solution. This was formerly explained by assuming the reduction to be brought about by very active hydrogen atoms ('nascent' hydrogen). Nowadays the reduction is attributed directly to the zinc:

$$Zn + 2Fe^{3+} \rightarrow Zn^{2+} + 2Fe^{2+}$$

**Table 17-3.** REDOX SERIES

(Redox potentials ($E^{\ominus}$) at 298 K)

| Half-cell reaction | $E^{\ominus}$/V |
|---|---|
| $Li \rightarrow e + Li^+$ | $-3\cdot04$ |
| $K \rightarrow e + K^+$ | $-2\cdot92$ |
| $Ca \rightarrow 2e + Ca^{2+}$ | $-2\cdot87$ |
| $Na \rightarrow e + Na^+$ | $-2\cdot71$ |
| $Mg \rightarrow 2e + Mg^{2+}$ | $-2\cdot37$ |
| $Al \rightarrow 3e + Al^{3+}$ | $-1\cdot66$ |
| $Mn \rightarrow 2e + Mn^{2+}$ | $-1\cdot18$ |
| $Zn \rightarrow 2e + Zn^{2+}$ | $-0\cdot76$ |
| $Cr \rightarrow 3e + Cr^{3+}$ | $-0\cdot51$ |
| $Fe \rightarrow 2e + Fe^{2+}$ | $-0\cdot44$ |
| $Co \rightarrow 2e + Co^{2+}$ | $-0\cdot28$ |
| $Ni \rightarrow 2e + Ni^{2+}$ | $-0\cdot25$ |
| $Sn \rightarrow 2e + Sn^{2+}$ | $-0\cdot14$ |
| $Pb \rightarrow 2e + Pb^{2+}$ | $-0\cdot13$ |
| $\frac{1}{2}H_2 \rightarrow e + H^+$ | $0\cdot00$ |
| $Cu^+ \rightarrow e + Cu^{2+}$ | $+0\cdot17$ |
| $Sn^{2+} \rightarrow 2e + Sn^{4+}$ | $+0\cdot20$ |
| $Cu \rightarrow 2e + Cu^{2+}$ | $+0\cdot34$ |
| $Fe(CN)_6^{4-} \rightarrow e + Fe(CN)_6^{3-}$ | $+0\cdot48$ |
| $I^- \rightarrow e + \frac{1}{2}I_2$ | $+0\cdot54$ |
| $Fe^{2+} \rightarrow e + Fe^{3+}$ | $+0\cdot76$ |
| $Hg_2^{2+} \rightarrow 2e + 2Hg^{2+}$ | $+0\cdot79$ |
| $Ag \rightarrow e + Ag^+$ | $+0\cdot80$ |
| $Hg \rightarrow 2e + Hg^{2+}$ | $+0\cdot85$ |
| $Br^- \rightarrow e + \frac{1}{2}Br_2$ | $+1\cdot07$ |
| $2Cr^{3+} + 7H_2O \rightarrow 6e + Cr_2O_7^{2-} + 14H^+$ | $+1\cdot33$ |
| $Cl^- \rightarrow e + \frac{1}{2}Cl_2$ | $+1\cdot36$ |
| $Mn^{2+} + 4H_2O \rightarrow 5e + MnO_4^- + 8H^+$ | $+1\cdot52$ |
| $MnO_2 + 2H_2O \rightarrow 3e + MnO_4^- + 4H^+$ | $+1\cdot69$ |
| $SO_4^{2-} \rightarrow e + \frac{1}{2}S_2O_8^{2-}$ | $+2\cdot01$ |
| $F^- \rightarrow e + \frac{1}{2}F_2$ | $+2\cdot85$ |

(Left side vertical label: Strength as reducing agent decreases ↓)

(Right side vertical label: Strength as oxidizing agent increases ↓)

## EXERCISE 17

**Electrochemical Series**

**1.** What is the basis of the classification called the *electrochemical series*? Arrange the following elements in their correct sequence in the electrochemical series: silver; copper; iron; hydrogen; zinc; potassium; calcium; tin.

Illustrate *three* ways in which the characteristic chemical properties of a metal or its compounds are closely related to its position in the series.

Describe, by means of a simple diagram only, a galvanic cell involving the use of a suitable pair of the above metals and write the equation for the net cell reaction.

Mention *very briefly one* simple chemical experiment to demonstrate the appropriate placing of hydrogen relative to copper in the series.    (W.J.E.C.)

2. (Part question.) Put into their relative positions in the electrochemical series the elements: aluminium, calcium, potassium, and strontium. Account for your answer by reference to the electronic structures of these elements.
(S.U.)

3. What is the electrochemical series? By reference to the metals magnesium, aluminium, tin, and mercury show how the position of a metal in the series influences its method of extraction and the action of heat on its oxides and sulphates(VI).    (Lond.)

4. (Part question.) Comment on the significance of the following observations: (i) sheet iron rusts in use less rapidly when galvanized than when tinned; (ii) copper will dissolve in dilute nitric(V) acid, but not in dilute sulphuric acid(VI).
(C.L.)

## Electrolysis

5. Describe, and explain in each case, the chemical changes which are observed when an electric current is passed through aqueous solutions of the following compounds (the electrodes being inert and separated sufficiently to prevent subsequent interaction of the products of electrolysis):

(a) potassium(I) chloride, (b) copper(II) sulplate(VI), (c) potassium(I) dicyanoargentate(I), $KAg (CN)_2$.

What relationships exist between the quantity of electricity passing through the solutions and the amounts of the resulting products?

What would be the effect in (a) if the electrodes were placed close together and the solution was (i) warmed, (ii) kept cool?    (J.M.B.)

6. (Part question.) Explain why it is possible to plate out only copper from a solution containing copper, zinc, and iron ions, leaving the zinc and iron ions in solution.

What would be expected to happen(i) at the cathode and (ii) at the anode on electrolysing solutions of (a) silver(I) nitrate(V) with platinum electrodes, (b) copper(II) chloride with platinum electrodes, and (c) copper(II) sulphate(VI) with copper electrodes?    (J.M.B.)

7. Explain how the ionic theory accounts for the following:

(i) Magnesium cannot be obtained by electrolysis of aqueous solutions of its salts.

(ii) If an aqueous solution of silver(I) nitrate(V) is electrolysed with silver electrodes the concentration of silver in solution remains unchanged.

(iii) Electrolysis of dilute hydrochloric acid with platinum electrodes yields oxygen at the anode, while electrolysis of concentrated hydrochloric acid yields chlorine.

## Oxidation and Reduction

8. What do you understand by *oxidation* and *reduction*? Explain or illustrate the following statements:

(a) Oxidation cannot take place without reduction.

(b) A metal high in the electrode potential series will be a good reducing agent.

(c) The oxidizing power of the halogens decreases from fluorine to iodine.

(d) Some compounds can be both oxidizing and reducing agents. (O. and C.)

9. (Part question.) The salt $X_2M_2O_7$ (where X and M are metallic elements of relative atomic mass $X = 39$, $M = 52$) can be used in acidic solution for reactions in which the overall reaction of the salt is

$$M_2O_7^{2-} + 14H^+ + 6e \rightarrow 2M^{3+} + 7H_2O$$

Deduce the concentration in gram $dm^{-3}$ of a decimolar solution of the salt. Write *ionic* equations for the reactions which occur when the acidic solution of the salt is mixed with (i) potassium(I) iodide solution, (ii) iron(II) sulphate(VI) solution. (J.M.B.)

10. Define oxidation and reduction in terms of the transfer of electrons. Explain the following reactions in aqueous solution in terms of electron transfers, pointing out whether each molecular or ionic species on the left-hand side of the equation is oxidized, reduced, or unchanged:

(a) $I_2 + 2 Na_2S_2O_3 \rightarrow 2 NaI + Na_2S_4O_6$
(b) $SnCl_2 + 2 HgCl_2 \rightarrow Hg_2Cl_2 + SnCl_4$
(c) $H_2SO_4 + Zn \rightarrow ZnSO_4 + H_2$

The oxidizing action of potassium(I) dichromate(VI) in acid solution is essentially

$$Cr_2O_7^{2-} + 14H^+ + 6e \rightarrow 7H_2O + 2 Cr^{3+}$$

Deduce the ionic equation for the oxidation of iron(II) sulphate(VI) solution by acidified dichromate (VI). (J.M.B.)

## MORE DIFFICULT QUESTIONS

11. What do you understand by the electrode potential series of the elements? Comment on the following:

(a) Iron is above hydrogen in the electrode potential series, but hydrogen can reduce iron(III) oxide.

(b) Sodium is above zinc in the electrode potential series, but when sodium is placed in a solution of a zinc(II) salt hydrogen is liberated.

(c) Hydrogen is above copper in the electrode potential series, but hydrogen will not displace copper from solutions of its salts. (O. and C.)

12. Describe and explain what takes place when an electric current passes between platinum electrodes immersed in:

(a) an aqueous solution of copper(II) sulphate(VI);
(b) a hot aqueous solution of potassium(I) chloride;
(c) fused sodium(I) ethanoate;
(d) a conc. aq. solution of potassium(I) hydrogensulphate(VI) at 0°C;
(e) a solution of silver(I) cyanide in potassium(I) cyanide solution. (J.M.B.)

13. Develop the concept of oxidation, giving examples to illustrate each point and introducing the idea of electron transfer. Comment on the following from the standpoint of oxidation and reduction:

(a) processes at electrodes during electrolysis;
(b) the conversion of hydrogen to sodium(I) hydride;
(c) the conversion of fluorine to difluorine oxide.

(Knowledge of the reaction in (c) is not necessary.)

Calculate the mass of iodine liberated when 107 g of potassium(I) iodate(V) are added to 1 $dm^3$ of 2·0 molar HCl solution containing excess of potassium(I) iodide. $H = 1·01$; $O = 16·0$; $Cl = 35·5$; $K = 39·0$; $I = 127·0$. (S.U.)

# CHAPTER 18

# Further applications of the ionic theory

## Ionic Dissociation and the Law of Mass Action

**Ostwald's Dilution Law.** If the dissociation of an electrolyte in aqueous solution is a simple reversible reaction it should be possible to apply the law of mass action to it in the same way that the law was applied to the general case at p. 359. If we start with a concentration of 1 mole of a binary electrolyte AB in a volume $v$ (volume in dm³), and $\alpha$ is the degree of dissociation, the concentrations of the different species present at equilibrium are as follows:

$$AB \rightleftharpoons A^+ + B^-$$

$$(1 - \alpha)/v \qquad \alpha/v \qquad \alpha/v$$

Applying the law of mass action to the equilibrium, we obtain

$$K = \frac{(\alpha/v)(\alpha/v)}{(1 - \alpha)/v} = \frac{\alpha^2}{(1 - \alpha)v}$$

where $K$ is a constant known as the *dissociation constant*.

The final expression above is *Ostwald's dilution law* for a binary electrolyte. It is a purely theoretical deduction, but it can be tested by determining the degree of dissociation of an electrolyte at various dilutions. This has been done for a large number of electrolytes. The law holds remarkably well for solutions of weak electrolytes, but not for those of strong electrolytes. We can see this if we compare the values of $K$ obtained with solutions of ethanoic acid (a weak electrolyte) and sodium(I) chloride (a strong electrolyte) at various dilutions at 25°C. The values of the degree of dissociation, $\alpha$, have been deduced from the molar conductivity ratio ($\Lambda_v/\Lambda^\infty$). The dilution, $v$, is expressed in dm³, not in cm³.

The fact that strong electrolytes do not obey Ostwald's dilution law

Table 18-1. APPLICATION OF OSTWALD'S DILUTION LAW
TO SOLUTIONS OF ETHANOIC ACID AND SODIUM(I) CHLORIDE

| Ethanoic acid | | | Sodium(I) chloride | | |
|---|---|---|---|---|---|
| Dilution $v/dm^3$ | $\alpha$ | $K = \dfrac{\alpha^2}{(1-\alpha)v}$ | Dilution $v/dm^3$ | $\alpha$ | $'K' = \dfrac{\alpha^2}{(1-\alpha)v}$ |
| 13·57 | 0·0157 | $1·845 \times 10^{-5}$ | 1 | 0·74 | 2·106 |
| 54·28 | 0·0312 | $1·851 \times 10^{-5}$ | 10 | 0·85 | 0·481 |
| 108·56 | 0·0438 | $1·849 \times 10^{-5}$ | 100 | 0·94 | 0·147 |
| 217·12 | 0·0614 | $1·851 \times 10^{-5}$ | 1000 | 0·98 | 0·048 |

was regarded for many years as a serious objection to the ionic theory. However, with increasing knowledge of the nature of strong electrolytes the difficulty has disappeared. The anomaly is due to the different meaning of degree of dissociation as regards strong and weak electrolytes p. 406). In a solution of a highly dissociated electrolyte, like sodium(I) chloride or hydrochloric acid, there is mutual attraction between oppositely charged ions and the apparent degree of dissociation, as determined by the molar conductivity ratio does not represent the proportion of the electrolyte present as ions. Aqueous ethanoic acid, on the other hand, contains so few ions that they are free from interference by one another, and the molar conductivity ratio gives the actual fraction of the acid in the form of ions. Although Ostwald's dilution law is thus limited in its application, it is of great value in connection with the strengths of weak acids and weak bases.

**Solubility Product of a Sparingly Soluble Electrolyte.** Although the law of mass action does not apply to most salts in solution, it does give consistent results when applied to solutions of sparingly soluble salts. This is explained as follows.

All salts consist of ions, which are uniformly arranged in a crystalline structure. When a sparingly soluble salt, such as silver(I) chloride or barium(II) sulphate(VI), is added to water a little dissolves and produces ions in the liquid. An equilibrium is established between the ions and the solid salt when the rate at which ions leave the crystal framework of the solid is equal to the rate at which they are deposited again. Since the solubility is very small, the ions present in the liquid are few in number and are so far apart that they are free from mutual interference. We can therefore say that in the very dilute saturated solution the salt is completely dissociated.

With a solution of a sparingly soluble binary electrolyte $A^+B^-$ in contact with the solid the dissociation can be represented as follows, where for simplicity the ions are regarded as anhydrous:

$$A^+B^- \rightleftharpoons A^+ + B^-$$
$$\text{solid} \qquad \text{ions}$$

Applying the law of mass action to this equilibrium we have

$$K = \frac{[A^+] [B^-]}{[A^+B^-]} \quad (K \text{ in mol dm}^{-3})$$

$$\text{or} \quad K[A^+B^-] = [A^+] [B^-]$$

Since the 'concentration' of a solid remains constant we have

$$[A^+] [B^-] = \text{a constant } (K_{sp})$$

The constant, $K_{sp}$, is the *solubility product* of the electrolyte $A^+B^-$.

*The **solubility product** of a sparingly soluble binary electrolyte is the product of the concentrations of the ions in a saturated solution.*

When an electrolyte furnishes more than one of a particular kind of ion the concentration of that ion must be raised to the corresponding power in expressing the solubility product. Thus the solubility product of lead(II) chloride, $PbCl_2$, is given by $[Pb^{2+}] [Cl^-]^2 = K_{sp}$. The general case can be written as follows:

$$A_xB_y \rightleftharpoons xA + yB$$
$$\text{and} \quad [A]^x [B]^y = K_{sp}$$

We can measure the solubility of sparingly soluble salts like silver(I) chloride by conductivity experiments (p. 414). The solubility product can then be calculated from the solubility.

**Example 1.** *The solubility of silver(I) chloride at $18°C$ is $1.46 \times 10^{-3}$ g dm$^{-3}$. What is the solubility product? ($Ag = 108$, $Cl = 35.5$.)*

1 mole of $AgCl = 108 + 35.5 = 143.5$ g

Solubility of $AgCl = 1.46 \times 10^{-3}$ g dm$^{-3}$

$$= \frac{1.46 \times 10^{-3}}{143.5} \text{ mol dm}^{-3}$$

$$= 1.0 \times 10^{-5} \text{ mol dm}^{-3} \text{ approximately}$$

Since one 'molecule' of silver(I) chloride furnishes on dissociation one $Ag^+$ ion and one $Cl^-$ ion, and since the dissolved silver(I) chloride is completely dissociated into ions, we have,

$$[Ag^+] [Cl^-] = (1 \times 10^{-5}) (1 \times 10^{-5})$$
$$= 1 \times 10^{-10} \text{ mol}^2 \text{ dm}^{-6}$$

**Example 2.** *The solubility of calcium(II) hydroxide in water at $20°C$ is $2.78$ g dm$^{-3}$. What is the solubility product? ($Ca = 40$, $O = 16$, $H = 1$.)*

1 mole of $Ca(OH)_2 = 40 + 34 = 74$ g
$2.78$ g of $Ca(OH)_2 = 2.78 \div 74 = 0.0376$ mol

Since one 'molecule' of calcium(II) hydroxide gives one $Ca^{2+}$ ion and two $OH^-$ ions,

$$[Ca^{2+}] \, [OH^-]^2 = (0\cdot0376) \times (0\cdot0752)^2$$
$$= 2\cdot13 \times 10^{-4} \, mol^3 \, dm^{-9}$$

Note that salts like sodium(I) chloride, of relatively high solubility, do not possess solubility products, because the law of mass action does not hold in their case.

**Precipitation by 'Common Ion' Action.** The effect of increasing or decreasing the concentration of one of the ions participating in a balanced reaction can be predicted *qualitatively* from Le Chatelier's principle. Thus, in the dissociation

$$A^+B^- \rightleftharpoons A^+ + B^-$$

if the concentration of one of the ions is increased the point of equilibrium will move to the left so as to decrease the concentration of the other ion in solution—that is, the degree of dissociation will be reduced. If the solution is already saturated with the solid $A^+B^-$ the displacement of the equilibrium will result in precipitation of solid. Precipitation can be brought about in this way by adding to the solution a solute which provides a 'common ion.' This is illustrated in the following examples.

*Purification of Sodium(I) Chloride*

$$Na^+Cl^- \rightleftharpoons Na^+ + Cl^-$$

Sodium(I) chloride often contains small quantities of calcium(II) chloride and magnesium(II) chloride, which cause it to be deliquescent. It can be purified by passing hydrogen chloride through the saturated solution or by adding concentrated hydrochloric acid. With either method there is a considerable increase in $[Cl^-]$, resulting in displacement of the equilibrium point in the above dissociation to the left. As the solution is already saturated with sodium(I) chloride, the decrease in degree of dissociation causes precipitation of the chloride. Decreasing the degree of dissociation of the salt in solution by 'common ion' action is equivalent to decreasing its solubility at the given temperature.

Note that the above precipitation is not entirely due to 'common ion' action. It is partly caused by some of the water present being used to hydrate $H^+$ and $Cl^-$ ions.

*'Salting Out' of Soap*

$$C_{17}H_{35}COO^-Na^+ \rightleftharpoons C_{17}H_{35}COO^- + Na^+$$

Soap is a mixture of sodium(I) salts of certain organic acids like sodium(I) octadecanoate, the dissociation of which is shown above. It is made by boiling fats and oils with aqueous sodium(I) hydroxide, and precipitated by adding brine, which increases $[Na^+]$. It is not an essential condition of precipitation that the solution should be saturated before addition of the common ion, but clearly the solution must become saturated before precipitation can occur. An increase in $[Na^+]$ decreases the degree of dissociation of the sodium(I) octadecanoate in solution; that is, it decreases its effective solubility. Ultimately, as the brine is added, the solution becomes saturated with respect to the solute and precipitation begins. Again this is helped by hydration of the added ions.

**Sparingly Soluble Electrolytes.** When an electrolyte dissolves only slightly in water the conception of solubility product enables us to express the effect of 'common ion' action *quantitatively*. In the case of a saturated solution of silver(I) chloride the concentrations of $Ag^+$ and $Cl^-$ ions are equal. Denoting this concentration by $c$ and the solubility product by $K_{sp}$, we have,

$$[Ag^+] [Cl^-] = c^2 = K_{sp}$$

If the concentration of one of the ions is increased to $c_1$ the concentration of the other ion must decrease to $c_2$, so that

$$c_1 \times c_2 = K_{sp}$$

$K_{sp}$ for silver(I) chloride at 18°C is $1 \times 10^{-10}$ mol² dm⁻⁶. Suppose we have silver(I) nitrate(V) solution in which $[Ag^+] = 1 \times 10^{-4}$ mol dm⁻³, and that we blow hydrogen chloride into the liquid or add sodium(I) chloride in very small amounts. Nothing happens until $[Cl^-]$ increases to $1 \times 10^{-6}$ mol dm⁻³, when $K_{sp}$ for silver chloride is reached. Any tendency for the solubility product to be exceeded by further increase in $[Cl^-]$ is prevented by combination of $Ag^+$ ions with some of the excess $Cl^-$ ions to form solid silver chloride. Whenever the concentrations of the ions in solution are such as to exceed solubility product of an electrolyte the electrolyte is precipitated. The extent to which precipitation occurs can be calculated if the solubility product of the sparingly soluble electrolyte and the concentration of the added ion are known.

**Example 1.** *The solubility product of silver(I) chloride at 18°C is $1 \times 10^{-10}$ mol² dm⁻⁶. What mass of the salt will be precipitated if 0·585 g of sodium(I) chloride is dissolved in 1 dm³ of a saturated solution of silver(I) chloride?* ($Na = 23$, $Cl = 35.5$.)

0·585 g of NaCl = 0·585 ÷ 58·5 = 0·01 mole

$$[Ag^+] [Cl^-] = 1 \times 10^{-10} \text{ mol}^2 \text{ dm}^{-6}$$

Assuming that the sodium(I) chloride is completely dissociated and that the $[Cl^-]$ from the silver(I) chloride is negligible in comparison with that from the sodium(I) chloride, we have,

$$\text{New } [Cl^-] = 1 \times 10^{-2} \text{ mol dm}^{-3}$$

therefore   new $[Ag^+] = 1 \times 10^{-8}$ mol dm⁻³

The $[Ag^+]$ is decreased from $1 \times 10^{-5}$ to $1 \times 10^{-8}$ mol $dm^{-3}$. Since 1 mole of silver gives 1 mole of silver(I) chloride, the mass of the latter precipitated

$$= (10^{-5} - 10^{-8}) \text{ mol}$$
$$= (10^{-5} - 10^{-8}) \times 143 \cdot 5 \text{ g}$$
$$= 0 \cdot 001434 \text{ g approximately}$$

That is, almost all the silver in solution is precipitated as chloride.

**Example 2.** The solubility product of magnesium(II) hydroxide at $18°C$ is $4 \cdot 2 \times 10^{-12}$ $mol^3$ $dm^{-9}$. What mass of magnesium(II) hydroxide can be dissolved in 1 $dm^3$ of $0 \cdot 01$ molar NaOH solution at $18°C$? ($Mg = 24, O = 16, H = 1$.)

$$[Mg^{2+}] [OH^-]^2 = 4 \cdot 2 \times 10^{-12} \text{ mol}^3 \text{ dm}^{-9}$$

Again assuming that the $[OH^-]$ is due only to NaOH, we have,

$$[OH^-] = 0 \cdot 01 \text{ mol dm}^{-3}$$
$$[OH^-]^2 = 1 \times 10^{-4} \text{ mol}^2 \text{ dm}^{-9}$$

Therefore    $[Mg^{2+}] = \dfrac{4 \cdot 2 \times 10^{-12}}{1 \times 10^{-4}}$

$$= 4 \cdot 2 \times 10^{-8} \text{ mol dm}^{-3}$$

Since 1 mole of $Mg^{2+}$ is equivalent to 1 mole of $Mg(OH)_2$, the mass of magnesium(II) hydroxide dissolved

$$= 4 \cdot 2 \times 10^{-8} \times 58 \text{ g dm}^{-3}$$
$$= 2 \cdot 436 \times 10^{-6} \text{ g dm}^{-3}$$

Note that deductions based on solubility product are valid only when (i) there is no combination of ions to form 'complex' ions, and (ii) *all* the ionic concentrations are small.

We should expect lead(II) chloride to be less soluble in concentrated than in dilute hydrochloric acid because the former has a larger concentration of $Cl^-$ ions. Actually lead(II) chloride is more soluble in the concentrated acid. Similarly silver(I) cyanide dissolves readily in potassium(I) cyanide solution. In both cases complex ions are formed:

$$PbCl_2 + 2Cl^- \rightleftharpoons PbCl_4^{2-} \quad \text{(tetrachloroplumbate(IV) ion)}$$
$$AgCN + CN^- \rightleftharpoons Ag(CN)_2^- \quad \text{(dicyanoargentate(I) ion)}$$

There is a tendency for the solubility of a sparingly soluble electrolyte to be increased if the solution contains a large concentration of other ions, whether or not these are the same as the ions of the electrolyte. This is because ions are attracted from the crystal framework into solution by the oppositely charged ions in the liquid. This effect may outweigh the tendency for the solubility to be decreased by 'common ion' action. Thus although the solubility of silver(I) chloride is *less* in a dilute solution of potassium(I) chloride than in pure water, the solubility in a concentrated solution of potassium(I) chloride is *larger* than in pure water.

### Determination of the Solubility Product of Calcium(II) Hydroxide.

The solubility product of calcium(II) hydroxide can be determined by leaving the hydroxide in contact with water and three or four standard solutions of sodium(I) hydroxide until equilibrium has been reached. The various solutions are then filtered and the $[OH^-]$ for each solution is found by titration with standard acid. It can be shown that $[Ca^{2+}]$ $[OH^-]^2$ is approximately constant. There is one difference, however, from the examples on solubility product in the last section. In these the concentration of the common ion provided by the sparingly soluble electrolyte was regarded as negligible compared with that due to the strong electrolyte. This is only the case if the solubility product of the former is very small. Calcium(II) hydroxide is soluble enough in water to contribute appreciably to the hydroxyl ion concentration at equilibrium.

Experiment.[1] Put into four 300-cm³ conical flasks 100 cm³ of (i) lime-water, (ii) M/40 NaOH solution, (iii) M/20 NaOH solution, and (iv) M/10 NaOH solution. Add to each flask a small spoonful of calcium(II) hydroxide, stopper the flasks, and leave them for several hours (shaking the flasks occasionally). In turn filter about 50 cm³ of each solution, and titrate two portions of 20 cm³ with M/10 HCl solution, using methyl orange as the indicator.

In each case the $[OH^-]$ is proportional to the volume of M/10 acid. The $[OH^-]$ due to calcium(II) hydroxide is obtained by subtracting from the total $[OH^-]$ that of the sodium(I) hydroxide. Since calcium(II) hydroxide gives one calcium(II) ion for every two hydroxyl ions, the $[Ca^{2+}]$ is one half of the $[OH^-]$ due to calcium(II) hydroxide. In the final expression $[Ca^{2+}] \times [OH^-]^2$ the hydroxyl ion concentration which must be used is the *total* concentration of the ion. Specimen results are shown in the Table 18–2.

### Table 18–2

| Solution in | Titration $v$/cm³ | Ionic concentrations/mol dm⁻³ | | | | $[Ca^{2+}] \times [OH^-]^2$ /mol³ dm⁻⁹ |
|---|---|---|---|---|---|---|
| | | Total $[OH^-]$ present | $[OH^-]$ due to NaOH | $[OH^-]$ due to $Ca(OH)_2$ | $[Ca^{2+}]$ | |
| Water | 9·5 | 0·0475 | — | 0·0475 | 0·0237 | $5·4 \times 10^{-5}$ |
| M/40 NaOH | 11·6 | 0·058 | 0·025 | 0·033 | 0·0165 | $5·6 \times 10^{-5}$ |
| M/20 NaOH | 14·2 | 0·071 | 0·050 | 0·021 | 0·0105 | $5·3 \times 10^{-5}$ |
| M/10 NaOH | 21·8 | 0·109 | 0·100 | 0·009 | 0·0045 | $5·4 \times 10^{-5}$ |

[1] This experiment and the results quoted are due to F. G. Mee. The table of results is taken by permission of the publishers from *The Science Masters' Book*, Series I, Part II (Murray).

## Solubility Products in Qualitative Analysis.

*Precipitation of Chlorides.* Group I of the classical qualitative analysis table consists of metal ions ($Ag^+$, $Pb^{2+}$, $Hg^+$) which are precipitated as chlorides by cold dilute hydrochloric acid. The chlorides have only a small solubility in water, so that even in a saturated solution they produce few ions and their solubility products are small.

$$Ag^+ + Cl^- \rightleftharpoons \underset{\text{(solid)}}{AgCl} \qquad [Ag^+][Cl^-] = K_{sp}$$

Therefore when dilute hydrochloric acid is added in excess to a solution containing ions of a Group I metal the number of ions left in solution is negligibly small.

Certain errors may arise in Group I unless suitable precautions are taken. If the solutions are not cold precipitation of lead(II) chloride is incomplete or may not take place at all. This is because the solubility, and therefore solubility product, of lead(II) chloride increases rapidly with rise of temperature. Again, concentrated hydrochloric acid must not be used, because the concentrated acid tends to keep lead in solution due to formation of the tetrachloroplumbate(IV) ion $PbCl_4^{2-}$, as explained previously. In this group, also, concentrated hydrochloric acid will precipitate barium(II) chloride from a strong solution of this substance, owing to 'common ion effect.' The use of concentrated acid in this group also leads to errors in Group II.

*Precipitation of Sulphides.* In Group II certain metal sulphides (*e.g.*, HgS, CuS, CdS, etc.) are precipitated by passing hydrogen sulphide into a solution of the metal ions in the presence of dilute hydrochloric acid. If the metal M is divalent we can write the precipitation as follows:

$$M^{2+} + S^{2-} \rightleftharpoons \underset{\text{(solid)}}{MS} \qquad [M^{2+}][S^{2-}] = K_{sp}$$

What is the effect produced by the dilute hydrochloric acid from Group I? In solution there are the following equilibria:

$$(1) \quad MS \rightleftharpoons M^{2+} + S^{2-}$$

$$(2) \quad H_2S \rightleftharpoons H^+ + HS^- \rightleftharpoons 2H^+ + S^{2-}$$

$$\text{and} \quad (3) \quad HCl \rightleftharpoons H^+ + Cl^-$$

The dissociations represented by (1) and (2) are only small, while that represented by (3) is large. Thus the hydrogen ion concentration in (2) is increased by the presence of the hydrochloric acid. Therefore by Le Chatelier's principle $[S^{2-}]$ is decreased. But decrease of $[S^{2-}]$ increases $[M^{2+}]$ in (1). It follows that the presence of hydrochloric acid *hinders* the precipitation of the sulphides. Only those sulphides which have very small solubility products are precipitated. In particular the sulphides of the Group IV metals, zinc, manganese, cobalt, and nickel remain in solution. Some typical (but approximate) values of solubility products at 18°C of Group II and Group IV metal sulphides are, in $mol^2\,dm^{-6}$:

$$CuS\ 1\times 10^{-40} \qquad ZnS\ 1\times 10^{-24}$$
$$SnS\ 1\times 10^{-28} \qquad MnS\ 1\times 10^{-16}$$
$$CdS\ 1\times 10^{-28} \qquad NiS\ 1\times 10^{-21}$$

Note that the values of solubility products of metal sulphides are very uncertain.

In the presence of concentrated hydrochloric acid $[H^+]$ is increased considerably and $[M^{2+}]$ correspondingly increased. This explains why concentrated hydrochloric acid prevents the complete precipitation of metal sulphides in Group II.

In Group IV the sulphides of zinc, manganese, cobalt, and nickel are precipitated by ammonium sulphide in *alkaline* solution (dilute ammonia is present). The ammonium sulphide can be regarded as furnishing ammonia and hydrogen sulphide in solution:

$$(NH_4)_2S \rightleftharpoons 2NH_3 + H_2S$$

As in Group II,

$$(1)\ ZnS \rightleftharpoons Zn^{2+} + S^{2-}$$
$$(2)\ H_2S \rightleftharpoons H^+ + HS^- \rightleftharpoons 2H^+ + S^{2-}$$

In ammoniacal solution there is an excess of $OH^-$ ions, which combine with the $H^+$ ions of (2) to form water. Thus $[S^{2-}]$ is increased. But increase of $[S^{2-}]$ in (1) decreases $[Zn^{2+}]$, so that the metal ions are removed from solution as metal sulphide. This explains why hydrogen sulphide will precipitate zinc as sulphide from an ammoniacal solution, but not from the acid solution used in Group II.

*Precipitation of Hydroxides.* Iron, aluminium, and chromium are precipitated as hydroxides by ammonia solution in the presence of ammonium chloride, *e.g.*

$$Fe^{3+} + 3OH^- \rightleftharpoons Fe(OH)_3 \qquad [Fe^{3+}]\,[OH^-]^3 = K_{sp}$$

The object of adding ammonium chloride is to keep the hydroxides of manganese and magnesium in solution (the hydroxides of zinc, cobalt, and nickel are soluble in excess of ammonia).

$$Mg(OH)_2 \rightleftharpoons Mg^{2+} + 2OH^-$$
$$NH_3 + H_2O \rightleftharpoons NH_4^+ + OH^-$$
$$NH_4Cl \rightleftharpoons NH_4^+ + Cl^-$$

The presence of ammonium chloride increases $[NH_4^+]$, thus decreasing $[OH^-]$ and increasing $[Mg^{2+}]$. In this way precipitation of magnesium(II) hydroxide is prevented. The precipitation of iron(III) hydroxide, etc., is also hindered, but not prevented, because these hydroxides have much smaller solubility products than magnesium(II) hydroxide and manganese (II) hydroxide. This can be seen from the values now given (for 18°C).

$Fe(OH)_3$ $1 \times 10^{-38}$ $mol^4$ $dm^{-12}$        $Mg(OH)_2$ $4 \times 10^{-12}$ $mol^3$ $dm^{-9}$

$Al(OH)_3$ $1 \times 10^{-33}$ $mol^4$ $dm^{-12}$        $Mn(OH)_2$ $1 \times 10^{-14}$ $mol^3$ $dm^{-9}$

Notice that the solubility product ($1 \times 10^{-15}$ $mol^3$ $dm^{-9}$) of iron(II) hydroxide is larger than that of iron(III) hydroxide, and the addition of ammonium chloride prevents its complete precipitation by ammonia. We must therefore have iron in the trivalent state before proceeding with the third group; this is ensured by boiling the solution with one or two drops of concentrated nitric(V) acid.

Actually, the explanations which we have given for precipitations in qualitative analysis are somewhat oversimplified. Solubility products apply to equilibrium conditions in very dilute solution, and these conditions are not usually present in actual working. Thus some sulphides (*e.g.*, ZnS) are precipitated in a metastable form, which passes relatively slowly into the stable crystalline form. In other cases (*e.g.*, $Cd^{2+}$) the simple ions form complex ions (*e.g.*, $CdCl_4^{2-}$) in concentrated solution. As a result of these and other factors little reliance can be placed on calculations of ionic concentrations necessary to bring about precipitation.[1]

**Use of Potassium(I) Chromate(VI) in Silver(I) Nitrate(V) Titrations.** When silver(I) nitrate(V) solution is titrated with sodium(I) chloride solution containing a little potassium(I) chromate(VI), only silver(I) chloride is precipitated as long as there are any $Cl^-$ ions in the liquid. Only when all the $Cl^-$ ions have reacted does a red precipitate of silver(I) chromate(VI) appear. This is explained by the solubility products of the two silver(I) salts. $[Ag^+][Cl^-] = 1 \times 10^{-10}$ $mol^2$ $dm^{-6}$, while $[Ag^+]^2$ $[CrO_4^{2-}] = 2 \cdot 5 \times 10^{-12}$ $mol^3$ $dm^{-9}$. Suppose that the concentrations of $Cl^-$ ions and $CrO_4^{2-}$ ions in the liquid are both $0 \cdot 1$ $mol$ $dm^{-3}$. Then

$[Ag^+]$ needed to precipitate silver(I) chloride

$$= \frac{1 \times 10^{-10}}{0 \cdot 1} = 1 \times 10^{-9} \text{ mol dm}^{-3}$$

$[Ag^+]$ needed to precipitate silver(I) chromate

$$= \sqrt{\left(\frac{2 \cdot 5 \times 10^{-12}}{0 \cdot 1}\right)} = 5 \times 10^{-6} \text{ mol dm}^{-3}$$

Thus, even if the concentrations of $Cl^-$ ions and $CrO_4^{2-}$ ions in the liquid are equal, it is easier by adding $Ag^+$ ions to reach the solubility product of

[1] For a critical account of the use of solubility products in qualitative analysis see *The Solubility Product Principle* by S. Lewin (Pitman).

the chloride than that of the chromate(VI) in spite of the lower solubility product of the latter. In practice $[Cl^-]$ is much larger than $[CrO_4^{2-}]$. Hence silver(I) chloride is precipitated even more readily.

### Manufacture of Sodium(I) Hydroxide by Gossage's Process

$$Ca^{2+} + 2OH^- + 2Na^+ + CO_3^{2-} \rightleftharpoons Ca^{2+} + CO_3^{2-} + 2Na^+ + 2OH^-$$

$$\underbrace{\qquad\qquad}\qquad\qquad\qquad\underbrace{\qquad\qquad}$$

Ca(OH)₂          CaCO₃
(solid)          (solid)

In this process sodium(I) hydroxide is made by heating aqueous sodium(I) carbonate with solid calcium(II) hydroxide. Calcium(II) carbonate is precipitated and sodium(I) hydroxide is left in solution. The reaction is an equilibrium one, but owing to the fact that the solubility product of calcium(II) carbonate is smaller than that of the hydroxide the equilibrium point is far to the right.

The efficiency of the process depends on the extent to which sodium(I) carbonate is converted into the hydroxide, and this in turn depends on the concentration of the carbonate. This can be shown as follows. The solubility product of calcium(II) hydroxide is $5{\cdot}4 \times 10^{-5} \, mol^3 \, dm^{-9}$, while that of calcium(II) carbonate is $1{\cdot}7 \times 10^{-8} \, mol^2 \, dm^{-6}$. Then

$$\frac{[Ca^{2+}] \, [OH^-]^2}{[Ca^{2+}] \, [CO_3^{2-}]} = \frac{5{\cdot}4 \times 10^{-5}}{1{\cdot}7 \times 10^{-8}} = \frac{3200}{1} \, mol \, dm^{-3} \text{ approximately}$$

Since there is only one $[Ca^{2+}]$ in the equilibrium mixture, this term is the same in both the numerator and denominator, so that,

$$\frac{[OH^-]^2}{[CO_3^{2-}]} = \frac{3200}{1} \, mol \, dm^{-3} \text{ approximately}$$

The extent to which sodium(I) carbonate is converted into sodium(I) hydroxide can be deduced from the ratio $[OH^-]:[CO_3^{2-}]$ in the equilibrium mixture. Let the $[CO_3^{2-}]$ in the mixture be $n \, mol \, dm^{-3}$. Then,

$$[OH^-]^2 = 3200 \times [CO_3^{2-}]$$
$$= 3200 \times n \, mol^2 \, dm^{-6}$$

Therefore $[OH^-] = \sqrt{(3200n)} \, mol \, dm^{-3}$

and $\dfrac{[OH^-]}{[CO_3^{2-}]} = \dfrac{\sqrt{(3200n)}}{n} = 40\sqrt{\dfrac{2}{n}}$

It follows that the smaller the value of $n$ the larger will be the conversion of carbonate to hydroxide. Clearly one method by which $n$ can be made smaller is to have a smaller initial concentration of sodium(I) carbonate; that is, with more dilute sodium(I) carbonate solution a relatively better yield is obtained. In practice, the concentration is adjusted so that there is about 90 per cent conversion to hydroxide.

## Acids and Bases

Arrhenius's theory of ionization focused attention on the importance of water in connection with the properties of acids and bases. Arrhenius attributed acidic properties to the combined effect of the acid and water and defined an acid as "a substance which on dissolving in water dissociates to produce hydrogen ions."

The term 'base' was introduced about 1774 by Rouelle, a compatriot of Lavoisier, to denote a substance which combined with an acid to form a salt. As the common bases (metal oxides) were earthy substances, they were regarded as the foundation or base on which the beautiful structure characteristic of a salt was built up. With the establishment of the ionic theory of Arrhenius it was natural that soluble bases (metal hydroxides) should be put into a class of their own, called alkalis, and that an alkali should be defined as "a substance which on dissolving in water dissociates to produce hydroxyl ions."

**Modern Theory of Acids and Bases.** In modern times the Arrhenius theory of acids and bases has become merged in a wider theory put forward in 1923 by Brönsted and Lowry. In the same way Newton's theory of gravitation has been absorbed in the more general theory of relativity due to Einstein. Although the new theory of acids and bases is derived principally from a study of non-aqueous solutions, it has important applications to aqueous solutions. Now there would be no point in inventing another theory unless it explained more facts, or explained the facts more convincingly, than the old theory. This is precisely what the Brönsted–Lowry theory does. It accounts for new discoveries which are at variance with the older theory. Some of the facts which are difficult to explain on the Arrhenius theory of acids and bases are the following:

* The ammonium ion has acidic properties, and certain metals will liberate hydrogen from aqueous solutions of ammonium salts—*e.g.*

$$Mg + 2NH_4^+ \rightarrow Mg^{2+} + 2NH_3 + H_2$$

* Some reactions which are catalysed in aqueous solution by acids (presumably due to $H^+(aq)$ ions) are catalysed better by the same acids in hydrocarbon solvents (*e.g.*, pentane) although there are no $H^+$ ions present (the solutions are non-conductors).
* Other solvents behave exactly like water in regard to electrolytes. Thus solutions of salts in liquid ammonia conduct a current and are electrolysed. Like water, liquid ammonia dissolves sodium and potassium with evolution of hydrogen, and the resulting solutions turn phenolphthalein pink.

Brönsted and Lowry defined an acid as *a substance which can give up a proton to a base*. This is not widely different from the Arrhenius definition (since a hydrogen ion and a proton can be regarded as the same thing), but there is no mention of a particular solvent, while there is mention of a base. A base is defined as *any substance which can combine with a proton.* In other words, an acid is a 'proton-donor' and a base is a 'proton-acceptor.' (Compare this with "A reducing agent is an 'electron-donor' and an oxidizing agent is an 'electron-acceptor' " (p. 438).) The relation between an acid and a base can be expressed by the reversible reaction:

conjugate pair

$$\text{Acid} \rightleftharpoons \text{H}^+ + \text{Base}$$

$$e.g. \quad \text{HCl} \rightleftharpoons \text{H}^+ + \text{Cl}^-$$

When an acid loses a proton it leaves a base, which is called the conjugate base to the acid, and when a base accepts a proton it forms an acid, which is called the conjugate acid to the base. Acid and base together are called a 'conjugate pair.' Acids and bases can be neutral molecules, positive ions, or negative ions, as will be seen from Table 18–3, in which each acid is listed with its conjugate base.

Table 18–3

| Acid | Base | Acid | Base |
|------|------|------|------|
| $HCl$ | $Cl^-$ | $CH_3COOH$ | $CH_3COO^-$ |
| $HNO_3$ | $NO_3^-$ | $H_2S$ | $HS^-$ |
| $H_2SO_4$ | $HSO_4^-$ | $HS^-$ | $S^{2-}$ |
| $HSO_4^-$ | $SO_4^{2-}$ | $NH_4^+$ | $NH_3$ |
| $H_3O^+$ | $H_2O$ | $NH_3$ | $NH_2^-$ |
| $H_2O$ | $OH^-$ | $HCO_3^-$ | $CO_3^{2-}$ |

The term 'base' has here been given a much wider interpretation than that given by the Arrhenius theory, which restricted it to solutions of $OH^-$ ions. The $OH^-$ ion is now only one of a large class of substances. We should mention that sodium(I) hydroxide and potassium(I) hydroxide are still often described as bases although, strictly speaking, the term should be applied to the $OH^-$ ions which they contain.

When a base combines with a proton it does so by giving a share in two of its electrons to the proton, thus forming a co-ordinate covalent linkage. The ability to do this depends on the base having a lone pair of unshared electrons, and no molecule or ion can act as a base unless it possesses a pair of unshared electrons. *E.g.*

$$\left[ \text{H}:\overset{..}{\underset{..}{\text{O}}}: \right]^- + \text{H}^+ \quad \rightarrow \quad \text{H}:\overset{..}{\underset{..}{\text{O}}}:\text{H}$$

$$H:\overset{..}{\underset{..}{N}}: + H^+ \rightarrow \left[ H:\overset{..}{\underset{..}{N}}:H \atop \phantom{x} \right]^+$$

The dissolving of an acid in water was formerly regarded as a purely physical change, but it is now obvious that the change is as much chemical as the dissolving of zinc in hydrochloric acid. When an acid dissolves in water the latter acts as a base. In general:

$$HA + H_2O \rightleftharpoons H_3O^+ + A^-$$

Notice that the loss of a proton by the acid to the water (the base) leaves a new acid and a new base. This is typical of the interaction of any acid and any base. Moreover, the new acid is the conjugate acid of the first base, and the new base is the conjugate base of the first acid. This can be represented as:

Acid (A) + base (B) $\rightleftharpoons$ new acid (B) + new base (A)

The classification of water and ammonia as both acids and bases may puzzle the reader. These substances can act as acids or bases according to the circumstances. Thus when a molecule of water dissociates to give a proton and a hydroxyl ion it is behaving as an acid. In the subsequent combination of the proton with a water molecule the latter functions as a base:

$$\underset{\text{(acid)}}{H_2O} \rightleftharpoons H^+ + OH^-$$

$$\underset{\text{(base)}}{H_2O} + H^+ \rightleftharpoons H_3O^+$$

Or, $\underset{\text{(acid)}}{H_2O} + \underset{\text{(base)}}{H_2O} \rightleftharpoons \underset{\text{(new acid)}}{H_3O^+} + \underset{\text{(new base)}}{OH^-}$

We can now understand more fully what happens when an aqueous solution of an acid is neutralized by an aqueous solution of an alkali. Fundamentally the reaction is one between oxonium ions and hydroxyl ions, and can be written

$$\underset{\text{(acid)}}{H_3O^+} + \underset{\text{(base)}}{OH^-} \rightarrow \underset{\text{(new acid)}}{H_2O} + \underset{\text{(new base)}}{H_2O}$$

As with water, the feeble conductivity of liquid ammonia is due to slight ionization followed by 'solvation' of the proton:

$$NH_3 + NH_3 \rightleftharpoons NH_4^+ + NH_2^-$$

The ammonium ion, $NH_4^+$, and amide ion, $NH_2^-$, stand in the same relation to liquid ammonia as the $H_3O^+$ ion and $OH^-$ ion to water. Like water, liquid ammonia has a high relative permittivity and is an ionizing solvent. In recent years striking analogies between reactions in liquid ammonia and reactions in water have been found. Sodium(I) amide $NaNH_2$, dissolves in liquid ammonia just as sodium hydroxide, NaOH, dissolves in water. Phenolphthalein added to both solutions turns pink. Ammonium chloride dissolved in liquid ammonia acts as an acid, like hydrochloric acid dissolved in water. The ammonium chloride solution can be titrated with sodium(I) amide solution in the presence of phenolphthalein as indicator in the same way that in aqueous solution hydrochloric acid can be titrated with sodium(I) hydroxide. It is interesting to compare the equations for the reactions:

$$(H_3O^+ + Cl^-) + (Na^+ + OH^-) \rightarrow 2H_2O + Na^+ + Cl^-$$

$$(NH_4^+ + Cl^-) + (Na^+ + NH_2^-) \rightarrow 2NH_3 + Na^+ + Cl^-$$

Experiments like these, involving the use of liquid ammonia and other non-aqueous solvents, have been largely responsible for the broadening of the concept of acids and bases.

**Use of Chemical Terms.** At this stage it is perhaps not out of place to pause and examine the use of chemical terms in general. The language used by scientists does not stand still any more than that used in everyday life and most scientific terms have changed in meaning in the course of time. The meaning of some terms becomes stabilized (*e.g.*, Boyle's definition of an 'element'), but this is to be regarded as exceptional. The general tendency is for the meaning of scientific terms to expand so as to cover a wider range of phenomena. The modern definition of 'acid' brings under one heading such diverse substances as hydrochloric acid, the 'neutral' liquid water, and the 'alkaline' gas ammonia. At first this may appear nonsensical. The apparent contradiction is due to the fact that certain properties, such as the action on indicators, have come to be associated with acids by custom. We are now required to discard the traditional conception of acids and substitute a deeper meaning. Only in this way can progress be made in science. If our ancestors had not been willing to accept new ideas we might still be thinking of 'elements' in the Greek sense of earth, air, fire, and water!

**Strengths of Acids in Aqueous Solution.** The description 'strong' is often applied to acids in two senses. Reference may be made to 'strong' sulphuric(VI) acid to distinguish it from the dilute solution. On the other hand sulphuric(VI) acid, nitric(V) acid, and hydrochloric acid are said to be 'strong' acids in comparison with ethanoic acid and carbonic acid, which are weak acids. It is advisable to use the term 'concentrated' in the first case and to restrict the use of 'strong' to those acids in which the so-called acid properties are most marked.

The strengths of acids vary in different solvents. The stronger the basic character of the solvent, the more readily will an acid give up its protons. Thus ethanoic acid is only a weak acid in aqueous solution, but in liquid ammonia, which is a stronger base than water, it is almost completely ionized and is therefore a strong acid.

$$CH_3COOH + NH_3 \rightleftharpoons CH_3COO^- + NH_4^+$$

The strength of an acid HA in water is determined by the extent to which the following reaction takes place:

$$H_2O + \overset{\delta+}{H} - \overset{\delta-}{A} \rightleftharpoons H_3O^+ + A^-$$

If in dilute solution the forward change takes place much more readily than the reverse one, HA is a strong acid; if recombination of the ions predominates HA is a relatively weak acid. As in other balanced reactions, the extents to which the forward and backward reactions occur depend on their relative activation energies.

The activation energy of the forward reaction is determined by the strength of the H—A bond (which has to be broken). We might expect that ionization of HA in water would be promoted by increase in the ionic character of the H—A bond, but this does not follow. The H—I bond has only about 5 per cent of ionic character as compared with about 43 per cent for the H—F bond, but hydrogen iodide is a stronger acid than hydrogen fluoride in aqueous solution. The energy required to break the H—I bond is, however, smaller than that needed for the H—F bond. At the same time the backward reaction takes place more easily with the $F^-$ ion, the smallest of the halogen anions, than with $I^-$, the largest. The two factors combine to give the following order of acid strengths for hydrogen halides: HI > HBr > HCl > HF. In connection with the activation energy of the backward reaction we must remember that the ions exist in solution, not as simple ions, but as hydrated ions.

The methods now described are used for comparing the strengths of acids under similar conditions.

METHOD 1. *By Degree of Dissociation.* The ionic theory ascribes acidic properties to the hydrogen ions (or, more accurately, oxonium ions) formed by the dissociation of the acid in aqueous solution. Of two acids the stronger one is therefore the acid which furnishes the greater concentration of hydrogen ions under the same conditions. This depends on the relative degrees of dissociation of the acids. Hence we can compare the strengths by the degrees of dissociation at the same dilution and temperature. In strong acids the degree of dissociation can be found (approximately) by the general methods previously described—that is, by the molar conductivity ratio or by the freezing-point method. For weak acids only the first method (and that indirectly) is available.

If a certain acid is found to be dissociated to twice the extent of another acid at one dilution it does not follow that the same ratio will be found at a different dilution. The degree of dissociation of all acids increases with dilution, but the effect of increasing the dilution is greater with some acids than with others. Therefore the relative strengths will depend on the dilution employed for comparison. Theoretically all acids become equal in strength at infinite dilution, since they are then completely dissociated. This tendency is seen in Table 18–4.

**Table 18–4.** DEGREES OF DISSOCIATION OF ACIDS AT 298 K

| Acid | M Solution | M/10 Solution | M/1000 Solution |
|------|-----------|---------------|-----------------|
| $HNO_3$ | 0·82 | 0·96 | 0·99 |
| HCl | 0·79 | 0·92 | 0·99 |
| $\frac{1}{2}H_2SO_4$ | 0·51 | 0·65 | 0·96 |
| $CH_3COOH$ | 0·004 | 0·013 | 0·13 |
| $H_3BO_3$ | 0·00002 | 0·00008 | 0·0008 |

METHOD 2. *By Dissociation Constants (Weak Acids)*. The strengths of weak acids can be compared more conveniently by means of their dissociation constants deduced from Ostwald's dilution law (strong acids do not obey the law). For a weak acid which dissociates into two ions we have

$$K_a = \frac{\alpha^2}{(1 - \alpha)v} \qquad (K_a \text{ in mol dm}^{-3})$$

$$\text{or} \quad \alpha^2 = K_a v(1 = \alpha)$$

The dissociation constant of a weak acid is very small (see Table 18–5), and the degree of dissociation is also small. Hence the term $K_a v \alpha$ can be neglected in comparison with $K_a v$, so that

$$\alpha = \sqrt{(K_a v)}$$

It follows that if $v = 1$ dm$^3$, $\alpha = \sqrt{K_a}$. This means that for a normal solution of a weak acid the degree of dissociation is equal to the square root of the dissociation constant. The strengths of weak acids can therefore be compared approximately by means of their dissociation constants.

**Table 18–5.** DISSOCIATION CONSTANTS OF ACIDS AT 298 K

| Acid | Formula | $K_a$/mol dm$^{-3}$ |
|------|---------|---------------------|
| Methanoic acid | HCOOH | $2 \times 10^{-4}$ |
| Ethanoic acid | $CH_3COOH$ | $1·8 \times 10^{-5}$ |
| Chloroethanoic acid | $CH_2ClCOOH$ | $1·6 \times 10^{-3}$ |
| Benzenecarboxylic acid | $C_6H_5COOH$ | $6·6 \times 10^{-5}$ |
| Nitric(III) acid (18°C) | $HNO_2$ | $4 \times 10^{-4}$ |
| 2-hydroxypropanoic acid | $C_3H_6O_3$ | $1·4 \times 10^{-4}$ |
| Chloric(I) acid | HOCl | $6·7 \times 10^{-10}$ |
| Hydrogen peroxide | $H_2O_2$ | $2·4 \times 10^{-12}$ |
| Hydrocyanic acid (18°C) | HCN | $1·3 \times 10^{-9}$ |
| Boric acid | $H_3BO_3$ | $6 \times 10^{-10}$ |
| Phenol | $C_6H_5OH$ | $1 \times 10^{-10}$ |

Weak polybasic acids dissociate in stages and have a dissociation constant corresponding to each stage of dissociation. For hydrogen sulphide in aqueous solution we have

(i) $$H_2S \rightleftharpoons H^+ + HS^-$$

(ii) $$HS^- \rightleftharpoons H^+ + S^{2-}$$

For the first dissociation at 25°C

$$K_a = \frac{[H^+][HS^-]}{[H_2S]} = 9 \cdot 1 \times 10^{-8} \text{ mol dm}^{-3}$$

For the second dissociation

$$K_a' = \frac{[H^+][S^{2-}]}{[HS^-]} = 1 \times 10^{-15} \text{ mol dm}^{-3}$$

From the values of $K_a$ and $K_a'$ we see that the second dissociation occurs to a much smaller extent than the first (which is itself small). Thus in an aqueous solution of hydrogen sulphide there are very few $S^{2-}$ ions present compared with $HS^-$ ions.

In the practical determination of the dissociation constant of a weak acid, *e.g.*, ethanoic acid, several solutions of the acid, about M/10, M/20, M/40, etc., are prepared and the exact concentrations are found by titration with standard alkali. The molar conductivity ($\Lambda_V$) for each solution is then determined (p. 405). $\Lambda^\infty$ for the acid is obtained by adding together $\Lambda^\infty$ for the $H^+(aq)$ ion and the $CH_3COO^-$ ion at the given temperature. The degree of dissociation for each solution is found from the molar conductivity ratio $\Lambda_V/\Lambda^\infty$, and $K_a$ is calculated from the formula $\alpha = \sqrt{(K_a v)}$. The average value for the dissociation constant is determined.

Another method of finding $\alpha$ and hence $K_a$ for a weak acid is by measurement of the hydrogen ion concentration of a solution of known concentration (see p. 469).

**Example.** *The molar conductivity of 0·093 M CH₃COOH solution at 293 K is 5·36 × 10⁻⁴ Sm² mol⁻¹. The molar conductivities at infinite dilution at 293 K of the H⁺(aq) and CH₃COO⁻ ions are 3·50 × 10⁻² and 0·41 × 10⁻² Sm² mol⁻¹. What is the dissociation constant of ethanoic acid?*

$$\Lambda^\infty(CH_3COOH, 293 \text{ K}) = (3 \cdot 50 + 0 \cdot 41) \times 10^{-2} = 3 \cdot 91 \times 10^{-2}$$
$$\text{Sm}^2 \text{ mol}^{-1}$$

$$\alpha - \frac{\Lambda_V}{\Lambda^\infty} = \frac{5 \cdot 36 \times 10^{-4}}{3 \cdot 91 \times 10^{-2}} = 1 \cdot 37 \times 10^{-2}$$

$$v = \frac{1}{0 \cdot 093} \text{ dm}^3 \text{ mol}^{-1}$$

$$K_a = \frac{\alpha^2}{v} = (0 \cdot 0137)^2 \times 0 \cdot 093$$

$$= 1 \cdot 75 \times 10^{-5} \text{ mol dm}^{-3}$$

METHOD 3. *By the Catalytic Effect of Acids.* The rate at which certain reactions proceed is increased by the presence of a mineral acid. The hydrolysis of an ester is an example of such a reaction. The catalytic effect depends on (i) the concentration of the acid, and (ii) its strength. The strengths of two acids can therefore be compared by measuring the rate of hydrolysis when equivalent concentrations of the acids are used as catalysts. The method is applicable only to strong acids (see p. 377).

Salts of weak acids—*e.g.*, calcium(II) carbonate and calcium(II) phosphate(V)—are often insoluble in water, but dissolve easily in a strong acid like hydrochloric acid. No salt is completely insoluble, and we must imagine that when such salts are added to water small amounts dissolve and produce ions in solution. For example, when hydrochloric acid is added to calcium(II) carbonate we have

$$Ca^{2+} + CO_3^{2-} + 2H^+ + 2Cl^- \rightleftharpoons H_2CO_3 + Ca^{2+} + 2Cl^-$$
$$\Updownarrow$$
$$H_2O + CO_2$$

As carbonic acid is only feebly ionized, the carbonate ions are largely removed from solution and the equilibrium is displaced to the right, causing more of the carbonate to dissolve. In the cases of carbonates, sulphides, etc., the process is aided by the ease with which the resulting weak acid forms volatile products which are continuously removed, thus causing further displacement of the equilibrium.

**Strength of 'Hydroxide' Bases.** The characteristic property of soluble bases of the sodium(I) hydroxide type is their ability to give hydroxyl ions in solution. Although the modern ionic theory applies the term 'base' strictly only to the OH⁻ ions produced, it is still quite common to describe the parent substances by this name. As in the case of acids, bases can be roughly divided into strong bases and weak bases, depending on their degree of dissociation in solution. Strong bases differ from strong acids, however, in being ionic, and not covalent, compounds, and when they can be fused without decomposition they conduct electricity and are electrolysed in the liquid state.

The same general methods used for comparing the strengths of acids can also be used for comparing the strengths of bases. Thus the degree of dissociation can be ascertained by conductivity measurements. The relative strengths of some common bases in M/10 solution at 298 K are given in Table 18–6.

**Table 18–6.** RELATIVE STRENGTHS OF BASES

|  | KOH | NaOH | $\frac{1}{2}Ba(OH)_2$ | Aq.NH₃ |
|---|---|---|---|---|
| Degree of dissociation | 0·95 | 0·93 | 0·90 | 0·013 |
| Strength | 100 | 98 | 95 | 1·51 |

The strengths of weak bases, like those of weak acids, can also be compared by means of their dissociation constants (Table 18–7).

**Table 18–7.** DISSOCIATION CONSTANTS ($K_b$) OF BASES AT 298 K

| Base | Ions formed | $K_b$/mol dm$^{-3}$ |
|---|---|---|
| Aqueous ammonia | $NH_4^+ + OH^-$ | $1.8 \times 10^{-5}$ |
| Aqueous methylamine | $CH_3NH_3^+ + OH^-$ | $5.0 \times 10^{-4}$ |
| Aqueous phenylamine | $C_6H_5NH_3^+ + OH^-$ | $3.5 \times 10^{-10}$ |

**Dissociation of Water.** Pure water is a bad conductor of electricity. The presence of small amounts of dissolved substances, however, increases the conductance considerably. Tap water usually contains appreciable quantities of dissolved substances, while distilled water rapidly acquires impurities from the air and from the walls of the containing vessel. For experiments on conductances the water used as a solvent has to be specially prepared so that its own conductance will be as small as possible. This water is known as *conductivity water*. It is usually obtained by 'demineralizing' ordinary water by the ion exchange process described in Chapter 13.

In spite of elaborate precautions to purify water, it always has a slight conductivity, due to the feeble dissociation into $H_3O^+$ and $OH^-$ ions. For simplicity the dissociation is usually written

$$H_2O \rightleftharpoons H^+ + OH^-$$

Applying the law of mass action we obtain

$$K = \frac{[H^+][OH^-]}{[H_2O]} \quad (1)$$

Since the degree of dissociation is very small $[H_2O]$ can be taken as constant, so that

$$K_W = [H^+][OH^-] \quad (2)$$

The constant $K_W$ is called the *ionic product* of water. At 25°C 1 dm$^3$ of water contains approximately $1 \times 10^{-7}$ mol of both hydrogen and hydroxyl ions. Hence from (2),

$$K_W = 10^{-7} \times 10^{-7} = 10^{-14} \text{ mol}^2 \text{ dm}^{-6}$$

The value of $K_W$ increases with rise of temperature (from $0.29 \times 10^{-14}$ mol$^2$ dm$^{-6}$ at 10°C to $5.47 \times 10^{-14}$ mol$^2$ dm$^{-6}$ at 50°C), but as 25°C is commonly used as a temperature for carrying out experiments, the value at this temperature is the one most frequently used.

The dissociation constant, $K_a$, for water as an acid is obtained from equation (1) by substituting the concentrations of ions or molecules in mol dm$^3$.

$$K_a = \frac{10^{-7} \times 10^{-7}}{1000/18} = 1.8 \times 10^{-16} \text{ mol dm}^{-3}$$

## Hydrogen Ion Concentration and pH Values

The hydrogen ion concentrations of solutions are often a matter of great practical importance. In many industrial operations the success of the process depends on a careful control being maintained over the degree of acidity of the solutions employed. As the hydrogen ion concentrations are usually small, it is more convenient to express them in terms of the *hydrogen ion exponent* (represented by pH). The pH value of a solution is defined as the logarithm to the base 10 of the reciprocal of the hydrogen ion concentration. Pure water contains $1 \times 10^{-7}$ mol dm$^{-3}$ of hydrogen ions. The pH value of water is therefore $\log \dfrac{1}{10^{-7}} = 7$.

Expressed otherwise the pH of a liquid is the logarithm of the hydrogen ion concentration with the sign reversed. Note that as the hydrogen ion concentration increases the pH value decreases, and vice versa. The pH of an aqueous solution of an acid or alkali depends on the concentration and on the degree of dissociation. The following examples will make this clear.

**pH of Acids.** A M/1000 solution of HCl may be said to be completely dissociated and therefore contains 0·001 mol dm$^{-3}$ of hydrogen ions. $[H^+] = 10^{-3}$ mol dm$^{-3}$ and the pH = 3. With more concentrated solutions the pH can be found if the concentration of the acid and the degree of dissociation are known. Thus the degree of dissociation in a 3M solution of hydrochloric acid is 0·55. Hence

$$[H^+] = 3 \times 0.55 = 1.65 = 10^{0.2175} \text{ mol dm}^{-3}$$
$$\text{pH of the solution} = -\log 1.65 = -0.2175$$

The hydrogen ion concentration of an incompletely dissociated acid is given by

$$[H^+] = \text{mol dm}^{-3} \text{ of replaceable hydrogen} \times \alpha$$

where $\alpha$ is the degree of dissociation.

The increase in hydrogen ion concentration resulting from increased concentration of the total acid is opposed by a smaller degree of dissociation. The curves in Fig. 18–1 show the variation of the degree of dissociation and the hydrogen ion concentration with the concentration of hydrochloric acid.[1] The degree of dissociation (curve A) increases as the concentration of the acid decreases. The hydrogen ion concentration (curve B) at first increases as the concentration of acid increases, but reaches a maximum (pH = −0·3) when the solution is approximately 6M. With more concentrated solutions the smaller degree of dissociation more than counterbalances the increase in the total amount of hydrogen present, and the hydrogen ion concentration falls.

[1] Reproduced from 'Reaction and the Determination of pH Values,' by T. Tusting Cocking, *Chemistry in Commerce* (Newnes).

Note that M sulphuric(VI) acid ($H_2SO_4$) is about 50 per cent dissociated into hydrogen ions. Hence $[H^+] = 1$ mol $dm^{-3}$, and the pH is 0.

Weak acids such as ethanoic acid are not completely dissociated even in very dilute solution. For these the hydrogen ion concentration (and hence the pH value) can be found from $[H^+] =$ mol $dm^{-3}$ of replaceable hydrogen $\times \alpha$, or it can be calculated from the concentration and the dissociation constant of the acid. For a weak acid we have

$$\frac{[H^+]\,[A^-]}{[HA]} = K_a$$

$$\text{Therefore} \quad [H^+] = K_a \times \frac{[HA]}{[A^-]}$$

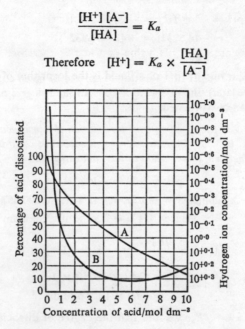

*Fig.* 18–1. Curve A shows dissociation of hydrochloric acid. Curve B shows corresponding hydrogen ion concentration

For a solution of 1 mole in a volume $v$ (volume in $dm^3$), $[HA] = 1/v$ approximately and $[A^-] = \alpha/v$, where $\alpha$ is the degree of dissociation. Hence

$$[H^+] = K_a \times \frac{1}{\alpha} = \frac{K_a}{\sqrt{(K_a v)}}$$

$$= \sqrt{\frac{K_a}{v}}$$

(The reader is advised to make a special note of the two expressions: $\alpha = \sqrt{(K_a v)}$ and $[H^+] = \sqrt{(K_a/v)}$. These are used frequently in working out problems connected with weak acids.)

**Example.** *Calculate the hydrogen ion concentration and pH of a* $0.01$ *molar solution of ethanoic acid,* $CH_3COOH$. $(K_a = 1.8 \times 10^{-5}$ *mol dm*$^{-3}$.)

$$[H^+] = \sqrt{\frac{K_a}{v}} = \sqrt{\left(\frac{1.8 \times 10^{-5}}{10^2}\right)}$$

$$= \sqrt{(1.8 \times 10^{-7})}$$

$$= 4.2 \times 10^{-4} \text{ mol dm}^3$$

$$pH = -\log[H^+] = -\log(4.2 \times 10^{-4})$$

$$= -(\bar{4}.6232) = +4 - 0.6232$$

$$= 3.38$$

Conversely, if the pH of a solution of an acid of given concentration is measured (see later), the dissociation constant and degree of dissociation of the acid can be calculated.

**Example.** *The pH of a* $0.001$ *molar solution if benzenecarboxylic acid,* $C_6H_5COOH$, *is* $3.59$. *Calculate* (i) *the dissociation constant of the acid, and* (ii) *the degree of dissociation at this concentration.*

(i) $$pH = 3.59 = -\log [H^+]$$

$$\therefore \quad \log[H^+] = -3.59 = \bar{4}.41 = \log 0.000\,257$$

$$\text{Hence} \quad [H^+] = 2.57 \times 10^{-4} \text{ mol dm}^{-3}$$

$$[H^+] = \sqrt{\frac{K_a}{v}} \quad \text{or} \quad K_a = [H^+]^2 \times v$$

$$\therefore \quad K_a = (2.57 \times 10^{-4})^2 \times 10^3$$

$$= 6.60 \times 10^{-5} \text{ mol dm}^{-3}$$

(ii) From $\alpha = \sqrt{K_a v}$ we find that the degree of dissociation is $0.257$.

**pH of Alkalis.** The pH values of alkalis in aqueous solution depend on the relation $[H^+] [OH^-] = K_W = 10^{-14} \text{ mol}^2 \text{ dm}^{-6}$. We must first find the hydroxyl ion concentration and then the hydrogen ion concentration and the pH can be calculated. Thus a $0.001$ molar solution of KOH is almost completely dissociated into ions, so that $[OH^-] = 10^{-3} \text{ mol dm}^{-3}$. It follows that $[H^+] = 10^{-14} \div 10^{-3} = 10^{-11} \text{ mol dm}^{-3}$. The pH of the solution is therefore 11.

The pH of solutions of weak alkalis of known concentration can be calculated from their dissociation constants. As in the case of weak acids, it can be shown that for weak alkalis

$$[OH^-] = \sqrt{\frac{K_b}{v}}$$

where $K_b$ is the dissociation constant and $v$ is the volume in dm$^3$ containing one mole of the base.

**Example.** *Calculate the hydrogen ion concentration and pH of a* 0·001 *molar solution of ammonia, NH₃, in water,* ($K_b = 1·8 \times 10^{-5}$ *mol dm³.*)

$$[OH^-] = \sqrt{\frac{K_b}{v}} = \sqrt{\frac{1·8 \times 10^{-5}}{10^3}}$$

$$= \sqrt{(1·8 \times 10^{-8})} = 1·34 \times 10^{-4} \text{ mol dm}^{-3}$$

$$[H^+] = \frac{10^{-14}}{[OH^-]} = \frac{10^{-14}}{1·34 \times 10^{-4}}$$

$$= 7·46 \times 10^{-11} \text{ mol dm}^{-3}$$

$$pH = -\log[H^+] = -\log(7·46 \times 10^{-11})$$

$$= -(\overline{11}·8722)$$

$$= 10·13$$

**The pH Scale.** No acid gives a solution of greater acidity than pH = −0·3. Similarly the smallest degree of acidity obtainable with any alkali is pH = 14·5. All solutions have pH values which fall within these limits. The range represented by these limits is therefore referred to as the *pH scale.*

**Buffer Solutions.** Theoretically, as we have seen, it is possible to prepare solutions of known pH by making up solutions of a strong acid or alkali of known concentration. For example, a 0·0001 molar solution of HCl in water has a pH of 4, while a 0·0001 molar solution of NaOH in water has a pH of 10. In practice, however, such solutions do not retain a constant hydrogen ion concentration for long, as they dissolve impurities from the air and the walls of the containing vessel. To obtain solutions of hydrogen ion concentration which will remain fairly constant, we use buffer solutions. A buffer solution can be made from a weak acid and the sodium(I) salt of the acid, or from a weak base and a salt of the base and a strong acid. Examples of the first are ethanoic acid, phosphoric(V) acid, boric(III) acid, and the corresponding sodium(I) salts; an illustration of the second is a mixture of aqueous ammonia and ammonium chloride.

As ethanoic acid is only slightly dissociated and sodium(I) ethanoate is highly dissociated, a mixture of the two in solution contains few hydrogen ions, but a large proportion of the anions of the acid:

$$CH_3COOH \rightleftharpoons CH_3COO^- + H^+$$
$$CH_3COONa \rightleftharpoons CH_3COO^- + Na^+$$

If a small amount of hydrochloric acid is added to the solution the hydrogen ions added largely unite with ethanoate ions to form molecules of ethanoic acid. Hence the increase in hydrogen ion concentration in the liquid is greatly reduced by the presence of sodium(I) ethanoate. Similarly, if a small amount of alkali is added to the solution the hydroxyl ions combine with the hydrogen ions to form water and further dissociation of the acid occurs. Again, the hydrogen ion concentration undergoes practically no change. In the same way a buffer solution made from ammonia solution and ammonium chloride retains a nearly constant hydrogen ion concentration. The second kind of buffer solution is used to cover a different range of pH from that covered by the first kind. Since buffer solutions tend to neutralize the effect of adding either hydroxyl or hydrogen ions, they are described as 'solutions of reserve acidity and alkalinity.'

The pH of a buffer solution can be calculated from the dissociation constant of the weak acid (or weak base) and the concentrations of acid (or base) and salt present in the mixture. In the case of a weak acid and its sodium(I) salt we have for the acid

$$\frac{[H^+] \, [A^-]}{[HA]} = K_a$$

Therefore $[H^+] = K_a \times \dfrac{[HA]}{[A^-]}$

Since the acid is only slightly dissociated and the salt highly dissociated, $[A^-]$ can be regarded as derived entirely from the salt. Then

$$[H^+] = K_a \times \frac{[\text{acid}]}{[\text{salt}]}$$

Since the ratio [acid]/[salt] remains the same on dilution, the pH of the solution is not affected by dilution.

**Example.** *3·28 g of sodium(I) ethanoate are dissolved in 1 dm³ of 0·01 molar $CH_3COOH$ solution. What is the pH of the resulting solution?* *($K_a$ for ethanoic acid = $1·84 \times 10^{-5}$ mol dm⁻³. H = 1, C = 12, O = 16, Na = 23.)*

1 mole of sodium(I) ethanoate, $CH_3COONa = 82$ g

$$3·28 \text{ g sodium(I) ethanoate} = \frac{3·28}{82} = 0·04 \text{ mole}$$

$$[H^+] = K_a \times \frac{[\text{acid}]}{[\text{salt}]}$$

$$= 1·84 \times 10^{-5} \times \frac{0·01}{0·04}$$

$$= 4·6 \times 10^{-6} \text{ mol dm}^{-3}$$

$$pH = -\log[H^+] = -\log(4·6 \times 10^{-6})$$

$$= -(\bar{6}·6628) = +6 - 0·6628$$

$$= 5·34$$

It is simple to calculate the change in pH caused by the addition of a known amount of strong acid or strong base to the buffer solution. Thus, let us suppose that 1 cm³ of a molar HCl solution is added to 1 dm³ of the solution in the example just given. The addition of 1 cm³ of molar hydrochloric acid to 1 dm³ of water would give a solution in which $[H^+] = 0.001$ mol dm⁻³ approximately. By combination between the added $H^+$ ions and the reserve of ethanoate ions in the buffer solution $[CH_3COO^-]$ will be reduced from 0·04 to 0·039 mol dm⁻³. $[CH_3COOH]$ will increase from 0·01 to 0·011 mol dm⁻³. Therefore,

$$[H^+] = K_a \times \frac{[acid]}{[salt]}$$

$$= 1.84 \times 10^{-5} \times \frac{0.011}{0.039}$$

$$= 5.19 \times 10^{-6} \text{ mol dm}^{-3}$$

$$pH = -\log[H^+] = 5.28$$

The change in pH is thus only 0·06 unit, whereas when the acid is added to 1 dm³ of water the pH changes from 7 to 3—that is, by 4 units.

Similarly by dissolving sufficient sodium(I) hydroxide in the solution to give 0·001 mol dm⁻³ of $OH^-$ the concentration of ethanoic acid will diminish by 0·001 mol dm⁻³ and $[CH_3COO^-]$ will increase by 0·001 mol dm⁻³. The new pH can be calculated as shown previously. Its value is 5·39 approximately.

$pK_a$ **Values.** We have seen that the relative strengths of weak acids can be expressed by their acid dissociation constants, $K_a$. A more convenient scale is provided by the *dissociation constant exponents*, $pK_a$, where $pK_a = -\log K_a$ (compare $pH = -\log [H^+]$). Thus, for ethanoic acid $K_a = 1.82 \times 10^{-5}$ mol dm⁻³ and $pK_a = 4.74$.

It is easily shown that $pK_a$ for a weak acid *is equal to the pH of an aqueous solution of the acid which has been half-neutralized by sodium(I) hydroxide*. We have seen at p. 472 that for a mixture of a weak acid and its sodium(I) salt in aqueous solution

$$[H^+] = K_a \times \frac{[acid]}{[salt]}$$

When the acid is half-neutralized $[acid] = [salt]$, and therefore

$$[H^+] = K_a$$

It follows that

$$pK_a = -\log K_a = -\log[H^+] = pH$$

In practice the volume of sodium(I) hydroxide solution required to neutralize a known volume of the acid solution is found by titration, using phenolphthalein as indicator. The concentrations of acid and alkali need not be known. Half the amount of alkali is then added to the same volume of acid and the pH of the liquid is measured (p. 485).

## Acid-Base Indicators

**Theory of Acid-base Indicators.** Acid-base indicators are substances which change colour according to the hydrogen ion concentration of the liquid in which they are placed. They are either weak acids or weak bases, and are therefore slightly dissociated when dissolved in water. The colour of the indicator depends on the colour of the undissociated molecules and the colour of the ions produced. Thus for methyl orange dissociation occurs as follows:[1]

$$\xleftarrow{\qquad \text{Acids} \qquad}$$
$$\underset{\text{red \quad colourless \quad yellow}}{HA \rightleftharpoons H^+ + A^-}$$
$$\xrightarrow{\qquad \text{Alkalis} \qquad}$$

When a drop of methyl orange is added to water the resultant colour is orange. If now an acid is added, the hydrogen ions of the acid drive back the ionization of the methyl orange, very few $A^-$ ions remain, and the indicator becomes pink. On the other hand, addition of an alkali provides a large concentration of hydroxyl ions, which combine with the hydrogen ions of the indicator to form water. More of the indicator ionizes and a relatively large concentration of $A^-$ ions is produced, giving a yellow colour. Similarly, litmus has red HA molecules and blue $A^-$ ions.

For phenolphthalein HA is colourless and $A^-$ red. Since phenolphthalein is colourless in water, we must assume that the degree of dissociation is so small that there are insufficient $A^-$ ions present to produce a visible colour. Hence phenolphthalein is a much weaker acid even than methyl orange. This is confirmed by the fact that the dissociation constant for phenolphthalein is only $7 \times 10^{-10}$ mol dm$^{-3}$, while that of methyl orange is $2 \times 10^{-4}$ mol dm$^{-3}$. The addition of hydroxyl ions, however, increases the number of $A^-$ ions sufficiently to show their red colour.

**Neutrality.** An indicator which is an acid will indicate neutrality when the numbers of HA molecules and $A^-$ ions present in solution are the same, for then the two colours of the indicator will be showing to an equal extent. This occurs at a different hydrogen ion concentration, or

---

[1] The formula of methyl orange is $(CH_3)_2N$⟨○⟩$—N_2—$⟨○⟩$SO_3H$.

pH, for each indicator, as can be shown by applying the law of mass action to the overall change. If the dissociation constant of the indicator is represented by $K_I$, we have

$$K_I = \frac{[H^+][A^-]}{[HA]}$$

$$\text{or} \quad [H^+] = K_I \times \frac{[HA]}{[A^-]}$$

When the colours of HA and $A^-$ are equal $[HA] = [A^-]$, so that $[H^+] = K_I$,

$$\text{or} \quad pH = -\log[H^+] = -\log K_I$$

The half-way colour of an indicator—that is, the point at which it indicates neutrality—thus occurs at a different hydrogen ion concentration for each indicator, depending on its dissociation constant. In the case of methyl orange $K_I = 2 \times 10^{-4}$ mol dm$^{-3}$, and the corresponding pH = 3·7. For phenolphthalein $K_I = 7 \times 10^{-10}$ mol dm$^{-3}$, and the half-way colour occurs at a pH of 9·1. Litmus has a half-way colour at a pH slightly below 7. Thus water, while roughly neutral with respect to litmus, is alkaline to methyl orange (in dilute solution) and acid to phenolphthalein.

**Range of an Indicator.** Every indicator has a definite range of hydrogen ion concentration, or pH, over which it changes colour. Methyl orange is pink at a pH of 2·9, orange when the pH = 3·7, and yellow when the pH = 4·6. Methyl orange is said to have a pH range of 2·9–4·6. The colours and pH ranges of some well-known indicators are given in Table 18–8.

Table 18–8

| Indicator | pH range | Colours (Acid—Alkali) |
|-----------|----------|------------------------|
| Thymol blue[1] | 1·2— 2·8 | Red—yellow |
| Methyl orange | 2·9— 4·6 | Pink—yellow |
| Congo red | 3·0— 5·0 | Blue—red |
| Methyl red | 4·2— 6·3 | Pink—yellow |
| Litmus | 5·0— 8·0 | Red—blue |
| Phenolphthalein | 8·3—10·0 | Colourless—red |

**Sharpness of End Point in Titrations.** To obtain a sharp end point in titrating acid or alkali, a small change in the volume of acid or alkali added near the end point must produce a large change in hydrogen ion concentration, or pH, and hence a big colour change. The extent to which a sharp end point can be obtained depends on whether the acid and alkali which are being titrated together are strong or weak. This is illustrated graphically in Figs. 18–3 and 18–4.

[1] Thymol blue has a further colour change from yellow to blue over the pH range 8·0–9·6.

Fig. 18–2 shows the effect on the pH value of adding to 50 cm³ of molar HCl solution (a strong acid) increasing quantities of, firstly, a molar solution of NaOH (a strong base) and, secondly, a molar solution of $NH_3$

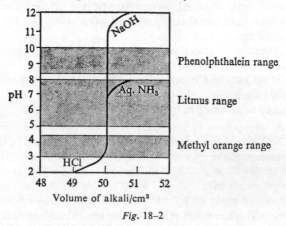

*Fig.* 18–2

(a weak base). With sodium(I) hydroxide the addition of a very little alkali near the end point produces a very large change in the hydrogen ion concentration, almost completely covering the pH ranges of methyl orange, litmus, and phenolphthalein. In practice, therefore, it is immaterial which of these indicators is employed in the titration of a strong acid and a strong base. With ammonia solution adding a small amount

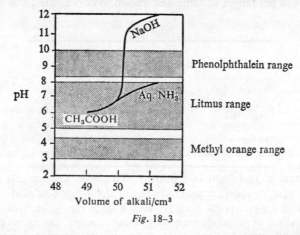

*Fig.* 18–3

of alkali near the end point produces a smaller change in hydrogen ion concentration, and the end point is less sharp, no matter what indicator is used.

Fig. 18–3 shows the effect on the pH value of adding to 50 cm³ of molar

$CH_3COOH$ (a weak acid) increasing quantities of, firstly, molar NaOH solution (a strong base) and, secondly, a molar solution of $NH_3$ (a weak base). In the first case a less sharp end point than with a strong acid and strong base is again obtained. When the acid and base are both weak the end point is so indefinite that in practice a weak acid and a weak base are never titrated against each other.

It is interesting to calculate the pH corresponding to the addition of known amounts of a molar solution of NaOH to 50 cm³ of a molar solution of HCl close to the end-point. The calculation is shown for the addition of 49·9 cm³ and 50·1 cm³ of the alkali (the volume of the resulting solution may be assumed to be 100 cm³).

(i) *Addition of 49·9 cm³ of molar NaOH solution*

Left over we have 0·1 cm³ of molar HCl solution

$$= 0 \cdot 0001 \text{ mole } H^+ \text{ ions in } 100 \text{ cm}^3$$

$$= 0 \cdot 001 \text{ mole } H^+ \text{ ions in } 1 \text{ dm}^3$$

Therefore $[H^+] = 10^{-3} \text{ mol dm}^{-3}$ and pH = 3

(ii) *Addition of 50·1 cm³ of molar NaOH solution*

Left over we have 0·1 cm³ of molar NaOH solution

$$= 0 \cdot 0001 \text{ mole } OH^- \text{ in } 100 \text{ cm}^3$$

$$= 0 \cdot 001 \text{ mole } OH^- \text{ in } 1 \text{ dm}^3$$

Therefore $[OH^-] = 10^{-3} \text{ mol dm}^{-3}$, $[H^+] = 10^{-11}$ and pH = 11

Similarly, the pH corresponding to the addition of other known amounts of alkali can be calculated. Some values are given here:

| Vol. of NaOH/cm³ | 49·0 | 49·9 | 49·95 | 50·05 | 50·1 | 51·0 |
|---|---|---|---|---|---|---|
| pH | 2·0 | 3·0 | 3·3 | 10·7 | 11·0 | 12·0 |

**Choice of an Indicator.** *The true point of neutralization in any titration occurs when the amounts of acid and alkali added together are chemically equivalent to each other.* The solution, however, may not have a pH equal to that of water at this point. With strong acids and alkalis the final pH is about 7. In other cases hydrolysis of the salt takes place, producing a pH greater or less than 7. Thus, according to Fig. 18–2 the pH is about 5 for 0·5 molar $NH_4Cl$ solution, and from Fig. 18–3 about 9 for a 0·5 molar solution of sodium(I) ethanoate, $CH_3COONa$. (Addition of 50 cm³ of molar $NH_3$ solution to 50 cm³ of molar HCl solution results in the formation of approximately 100 cm³ of 0·5 molar ammonium chloride, $NH_4Cl$, solution.) The indicator to be selected is the one which has its half-way colour nearest to the pH of the neutralized solution. For a strong acid and a strong alkali the theoretically correct indicator (of the indicators listed in Table 18–8) is therefore litmus, although in practice either methyl orange or phenolphthalein can be employed. Similarly, methyl orange (or, more accurately, methyl red) is the correct indicator for titrating aqueous ammonia and hydrochloric acid, and phenolphthalein for sodium(I) hydroxide solution and ethanoic acid. The general rules for indicators can be summarized as follows:

> Strong acid and strong base—any indicator
> Strong acid and weak base—methyl orange
> Weak acid and strong base—phenolphthalein
> (Weak acid and weak base—not titrated)

In addition, there are some titrations for which a particular indicator has a special application.

(i) *Titration of Disodium(I) Tetraborate(III) Solution*

$$Na_2B_4O_7 + 5H_2O + 2HCl \rightarrow 2NaCl + 4H_3BO_3$$

A solution of 'borax' can be regarded as a solution of sodium(I) hydroxide of corresponding molarity providing that methyl orange is used as indicator. Boric(III) acid is such a weak acid that any hydrogen ion concentration which it gives is too small to reach the colour range of methyl orange. The latter is therefore unaffected by the presence of the acid.

(ii) *Titration of Alkali Carbonates*

$$Na_2CO_3 + 2HCl \rightarrow 2NaCl + H_2CO_3$$

A cold solution of sodium(I) carbonate can be titrated with a strong acid if methyl orange is the indicator, since the latter is not affected by carbonic acid. Litmus and phenolphthalein are both sensitive to carbonic acid and can be used only when the solution is kept boiling to decompose carbonic acid. If phenolphthalein is used without boiling it changes from red to colourless when the sodium(I) carbonate is half-neutralized —that is, when sufficient acid is added to complete the following reaction:

$$Na_2CO_3 + HCl \rightarrow NaHCO_3 + NaCl \quad (pH = 8\cdot4)$$

This is used in estimating the amounts of sodium(I) hydroxide and sodium(I) carbonate in a mixed solution. An aliquot portion of solution is first titrated with acid by using phenolphthalein, which changes from red to colourless when *all* the hydroxide is neutralized and the carbonate is half-neutralized. Methyl orange is then added, and more acid is run in to complete the reaction:

$$NaHCO_3 + HCl \rightarrow NaCl + H_2O + CO_2$$

If a volume $v_1$ of acid has been used at the first end point, and a further volume $v_2$ at the second end point the sodium(I) carbonate is equivalent to a volume $2v_2$ of acid and the sodium(I) hydroxide to a volume $(v_1 - v_2)$ of acid.

Similarly, phenolphthalein and methyl orange can be used to estimate the amounts of $Na_2CO_3$ and $NaHCO_3$ in a mixture of the two.

(iii) *Titration of Phosphoric(V) Acid.* If sodium(I) hydroxide solution is added to phosphoric(V) acid solution in the presence of methyl orange the latter just turns yellow when the following reaction is completed:

$$H_3PO_4 + NaOH \rightarrow NaH_2PO_4 + H_2O$$

A solution of sodium(I) dihydrogenphosphate(V) (pH = 4·4 for a 0·1 molar solution) is thus roughly neutral to methyl orange, but acidic to litmus and phenolphthalein. The other two sodium(I) phosphates(V) are prepared by adding the same amount of acid used in the titration to two, and to three, times the amount of alkali used. The resulting solutions are then crystallized out. pH values for 0·1 molar solutions are given below.

$$H_3PO_4 + 2NaOH \rightarrow Na_2HPO_4 + 2H_2O \qquad (pH = 9·5)$$
$$H_3PO_4 + 3NaOH \rightarrow Na_3PO_4 + 3H_2O \qquad (pH = 12·0)$$

Since a solution of the disodium(I) salt has a pH which falls within the range of phenolphthalein, the amount of alkali required to prepare this salt can also be determined by titration in the presence of phenolphthalein.

## Salts

The classical definition of a *salt* is that it is a substance formed from an acid by replacing its replaceable hydrogen, either wholly or partially, by a metal or an equivalent radical (such as $NH_4$). In modern ionic theory, a salt is a compound composed of oppositely charged ions. From this point of view there is no difference between sodium(I) chloride, $Na^+Cl^-$, and sodium(I) hydroxide, $Na^+(OH^-)$. Anhydrous aluminium(III) chloride is not a salt, since the bonds are essentially covalent. The hydrated solid, however, consists of $Al(H_2O)_6^{3+}$ and $Cl^-$ ions and is therefore classed as a salt.

**Normal Salts.** These are obtained when the replaceable hydrogen of the acid has been completely replaced by a metal or equivalent radical —*e.g.*, normal sodium(I) sulphate(VI), $Na_2SO_4$.

**Acid Salts.** This type of salt is formed when the replaceable hydrogen of the acid has been only partially replaced by a metallic radical or an equivalent radical. An example is sodium(I) hydrogensulphate(VI), $NaHSO_4$. As an acid salt still contains replaceable hydrogen, it possesses the properties of a salt and those of an acid. In aqueous solution, acid salts dissociate in stages—*e.g.*

$$NaHSO_4 \rightleftharpoons Na^+ + HSO_4^- \rightleftharpoons Na^+ + H^+ + SO_4^{2-}$$

The point of equilibrium depends not only on the dilution, but on the strength of the acid. Increasing the dilution moves the point of equilibrium to the right, but the tendency is less marked when the acid is weak, as in the case of $NaHCO_3$. Hydrolysis frequently occurs with acid salts, so that the solution may actually give an alkaline reaction with litmus. This happens with sodium(I) hydrogencarbonate (see p. 484).

**Basic Salts.** These can be regarded in two ways. They can be considered as formed from normal salts by replacing part of the acid group by oxygen or hydroxyl. Antimony(III) chloride oxide is formed when antimony(III) chloride is added to water; thus

$$\text{Sb}\underset{\text{Cl}}{\overset{\text{Cl}}{\underset{|}{\diagup}}}\text{Cl} + H_2O \rightarrow \text{Sb}\overset{\text{Cl}}{\underset{\text{O}}{\diagup}} + 2HCl$$

Bismuth(III) nitrate(V) oxide, $BiO(NO_3)$, is similarly formed from bismuth(III) nitrate(V), $Bi(NO_3)_3$. Alternatively, a basic salt may be regarded as a compound of a normal salt and a base in stoichiometric proportions. The above basic salts can be represented as $SbCl_3.Sb_2O_3$ and $Bi(NO_3)_3.Bi_2O_3$. The second method is simpler when a series of basic salts occurs. For example, there are several basic copper(II) sulphates(VI) of the type $xCuSO_4.yCu(OH)_2$.

**Double Salts.** A double salt is one which is formed by the combination of two simple salts in stoichiometric ratio but which in solution gives the reactions of the constituent single salts. Thus if stoichiometric proportions of iron(II) sulphate(VI) and ammonium sulphate(VI) are dissolved in water, and the solutions added to each other, the crystals obtained on evaporation have the composition $FeSO_4.(NH_4)_2SO_4.6H_2O$. (This does *not* mean that the solid double salt actually exists in the composite form shown.) Double salts prepared in this way differ in certain properties, such as colour and crystalline form, from the constituent single salts. Iron(II) sulphate(VI) undergoes oxidation on exposure to air, whereas iron(II) ammonium sulphate(VI) is unaffected.

**Complex Salts.** A complex salt is one containing a complex ion, and a complex ion is an ion consisting of a charged group of atoms. Examples of complex ions include $NH_4^+$, $NO_3^-$, and $SO_4^{2-}$. In contrast $Na^+$, $Cl^-$, etc., are 'simple' ions. Complex ions may be produced in various ways, as now illustrated.

*Combination of simple ions with neutral molecules.* Examples are the formation of hydrated metal ions such as $Ca(H_2O)_6^{2+}$ and $Cu(H_2O)_4^{2+}$. Ammonia often gives rise to complex ions of this type, *e.g.*, $Cu(NH_3)_4^{2+}$ (tetraamminecopper(II) ion) and $Ag(NH_3)_2^+$ (diamminesilver(I) ion). The latter is formed when silver(I) chloride dissolves in ammonia solution.

$$Ag^+Cl^- + 2NH_3 \rightleftharpoons Ag(NH_3)_2^+ + Cl^-$$

*Combination of simple or complex ions with salts.* Lead(II) chloride dissolves in concentrated hydrochloric acid with the formation of the tetrachloroplumbate(IV) ion.

$$PbCl_2 + 2Cl^- \rightleftharpoons PbCl_4^{2-}$$

Similarly iron(II) cyanide dissolves in aqueous potassium(I) cyanide to give the hexacyanoferrate(II) ion.

$$Fe(CN)_2 + 4CN^- \rightleftharpoons Fe(CN)_6^{4-}$$

*Combination of oppositely charged ions.* In the fixing process in photography silver(I) bromide is dissolved from the plate or film by a solution of sodium(I) thiosulphate(VI) ('hypo'), forming the complex thiosulphatoargentate(I) ion.

$$Ag^+Br^- + S_2O_3^{2-} \rightleftharpoons AgS_2O_3^- + Br^-$$

The formation of complex anions or cations is often used to get 'insoluble' substances into solution, as many complex salts have a high solubility in water. The dissolving of silver(I) chloride in aqueous ammonia and silver(I) bromide in sodium(I) thiosulphate(VI) solution are examples. In qualitative analysis the sulphides As₂S₃, Sb₂S₃, and SnS are dissolved in yellow ammonium sulphide to separate them from the sulphides of other Group II metals. The dissolving of tin(II) sulphide to form the trithiostannate(IV) ion may be taken as typical. Yellow ammonium sulphide can be regarded as a solution of sulphur in ammonium sulphide:

$$SnS + (NH_4)_2S + S \rightleftharpoons (NH_4)_2SnS_3$$
$$\text{or} \quad SnS + S^{2-} + S \rightleftharpoons SnS_3^{2-}$$

The incorporation of the additional sulphur in the complex ion results in the *higher* sulphide being precipitated when the solution is acidified with hydrochloric acid:

$$SnS_3^{2-} + 2H^+ \rightarrow SnS_2 + H_2S$$

**Investigation of Complex Ions.** There are several methods for determining the composition of complex ions. The most important are the following:

*By Solubility.* When solid mercury(II) iodide is added to a solution of potassium(I) iodide of known concentration the solid can be dissolved until the ratio of mercury(II) iodide to potassium(I) iodide is approximately that represented by HgI₂:2KI. Hence it is deduced that the composition of the complex ion is HgI₄²⁻. This is the tetraiodomercurate(II) ion.

*Freezing point Method.* The freezing point of a dilute solution of potassium(I) iodide shows that the salt is almost completely dissociated. If a small amount of mercury(II) iodide is dissolved in the solution the freezing point depression is not as large as previously. This shows that, although mercury(II) iodide is being added, the number of particles in solution has decreased. The decrease is due to combination between the iodide ion and mercury(II) iodide to form a complex ion. Two ways in which this combination could occur are:

$$K^+ + I^- + HgI_2 \rightarrow K^+ + HgI_3^-$$

$$\text{and} \quad 2K^+ + 2I^- + HgI_2 \rightarrow 2K^+ + HgI_4^{2-}$$

If the combination occurs according to the first equation the freezing point of the solution will not alter when mercury(II) iodide is added, because the number of particles in solution is the same after the addition of mercury(II) iodide as before. According to the second equation, if all the mercury(II) iodide is converted into the complex ion the freezing point of the solution must rise, because the number of particles in solution decreases. As mercury(II) iodide is added, the freezing point must continue to rise until the amounts of potassium(I) iodide and mercury(II) iodide present are in the ratio represented by 2KI:HgI₂. Furthermore, when the maximum value for the freezing point is reached the freezing point depression should be approximately three quarters of that corresponding to complete dissociation of the potassium(I) salt, since if the complex salt is completely dissociated there will be three quarters as many particles in solution. In practice both these deductions are found to hold, so that the second equation given above represents the action.

The dissolving of iodine in a dilute solution of potassium(I) iodide does not affect the freezing point of the solution, because the number of particles in solution is unaltered:

$$K^+ + I^- + I_2 \rightleftharpoons K^+ + I_3^-$$

*Partition Coefficient Method.* This method is described at p. 291.

**Hydrolysis of Salts.** The term 'hydrolysis' is used in general to describe certain chemical reactions of double decomposition brought about by water. In salt hydrolysis there is a reaction between a salt and water whereby the equilibrium existing between the hydrogen and hydroxyl ions in water is disturbed. We have seen previously that for water $[H^+][OH^-] = K_W = 10^{-14} \, mol^2 \, dm^{-6}$. Any disturbance of the equilibrium will result in a change of $[H^+]$, so that the pH of the liquid will be greater, or less, than 7. Any liquid which has a pH less than 7 gives an acid reaction with litmus, and any liquid with a pH greater than 7 gives an alkaline reaction with litmus. The mechanism and extent of hydrolysis depend on the nature of the salt as now described.

(i) *Salts Containing a Metal Cation and an Anion of a Weak Acid.* When dissolved in water salts of this type give an alkaline reaction with litmus. Examples are potassium(I) cyanide, sodium(I) ethanoate, and sodium(I) carbonate. Thus, potassium(I) cyanide, being a salt, is highly dissociated, while water is slightly dissociated. Reaction occurs in the solution as follows:

$$K^+ + \underbrace{CN^- + H^+} + OH^- \rightleftharpoons HCN + K^+ + OH^-$$

Potassium(I) hydroxide is almost wholly dissociated in dilute solution, whereas hydrocyanic acid, being a very weak acid, is scarcely dissociated at all. $H^+$ and $CN^-$ ions are therefore largely removed from solution as hydrocyanic acid molecules. The removal of $H^+$ ions disturbs the equilibrium $H_2O \rightleftharpoons H^+ + OH^-$, and further dissociation of water molecules occurs until equilibrium is established between the ions and hydrocyanic acid molecules. There then remains in solution an excess of $OH^-$ ions (pH > 7), which cause red litmus to turn blue. In general this type of hydrolysis can be represented as follows:

$$A^- + H^+ + OH^- \rightleftharpoons HA + OH^-$$

(ii) *Ammonium Salts.* Ammonium chloride may be taken as typical of ammonium salts of strong acids. The $NH_4^+$ ion is quite strongly acidic and readily gives up a proton to a strong base such as $OH^-$ to form ammonia and water:

$$H^+ + \underbrace{OH^- + NH_4^+} + Cl^- \rightleftharpoons NH_3 + H_2O + H^+ + Cl^-$$

In this case $OH^-$ ions are withdrawn from the liquid and by further dissociation of water molecules an excess of $H^+$ ions is produced in solution (pH < 7).

Ammonium chloride solution is only faintly acidic to litmus. The phenylammonium ion, $C_6H_5NH_3^+$, is a stronger acid than $NH_4^+$ because of the presence of the benzene nucleus. Thus an aqueous solution of a phenylammonium salt such as the chloride, $C_6H_5NH_3^+$ $Cl^-$, immediately turns blue litmus red. Again, when aqueous ammonium sulphate(VI) is boiled ammonia gas is expelled from the liquid, while the less volatile sulphuric(VI) acid remains in solution. The equilibrium point in the hydrolysis reaction is thus displaced to the right, and after a time (20 minutes) the solution turns blue litmus paper red.

If the ammonium salt contains the anion of a weak acid the situation is different, for both $H^+$ ions and $OH^-$ ions are withdrawn from solution. Thus ammonium ethanoate undergoes considerable hydrolysis with water:

$$H^+ + OH^- + NH_4^+ + CH_3COO^- \rightleftharpoons NH_3 + H_2O + CH_3COOH$$

The behaviour of the solution towards litmus depends on the relative extents to which $H^+$ ions and $OH^-$ ions are withdrawn from solution. This in turn depends on the dissociation constants of ammonia solution and the weak acid. If the dissociation constants are about equal, as in the case of ethanoic acid, the liquid is neutral (pH = 7) in spite of hydrolysis. If the weak acid has a smaller dissociation constant, as with carbonic acid, the solution is alkaline.

The strong smell of ammonium sulphide solution is due to large hydrolysis and the escape of ammonia and hydrogen sulphide. Similarly the giving-off of ammonia from ammonium carbonate in 'smelling salts' depends on hydrolysis.

(iii) *Acid Salts*. When an acid salt is dissolved in water the pH of the solution depends on the extent to which two opposing reactions take place. These are dissociation of the acid anion, and hydrolysis. Thus, with sodium(I) hydrogencarbonate solution the two reactions are

$$HCO_3^- \rightleftharpoons H^+ + CO_3^{2-}$$

$$HCO_3^- + H^+ + OH^- \rightleftharpoons H_2CO_3 + OH^-$$

Both the hydrogencarbonate ion and carbonic acid are weak acids and only feebly dissociated. The hydrolysis reaction therefore predominates and the solution is alkaline to litmus.

On the other hand, a solution of sodium(I) hydrogensulphate(IV), $NaHSO_3$, turns blue litmus red:

$$HSO_3^- \rightleftharpoons H^+ + SO_3^{2-}$$

$$HSO_3^- + H^+ + OH^- \rightleftharpoons H_2SO_3 + OH^-$$

Here, the first reaction occurs to the greater extent and an acidic solution is formed.

(iv) *Salts containing Hydrated Cations*. A number of normal salts when dissolved in water give an acidic reaction with litmus paper. This is caused by the hydrated metal ions behaving as an acid in the presence of the base water. Thus in iron(III) chloride solution the hydrated iron (III) ions have the formula $Fe(H_2O)_6^{3+}$, the six water molecules having an octahedral distribution about the central iron(III) ion (see p. 184). Each water molecule is attached by a co-ordinate covalent bond formed from a lone pair of electrons donated by the oxygen atom. However, the strong attraction of the trebly charged iron(III) ion extends beyond the electrons of the oxygen atoms to those in the O—H bonds. These electrons also are displaced to some extent towards the iron(III) ion. As a result the O—H bonds are weakened, and in aqueous solution the hydrated iron(III) ion loses a proton to a water molecule.

$$\left[ \begin{array}{c} H \quad\quad H \\ \searrow \quad \swarrow \\ O \\ H_2O \quad | \quad OH_2 \\ \searrow \downarrow \swarrow \\ Fe \\ \nearrow \uparrow \nwarrow \\ H_2O \quad | \quad OH_2 \\ H_2O \end{array} \right]^{3+} + H_2O \rightleftharpoons H_3O^+ + Fe(H_2O)_5(OH)^{2+}$$

If sufficient water is present the hydrolysis reaction may proceed further:

$$Fe(H_2O)_5(OH)^{2+} + H_2O \rightleftharpoons H_3O^+ + Fe(H_2O)_4(OH)_2^+$$
$$Fe(H_2O)_4(OH)_2^+ + H_2O \rightleftharpoons H_3O^+ + Fe(H_2O)_3(OH)_3$$

Thus, when iron(III) chloride is poured into a large amount of water a colloidal dispersion of (hydrated) iron(III) hydroxide is produced. With iron(III) sulphate(VI) a precipitate of basic sulphate(VI) is formed. This can be represented as $xFe_2(SO_4)_3.yFe(OH)_3$, where $x$ and $y$ vary with the amount of water.

Similar hydrolysis reactions take place with other hydrated metal ions, the extent to which they occur depending partly on the charge on the ion. Hydrolysis is most marked with trivalent ions (*e.g.*, $Fe(H_2O)_6^{3+}$ and $Al(H_2O)_6^{3+}$). It occurs with some divalent ions (*e.g.*, $Zn(H_2O)_4^{2+}$) and $Cu(H_2O)_4^{2+}$), but not with others (*e.g.*, $Mg(H_2O)_6^{2+}$ and $Fe(H_2O)_6^{2+}$), which therefore give neutral solutions. It does not occur with monovalent ions (*e.g.*, silver(I) ions).

## Measurements of pH Values

Two methods are in general use for the direct measurement of the pH of solutions. The accurate one is an electrical method involving the use of the potentiometer, but is not as convenient or as rapid as the other method which involves the use of indicators. The latter method will be described first.

**Indicator Method.** The B.D.H. Universal Indicator shows the pH value of solutions between pH = 3 and pH = 11. As the pH increases from 3 to 11 the colour of the indicator passes in order through the colours of the spectrum from red to violet. By the use of test papers impregnated with the indicator the approximate pH is given by the resulting colour as in Table 18-9.

**Table 18-9**

| Colour | pH | Colour | pH |
|---|---|---|---|
| Red | 3·0 or less | Green | 8·0 |
| Deep red | 4·0 | Greenish blue | 9·0 |
| Orange red | 5·0 | Blue | 9·5 |
| Orange yellow | 6·0 | Violet | 10·0 |
| Yellow | 6·5 | Reddish violet | 11·0 |
| Greenish yellow | 7·0 | | |

To obtain the value more accurately an indicator is chosen which has a range covering the approximate value of the pH. A standard amount of this indicator is added to a standard amount of solution, and the colour compared with a series of prepared buffer solutions of known pH containing the same amount of indicator. Thus if the solution under investigation has a pH value of from 7 to 8 use is made of phenol red, which has a range from 6·8 to 8·4 with a colour change from yellow to red. A set of standard buffer tubes is available containing this indicator in solutions of pH 6·8, 7·0, 7·2, . . . 8·4. Similar sets are available for other indicators. The pH is then equal to that of the solution which it matches in colour.

**Potentiometer Method.** The liquid of which the pH is required is placed in a vessel A (Fig. 18–4), into which a 'hydrogen electrode' is inserted. The latter consists of a 'platinized' strip of platinum foil (*i.e.*, platinum

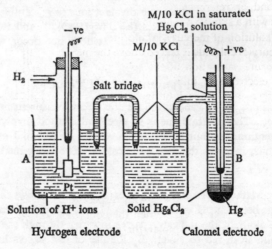

*Fig.* 18–4. Cell composed of hydrogen and calomel electrodes

on which a layer of spongy platinum has been deposited by electrolysis).

Pure hydrogen at atmospheric pressure is kept bubbling over the surface of the platinum so that the latter is about half in, and half out of, the liquid. Hydrogen is adsorbed on the platinum surface, and in these circumstances the electrode behaves as a solid hydrogen electrode, equilibrium being maintained between the hydrogen ions in the liquid and hydrogen on the platinum surface as follows:

$$H_2 - 2e \rightleftharpoons 2H^+$$

The hydrogen electrode is one half of a cell. Theoretically the other half of the cell could be a 'molar hydrogen electrode'—that is, a hydrogen electrode placed in a molar solution of $H^+$ ions. (More accurately, the liquid should contain hydrogen ions at unit 'activity.' At 298 K this occurs with a 1·2 molar HCl solution.) The e.m.f. of this combination would then be given by the Nernst equation

$$E = \frac{RT}{zF} \log_e [H^+] = \frac{2 \cdot 303RT}{zF} \log_{10} [H^+]$$

where $R$ is the gas constant, $T$ the kelvin temperature, $z$ the valency of the $H^+$ ion, and $F$ is the Faraday constant (approximately 96 500 C mol$^{-1}$). At 298 K the expression becomes

$$E = 0 \cdot 059 \log_{10} [H^+] \text{ V}$$

In practice it is easier to use a mercury(I) chloride ('calomel') electrode as the second, or 'reference', electrode. In this a tube B contains mercury in contact with a standard solution of potassium(I) chloride made up in a saturated solution of mercury(I) chloride. Some solid mercury(I) chloride is placed above the mercury to ensure saturation of the solution. The reference electrode is joined to the hydrogen electrode by means of a 'salt bridge' consisting of an inverted U-tube full of standard potassium(I) chloride solution and plugged at the ends with filter paper or plaster of Paris.

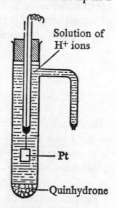

Solution of H$^+$ ions

Pt

Quinhydrone

*Fig.* 18–5. The quin-hydrone electrode

The standard KCl solutions may be molar, 0·1 molar, or a saturated solution of the salt. If the concentration is 0·1 molar a calomel electrode develops an e.m.f. of +0·334 V with respect to a *molar* hydrogen electrode. When the concentration of H$^+$ ions is not molar the e.m.f. obtained at 298 K has a value $E$ in volt given by the equation

$$0 \cdot 334 \text{ V} - E = 0 \cdot 059 \log_{10} [H^+] \text{ V}$$

Hence $\quad \log_{10}[H^+] = \dfrac{0 \cdot 334 \text{ V} - E}{0 \cdot 059 \text{ V}}$

or $\quad \text{pH} = -\log [H^+] = \dfrac{E - 0 \cdot 334 \text{ V}}{0 \cdot 059 \text{ V}}$

Instead of the hydrogen electrode a 'quinhydrone' electrode may be used as a half-cell. This electrode, which is much easier to set up, simply consists of a strip of platinum foil immersed in the solution of hydrogen ions, the solution also being saturated with the sparingly soluble substance 'quinhydrone' (Fig. 18–5.) The latter is a double compound consisting of cyclohexadiene-1,4-dione ('quinone'), $C_6H_4O_2$, and its reduction compound benzene-1,4-diol ('hydroquinone'), $C_6H_4(OH)_2$, in stoichiometric proportions. The two compounds form a redox couple in solution, and in a solution containing $H^+(aq)$ ions an equilibrium is established as follows:

$$C_6H_4O_2 + 2H^+ + 2e \rightleftharpoons C_6H_4(OH)_2$$

The position of equilibrium and hence the reversible electric potential vary with the pH of the solution. The electrode is used in conjunction with a calomel electrode or some other form of reference electrode. At 298 K the potential of a quinhydrone electrode with a molar solution of $H^+$ ions is $+0.699$ V with respect to a standard hydrogen electrode. The electrode potential of a calomel electrode (that is, with respect to a standard hydrogen electrode) is $+0.334$ V. Therefore under standard conditions the quinhydrone electrode will have a potential of $0.699 - 0.334 (= +0.365)$ V with respect to the calomel electrode. When the concentration of $H^+$ ions in the quinhydrone half-cell is not unity the observed e.m.f. of the combination will differ from $0.365$ V. If the observed e.m.f. (found by potentiometer) is $E$ we have

$$E = E^{\ominus} + 0.059 \log [H^+] \text{ V}$$

$$= 0.365 + 0.059 \log [H^+] \text{ V}$$

Instead of the calomel electrode a silver-silver(I) chloride electrode is often used as a reference electrode, as it is more easily assembled. It consists of a strip of silver foil coated with silver chloride by electrolysis. It is immersed in a $0.1$ molar KCl solution. Its e.m.f. with respect to a standard hydrogen electrode is $0.2875$ V at 298 K.

**Potentiometric Titration.** When a solution of an alkali is gradually added from a burette to a solution of an acid the change in pH can be represented graphically as shown in Figs. 18–3 and 18–4. If the acid contains a hydrogen electrode used in conjunction with a reference electrode (calomel or silver-silver(I) chloride) the course of the titration can be followed by the changes in e.m.f. caused by the changes in hydrogen ion concentration. Thus, if 50 $cm^3$ of molar HCl solution are titrated with molar NaOH solution, the pH for each addition of alkali can be measured and the graphs shown in Figs. 18–2 and 18·3 can be constructed.

Near the end point in an acid-alkali titration the pH of the solution (or e.m.f. of the cell) undergoes a large change. The gradient of the pH curve has a maximum at the point of neutralization (the point at which the gradient of a curve has a maximum or minimum is called a 'point of inflection'). The point of inflection is found accurately, as now described.

Suppose that 20 $cm^3$ of an acid of unknown concentration are being titrated with a $0.1$ molar solution of NaOH. Since the pH of the solution is proportional to the e.m.f. developed, it is sufficient to plot the e.m.f. ($E$) against the number of $cm^3$ of the alkali added. A graph of the form shown in Fig. 18–6(*a*) is obtained, showing a point of inflection between 24 cm³ and 25 cm³ of alkali. To find the exact point of inflection the slope of the curve is determined at a number of points by drawing tangents. This gives $dE/dV$ the rate of change of e.m.f. with volume of alkali added,

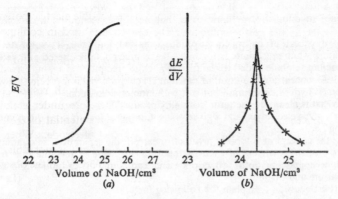

*Fig.* 18–6

on both sides of the neutralization point. The values of $dE/dV$ are now plotted against the volume, $V$, of alkali added (Fig. 18–6(*b*)). The exact point of neutralization is readily ascertained, since it is the point at which $dE/dV$ has its maximum value.

When a polybasic acid is titrated electrometrically with alkali as described above there is usually more than one point of inflection in the pH, or e.m.f., curve. Carbonic acid and ethanedioic acid give two such points, corresponding to the two stages of dissociation of these acids. We might expect that the number of points of inflection would always correspond to the basicity of the acid, but in practice this does not hold. Sulphuric(VI) acid (dibasic) has only one point of inflection, and phosphoric(V) acid (tribasic) two. Those for phosphoric(V) acid occur

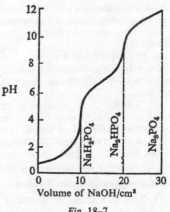

*Fig.* 18–7

when sufficient alkali has been added to form the salts $NaH_2PO_4$ and $Na_2HPO_4$. This is illustrated in Fig. 18–7, which shows the pH curve obtained when 0·1 molar NaOH solution is added to 10 cm³ of a 0·1 molar solution of phosphoric(V) acid ($H_3PO_4$). There is no point of inflection corresponding to the replacement of the third hydrogen atom from the acid.

## EXERCISE 18
*(Relative atomic masses are given at the end of the book)*

**1.** Explain the following by reference to the ionic theory:

(i) In qualitative analysis barium(II) chloride may be precipitated from a concentrated solution by hydrochloric acid in Group 1.

(ii) Antimony(III) sulphide is precipitated by hydrogen sulphide in the presence of dilute hydrochloric acid, but not of concentrated hydrochloric acid.

(iii) If excess of hydrogen sulphide is passed into zinc(II) sulphate(VI) solution a white precipitate is formed. If the precipitate is filtered and yellow ammonium sulphide added to the filtrate a white precipitate is again formed.

(iv) A concentrated solution of calcium(II) chloride yields a precipitate with sodium(I) hydroxide solution but not with ammonia solution.

**2.** What is meant by the term 'solubility product'?

For a given electrolyte, and a given solvent, on what factors does this quantity depend?

Use the concept of solubility product to explain the separation, in qualitative analysis, of the members of each of the following pairs of metallic ions from solutions containing both: (i) copper and cadmium, (ii) cadmium and zinc, (iii) barium and calcium. (Lond.)

**3.** (Part question.) Explain the following facts:

(*a*) Mercury(II) sulphide is precipitated by hydrogen sulphide in both dilute and concentrated hydrochloric acid; zinc(II) sulphide is precipitated by hydrogen sulphide in ethanoic acid solution, but not in molar HCl solution.

(*b*) Ammonium chloride solution prevents precipitation of manganese(II) hydroxide by ammonium hydroxide solution.

(*c*) Lead(II) chloride is soluble in concentrated hydrochloric acid. (O. and C.)

**4.** State the law of mass action, and explain how it may be applied to the dissociation of a weak electrolyte in solution. What do you understand by (*a*) dissociation constant (of such an electrolyte), (*b*) solubility product? Show how the dissociation constant of a weak monobasic acid may be calculated approximately if the concentration of hydrogen ions in a decimolar solution of the acid is known. (J.M.B.)

**5.** Explain the following:

(*a*) Gradual addition of a solution of ammonia to a solution of a copper(II) salt gives at first a pale blue precipitate, which then dissolves forming a deep blue solution.

(*b*) If some tin is dissolved in excess of concentrated hydrochloric acid, the resulting solution gives no precipitate with hydrogen sulphide, but on the addition of distilled water, a dark brown precipitate is eventually deposited.

(*c*) A solution of iron(III) chloride is acid to litmus.

(*d*) Zinc(II) hydroxide dissolves in excess of dilute hydrochloric acid and also in excess of sodium(I) hydroxide solution. (O.L.)

### Calculations on Solubility Product

**6.** The solubilities (in $g\ dm^{-3}$) of silver(I) bromide, lead(II) iodide, and silver(I) carbonate at 15°C are as follows: (i) $AgBr:0.000113$, (ii) $PbI_2:0.6$, (iii) $Ag_2CO_3:0.03$. Calculate the solubility products of these substances.

**7.** The solubility products of calcium(II) sulphate and lead(II) sulphide at 15°C are: (i) $CaSO_4:2.24 \times 10^{-4}\ mol^2\ dm^{-6}$, (ii) $PbS:3.4 \times 10^{-28}\ mol^{-2}\ dm^{-6}$.

Calculate the solubilities (in $g\ dm^{-3}$) of these substances at 15°C.

**8.** (Part question.) (*a*) The solubility product of silver(I) chloride is $1.44 \times 10^{-10}\ mol^2\ dm^{-6}$ at room temperature. Calculate the mass of silver(I) chloride which dissolves at room temperature in (i) 1 $dm^3$ of water, (ii) 1 $dm^3$ of 0.1 molar HCl solution.

(*b*) The solubility product of copper(II) sulphide is $1 \times 10^{-44}$ mol$^2$ dm$^{-6}$. If the concentration of the sulphide ion in a saturated solution of hydrogen sulphide acidified with hydrochloric acid is $2 \times 10^{-22}$ mol dm$^{-3}$, what mass of copper(II) ion will remain in 1 dm$^3$ of solution after precipitation by excess hydrogen sulphide?    (O.L.)

9. The solubility of barium(II) sulphate(VI) in water at 15°C is 0·0025 g dm$^{-3}$. What mass of barium(II) sulphate(VI) can be dissolved in 1 dm$^3$ of 0·05 molar H$_2$SO$_4$ at 15°C? (Assume complete dissociation of both substances.)

10. (Part question.) An aqueous solution contained Na$_2$CO$_3$ and Na$_2$SO$_4$, each of concentration 0·1 mol dm$^{-3}$. To 1 dm$^3$ of this solution was added slowly 0·1 mol of CaCl$_2$. Deduce what would be precipitated and calculate the quantity (in mole) of any precipitate, given that the solubility product for calcium(II) carbonate is $1·69 \times 10^{-8}$ mol$^2$ dm$^{-6}$ and for calcium(II) sulphate(VI) $2·3 \times 10^{-4}$ mol$^2$ dm$^{-6}$.    (J.M.B.)

**Dissociation Constants, [H⁺], and pH.** (Use the dissociation constants listed at pp. 464 and 467, and take $K_W$ to be $10^{-14}$ mol$^2$ dm$^{-6}$.)

11. The degrees of dissociation of M/10 and M/100 HCOOH at 25°C are 0·045 and 0·14 respectively. Use Ostwald's dilution law to calculate the corresponding values of the dissociation constant of methanoic acid.

12. What is the degree of dissociation of CH$_3$COOH in (i) M/4 and (ii) M/20 solution?

13. The degree of dissociation of a weak electrolyte in M/100 solution is 0·05. What is the degree of dissociation in M/1 000 solution?

14. Calculate the hydrogen ion concentrations and pH values of the following solutions: (i) M/1 000 HCOOH, (ii) M/50 C$_6$H$_5$COOH, (iii) M/100 CH$_3$NH$_2$ solution, (iv) M/25 NH$_3$ solution.

15. (Part question.) Derive an expression relating the degree of dissociation of a *very weak* monobasic acid to its dissociation constant. Without giving experimental details, state briefly how the validity of the expression might be tested.

The dissociation constant of a very weak monobasic acid is $2 \times 10^{-5}$ mol dm$^{-3}$ at 25°C. For a 0·05M solution of it, calculate:

(*a*) the degree of dissociation; (*b*) the hydrogen ion concentration; (*c*) the pH.    (S.U.)

16. (Part question.) Calculate the basic constant

$$\frac{[NH_4^+]\,[OH^-]}{[NH_3]}$$

for ammonia, given that the pH of a M/100 solution is 10·6.

$$[K_W = 10^{-14} \text{ mol}^2 \text{ dm}^{-6}] \tag{C.L.}$$

17. What are the hydrogen ion concentrations of the solutions obtained by adding to 50 cm$^3$ of M/10 hydrochloric acid, HCl: (i) 49·5 cm$^3$ of M/10 sodium(I) hydroxide solution, NaOH, (ii) 50·5 cm$^3$ of M/10 sodium(I) hydroxide solution? (Assume that the final volume is 100 cm$^3$ in each case and that the substances are completely dissociated.)

**Buffer Solutions**

18. 6·8 g of sodium(I) methanoate are dissolved in 1 dm$^3$ of M/100 HCOOH. What is the approximate hydrogen ion concentration of the resulting solution?

19. 0·02 mol of NH$_4$Cl is dissolved in 1 dm$^3$ of M/10 NH$_3$ solution. Calculate the [OH⁻] (i) before, (ii) after, the addition of the salt.

**20.** (Part question.) Calculate the mass of sodium(I) ethanoate that must be dissolved in 1 dm³ of decimolar $CH_3COOH$ in order to produce a solution of pH = 4. Explain how you arrive at your answer. H = 1; C = 12; O = 16; Na = 23. The dissociation constant of ethanoic acid, $K_a = 1.8 \times 10^{-5}$ mol dm⁻³.

(W.J.E.C.)

## MORE DIFFICULT QUESTIONS

**21.** What is your definition of (i) an acid, (ii) a base, (iii) a salt? Discuss the application of your definitions to the cases of (a) zinc(II) hydroxide, (b) sodium(I) carbonate, (c) ethanol, (d) water. (O. and C.)

**22.** The solubility of lead(II) sulphate(VI) in water at 15°C is 0.04 g dm⁻³. Calculate the mass of lead(II) sulphate(VI) precipitated when 1 g of potassium(I) sulphate(VI) is dissolved in 1 dm³ of saturated $PbSO_4$ solution at 15°C.

**23.** (Part question.) Mercury(II) iodide is insoluble in water but dissolves in potassium(I) iodide solution to form a complex anion. A solution of 3.32 g of potassium(I) iodide in 100 g of water begins to freeze at −0.745°C. A saturated solution of mercury(II) iodide in the above potassium(I) iodide solution contains 4.60 g of mercury(II) iodide per 100 g of water and it begins to freeze at −0.558°C. Give a reasoned deduction of the formula of the complex anion from these data. Mention any simplifying assumptions which you make. K = 39.0; I = 127; Hg = 200.6.

(W.J.E.C.)

**24.** What do you understand by the solubility product of a sparingly soluble salt?

In saturated solutions of zinc(II) hydroxide there are two solubility products (*P*) corresponding to the equilibria :

$$Zn(OH)_2 \ (s) \rightleftharpoons Zn^{2+} + 2\,OH^- \qquad P_1 = 4 \times 10^{-17} \ \text{mol}^3 \ \text{dm}^{-9}$$

$$Zn(OH)_2 \ (s) \rightleftharpoons ZnO_2^{2-} + 2H^+ \qquad P_2 = 1 \times 10^{-29} \ \text{mol}^3 \ \text{dm}^{-9}$$

Derive an expression relating the total concentration of zinc in solution ($[Zn^{2+}] + [ZnO_2^{2-}]$) to the hydrogen ion concentration, and calculate the total concentration of zinc in a saturated solution at pH = 10. $K_W = 10^{-14}$ mol² dm⁻⁶.

(C.L.)

**25.** What is the pH value of a solution which is 0.1 molar with respect to methanoic acid, HCOOH, and 0.2 molar with respect to sodium(I) methanoate, HCOONa? Calculate the new pH when there is added to 1 dm³ of this solution (i) 1 cm³ of 10M hydrochloric acid, HCl, (ii) 1 g of sodium(I) hydroxide. (Assume the volume to be 1 dm³ in each case.)

## ANSWERS TO NUMERICAL QUESTIONS

### EXERCISE 1 (p. 37)

**5.** 26·0 or 52·0.  
**6.** 39·0 g.  
**7.** 31·75 g.  
**8.** 56·0 g.  
**9.** $M_2O_3$.  
**10.** 137·4.  
**11.** 137·4.  
**12.** 118·7; XO and $XO_2$; $XCl_4$.  
**13.** (a) 85·49; (b) 56·01.  
**14.** 137·3.  
**15.** 5 cm³ $CH_4$, 10 cm³ $C_2H_6$.  
**16.** 64·7 per cent steam, 27·1 per cent carbon dioxide, 5·9 per cent oxygen, 2·3 per cent nitrogen.  
**19.** (i) $x = 3$; (ii) 87·8.  
**21.** Equivalent mass 30·62 g, relative atomic mass 183·7, $XO_3$.  
**22.** 24·40.  

### EXERCISE 2 (p. 58)

**1.** (i) 1 384 m³; (ii) 224 m³.  
**2.** 82·3 m³.  
**3.** (i) 111 000 Pa; (ii) 315°C.  
**4.** 85 600 Pa.  
**5.** 12·2 g.  
**6.** 1·17 times.  
**7.** $1·8 \times 10^2$ g/m³.  
**9.** 0·25 atm.  
**10.** 8·31 J mol⁻¹ K⁻¹.  
**11.** (i) 122 cm³; (ii) 151 cm³.  

### EXERCISE 3 (p. 78)

**3.** 1·96 g.  
**4.** 28·77.  
**5.** 91·8.  
**6.** 154.  
**7.** 24·3 cm³.  
**8.** 55·95; $M_2Cl_6$.  
**9.** 4·00.  
**10.** 29·7 cm.  
**11.** $C_2N_2$; 54·2 cm³.  
**13.** $AH_3$.  
**15.** $C_2H_2$.  
**16.** $C_3H_8$.  
**17.** 7·67 dm³.  
**18.** 81 per cent.  
**19.** 91 per cent.  
**20.** Relative molecular mass = 119; molecular formula = $SOCl_2$.  
**21.** (i) 0·586; (ii) 261 cm³ $N_2O_4$, 739 cm³ $NO_2$.  
**22.** (a) $C_3H_6$; (b) 12 cm³ $C_3H_8$, 12 cm³ $C_3H_6$.  
**23.** 40 per cent $CH_4$, 30 per cent $C_2H_4$, 30 per cent $C_2H_2$.  

### EXERCISE 4 (p. 110)

**1.** 60·6.  
**2.** 120.  
**3.** −7·1°C.  
**4.** 39·95°C mol⁻¹ kg⁻¹.  
**5.** 295 g.  
**6.** 5·08°C mol⁻¹ kg⁻¹; 243.  
**7.** $P_4$.  
**8.** 78·70°C.  
**9.** 364.  
**10.** 182·4.  
**11.** (i) 251; (ii) $S_8$.  
**12.** 32 160 Pa.  
**13.** 182.  
**14.** 138 000 Pa.  
**15.** 27,500.  
**16.** 13·6°C.  
**17.** (a) −0·103°C; (b) 135 300 Pa.  
**18.** −0·053°C.  
**19.** 2:1.  
**20.** (a) −0·72°C; (b) 1·80 degC mol⁻¹ kg⁻¹.  
**21.** 73·3.  
**22.** $248 \times 10^3$ Pa.

## EXERCISE 10 (p. 257)

6. Composition: $N_2 = 95 \cdot 53$, $CO_2 = 3 \cdot 98$, $H_2O = 0 \cdot 49$ per cent. Partial pressures: $N_2 = 95\ 790$ Pa, $CO_2 = 3987$ Pa, $H_2O = 493$ Pa.

## EXERCISE 12 (p. 300)

2. $3 \cdot 31 \times 10^{-2}$ mol kg$^{-1}$.
3. $39 \cdot 2$ g per 100 g and $48 \cdot 7$ g per 100 g.
6. $H_2O$, $3H_2O$, and $5H_2O$.
7. 122.
8. 23 g.
9. 72 700 Pa (water) and 27 600 Pa (chlorobenzene).
10. $0 \cdot 375$ g.
11. $90 \cdot 0$ $(CCl_4):1$ $(H_2O)$.
12. 120.
13. (i) $47 \cdot 3$; (ii) $49 \cdot 9$.
14. $333 \cdot 3$ cm$^3$; final pressure $= 67\ 550$ Pa.
15. $5 \cdot 60$ cm$^3$ X, $5 \cdot 60$ cm$^3$ Y.
16. $13 \cdot 8$ cm$^3$ oxygen, $19 \cdot 6$ cm$^3$ methane.
18. $98 \cdot 7°C$ (by graph); 1 phenylamine: $3 \cdot 2$ water (approx.).
20. $25 \cdot 2$.

## EXERCISE 14 (p. 346)

1. $\Delta H_f^\ominus = -110$ kJ mol$^{-1}$.
2. $\Delta H^\ominus = -1\ 410 \cdot 3$ kJ mol$^{-1}$.
3. $\Delta H = -3\ 265 \cdot 3$ kJ mol$^{-1}$.
4. $\Delta H_f^\ominus = -251$ kJ mol$^{-1}$.
5. $\Delta H^\ominus = -1\ 370$ kJ mol$^{-1}$.
6. $\Delta H_f^\ominus = -49 \cdot 5$ kJ mol$^{-1}$.
7. $\Delta H_f^\ominus = +73 \cdot 2$ kJ mol$^{-1}$.
8. (a) $597 \cdot 5$ kJ mol$^{-1}$ evolved; (b) $1\ 255 \cdot 2$ kJ mol$^{-1}$ evolved.
9. $\Delta H = +229$ kJ mol$^{-1}$; $\Delta H = -640$ kJ mol$^{-1}$.
10. 774 kJ mol$^{-1}$ evolved.
11. $\Delta H = -18 \cdot 0$ kJ mol$^{-1}$.
12. $7 \cdot 963$ kJ mol$^{-1}$.
13. $\Delta H = -787 \cdot 5$ kJ mol$^{-1}$.
14. $135 \cdot 5$ kJ mol$^{-1}$ absorbed; $148 \cdot 5$ kJ mol$^{-1}$ evolved.
15. $\Delta H = -45$ kJ mol$^{-1}$.
16. $\Delta H = -243$ kJ mol$^{-1}$.

## EXERCISE 15 (p. 389)

7. $1 \cdot 05 \times 10^{-2}$ mol dm$^{-3}$ min$^{-1}$.
8. $0 \cdot 5$.
10. $4 \cdot 2$ per cent $N_2$, $84 \cdot 7$ per cent $H_2$, $11 \cdot 1$ per cent $NH_3$.
11. $K_c = 3 \cdot 99$.
12. $K_o = 48 \cdot 2$.
13. $0 \cdot 642$ mole.
14. (i) CO and steam both $22 \cdot 85$ per cent, $H_2$ and $CO_2$ both $27 \cdot 15$ per cent.
(ii) $9 \cdot 5$ per cent CO, $42 \cdot 8$ per cent steam, $23 \cdot 8$ per cent $CO_2$, $23 \cdot 8$ per cent $H_2$.
15. $2 \cdot 91$ per cent; $2 \cdot 91$ per cent.
17. Rate $\propto [H^+]^2$ $[Br^-]$ $[BrO_3^-]$.
18. $67 \cdot 5$ hours.
19. $K_p = 3 \cdot 23$ atm; $78 \cdot 6$ per cent. Partial pressures: $PCl_5 = 0 \cdot 24$ atm; $PCl_3 = 0 \cdot 88$ atm; $Cl_2 = 0 \cdot 88$ atm.
20. $K_p = 7\ 496$ Pa.
21. 43 per cent; $K_p = 0 \cdot 908$ atm; (i) 19 per cent; (ii) $0 \cdot 13$ atm.

## EXERCISE 16 (p. 418)

3. 19 160 C; $31 \cdot 9$ min.
4. $121 \cdot 8$.
5. $11 \cdot 49$ g.
6. (a) $89 \cdot 3$ mA; (b) $0 \cdot 360$ g; (c) $59 \cdot 85$ cm$^3$.

**7.** 107·5 g.                    **8.** 1·592 × 10⁻¹⁹ C.
**10.**

|  | *Molar conductivity/Sm² mol⁻¹* | *Degree of dissociation* |
|---|---|---|
| (i) | 1·29 × 10⁻² | 0·86 |
| (ii) | 4·07 × 10⁻² | 0·96 |
| (iii) | 1·28 × 10⁻² | 0·96 |

**11.** 0·012.
**12.** (i) 0·91; (ii) 0·81; (iii) 0·86; (iv) 0·71.
**13.** (i) −3·40°C; (ii) −3·98°C; (iii) −1·90°C.
**14.** Magnesium(II) sulphate(VI) is 50 per cent dissociated.
**16.** 0·035.                    **17.** 0·87.
**18.** 4·3 × 10⁻⁷ mol kg⁻¹.
**19.** (i) 3·92 × 10⁻² g kg⁻¹; (ii) 4·26 × 10⁻² g kg⁻¹.

## EXERCISE 17 (p. 445)

**9.** 29·4 g dm⁻³.                    **13.** 254 g.

## EXERCISE 18 (p. 490)

**6.** (i) 3·61 × 10⁻¹³ mol² dm⁻⁶; (ii) 8·82 × 10⁻⁹ mol³ dm⁻⁹; (iii) 5·14 × 10⁻¹²
mol³ dm⁻⁹.
**7.** 2·04 g CaSO₄ per dm³;  4·40 × 10⁻¹² g PbS per dm³.
**8.** (a) (i) 1·72 × 10⁻³ g; (ii) 2·07 × 10⁻⁷ g. (b) 3·175 × 10⁻²¹ g.
**9.** 5·4 × 10⁻⁷ g.                    **10.** 0·0999 mol CaCO₃.
**11.** 2·12 × 10⁻⁴ mol dm⁻³;  2·28 × 10⁻⁴ mol dm⁻³.
**12.** (i) 0·0085; (ii) 0·019.                    **13.** 0·158.
**14.** [H⁺] (i) 4·47 × 10⁻⁴; (ii) 1·15 × 10⁻³; (iii) 4·47 × 10⁻¹²;
    (iv) 1·18 × 10⁻¹¹ mol dm⁻³.
    pH (i) 3·35; (ii) 2·94; (iii) 11·35; (iv) 10·93.
**15.** (a) 0·02. (b) 1 × 10⁻³ mol dm⁻³. (c) pH = 3.
**16.** 1·6 × 10⁻⁵ mol dm⁻³.
**17.** (i) 5 × 10⁻⁴ mol dm⁻³; (ii) 2 × 10⁻¹¹ mol dm⁻³.
**18.** 2 × 10⁻⁵ mol dm⁻³.
**19.** (i) 1·34 × 10⁻³ mol dm⁻³; (ii) 9 × 10⁻⁵ mol dm⁻³.
**20.** 1·476 g.                    **22.** 0·0391 g.
**23.** HgI₄²⁻.                    **24.** 5 × 10⁻⁹ mol dm⁻³.
**25.** 4·00; (i) 3·94; (ii) 4·18.

# INDEX

Absolute zero, 41, 46
Absorption coefficient, 293
Acceptor atom, 172
Acids,
  Brönsted-Lowry theory, 459–462
  catalytic effect, 377, 466
  degree of dissociation, 463, 464
  dissociation constants, 464
  ionization, 173, 397, 407
  neutralization, 461, 477
  strengths, 462–466
Actinides, 133, 193, 207
Activated complex, 351, 366
Activation energy, 350–353, 356
  and catalysis, 366
Active mass, 357
Activity, 407
  coefficient, 408
Adiabatic expansion, 247
Adsorption, 319–326
  and catalysis, 365
  by colloids, 314
  indicators, 320
  of gases by solids, 319, 320
Aerosols, 307
Air, liquefaction, 248
Alkali, 459
Alkali carbonates, titration, 478
Alkali metals, and Periodic Table, 209
Alkaline-earth metals, and Periodic Table, 210
Allotropy, 261–268
  carbon, 265, 266
  helium, 267, 268
  hydrogen, 268
  oxygen, 267
  phosphorus, 263–265
  sulphur, 261–263
  tin, 266, 267
Alloys, 237, 238
Alpha-rays, 117
Aluminium, and Periodic Table, 199
  anodizing, 442
  chloride, 172, 174
Aluminium (III) ion, hydrated, 174
  hydrolysis, 485
  shape, 184
Alums, 35
Amagat, E. H., 240
Ammines, 206

Ammonia,
  dissociation constant, 467
  electronic formula, 163, 168
  liquid, 461, 462
  manufacture, 368
  molecular formula, 72
  reaction with water, 173, 397
  shape of molecule, 182, 183
Ammonium chloride,
  hydrolysis, 483
  thermal dissociation, 74
  use in qualitative analysis, 456
Amorphous solids, 52
Andrews, T., 244
Anions, 402
Anodic oxidation, 442
Aragonite, 260
Arcton, 246
Argon,
  electronic structure, 157
  ionization energies, 144, 145
  separation from liquid oxygen, 248
Arrhenius's ionic theory, 395, 396, 459
Association, 74, 250, 255
  in solution, 86, 105, 288
Aston, F. W., 125
Atom, Rutherford's nuclear model, 121
Atomic bomb, 135
Atomic mass, relative, definition, 19, 31, 37
Atomic masses, determination, 125, 126
Atomic masses, relative,
  after Cannizzaro, 27–37
  carbon standard, 37
  determination of, 23–26, 29, 33–37, 128
  hydrogen standard, 19
  oxygen standard, 31, 129
  problem of, 19, 23
  use in finding empirical formulae, 27
Atomic nucleus, 121, 123–125
Atomic number, 123
  and classification of elements, 189, 190
  determination, 122, 123, 144
Atomic orbitals, 150
  hybridization, 178–181, 266
Atomic radii, 160
  and Periodic Table, 193–195

## TABLE OF THE PRINCIPAL ELEMENTS

| Element | Symbol | Atomic number | Relative atomic mass | |
| --- | --- | --- | --- | --- |
| | | | Approximate | Accurate |
| Aluminium | Al | 13 | 27 | 26·98 |
| Antimony | Sb | 51 | 122 | 121·75 |
| Argon | Ar | 18 | 40 | 39·95 |
| Arsenic | As | 33 | 75 | 74·92 |
| Barium | Ba | 56 | 137 | 137·34 |
| Beryllium | Be | 4 | 9 | 9·01 |
| Bismuth | Bi | 83 | 209 | 208·98 |
| Boron | B | 5 | 11 | 10·81 |
| Bromine | Br | 35 | 80 | 79·91 |
| Cadmium | Cd | 48 | 112 | 112·40 |
| Caesium | Cs | 55 | 133 | 132·90 |
| Calcium | Ca | 20 | 40 | 40·08 |
| Carbon | C | 6 | 12 | 12·01 |
| Chlorine | Cl | 17 | 35·5 | 35·45 |
| Chromium | Cr | 24 | 52 | 52·00 |
| Cobalt | Co | 27 | 59 | 58·93 |
| Copper | Cu | 29 | 63·5 | 63·54 |
| Fluorine | F | 9 | 19 | 19·00 |
| Gallium | Ga | 31 | 70 | 69·72 |
| Germanium | Ge | 32 | 72·5 | 72·59 |
| Gold | Au | 79 | 197 | 196·97 |
| Helium | He | 2 | 4 | 4·00 |
| Hydrogen | H | 1 | 1 | 1·007 97 |
| Iodine | I | 53 | 127 | 126·90 |
| Iridium | Ir | 77 | 192 | 192·2 |
| Iron | Fe | 26 | 56 | 55·85 |
| Krypton | Kr | 36 | 84 | 83·80 |
| Lanthanum | La | 57 | 139 | 138·91 |

| Element | Symbol | Atomic number | Relative atomic mass | |
|---------|--------|---------------|-----------|----------|
| | | | Approximate | Accurate |
| Lead | Pb | 82 | 207 | 207·19 |
| Lithium | Li | 3 | 7 | 6·94 |
| Magnesium | Mg | 12 | 24 | 24·31 |
| Manganese | Mn | 25 | 55 | 54·94 |
| Mercury | Hg | 80 | 200·5 | 200·59 |
| Molybdenum | Mo | 42 | 96 | 95·94 |
| Neon | Ne | 10 | 20 | 20·18 |
| Nickel | Ni | 28 | 59 | 58·71 |
| Nitrogen | N | 7 | 14 | 14·01 |
| Oxygen | O | 8 | 16 | 15·999 4 |
| Phosphorus | P | 15 | 31 | 30·97 |
| Platinum | Pt | 78 | 195 | 195·09 |
| Potassium | K | 19 | 39 | 39·10 |
| Radium | Ra | 88 | 226 | |
| Rubidium | Rb | 37 | 85·5 | 85·47 |
| Selenium | Se | 34 | 79 | 78·96 |
| Silicon | Si | 14 | 28 | 28·09 |
| Silver | Ag | 47 | 108 | 107·87 |
| Sodium | Na | 11 | 23 | 22·99 |
| Strontium | Sr | 38 | 87·5 | 87·62 |
| Sulphur | S | 16 | 32 | 32·06 |
| Tin | Sn | 50 | 119 | 118·69 |
| Titanium | Ti | 22 | 48 | 47·90 |
| Tungsten | W | 74 | 184 | 183·85 |
| Uranium | U | 92 | 238 | 238·03 |
| Vanadium | V | 23 | 51 | 50·94 |
| Xenon | Xe | 54 | 131 | 131·30 |
| Zinc | Zn | 30 | 65 | 65·37 |